AMSCO'S

GEOMETRY

Ann Xavier Gantert

AMSCO SCHOOL PUBLICATIONS, INC.
315 HUDSON STREET, NEW YORK, N.Y. 10013

Dedication

This book is dedicated to Patricia Sullivan whose friendship and support has always been the mainstay of the author's life and work.

Ann Xavier Gantert

The author has been associated with mathematics education in New York State as a teacher and an author throughout the many changes of the past fifty years. She has worked as a consultant to the Mathematics Bureau of the Department of Education in the development and writing of Sequential Mathematics and has been a coauthor of Amsco's *Integrated Mathematics* series, which accompanied that course of study.

Reviewers:

Steven J. Balasiano
Assistant Principal,
Supervision Mathematics
Canarsie High School
Brooklyn, NY

Ronald Hattar
Mathematics Chairperson
Eastchester High School
Eastchester, NY

Sal Sutera
Mathematics Teacher
New Utrecht High School
Brooklyn, NY

Debbie Calvino
Mathematics Supervisor,
Grades 7–12
Valley Central High School
Montgomery, NY

Juanita Maltese
Mathematics, Business, and
Technology Chairperson
Carle Place Middle School/
High School
Carle Place, NY

Domenic D'Orazio
Mathematics Teacher
Midwood High School
Brooklyn, NY

Raymond Scacalossi Jr.
Mathematics Coordinator
Manhasset High School
Manhasset, NY

Text Designer: Nesbitt Graphics, Inc.
Compositor: Compset, Inc.
Cover Design by Meghan J. Shupe
Cover Art by Brand X Pictures (RF)

Please visit our Web site at: *www.amscopub.com*

When ordering this book, please specify:
R 80 P *or* GEOMETRY, *Paperback*   *or*   **R 80 H** *or* GEOMETRY, *Hardbound*

ISBN 978-1-56765-596-4 (Paperback edition)   ISBN 978-1-56765-595-7 (Hardcover edition)
NYC Item 56765-596-3 (Paperback edition)   NYC Item 56765-595-6 (Hardcover edition)

PREFACE

Geometry is a new text for high school geometry that continues the approach that has made Amsco a leader in presenting mathematics in a contemporary, integrated manner. Over the past decades, this approach has undergone many changes and refinements to keep pace with the introduction and expansion of technology in the classroom.

Amsco texts parallel the integrated approach to the teaching of high school mathematics promoted by the National Council of Teachers of Mathematics in its *Curriculum and Evaluation Standards for School Mathematics*. In addition, the content of the book follows the guidelines mandated by the New York State Board of Regents in the *Mathematics Core Curriculum*. This book presents a range of materials and explanations to enable students to achieve a high level of excellence in their understanding of mathematics.

In this book:

✔ **Formal logic** is presented as the basis for geometric reasoning. Most of the geometric facts presented in this text are already familiar to the student. The purpose of this text is to help the student to use the principles of logic to understand the interdependence among these geometric and algebraic concepts.

✔ **Coordinate geometry** is presented with a postulational approach and used when appropriate to enhance and clarify synthetic proof.

✔ **Transformations** are introduced to further expand the students understanding of function and to relate that concept to geometry.

✔ **The concurrence theorems** for the altitudes, angle bisectors, medians and perpendicular bisectors of triangles are proved using a variety of approaches.

✔ **Solid geometry** is introduced and students are encouraged to expand their understanding of the three-dimensional world, particularly through the study of perpendicular and parallel lines and planes.

✔ **Algebraic skills** from *Integrated Algebra 1* are maintained, strengthened, and expanded as a bridge to *Algebra 2 and Trigonometry*.

✔ **Writing About Mathematics** encourages students to reflect on and justify mathematical conjectures, to discover counterexamples, and to express mathematical ideas in their own language.

✔ **Enrichment** is stressed both in the text and in the Teacher's Manual where many suggestions are given for teaching strategies and alternative assessment. The Manual provides opportunities for *Extended Tasks* and *Hands-On Activities*. Reproducible *Enrichment Activities* that challenge students to explore topics in greater depth are provided in each chapter of the Manual.

While *Integrated Algebra 1* is concerned with an intuitive approach to mathematics, the emphasis in *Geometry* is proof. In this text, geometry is developed as a postulational system of reasoning beginning with definitions, postulates, and the laws of reasoning. A unique blending occurs when students learn to apply the laws of logic to traditional deductive proof in geometry, both direct and indirect. The integration of traditional synthetic geometry, coordinate geometry, and transformational geometry is seen throughout the text and students learn to appreciate the interdependence of those branches of mathematics.

The intent of the author is to make the book of greatest service to the average student using thorough explanations and multiple examples. Each section provides careful step-by-step procedures for solving routine exercises as well as the non-routine applications of the material. Sufficient enrichment material is included to challenge students of all abilities. Specifically:

✔ Concepts are carefully developed using appropriate language and mathematical symbolism. General principles are stated clearly and concisely.

✔ Numerous examples are solved as models for students with detailed explanations of the mathematical concepts that underlie the solution. Where appropriate, alternative approaches are suggested.

✔ Varied and carefully graded exercises are given in abundance to develop skills and to encourage the application of those skills. Additional enrichment materials challenge the most capable students.

CONTENTS

ESSENTIALS OF GEOMETRY

For thousands of years, civilized people have used mathematics to investigate sizes, shapes, and the relationships among physical objects. Ancient Egyptians used geometry to solve many practical problems involving boundaries and land areas.

The work of Greek scholars such as Thales, Eratosthenes, Pythagoras, and Euclid for centuries provided the basis for the study of geometry in the Western world. About 300 B.C., Euclid and his followers organized the geometry of his day into a logical system contained in thirteen "books" known today as Euclid's *Elements*.

Euclid began his *Elements* with definitions:

- A *point* is that which has no part.
- A *line* is breadthless length.
- The *extremities of a line* are points.
- A *straight line* is a line that lies evenly with the points on itself.
- A *surface* is that which has length and breadth only.

Today mathematicians consider the words *point*, *line*, and *straight line* to be undefined terms.

Euclid's *Elements* was, of course, written in Greek. One of the most commonly used translations is that of Sir Thomas L. Heath.

1-1 UNDEFINED TERMS

Geometry is the branch of mathematics that defines and relates the basic properties and measurement of line segments and angles. When we study mathematics, we reason about numbers and shapes. Most of the geometric and algebraic facts that are presented in this book you have already learned in previous mathematics courses. In this course we will investigate how these ideas interact to form a logical system. That logical system must start with generally accepted terms and concepts that are building blocks of the system.

There are four terms that we will accept without definition: *set*, *point*, *line*, and *plane*. These are called **undefined terms** because their meaning is accepted without definition. While we do not define these words, we will try to identify them in such a way that their meaning will be accepted by all.

A **set** is a collection of objects such that it is possible to determine whether a given object belongs to the collection or not. For example, the set of students in your mathematics class is determined by your teacher's class list. That list enables us to determine whether or not a student belongs to the set.

•P A **point** may be represented by a dot on a piece of paper, and is usually named by a capital letter. The dot shown, for example, represents a point, which we call point P. A geometric point has no length, width, or thickness; it only indicates place or position.

A **line** is a set of points. A set of points may form a curved line or a straight line as shown. Unless otherwise stated, the term *line* will mean a **straight line**.

Curved Line

Straight Line

We understand what a straight line is by considering the set of points that can be arranged along a stretched string or along the edge of a ruler. While a stretched string or the edge of a ruler is limited in length, a line is an infinite set of points extending endlessly in both directions. Arrowheads are sometimes used to emphasize the fact that the line has no endpoints.

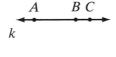

To name a line, we usually use two capital letters that name two points on the line. The two letters may be written in either order. The line shown here may be named line AB, written as $\overleftrightarrow{AB}$. It may also be named line AC or line BC, written as $\overleftrightarrow{AC}$ or $\overleftrightarrow{BC}$. The line may also be named by a single lowercase letter. Thus, the line on the left can also be called line k or simply k.

A **plane** is a set of points that form a flat surface extending indefinitely in all directions. A plane is suggested by the surface of the floor of a room or the surface of your desk. While a floor or your desk have boundaries, a plane extends endlessly in all directions. The figure here represents a plane, called plane p. This four-sided figure shows only part of a plane, however, because a plane has no boundaries.

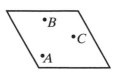

A plane may also be named by using letters that identify three points in the plane that are not on the same line. The plane shown on the left is plane ABC.

Exercises

Writing About Mathematics

1. A section of the United States west of the Mississippi river is called the *Great Plains*. Explain why this area may be called a plane and how it is unlike a mathematical plane.

2. Explain why a stretched string is not a line.

Developing Skills

In 3–8, determine whether each statement is true or false.

3. A point has no length, width, or thickness.

4. A line is limited in length.

5. A plane has boundaries that are lines.

6. $\overleftrightarrow{AB}$ and $\overleftrightarrow{BA}$ name the same line.

7. A line is unlimited in length but has a definite thickness.

8. A point indicates place or position.

Applying Skills

9. Name three things that can be represented by a point.

10. Name three things that can be represented by a line.

11. Name three things that can be represented by a plane.

1-2 THE REAL NUMBERS AND THEIR PROPERTIES

As young children, our first mathematical venture was learning to count. As we continued to grow in our knowledge of mathematics, we expanded our knowledge of numbers to the set of integers, then to the set of rational numbers, and finally to the set of real numbers. We will use our understanding of the real numbers and their properties in our study of geometry.

In *Integrated Algebra 1*, we learned that every real number corresponds to a point on a **number line** and every point on the number line corresponds to a real number. The real number line below shows this correspondence for a part of the real numbers.

The number that corresponds to a point on a line is called the **coordinate** of the point. The coordinate of point A on the number line shown above is 3, and the coordinate of point B is $-\frac{1}{2}$. The point to which a number corresponds is called the **graph** of the number. The graph of 5 is C, and the graph of -2 is D.

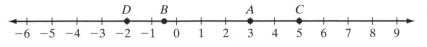

Properties of the Real Number System

A **numerical operation** assigns a real number to every pair of real numbers. The following properties of the real numbers govern the operations of addition and multiplication.

1. *Closure*

   The sum of two real numbers is a real number. This property is called the **closure property of addition**. For all real numbers a and b:

   $$a + b \text{ is a real number.}$$

   The product of two real numbers is a real number. This property is called the **closure property of multiplication**. For all real numbers a and b:

   $$a \cdot b \text{ is a real number.}$$

2. *Commutative Property*

   When we add real numbers, we can change the **order** in which two numbers are added without changing the sum. This property is called the **commutative property of addition**. For all real numbers a and b:

   $$a + b = b + a$$

   When we multiply real numbers, we can change the **order** of the factors without changing the product. This property is called the **commutative property of multiplication**. For all real numbers a and b:

   $$a \cdot b = b \cdot a$$

3. *Associative Property*

   When three numbers are added, two are added first and then their sum is added to the third. The sum does not depend on which two numbers are added first. This property is called the **associative property of addition**. For all real numbers a, b, and c:

   $$(a + b) + c = a + (b + c)$$

When three numbers are multiplied, two are multiplied first and then their product is multiplied by the third. The product does not depend on which two numbers are multiplied first. This property is called the **associative property of multiplication**. For all real numbers a, b, and c:

$$a \times (b \times c) = (a \times b) \times c$$

4. Identity Property

When 0 is added to any real number a, the sum is a. The real number 0 is called the **additive identity**. For every real number a:

$$a + 0 = a \quad \text{and} \quad 0 + a = a$$

When 1 is multiplied by any real number a, the product is a. The real number 1 is called the **multiplicative identity**. For every real number a:

$$a \cdot 1 = a \quad \text{and} \quad 1 \cdot a = a$$

5. Inverse Property

Two real numbers are called **additive inverses** if their sum is the additive identity, 0. For every real number a, there exists a real number $-a$ such that:

$$a + (-a) = 0$$

Two real numbers are called **multiplicative inverses** if their product is the multiplicative identity, 1. For every real number $a \neq 0$, there exists a real number $\frac{1}{a}$ such that:

$$a \cdot \frac{1}{a} = 1$$

6. Distributive Property

The **distributive property** combines the operations of multiplication and addition. Multiplication distributes over addition. For all real numbers a, b, and c:

$$a(b + c) = ab + ac \quad \text{and} \quad (a + b)c = ac + bc$$

7. Multiplication Property of Zero

Zero has no multiplicative inverse. The **multiplication property of zero** defines multiplication by zero. For all real numbers a and b:

$$ab = 0 \text{ if and only if } a = 0 \text{ or } b = 0.$$

EXAMPLE 1

In **a–c**, name the property illustrated in each equality.

	Answers

a. $4(7 \cdot 5) = (4 \cdot 7)5$ — Associative property of multiplication

b. $\sqrt{3} + 3 = 3 + \sqrt{3}$ — Commutative property of addition

c. $\frac{1}{9}(4 + \sqrt{2}) = \frac{1}{9}(4) + \frac{1}{9}(\sqrt{2})$ — Distributive property of multiplication over addition

Exercises

Writing About Mathematics

1. a. Explain why the set of positive real numbers is not closed under subtraction.

   b. Is the set of real numbers closed under subtraction? Explain your answer.

2. Explain why the set of negative real numbers is not closed under multiplication.

Developing Skills

3. Name the number that is the additive identity for the real numbers.

4. Name the number that is the multiplicative identity for the real numbers.

5. When $a = -11$, what does $-a$ represent?

6. Name the real number that has no multiplicative inverse.

In 7–16: **a.** Replace each question mark with a number that makes the statement true. **b.** Name the property illustrated in the equation that is formed when the replacement is made.

7. $7 + 14 = 14 + ?$

8. $7 + (?) = 0$

9. $3(4 \cdot 6) = (3 \cdot 4)?$

10. $11(9) = ?(11)$

11. $1x = ?$

12. $6(4 + b) = 6(4) + ?$

13. $17(?) = 1$

14. $\frac{1}{4}(24) = 24(?)$

15. $12(? + 10) = 12(10 + 4)$

16. $\pi\left(\frac{1}{\pi}\right) = ?$

17. For what value of x does $3x - 1$ have no multiplicative inverse?

18. If $ab = 0$ and $a \neq 0$, explain why $a + b = a$.

1-3 DEFINITIONS, LINES, AND LINE SEGMENTS

Qualities of a Good Definition

A **definition** is a statement of the meaning of a term. When we state a definition, we will use only undefined terms or terms that we have previously defined.

The definition of a term will state the set to which the term belongs and the qualities that makes it different from others in the set.

Collinear Points

DEFINITION
A **collinear set of points** is a set of points all of which lie on the same straight line.

This definition has the properties of a good definition:

1. It is expressed in undefined words.

Set, *point*, and *line* are undefined words.

2. It states the set to which the defined term belongs.

The set is a set of points.

3. It distinguishes the term from other members of the class.

The points must be on the same straight line.

A, *B*, and *C* are collinear points.

DEFINITION
A **noncollinear set of points** is a set of three or more points that do not all lie on the same straight line.

D, *E*, and *F* are noncollinear points.

The Distance Between Two Points

Every point on a line corresponds to a real number called its *coordinate*. To find the distance between any two points, we find the absolute value of the difference between the coordinates of the two points. For example, the distance between *G* and *C*, written as *GC*, is $|5 - 1| = 4$. We use absolute value so that the distance between any two points is independent of the order that we subtract the coordinates.

$$GC = |5 - 1| \qquad\qquad CG = |1 - 5|$$
$$= |4| \qquad\qquad\qquad = |-4|$$
$$= 4 \qquad\qquad\qquad\quad = 4$$

Recall that:

1. The absolute value of a positive number p is p. Thus, $|12| = 12$.

2. The absolute value of a negative number n is $-n$, the opposite of n. If $n = -7$, then $-n = -(-7) = 7$. Thus, $|-7| = 7$.

3. The absolute value of 0 is 0. Thus, $|0| = 0$.

DEFINITION _____

The **distance between two points on the real number line** is the absolute value of the difference of the coordinates of the two points.

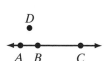

On the real number line, the distance between point A, whose coordinate is a, and B, whose coordinate is b, can be represented as

$$AB = |a - b| = |b - a|.$$

Order of Points on a Line

When we learned to count, we placed the natural numbers in order, starting with the smallest. We count "one, two, three." We are saying that $1 < 2$ and $2 < 3$ or that $1 < 3$. When we locate the graphs of numbers on the number line, a number m is to the left of a number n if and only if $m < n$. In general, if the numbers m, n, and p occur in that order from left to right on the number line, then $m < n < p$. If $m < n$ and $n < p$, then n is between m and p on the number line.

In the figure, note that the position of point B with respect to A and C is different from the position of D with respect to A and C because A, B, and C are collinear but A, D, and C are noncollinear.

DEFINITION _____

B **is between** A **and** C if and only if A, B, and C are distinct collinear points and $AB + BC = AC$. This is also called **betweenness**.

The symbol $\overleftrightarrow{ABC}$ represents a line on which B is between A and C. Let a be the coordinate of A, b be the coordinate of B, and c be the coordinate of C.

In the figure, B is between A and C and $a < b < c$.

- Since $a < b$, $b - a$ is positive and $AB = |b - a| = b - a$.
- Since $b < c$, $c - b$ is positive and $BC = |c - b| = c - b$.
- Since $a < c$, $c - a$ is positive and $AC = |c - a| = c - a$.

Therefore,

$$AB + BC = (b - a) + (c - b)$$
$$= c - a$$
$$= AC$$

Line Segments

A line segment is a subset, or a part of, a line. In the figure, A and B are two points of line m. Points A and B determine segment AB, written in symbols as $\overline{AB}$.

DEFINITION

A **line segment**, or a **segment**, is a set of points consisting of two points on a line, called endpoints, and all of the points on the line between the endpoints.

A line may be named by any two of the points on the line. A line segment is always named by its two endpoints. In the diagram, $\overleftrightarrow{AB}$, $\overleftrightarrow{AC}$, and $\overleftrightarrow{CB}$ all name the same line. However, $\overline{AB}$, $\overline{AC}$, and $\overline{CB}$ name different segments. To indicate that C is a point on $\overline{AB}$ but not one of its endpoints, we use the symbol $\overline{ACB}$.

DEFINITION

The **length** or **measure of a line segment** is the distance between its endpoints.

The length or measure of $\overline{AB}$ is written in symbols as AB. If the coordinate of A is a and the coordinate of B is b, then $AB = |b - a|$.

The line segments shown have the same length. These segments are said to be congruent.

DEFINITION

Congruent segments are segments that have the same measure.

The symbol $\cong$ is used to state that two segments are congruent. For example, since $\overline{AB}$ and $\overline{CD}$ are congruent, we write $\overline{AB} \cong \overline{CD}$. Because congruent segments have the same measure, when $\overline{AB} \cong \overline{CD}$, $AB = CD$. Note the correct symbolism. We write $\overline{AB} = \overline{CD}$ only when $\overline{AB}$ and $\overline{CD}$ are two different names for the same set of points.

In the figure, each segment, $\overline{AB}$ and $\overline{CD}$, is marked with two tick marks. When segments are marked with the same number of tick marks, the segments are known to be congruent. Therefore, $\overline{AB} \cong \overline{CD}$.

EXAMPLE 1

On a number line, the coordinate of A is -7 and the coordinate of B is 4. Find AB, the distance between A and B.

Solution   The distance between A and B is the absolute value of the difference of their coordinates.

$$AB = |(-7) - 4|$$
$$= |-11|$$
$$= 11 \quad \textit{Answer}$$

EXAMPLE 2

R, T, and S are three points on a number line. The coordinate of R is 3, the coordinate of T is -7, and the coordinate of S is 9. Which point is between the other two?

Solution   Since $-7 < 3 < 9$, R is between T and S. *Answer*

Note: We can state in symbols that R is between T and S by writing $\overline{TRS}$.

Exercises

Writing About Mathematics

1. Explain why "A hammer is a tool" is not a good definition.

2. Explain why "A hammer is used to drive nails" is not a good definition.

Developing Skills

In 3–7: **a.** Write the definition that each phrase describes. **b.** State the class to which each defined term belongs. **c.** Tell what distinguishes each defined term from other members of the class.

3. noncollinear set of points

4. distance between any two points on the number line

5. line segment

6. measure of a line segment

7. congruent segments

In 8–13, use the number line to find each measure.

8. *AB*   **9.** *BD*

10. *CD*   **11.** *FH*

12. *GJ*   **13.** *EJ*

14. Name three segments on the number line shown above that are congruent.

15. Three segments have the following measures: $PQ = 12$, $QR = 10$, $PR = 18$. What conclusion can you draw about P, Q, and R? Draw a diagram to justify your answer.

Applying Skills

16. A carpenter is measuring the board with which he is working.

   a. Because one end of the measuring tape is unreadable, the carpenter places the measuring tape so that one end of the board is at 3 inches and the other at 15 inches. How long is the board?

   b. The carpenter wants to cut the board so that he has a piece 10 inches long. If he leaves the measuring tape in the position described in **a**, at what point on the tape should he mark the board to make the cut? (Two answers are possible.)

17. A salesperson drove from Troy to Albany, a distance of 8 miles, and then from Albany to Schenectady, a distance of 16 miles. The distance from Schenectady back to Troy is 18 miles. Are Troy, Albany, and Schenectady in a straight line? Explain your answer.

1-4 MIDPOINTS AND BISECTORS

Midpoint of a Line Segment

A line segment can be divided into any number of congruent parts. The point that divides the line segment into two equal parts is called the **midpoint** of the segment.

DEFINITION

The **midpoint of a line segment** is a point of that line segment that divides the segment into two congruent segments.

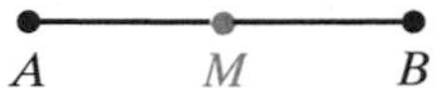

If M is the midpoint of $\overline{AB}$, then $\overline{AM} \cong \overline{MB}$ or $AM = MB$. It is also true that $AM = \frac{1}{2}AB$, that $MB = \frac{1}{2}AB$, that $AB = 2AM$, and that $AB = 2MB$.

On the number line, the coordinate of A is -2 and the coordinate of B is 4. Therefore,

$$AB = |-2 - 4|$$
$$= |-6|$$
$$= 6$$

If M is the midpoint of $\overline{AB}$, then $AM = 3$ and $MB = 3$. The coordinate of M can be found by adding 3 to the smaller coordinate, -2. The coordinate of M is $-2 + 3 = 1$. The coordinate of M can also be found by subtracting 3 from the larger coordinate, 4. The coordinate of M is $4 - 3 = 1$.

This example suggests the following relationship for finding the midpoint of a segment:

▶ **For any line segment $\overline{AB}$, if the coordinate of A is a and the coordinate of B is b, then the coordinate of the midpoint of $\overline{AB}$ is $\frac{a + b}{2}$.**

Bisector of a Line Segment

DEFINITION

The **bisector of a line segment** is any line, or subset of a line, that intersects the segment at its midpoint.

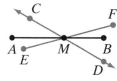

The line segment $\overline{AB}$ is bisected at M if M is the midpoint of $\overline{AB}$. Any line such as $\overleftrightarrow{CD}$, any segment such as $\overline{EF}$, and any ray such as $\overrightarrow{MC}$ are bisectors of $\overline{AB}$ because they intersect $\overline{AB}$ at its midpoint M.

Note: A line, unlike a line segment, cannot be bisected because it extends indefinitely in two directions.

Adding and Subtracting Line Segments

DEFINITION

A line segment, $\overline{RS}$ is the **sum of two line segments,** $\overline{RP}$ and $\overline{PS}$, if P is between R and S.

For three collinear points R, P, and S, with P between R and S, $\overline{RP} + \overline{PS}$ indicates the line segment formed by R, P, and S, that is, $\overline{RPS}$. Then $\overline{RP} + \overline{PS}$ is the *union* of segments $\overline{RP}$ and $\overline{PS}$. The expression $\overline{RP} + \overline{PS}$ represents a line segment only when P is between R and S.

We can also refer to the *lengths* of the line segments. Thus, $RS = RP + PS$. Recall that if P is between R and S, then R, P, and S are collinear. Observe that P divides $\overleftrightarrow{RS}$ into two segments, $\overline{RP}$ and $\overline{PS}$. The following equalities are true for the lengths of the segments.

$$RS = RP + PS \qquad RP = RS - PS \qquad PS = RS - RP$$

EXAMPLE I

The points P, Q, R, and S are shown as points on a number line.

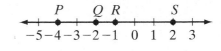

a. Find PQ, PR, QR, RS, PR, QS, and PS.

b. What point is the midpoint of $\overline{PS}$?

c. Show that $PQ + QS = PS$.

Solution　**a.** $PQ = |-2 - (-4)| = |-2 + 4| = 2$

$PR = |-1 - (-4)| = |-1 + 4| = 3$

$QR = |-1 - (-2)| = |-1 + 2| = 1$

$RS = |2 - (-1)| = |2 + 1| = 3$

$QS = |2 - (-2)| = |2 + 2| = 4$

$PS = |2 - (-4)| = |2 + 4| = 6$

b. Since $PR = RS$ and $PR + RS = PS$, R is the midpoint of $\overline{PS}$.

c. $PQ + QS = 2 + 4 = 6$ and $PS = 6$. Therefore, $PQ + QS = PS$.

Exercises

Writing About Mathematics

1. Is it ever correct to write $\overleftrightarrow{AB} = \overleftrightarrow{CD}$? If so, what does that imply about A, B, C, and D?

2. If $AM = MB$, does this necessarily mean that M is the midpoint of $\overline{AB}$? Explain why or why not.

Developing Skills

In 3–6, state two conclusions that can be drawn from the given data. In one conclusion, use the symbol $\cong$; in the other, use the symbol $=$.

3. T is the midpoint of $\overline{AC}$.

4. $\overleftrightarrow{MN}$ bisects $\overline{RS}$ at N.

5. C is the midpoint of $\overline{BD}$.

6. $\overleftrightarrow{LM}$ bisects $\overline{ST}$ at P.

7. A, B, C, D are points on $\overleftrightarrow{AD}$, B is between A and C, C is between B and D, and $\overline{AB} \cong \overline{CD}$.

    a. Draw $\overleftrightarrow{AD}$ showing B and C.   **b.** Explain why $\overline{AC} \cong \overline{BD}$.

8. What is the coordinate of the midpoint of $\overline{DE}$ if the coordinate of D is -4 and the coordinate of E is 9?

9. $S, M,$ and T are points on the number line. The coordinate of M is 2, and the coordinate of T is 14. What is the coordinate of S if M is the midpoint of $\overline{ST}$?

10. $P, Q,$ and R are points on the number line. The coordinate of P is -10, the coordinate of Q is 6, and the coordinate of R is 8.

    a. Does $PQ + QR = PR$?

    b. Is Q the midpoint of $\overline{PR}$? If not, what is the coordinate of the midpoint?

Applying Skills

11. Along the New York State Thruway, there are distance markers that show the distance in miles from the beginning of the Thruway. Alicia is at marker 215 and will exit the Thruway at marker 395. She wants to stop to rest at the midpoint of her trip. At what distance marker should she stop?

12. A carpenter is measuring the board with which she is working.

    a. The carpenter places a measuring tape so that one end of the board is at 24 inches and the other at 8 inches. How long is the board?

    b. The carpenter wants to cut the board so that she has two pieces of equal length. If she leaves the measuring tape in the position described in **a**, at what point on the tape should she mark the board to make the cut?

Hands-On Activity

For this activity, use a sheet of wax paper, patty paper, or other paper thin enough to see through. You may work alone or with a partner.

1. Draw a line segment $\overline{AB}$.

2. Fold the paper so that A coincides with B. Crease the paper. Label the crease $\overline{CDE}$ with D the point at which $\overline{CDE}$ intersects $\overline{AB}$.

3. What is true about $\overline{AB}$ and D? $\overline{AD}$ and $\overline{DB}$?

1-5 RAYS AND ANGLES

Half-Lines and Rays

DEFINITION

Two points, A and B, are **on one side of a point** P if $A, B,$ and P are collinear and P is not between A and B.

Every point on a line divides the line into two opposite sets of points called half-lines. A **half-line** consists of the set of all points on one side of the point of division. The point of division itself does not belong to the half-line. In the figure, point P divides $\overleftrightarrow{MB}$ into two half-lines.

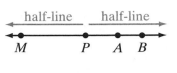

DEFINITION

A **ray** consists of a point on a line and all points on one side of the point.

This definition tells us that a ray is a set of all the points in a half-line and the dividing point. The dividing point is called the **endpoint** of the ray.

A ray is named by placing an arrow that points to the right over two capital letters. The first letter must name the endpoint of the ray. The second letter may be the name of any other point on the ray. The figure shows ray AB, which starts at endpoint A, contains point B, and extends indefinitely in one direction. Ray AB is written as $\overrightarrow{AB}$.

Note in the figure that the ray whose endpoint is O may be named $\overrightarrow{OA}$ or $\overrightarrow{OB}$. However, it may not be named $\overrightarrow{AB}$ since the first capital letter in the name of a ray must represent its endpoint.

A point on a line creates two different rays, each with the same endpoint. In the figure, $\overrightarrow{AB}$ and $\overrightarrow{AC}$ are opposite rays because points A, B, and C are collinear and points B and C are not on the same side of point A, but are on opposite sides of point A.

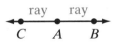

DEFINITION

Opposite rays are two rays of the same line with a common endpoint and no other point in common.

Angles

DEFINITION

An **angle** is a set of points that is the union of two rays having the same endpoint.

In the figure, $\overrightarrow{AB}$ and $\overrightarrow{AC}$, which form an angle, are called the **sides** of the angle. The endpoint of each ray, A, is called the **vertex** of the angle. The symbol for angle is $\angle$.

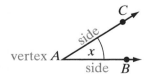

An angle, such as the one illustrated, may be named in any of the following ways:

1. By a capital letter that names its vertex.

Example: $\angle A$

2. By a lowercase letter or number placed inside the angle.

Example: $\angle x$

3. By three capital letters, the middle letter naming the vertex and each of the other letters naming a point on a different ray. Example: $\angle BAC$ or $\angle CAB$

When a diagram shows several angles with the same vertex, we avoid confusion by using three letters to name each angle. For example, in the figure, the smaller angles are $\angle RPS$ and $\angle SPT$; the larger angle is $\angle RPT$.

Straight Angles

DEFINITION

A **straight angle** is an angle that is the union of opposite rays.

The sides of a straight angle belong to the same straight line. In the figure, $\overrightarrow{OA}$ and $\overrightarrow{OB}$ are opposite rays of $\overleftrightarrow{AB}$ because they have a common endpoint, O, and no other points in common. Thus, $\angle AOB$ is a straight angle because it is the union of the opposite rays $\overrightarrow{OA}$ and $\overrightarrow{OB}$.

An angle divides the points of a plane that are not points of the angle into two sets or regions. One region is called the **interior of the angle**; the other region is called the **exterior of the angle**. In the figure, $\angle A$ is not a straight angle. M is any point on one side of $\angle A$ and N is any point on the other side of the angle.

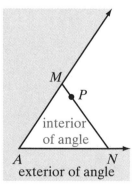

Neither M nor N is the vertex of the angle. Any point P on $\overline{MN}$ between points M and N belongs to the interior region of the angle. The region consisting of all points such as P is called the *interior of the angle*. All other points of the plane, except the points of the angle itself, form the region called the *exterior of the angle*.

The Measure of an Angle

To measure an angle, we must select a unit of measure and know how this unit is to be applied. In geometry, the unit used to measure an angle is usually a *degree*. The number of degrees in an angle is called its *degree measure*.

The degree measure of a straight angle is 180. A **degree** is, therefore, the measure of an angle that is $\frac{1}{180}$ of a straight angle. It is possible to use units other than the degree to measure angles. In this book, when we write m$\angle HRC = 45$, we mean that the measure of $\angle HRC$ is 45 degrees.

Classifying Angles According to Their Measures

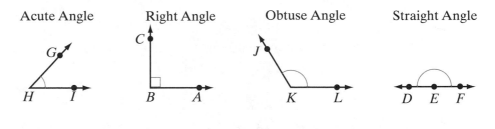

Acute Angle   Right Angle   Obtuse Angle   Straight Angle

DEFINITION _____
An **acute angle** is an angle whose degree measure is greater than 0 and less than 90.

In the figure above, $\angle GHI$ is an acute angle.

DEFINITION _____
A **right angle** is an angle whose degree measure is 90.

In the figure above, $\angle ABC$ is a right angle. Therefore, we know that $m\angle ABC = 90$. The symbol ⌐ at B is used to show that $\angle ABC$ is a right angle.

DEFINITION _____
An **obtuse angle** is an angle whose degree measure is greater than 90 and less than 180.

In the figure above, $\angle JKL$ is an obtuse angle.

A **straight angle**, previously defined as the union of opposite rays, is an angle whose degree measure is 180. In the figure above, DEF is a straight angle. Thus, $m\angle DEF = 180$. Note that $\overrightarrow{ED}$ and $\overrightarrow{EF}$, the sides of $\angle DEF$, are opposite rays and form a straight line.

EXAMPLE I ▨▨▨▨▨▨▨▨▨▨▨▨▨▨▨▨▨▨▨▨▨▨

D is a point in the interior of $\angle ABC$, $m\angle ABD = 15$, and $m\angle DBC = 90$.

a. Find $m\angle ABC$.

b. Name an acute angle.

c. Name a right angle.

d. Name an obtuse angle.

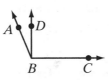

Solution   **a.** m∠ABC = m∠ABD + m∠DBC

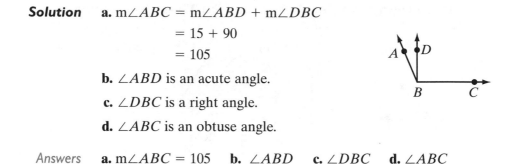

$$= 15 + 90$$

$$= 105$$

b. ∠ABD is an acute angle.

c. ∠DBC is a right angle.

d. ∠ABC is an obtuse angle.

Answers   **a.** m∠ABC = 105   **b.** ∠ABD   **c.** ∠DBC   **d.** ∠ABC

Exercises

Writing About Mathematics

1. Explain the difference between a half-line and a ray.

2. Point R is between points P and S. Are $\overrightarrow{PR}$ and $\overrightarrow{PS}$ the same ray, opposite rays, or neither the same nor opposite rays? Explain your answer.

Developing Skills

3. For the figure shown:

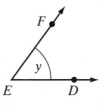

   a. Name two rays.

   b. Name the vertex of the angle.

   c. Name the sides of the angle.

   d. Name the angle in four ways.

4. a. Name the vertex of ∠BAD in the figure.

   b. Name the sides of ∠BAD.

   c. Name all the angles with A as vertex.

   d. Name the angle whose sides are $\overrightarrow{AB}$ and $\overrightarrow{AC}$.

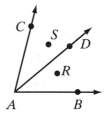

   e. Name the ray that is a side of both ∠BAD and ∠BAC.

   f. Name two angles in whose interior regions point R lies.

   g. Name the angle in whose exterior region point S lies.

   h. Are $\overrightarrow{AB}$ and $\overrightarrow{AC}$ opposite rays? Explain your answer.

   i. Is ∠BAC a straight angle? Explain your answer.

5. For the figure shown:

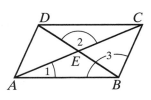

  a. Name angle 1 in four other ways.

  b. Name angle 2 in two other ways.

  c. Name angle 3 in two other ways.

  d. Name the point of intersection of $\overline{AC}$ and $\overline{BD}$.

  e. Name two pairs of opposite rays.

  f. Name two straight angles each of which has its vertex at E.

  g. Name two angles whose sum is $\angle ABC$.

1-6 MORE ANGLE DEFINITIONS

Congruent Angles

DEFINITION

  Congruent angles are angles that have the same measure.

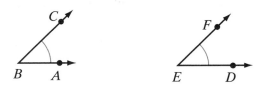

In the figures, $\angle ABC$ and $\angle DEF$ have the same measure. This is written as $m\angle ABC = m\angle DEF$. Since congruent angles are angles that have the same measure, we may also say that $\angle ABC$ and $\angle DEF$ are congruent angles, symbolized as $\angle ABC \cong \angle DEF$. We may use either notation.

    $\angle ABC \cong \angle DEF$ The angles are congruent.
    $m\angle ABC = m\angle DEF$ The measures of the angles are equal.

Note: We do not write $\angle ABC = \angle DEF$. Two angles are equal only if they name the union of the same rays. For example, $\angle ABC = \angle CBA$.

When two angles are known to be congruent, we may mark each one with the same number of arcs, as shown in the figures above, where each angle is marked with a single arc.

Bisector of an Angle

DEFINITION _____

A **bisector of an angle** is a ray whose endpoint is the vertex of the angle, and that divides that angle into two congruent angles.

For example, if $\overrightarrow{OC}$ is the bisector of $\angle AOB$, then $\angle AOC \cong \angle COB$ and m$\angle AOC$ = m$\angle COB$. If $\overrightarrow{OC}$ bisects $\angle AOB$, we may say that m$\angle AOC$ = $\frac{1}{2}$m$\angle AOB$, m$\angle COB$ = $\frac{1}{2}$m$\angle AOB$, m$\angle AOB$ = 2m$\angle AOC$, and m$\angle AOB$ = 2m$\angle COB$.

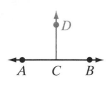

Let C be any point on $\overleftrightarrow{AB}$, $\angle ACB$ is a straight angle, and m$\angle ACB$ = 180. If $\overrightarrow{CD}$ is the bisector of $\angle ACB$, each of the two congruent angles, $\angle ACD$ and $\angle DCB$, has degree measure 90 and is a right angle. Any line, ray, or line segment that is a subset of $\overrightarrow{CD}$, and contains C, is said to be perpendicular to $\overleftrightarrow{AB}$ at C. For example, $\overline{CD}$ is perpendicular to $\overleftrightarrow{AB}$.

DEFINITION _____

Perpendicular lines are two lines that intersect to form right angles.

Since rays and line segments are contained in lines, any rays or line segments that intersect to form right angles are also perpendicular.

DEFINITION _____

The **distance from a point to a line** is the length of the perpendicular from the point to the line.

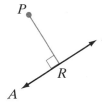

P is a point not on $\overleftrightarrow{AB}$, and $\overline{PR} \perp \overleftrightarrow{AB}$. The segment $\overline{PR}$ is called the perpendicular from P to $\overleftrightarrow{AB}$. The point R at which the perpendicular meets the line is called the **foot** of the perpendicular. The distance from P to $\overleftrightarrow{AB}$ is PR (the length of $\overline{PR}$).

Adding and Subtracting Angles

DEFINITION _____

If point P is a point in the interior of $\angle RST$ and $\angle RST$ is not a straight angle, or if P is any point not on straight angle RST, then $\angle RST$ is the **sum of two angles**, $\angle RSP$ and $\angle PST$.

In the figure, $\overrightarrow{SP}$ separates $\angle RST$ into two angles, $\angle RSP$ and $\angle PST$, such that the following relations are true for the measures of the angles:

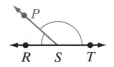

$$m\angle RST = m\angle RSP + m\angle PST$$
$$m\angle RSP = m\angle RST - m\angle PST$$
$$m\angle PST = m\angle RST - m\angle RSP$$

EXAMPLE 1

In the figure, $\overrightarrow{CD}$ bisects $\angle ACB$.

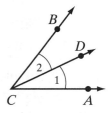

a. Name $\angle 1$ in two other ways.

b. Write a conclusion that states that two angles are congruent.

c. Write a conclusion that states that two angles have equal measures.

Solution   **a.** $\angle ACD$ and $\angle DCA$. *Answer*

b. $\angle ACD \cong \angle BCD$ or $\angle 2 \cong \angle 1$. *Answer*

c. $m\angle ACD = m\angle BCD$ or $m\angle 2 = m\angle 1$. *Answer*

Exercises

Writing About Mathematics

1. Tanya said that if $\overleftrightarrow{RST}$ separates $\angle PSQ$ into two congruent angles, then $\overleftrightarrow{RST}$ is the bisector of $\angle PSQ$. Do you agree with Tanya? Explain why or why not.

2. If an obtuse angle is bisected, what kind of an angle is each of the two angles formed?

Developing Skills

In 3–10, if an angle contains the given number of degrees, **a.** state whether the angle is an acute angle, a right angle, an obtuse angle, or a straight angle, **b.** find the measure of each of the angles formed by the bisector of the given angle.

3. 24   **4.** 98   **5.** 126   **6.** 90

7. 82   **8.** 180   **9.** 57   **10.** 3

In 11–13, find the measure of each of the following:

11. $\frac{1}{2}$ of a right angle   **12.** $\frac{1}{3}$ of a right angle   **13.** $\frac{4}{5}$ of a straight angle

In 14 and 15, in each case, use the information to:
 a. Draw a diagram.
 b. Write a conclusion that states the congruence of two angles.
  c. Write a conclusion that states the equality of the measures of two angles.

14. $\overrightarrow{CD}$ bisects $\angle ACB$.

15. $\overrightarrow{AC}$ is the bisector of $\angle DAB$.

16. If a straight angle is bisected, what types of angles are formed?

17. If a right angle is bisected, what types of angles are formed?

In 18–20, complete each statement, which refers to the figure shown.

18. $m\angle LMN = m\angle LMP + m\angle$ _____

19. $m\angle LMP = m\angle LMN - m\angle$ _____

20. $m\angle LMN - m\angle$ _____ $= m\angle NMP$

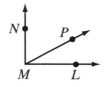

In 21–24, use the figure to answer each question.

21. $m\angle ABE + m\angle EBC = m\angle$ _____

22. $m\angle BEC + m\angle CED = m\angle$ _____

23. $m\angle ADC - m\angle CDE = m\angle$ _____

24. $m\angle AEC - m\angle AEB = m\angle$ _____

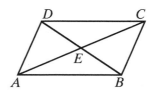

25. Every horizontal line is perpendicular to every vertical line. Draw, using pencil and paper or geometry software, a horizontal line ACB and a vertical line DCF.

 a. Name four congruent angles in your diagram.

 b. What is the measure of each of the angles that you named in part **a**?

 c. Name two rays that bisect $\angle ACB$.

 d. Name two rays that bisect $\angle DCF$.

Applying Skills

26. $\overrightarrow{PQ}$ and $\overrightarrow{PR}$ are opposite rays and $\overrightarrow{PS}$ bisects $\angle QPR$. If $m\angle QPS$ is represented by $4x + 30$, what is the value of x?

27. $\overrightarrow{BD}$ bisects $\angle ABC$. If $m\angle ABD$ can be represented by $3a + 10$ and $m\angle DBC$ can be represented by $5a - 6$, what is $m\angle ABC$?

28. $\overrightarrow{RT}$ bisects $\angle QRS$. If $m\angle QRS = 10x$ and $m\angle SRT = 3x + 30$, what is $m\angle QRS$?

29. $\overrightarrow{BD}$ bisects $\angle ABC$ and $\overrightarrow{MP}$ bisects $\angle LMN$. If $m\angle CBD = m\angle PMN$, is $\angle ABC \cong \angle LMN$? Justify your answer.

Hands-On Activity

For this activity, use a sheet of wax paper, patty paper, or other paper thin enough to see through. You may work alone or with a partner.

1. Draw a straight angle, $\angle ABC$.

2. Fold the paper through the vertex of the angle so that the two opposite rays correspond, and crease the paper. Label two points, D and E, on the crease with B between D and E.

3. Measure each angle formed by the crease and one of the opposite rays that form the straight angle. What is true about these angles?

4. What is true about $\overline{DE}$ and $\overline{AC}$?

1-7 TRIANGLES

DEFINITION _____

A **polygon** is a closed figure in a plane that is the union of line segments such that the segments intersect only at their endpoints and no segments sharing a common endpoint are collinear.

A polygon consists of three or more line segments, each of which is a **side** of the polygon. For example, a triangle has three sides, a quadrilateral has four sides, a pentagon has five sides, and so on.

In the definition of a polygon, we used the word *closed*. We understand by the word *closed* that, if we start at any point on the figure and trace along the sides, we will arrive back at the starting point. In the diagram, we see a figure that is not closed and is, therefore, not a polygon.

DEFINITION _____

A **triangle** is a polygon that has exactly three sides.

The polygon shown is triangle ABC, written as $\triangle ABC$. In $\triangle ABC$, each of the points A, B, and C is a vertex of the triangle. $\overline{AB}$, $\overline{BC}$, and $\overline{CA}$ are the sides of the triangle.

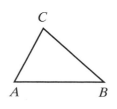

Included Sides and Included Angles of a Triangle

If a line segment is the side of a triangle, the endpoints of that segment are the vertices of two angles. For example, in $\triangle ABC$, the endpoints of $\overline{AB}$ are the vertices of $\angle A$ and $\angle B$. We say that the side, $\overline{AB}$, is **included** between the angles, $\angle A$ and $\angle B$. In $\triangle ABC$:

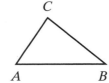

1. $\overline{AB}$ is included between $\angle A$ and $\angle B$.

2. $\overline{BC}$ is included between $\angle B$ and $\angle C$.

3. $\overline{CA}$ is included between $\angle C$ and $\angle A$.

In a similar way, two sides of a triangle are subsets of the rays of an angle, and we say that the angle is **included** between those sides. In $\triangle ABC$:

1. $\angle A$ is included between $\overline{AC}$ and $\overline{AB}$.

2. $\angle B$ is included between $\overline{BA}$ and $\overline{BC}$.

3. $\angle C$ is included between $\overline{CA}$ and $\overline{CB}$.

Opposite Sides and Opposite Angles in a Triangle

For each side of a triangle, there is one vertex of the triangle that is not an endpoint of that side. For example, in $\triangle ABC$, C is not an endpoint of $\overline{AB}$. We say that $\overline{AB}$ is the side **opposite** $\angle C$ and that $\angle C$ is the angle **opposite** side $\overline{AB}$. Similarly, $\overline{AC}$ is opposite $\angle B$ and $\angle B$ is opposite $\overline{AC}$; also $\overline{BC}$ is opposite $\angle A$ and $\angle A$ is opposite $\overline{BC}$.

The length of a side of a triangle may be represented by the lowercase form of the letter naming the opposite vertex. For example, in $\triangle ABC$, $BC = a$, $CA = b$, and $AB = c$.

Classifying Triangles According to Sides

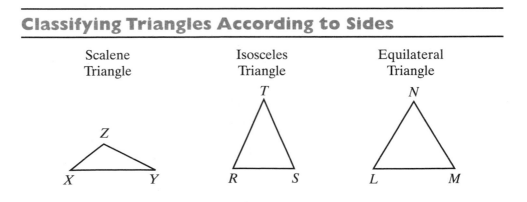

Scalene
Triangle

Isosceles
Triangle

Equilateral
Triangle

DEFINITION _____

A **scalene triangle** is a triangle that has no congruent sides.

An **isosceles triangle** is a triangle that has two congruent sides.

An **equilateral triangle** is a triangle that has three congruent sides.

Parts of an Isosceles Triangle

In isosceles triangle RST, the two congruent sides, $\overline{TR}$ and $\overline{TS}$, are called the **legs** of the triangle. The third side, $\overline{RS}$, is called the **base**.

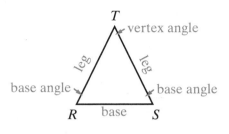

The angle formed by the two congruent sides of the isosceles triangle, $\angle T$, is called the **vertex angle** of the isosceles triangle. The vertex angle is the angle opposite the base.

The angles whose vertices are the endpoints of the base of the triangle, $\angle R$ and $\angle S$, are called the **base angles** of the isosceles triangle. The base angles are opposite the legs.

Classifying Triangles According to Angles

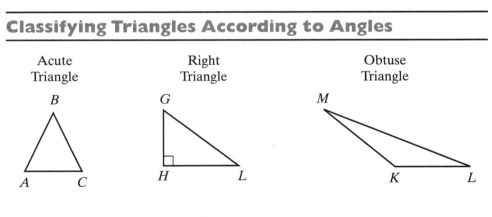

| Acute Triangle | Right Triangle | Obtuse Triangle |

DEFINITION _____

An **acute triangle** is a triangle that has three acute angles.

A **right triangle** is a triangle that has a right angle.

An **obtuse triangle** is a triangle that has an obtuse angle.

In each of these triangles, two of the angles may be congruent. In an acute triangle, all of the angles may be congruent.

DEFINITION _____

An **equiangular triangle** is a triangle that has three congruent angles.

Right Triangles

In right triangle GHL, the two sides of the triangle that form the right angle, $\overline{GH}$ and $\overline{HL}$, are called the **legs** of the right triangle. The third side of the triangle, $\overline{GL}$, the side opposite the right angle, is called the **hypotenuse**.

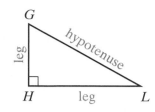

EXAMPLE 1

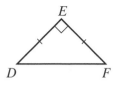

In $\triangle DEF$, $m\angle E = 90$ and $EF = ED$.

Answers

a. Classify the triangle according to sides.

The triangle is an isosceles triangle.

b. Classify the triangle according to angles.

The triangle is a right triangle.

c. What sides are the legs of the triangle?

The legs of the triangle are $\overline{EF}$ and $\overline{ED}$.

d. What side is opposite the right angle?

$\overline{FD}$ is opposite the right angle.

e. What angle is opposite $\overline{EF}$?

Angle D is opposite $\overline{EF}$.

f. What angle is included between $\overline{EF}$ and $\overline{FD}$?

Angle F is between $\overline{EF}$ and $\overline{FD}$.

Using Diagrams in Geometry

In geometry, diagrams are used to display and relate information. Diagrams are composed of line segments that are parallel or that intersect to form angles.

From a diagram, we may assume that:

- A line segment is a part of a straight line.
- The point at which two segments intersect is a point on each segment.
- Points on a line segment are between the endpoints of that segment.
- Points on a line are collinear.
- A ray in the interior of an angle with its endpoint at the vertex of the angle separates the angle into two adjacent angles.

From a diagram, we may NOT assume that:

- The measure of one segment is greater than, or is equal to, or is less than that of another segment.
- A point is a midpoint of a line segment.

- The measure of one angle is greater than, is equal to, or is less than that of another angle.
- Lines are perpendicular or that angles are right angles (unless indicated with an angle bracket).
- A given triangle is isosceles or equilateral (unless indicated with tick marks).
- A given quadrilateral is a parallelogram, rectangle, square, rhombus, or trapezoid.

EXAMPLE 2

Tell whether or not each statement may be assumed from the given diagram.

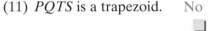

(1) $\overleftrightarrow{PR}$ is a straight line.   Yes

(2) S, T, and R are collinear.   Yes

(3) $\angle PSR$ is a right angle.   No

(4) $\triangle QTR$ is isosceles.   No

(5) $\angle QRW$ is adjacent to $\angle WRU$.   Yes

(6) Q is between V and T.   Yes

(7) $\overline{PQR}$ and $\overline{VQT}$ intersect.   Yes

(8) R is the midpoint of $\overline{TU}$.   No

(9) $QT < PS$ No

(10) m$\angle QRW >$ m$\angle VQP$ No

(11) $PQTS$ is a trapezoid.   No

Exercises

Writing About Mathematics

1. Is the statement "A triangle consists of three line segments" a good definition? Justify your answer.

2. Explain the difference between the legs of an isosceles triangle and the legs of a right triangle.

Developing Skills

In 3 and 4, name the legs and the hypotenuse in each right triangle shown.

3.

4.

In 5 and 6, name the legs, the base, the vertex angle, and the base angles in each isosceles triangle shown.

5.

6.

In 7–10, sketch, using pencil and paper or geometry software, each of the following. Mark congruent sides and right angles with the appropriate symbols.

7. An acute triangle that is isosceles

8. An obtuse triangle that is isosceles

9. A right triangle that is isosceles

10. An obtuse triangle that is scalene

11. The vertex angle of an isosceles triangle is $\angle ABC$. Name the base angles of this triangle.

12. In $\triangle DEF$, $\overline{DE}$ is included between which two angles?

13. In $\triangle RST$, $\angle S$ is included between which two sides?

14. a. Name three things that can be assumed from the diagram to the right.

 b. Name three things that can *not* be assumed from the diagram to the right.

Applying Skills

15. The degree measures of the acute angles of an obtuse triangle can be represented by $5a + 12$ and $3a - 2$. If the sum of the measures of the acute angles is 82, find the measure of each of the acute angles.

16. The measures of the sides of an equilateral triangle are represented by $x + 12, 3x - 8$, and $2x + 2$. What is the measure of each side of the triangle?

17. The lengths of the sides of an isosceles triangle are represented by $2x + 5, 2x - 6$, and $3x - 3$. What are the lengths of the sides of the triangle? (*Hint:* the length of a line segment is a positive quantity.)

18. The measures of the sides of an isosceles triangle are represented by $x + 5, 3x + 13$, and $4x + 11$. What are the measures of each side of the triangle? Two answers are possible.

CHAPTER SUMMARY

Undefined Terms
- set, point, line, plane

Definitions to Know
- A **collinear set of points** is a set of points all of which lie on the same straight line.
- A **noncollinear set of points** is a set of three or more points that do not all lie on the same straight line.
- The **distance between two points on the real number line** is the absolute value of the difference of the coordinates of the two points.
- *B* **is between** *A* **and** *C* if and only if *A*, *B*, and *C* are distinct collinear points and $AB + BC = AC$.
- A **line segment**, or a **segment**, is a set of points consisting of two points on a line, called endpoints, and all of the points on the line between the endpoints.
- The **length** or **measure of a line segment** is the distance between its endpoints.
- **Congruent segments** are segments that have the same measure.
- The **midpoint of a line segment** is a point of that line segment that divides the segment into two congruent segments.
- The **bisector of a line segment** is any line, or subset of a line, that intersects the segment at its midpoint.
- A line segment, $\overline{RS}$, is the **sum of two line segments**, $\overline{RP}$ and $\overline{PS}$, if *P* is between *R* and *S*.
- Two points, *A* and *B*, are **on one side of a point** *P* if *A*, *B*, and *P* are collinear and *P* is not between *A* and *B*.
- A **half-line** is a set of points on one side of a point.
- A **ray** is a part of a line that consists of a point on the line, called an endpoint, and all the points on one side of the endpoint.
- **Opposite rays** are two rays of the same line with a common endpoint and no other point in common.
- An **angle** is a set of points that is the union of two rays having the same endpoint.
- A **straight angle** is an angle that is the union of opposite rays and whose degree measure is 180.
- An **acute angle** is an angle whose degree measure is greater than 0 and less than 90.
- A **right angle** is an angle whose degree measure is 90.
- An **obtuse angle** is an angle whose degree measure is greater than 90 and less than 180.

- **Congruent angles** are angles that have the same measure.
- A **bisector of an angle** is a ray whose endpoint is the vertex of the angle, and that divides that angle into two congruent angles.
- **Perpendicular lines** are two lines that intersect to form right angles.
- The **distance from a point to a line** is the length of the perpendicular from the point to the line.
- If point P is a point in the interior of $\angle RST$ and $\angle RST$ is not a straight angle, or if P is any point not on straight angle RST, then $\angle RST$ is the **sum of two angles**, $\angle RSP$ and $\angle PST$.
- A **polygon** is a closed figure in a plane that is the union of line segments such that the segments intersect only at their endpoints and no segments sharing a common endpoint are collinear.
- A **triangle** is a polygon that has exactly three sides.
- A **scalene triangle** is a triangle that has no congruent sides.
- An **isosceles triangle** is a triangle that has two congruent sides.
- An **equilateral triangle** is a triangle that has three congruent sides.
- An **acute triangle** is a triangle that has three acute angles.
- A **right triangle** is a triangle that has a right angle.
- An **obtuse triangle** is a triangle that has an obtuse angle.
- An **equiangular triangle** is a triangle that has three congruent angles.

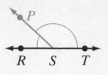

Properties of the Real Number System

	Addition	**Multiplication**
Closure	$a + b$ is a real number.	$a \cdot b$ is a real number.
Commutative Property	$a + b = b + a$	$a \cdot b = b \cdot a$
Associative Property	$(a + b) + c = a + (b + c)$	$a \times (b \times c) = (a \times b) \times c$
Identity Property	$a + 0 = a$ and $0 + a = a$	$a \cdot 1 = a$ and $1 \cdot a = a$
Inverse Property	$a + (-a) = 0$	$a \neq 0, a \cdot \frac{1}{a} = 1$
Distributive Property	$a(b + c) = ab + ac$ and $ab + ac = a(b + c)$	
Multiplication Property of Zero	$ab = 0$ if and only if $a = 0$ or $b = 0$	

VOCABULARY

1-1 Undefined term • Set • Point • Line • Straight line • Plane

1-2 Number line • Coordinate • Graph • Numerical operation • Closure property of addition • Closure property of multiplication • Commutative property of addition • Commutative property of multiplication • Associative property of addition • Associative property of multiplication • Additive identity • Multiplicative identity • Additive inverses • Multiplicative inverses • Distributive property • Multiplication property of zero •

1-3 Definition • Collinear set of points • Noncollinear set of points • Distance between two points on the real number line • Betweenness • Line segment • Segment • Length (Measure) of a line segment • Congruent segments

1-4 Midpoint of a line segment • Bisector of a line segment • Sum of two line segments

1-5 Half-line • Ray • Endpoint • Opposite rays • Angle • Sides of an angle • Vertex • Straight angle • Interior of an angle • Exterior of an angle • Degree • Acute angle • Right angle • Obtuse angle • Straight angle

1-6 Congruent angles • Bisector of an angle • Perpendicular lines • Distance from a point to a line • Foot • Sum of two angles

1-7 Polygon • Side • Triangle • Included side • Included angle • Opposite side • Opposite angle • Scalene triangle • Isosceles triangle • Equilateral triangle • Legs • Base • Vertex angle • Base angles • Acute triangle • Right triangle • Obtuse triangle • Equiangular triangle • Hypotenuse

REVIEW EXERCISES

1. Name four undefined terms.

2. Explain why "A line is like the edge of a ruler" is not a good definition.

In 3–10, write the term that is being defined.

3. The set of points all of which lie on a line.

4. The absolute value of the difference of the coordinates of two points.

5. A polygon that has exactly three sides.

6. Any line or subset of a line that intersects a line segment at its midpoint.

7. Two rays of the same line with a common endpoint and no other points in common.

8. Angles that have the same measure.

9. A triangle that has two congruent sides.

10. A set of points consisting of two points on a line and all points on the line between these two points.

11. In right triangle LMN, m$\angle M$ = 90. Which side of the triangle is the hypotenuse?

12. In isosceles triangle RST, $RS = ST$.

    a. Which side of the triangle is the base?

    b. Which angle is the vertex angle?

13. Points D, E, and F lie on a line. The coordinate of D is -3, the coordinate of E is 1, and the coordinate of F is 9.

    a. Find DE, EF, and DF.

    b. Find the coordinate of M, the midpoint of $\overline{DF}$.

    c. $\overleftrightarrow{AB}$ intersects $\overline{EF}$ at C. If $\overleftrightarrow{AB}$ is a bisector of $\overline{EF}$, what is the coordinate of C?

14. Explain why a line segment has only one midpoint but can have many bisectors.

In 15–18, use the figure shown. In polygon $ABCD$, $\overline{AC}$ and $\overline{BD}$ intersect at E, the midpoint of $\overline{BD}$.

15. Name two straight angles.

16. What angle is the sum of $\angle ADE$ and $\angle EDC$?

17. Name two congruent segments.

18. What line segment is the sum of $\overline{AE}$ and $\overline{EC}$?

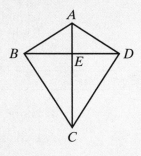

In 19–22, use the figure shown. $\overline{AB} \perp \overline{BC}$ and $\overrightarrow{BD}$ bisects $\angle ABC$.

19. Name two congruent angles.

20. What is the measure of $\angle ABD$?

21. Name a pair of angles the sum of whose measures is 180.

22. Does $AB + BC = AC$? Justify your answer.

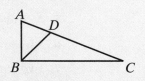

Exploration

Euclidean geometry, which we have been studying in this chapter, focuses on the plane. **Non-Euclidian geometry** focuses on surfaces other than the plane. For instance, *spherical geometry* focuses on the sphere. In this Exploration, you will draw on a spherical object, such as a grapefruit or a Styrofoam ball, to relate terms from Euclidean geometry to spherical geometry.

1. Draw a point on your sphere. How does this point compare to a point in Euclidean geometry?

2. The shortest distance between two points is called a **geodesic**. In Euclidean geometry, a line is a geodesic. Draw a second point on your sphere. Connect the points with a geodesic and extend it as far as possible. How does this geodesic compare to a line in Euclidean geometry?

3. Draw a second pair of points and a second geodesic joining them on your sphere. The intersection of the geodesics forms angles. How do these angles compare to an angle in Euclidean geometry?

4. Draw a third pair of points and a third geodesic joining them on your sphere. This will form triangles, which by definition are polygons with exactly three sides. How does a triangle on a sphere compare to a triangle in Euclidean geometry? Consider both sides and angles.

CHAPTER

2

LOGIC

Mathematicians throughout the centuries have used logic as the foundation of their understanding of the relationships among mathematical ideas. In the late 17th century, Gottfried Leibniz (1646–1716) organized logical discussion into a systematic form, but he was ahead of the mathematical thinking of his time and the value of his work on logic was not recognized.

It was not until the 19th century that George Boole (1815–1864), the son of an English shopkeeper, developed logic in a mathematical context, representing sentences with symbols and carefully organizing the possible relationships among those sentences.

Boole corresponded with Augustus DeMorgan (1806–1871) with whom he shared his work on logic. Two important relationships of logic are known today as DeMorgan's Laws.

$$\text{not } (p \text{ and } q) = (\text{not } p) \text{ or } (\text{not } q)$$
$$\text{not } (p \text{ or } q) = (\text{not } p) \text{ and } (\text{not } q)$$

Boolean algebra is key in the development of computer science and circuit design.

2-1 SENTENCES, STATEMENTS, AND TRUTH VALUES

Logic is the science of reasoning. The principles of logic allow us to determine if a statement is true, false, or uncertain on the basis of the truth of related statements.

We solve problems and draw conclusions by reasoning from what we know to be true. All reasoning, whether in mathematics or in everyday living, is based on the ways in which we put sentences together.

Sentences and Their Truth Values

When we can determine that a statement is true or that it is false, that statement is said to have a **truth value**. In this chapter we will study the ways in which statements with known truth values can be combined by the *laws of logic* to determine the truth value of other statements.

In the study of logic, we use simple declarative statements that state a fact. That fact may be either true or false. We call these statements **mathematical sentences**. For example:

1. Congruent angles are angles that have the same measure.　　True mathematical sentence

2. $17 - 5 = 12$　　True mathematical sentence

3. The Brooklyn Bridge is in New York.　　True mathematical sentence

4. $17 + 3 = 42$　　False mathematical sentence

5. The Brooklyn Bridge is in California.　　False mathematical sentence

Nonmathematical Sentences and Phrases

Sentences that do not state a fact, such as questions, commands, or exclamations, are not sentences that we use in the study of logic. For example:

1. Did you have soccer practice today?
This is not a mathematical sentence because it asks a question.

2. Go to your room.
This is not a mathematical sentence because it gives a command.

An incomplete sentence or part of a sentence, called a **phrase**, is not a mathematical sentence and usually has no truth value. For example:

1. Every parallelogram
This is not a mathematical sentence.

2. $19 - 2$
This is not a mathematical sentence.

Some sentences are true for some persons and false for others. For example:

1. I enjoy reading historical novels.

2. Summer is the most pleasant season.

3. Basketball is my favorite sport.

Conclusions based on sentences such as these do not have the same truth value for all persons. We will not use sentences such as these in the study of logic.

Open Sentences

In the study of algebra, we worked with **open sentences**, that is, sentences that contain a variable. The truth value of the open sentence depended on the value of the variable. For example, the open sentence $x + 2 = 5$ is true when $x = 3$ and false for all other values of x. In some sentences, a pronoun, such as *he*, *she*, or *it*, acts like a variable and the name that replaces the pronoun determines the truth value of the sentence.

1. $x \div 2 = 8$ Open sentence: the variable is x.

2. He broke my piggybank.   Open sentence: the variable is *he*.

3. Jenny found it behind the sofa.   Open sentence: the variable is *it*.

In previous courses, we learned that the **domain** or **replacement set** is the set of all elements that are possible replacements for the variable. The element or elements from the domain that make the open sentence true is the **solution set** or **truth set**. For instance:

Open sentence: $14 - x = 9$
Variable: x
Domain: $\{1, 2, 3, 4, 5\}$
Solution set: $\{5\}$
When $x = 5$, then $14 - 5 = 9$ is a true sentence.

The method we use for sentences in mathematics is the same method we apply to sentences in ordinary conversation. Of course, we would not use a domain like $\{1, 2, 3, 4\}$ for the open sentence "It is the third month of the year." Common sense tells us to use a domain consisting of the names of months. The following example compares this open sentence with the algebraic sentence used above. Open sentences, variables, domains, and solution sets behave in exactly the same way.

Open sentence: It is the third month of the year.
Variable: It
Domain: {Names of months}
Solution set: {March}
When "It" is replaced by "March," then "March is the third month of the year" is a true sentence.

Sometimes a solution set contains more than one element. If Elaine has two brothers, Ken and Kurt, then for her, the sentence "He is my brother" has the solution set {Ken, Kurt}. Here the domain is the set of boys' names. Some people have no brothers. For them, the solution set for the open sentence "He is my brother" is the empty set, $\emptyset$ or { }.

EXAMPLE 1

Identify each of the following as a true sentence, a false sentence, an open sentence, or not a mathematical sentence at all.

Answers

a. Football is a water sport.   False sentence

b. Football is a team sport.   True sentence

c. He is a football player.   Open sentence: the variable is *he*.

d. Do you like football?   Not a mathematical sentence: this is a question.

e. Read this book.   Not a mathematical sentence: this is a command.

f. $3x - 7 = 11$ Open sentence: the variable is *x*.

g. $3x - 7$ Not a mathematical sentence: this is a phrase or a binomial.

EXAMPLE 2

Use the replacement set $\left\{2, \frac{2\pi}{3}, 2.5, 2\sqrt{2}\right\}$ to find the truth set of the open sentence "It is an irrational number."

Solution   Both 2 and 2.5 are rational numbers.

Both π and $\sqrt{2}$ are irrational numbers. Since the product or quotient of a rational number and an irrational number is an irrational number, $\frac{2\pi}{3}$ and $2\sqrt{2}$ are irrational.

Answer $\left\{\frac{2\pi}{3}, 2\sqrt{2}\right\}$

Statements and Symbols

A sentence that can be judged to be true or false is called a **statement** or a **closed sentence**. In a statement, there are no variables.

A closed sentence is said to have a *truth value*. The truth values, true and false, are indicated by the symbols **T** and **F**.

Negations

In the study of logic, you will learn how to make new statements based upon statements that you already know. One of the simplest forms of this type of reasoning is negating a statement.

The **negation** of a statement always has the opposite truth value of the given or original statement and is usually formed by adding the word *not* to the given statement. For example:

1. Statement: Neil Armstrong walked on the moon.   (True)
  Negation: Neil Armstrong did *not* walk on the moon.   (False)

2. Statement: A duck is a mammal.   (False)
  Negation: A duck is *not* a mammal.   (True)

There are other ways to insert the word *not* into a statement to form its negation. One method starts the negation with the phrase "It is not true that . . ." For example:

3. Statement: A carpenter works with wood.   (True)
  Negation: It is *not* true that a carpenter works with wood.   (False)
  Negation: A carpenter does *not* work with wood.   (False)

Both negations express the same false statement.

Logic Symbols

The basic element of logic is a simple declarative sentence. We represent this basic element by a single, lowercase letter. Although any letter can be used to represent a sentence, the use of p, q, r, and s are the most common. For example, p might represent "Neil Armstrong walked on the moon."

The symbol that is used to represent the negation of a statement is the symbol $\sim$ placed before the letter that represents the given statement. Thus, if p represents "Neil Armstrong walked on the moon," then $\sim p$ represents "Neil Armstrong did *not* walk on the moon." The symbol $\sim p$ is read "not p."

Symbol	Statement in Words	Truth Value
p	There are 7 days in a week.	True
$\sim p$	There are not 7 days in a week.	False
q	$8 + 9 = 16$	False
$\sim q$	$8 + 9 \neq 16$	True

When p is true, then its negation $\sim p$ is false. When q is false, then its negation $\sim q$ is true.

▶ **A statement and its negation have opposite truth values.**

It is possible to use more than one negation in a sentence. Each time another negation is included, the truth value of the statement will change. For example:

Symbol	*Statement in Words*	*Truth Value*
r	A dime is worth 10 cents.	True
$\sim r$	A dime is not worth 10 cents.	False
$\sim(\sim r)$	It is not true that a dime is not worth 10 cents.	True

We do not usually use sentences like the third. Note that just as in the set of real numbers, $-(-a) = a$, $\sim(\sim r)$ always has the same truth value as r. We can use r in place of $\sim(\sim r)$. Therefore, we can negate a sentence that contains the word *not* by omitting that word.

$\sim r$: A dime is not worth 10 cents.
$\sim(\sim r)$: A dime is worth 10 cents.

The relationship between a statement p and its negation $\sim p$ can be summarized in the table at the right. When p is true, $\sim p$ is false. When p is false, $\sim p$ is true.

p	$\sim p$
T	F
F	T

EXAMPLE 3

In this example, symbols are used to represent statements. The truth value of each statement is given.

k: Oatmeal is a cereal.   (True)
m: Massachusetts is a city.   (False)

For each sentence given in symbolic form:

a. Write a complete sentence in words to show what the symbols represent.

b. Tell whether the statement is true or false.

Answers

(1) $\sim k$ **a.** Oatmeal is not a cereal.   **b.** False
(2) $\sim m$ **a.** Massachusetts is not a city.   **b.** True

Exercises

Writing About Mathematics

1. Explain the difference between the use of the term "sentence" in the study of grammar and in the study of logic.

2. a. Give an example of a statement that is true on some days and false on others.

   b. Give an example of a statement that is true for some people and false for others.

   c. Give an example of a statement that is true in some parts of the world and false in others.

Developing Skills

In 3–10, tell whether or not each of the following is a mathematical sentence.

3. Thanksgiving is on the fourth Thursday in November.

4. Albuquerque is a city in New Mexico.

5. Where did you go?

6. Twenty sit-ups, 4 times a week

7. Be quiet.

8. If Patrick leaps

9. $y - 7 = 3y + 4$

10. Tie your shoe.

In 11–18, all of the sentences are open sentences. Find the variable in each sentence.

11. She is tall.

12. We can vote at age 18.

13. $2y \geq 17$

14. $14x \div 8 = 9$

15. This country has the third largest population.

16. He hit a home run in the World Series.

17. It is my favorite food.

18. It is a fraction.

In 19–26: **a.** Tell whether each sentence is true, false, or open. **b.** If the sentence is an open sentence, identify the variable.

19. The Statue of Liberty was given to the United States by France.

20. They gained custody of the Panama Canal on December 31, 1999.

21. Tallahassee is a city in Montana.

22. A pentagon is a five-sided polygon.

23. $6x + 4 = 16$

24. $6(10) + 4 = 16$

25. $6(2) + 4 = 16$

26. $2^3 = 3^2$

In 27–31, find the truth set for each open sentence using the replacement set {Nevada, Illinois, Massachusetts, Alaska, New York}.

27. Its capital is Albany.

28. It does not border on or touch an ocean.

29. It is on the east coast of the United States.

30. It is one of the states of the United States of America.

31. It is one of the last two states admitted to the United States of America.

In 32–39, use the domain {square, triangle, rectangle, parallelogram, rhombus, trapezoid} to find the truth set for each open sentence.

32. It has three and only three sides.

33. It has exactly six sides.

34. It has fewer than four sides.

35. It contains only right angles.

36. It has four sides that are all equal in measure.

37. It has two pairs of opposite sides that are parallel.

38. It has exactly one pair of opposite sides that are parallel.

39. It has interior angles with measures whose sum is 360 degrees.

In 40–47, write the negation of each sentence.

40. The school has an auditorium.

41. A stop sign is painted red.

42. The measure of an obtuse angle is greater than 90°.

43. There are 1,760 yards in a mile.

44. Michigan is not a city.

45. $14 \times 2 - 16 = 12$

46. $3 + 4 + 5 \neq 6$

47. Today is not Wednesday.

In 48–56, for each given sentence: **a.** Write the sentence in symbolic form using the symbols shown below. **b.** Tell whether the sentence is true, false, or open.

> Let p represent "A snake is a reptile."
> Let q represent "A frog is a snake."
> Let r represent "Her snake is green."

48. A snake is a reptile.

49. A snake is not a reptile.

50. A frog is a snake.

51. A frog is not a snake.

52. Her snake is green.

53. Her snake is not green.

54. It is not true that a frog is a snake.

55. It is not the case that a snake is not a reptile.

56. It is not the case that a frog is not a snake.

In 57–64, the symbols represent sentences.

> p: Summer follows spring.
> q: August is a summer month.
> r: A year has 12 months.
> s: She likes spring.

For each sentence given in symbolic form: **a.** Write a complete sentence in words to show what the symbols represent. **b.** Tell whether the sentence is true, false, or open.

57. $\sim p$

58. $\sim q$

59. $\sim r$

60. $\sim s$

61. $\sim(\sim p)$

62. $\sim(\sim q)$

63. $\sim(\sim r)$

64. $\sim(\sim s)$

2-2 CONJUNCTIONS

We have identified simple sentences that have a truth value. Often we wish to use a connecting word to form a compound sentence. In mathematics, sentences formed by connectives are also called **compound sentences** or **compound statements**. One of the simplest compound statements that can be formed uses the connective *and*.

In logic, a **conjunction** is a compound statement formed by combining two simple statements using the word *and*. Each of the simple statements is called a **conjunct**. When p and q represent simple statements, the conjunction **p and q** is written in symbols as $p \wedge q$. For example:

> p: A dog is an animal.
> q: A sparrow is a bird.
> $p \wedge q$: A dog is an animal and a sparrow is a bird.

This compound sentence is true because both parts are true. "A dog is an animal is true and "A sparrow is a bird" is true. When one or both parts of a conjunction are false, the conjunction is false. For example,

- "A dog is an animal and a sparrow is not a bird" is false because "A sparrow is not a bird" is the negation of a true statement and is false.

- "A dog is not an animal and a sparrow is a bird" is false because "A dog is not an animal" is the negation of a true statement and is false.

- "A dog is not an animal and a sparrow is not a bird" is false because both "A dog is not an animal" and "A sparrow is not a bird" are false.

We can draw a diagram, called a **tree diagram**, to show all possible combinations of the truth values of p and q that are combined to make the compound statement $p \wedge q$.

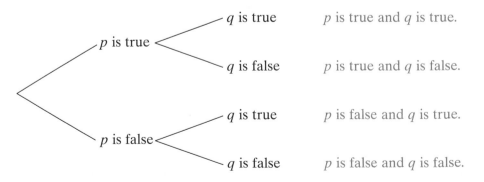

These four possible combinations of the truth values of p and q can be displayed in a chart called a **truth table**. The truth table can be used to show the possible truth values of a compound statement that is made up of two simple statements.

For instance, write a truth table for $p \wedge q$.

STEP 1. In the first column, we list the truth values of p. For each possible truth value of p, there are two possible truth values for q. Therefore, we list T twice and F twice.

p		
T		
T		
F		
F		

STEP 2. In the second column, we list the truth values of q. In the two rows in which p is true, list q as true in one and false in the other. In the two rows in which p is false, list q as true in one and false in the other.

p	q	
T	T	
T	F	
F	T	
F	F	

STEP 3. In the last column, list the truth values of the conjunction, $p \wedge q$. The conjunction is true only when both p and q are true. The conjunction is false when one or both conjuncts are false.

p	q	$p \wedge q$
T	T	T
T	F	F
F	T	F
F	F	F

▶ **The conjunction, p and q, is true only when both parts are true: p must be true and q must be true.**

For example, let p represent "It is spring," and let q represent "It is March."

CASE 1 *Both p and q are true.*

On March 30, "It is spring" is true and "It is March" is true. Therefore, "It is spring and it is March" is true.

CASE 2 *p is true and q is false.*

On April 30, "It is spring" is true and "It is March" is false. Therefore, "It is spring and it is March" is false.

CASE 3 *p is false and q is true.*

On March 10, "It is spring" is false and "It is March" is true. Therefore, "It is spring and it is March" is false.

CASE 4 *Both p and q are false.*

On February 28, "It is spring" is false and "It is March" is false. Therefore, "It is spring and it is March" is false.

A compound sentence may contain both negations and conjunctions at the same time. For example:

Let *p* represent "Ten is divisible by 2."
Let *q* represent "Ten is divisible by 3."

Then $p \wedge \sim q$ represents "Ten is divisible by 2 and ten is not divisible by 3." Here *p* is true, *q* is false, and $\sim q$ is true. Then for $p \wedge \sim q$, both parts are true so the conjunction is true. This can be summarized in the following table.

p	*q*	$\sim q$	$p \wedge \sim q$
T	F	T	T

EXAMPLE I

Let *p* represent "Albany is the capital of New York State." (True)

Let *q* represent "Philadelphia is the capital of Pennsylvania." (False)

For each given sentence:

a. Write the sentence in symbolic form.

b. Tell whether the statement is true or false.

(1) Albany is the capital of New York State and Philadelphia is the capital of Pennsylvania.

(2) Albany is the capital of New York State and Philadelphia is not the capital of Pennsylvania.

(3) Albany is not the capital of New York State and Philadelphia is the capital of Pennsylvania.

(4) Albany is not the capital of New York State and Philadelphia is not the capital of Pennsylvania.

(5) It is not true that Albany is the capital of New York State and Philadelphia is the capital of Pennsylvania.

Solution (1) The statement is a conjunction. Since *p* is true and *q* is false, the statement is false.

p	*q*	$p \wedge q$
T	F	F

Answers: **a.** $p \wedge q$ **b.** False

(2) The statement is a conjunction. Since *p* is true and $\sim q$ is true, the statement is true.

p	*q*	$\sim q$	$p \wedge \sim q$
T	F	T	T

Answers: **a.** $p \wedge \sim q$ **b.** True

(3) The statement is a conjunction. Since ~p is false and q is false, the statement is false.

p	~p	q	~$p \wedge q$
T	F	F	F

Answers: **a.** ~$p \wedge q$ **b.** False

(4) The statement is a conjunction. Since ~p is false and ~q is true, the statement is false.

p	~p	q	~q	~$p \wedge$ ~q
T	F	F	T	F

Answers: **a.** ~$p \wedge$ ~q **b.** False

(5) The phrase "It is not true that" applies to the entire conjunction. Since p is true and q is false, $(p \wedge q)$ is false and the negation of $(p \wedge q)$ is true.

p	q	$p \wedge q$	~$(p \wedge q)$
T	F	F	T

Answers: **a.** ~$(p \wedge q)$ **b.** True

EXAMPLE 2

Use the domain {1, 2, 3, 4} to find the truth set for the open sentence

$$(x < 3) \wedge (x \text{ is a prime})$$

Solution

Let $x = 1$	$(1 < 3) \wedge (1$ is a prime$)$	T $\wedge$ F = False
Let $x = 2$	$(2 < 3) \wedge (2$ is a prime$)$	T $\wedge$ T = True
Let $x = 3$	$(3 < 3) \wedge (3$ is a prime$)$	F $\wedge$ T = False
Let $x = 4$	$(4 < 3) \wedge (4$ is a prime$)$	F $\wedge$ F = False

A conjunction is true only when both simple sentences are true. This condition is met here when $x = 2$. Thus, the truth set or solution set is {2}.

Answer   {2}

EXAMPLE 3

Three sentences are written below. The truth values are given for the first two sentences. Determine whether the third sentence is true, is false, or has an uncertain truth value.

Today is Friday and I have soccer practice.	(False)
Today is Friday.	(True)
I have soccer practice.	(?)

Solution   Since the conjunction is false, at least one of the conjuncts must be false. But "Today is Friday" is true. Therefore, "I have soccer practice" must be false.

Answer   "I have soccer practice" is false.

EXAMPLE 4

Three sentences are written below. The truth values are given for the first two sentences. Determine whether the third sentence is true, is false, or has an uncertain truth value.

Today is Monday and the sun is shining.	(False)
Today is Monday.	(False)
The sun is shining.	(?)

Solution (1) Use symbols to represent the sentences. Indicate their truth values.

p: Today is Monday.	(False)
q: The sun is shining.	(?)
$p \wedge q$: Today is Monday and the sun is shining.	(False)

(2) Construct a truth for the conjunction. Study the truth values of p and $p \wedge q$.
Since $p \wedge q$ is false, the last three rows apply. Since p is false, the choices are narrowed to the last two rows. Therefore q could be either true or false.

p	q	$p \wedge q$
T	T	T
T	F	F
F	T	F
F	F	F

Answer The truth value of "The sun is shining" is uncertain.

Exercises

Writing About Mathematics

1. Is the negation of a conjunction, $\sim(p \wedge q)$, the same as $\sim p \wedge \sim q$? Justify your answer.

2. What must be the truth values of p, q, and r in order for $(p \wedge q) \wedge r$ to be true? Explain your answer.

Developing Skills

In 3–12, write each sentence in symbolic form, using the given symbols.

Let p represent "It is hot."
Let q represent "It is raining."
Let r represent "The sky is cloudy."

3. It is hot and it is raining.

4. It is hot and the sky is cloudy.

5. It is not hot.

6. It is not hot and the sky is cloudy.

7. It is raining and the sky is not cloudy.

8. It is not hot and it is not raining.

9. The sky is not cloudy and it is not hot.

10. The sky is not cloudy and it is hot.

11. It is not the case that it is hot and it is raining.

12. It is not the case that it is raining and it is not hot.

In 13–20, using the truth value for each given statement, tell if the conjunction is true or false.

A piano is a percussion instrument. (True)
A piano has 88 keys. (True)
A flute is a percussion instrument. (False)
A trumpet is a brass instrument. (True)

13. A flute is a percussion instrument and a piano is a percussion instrument.

14. A flute is a percussion instrument and a trumpet is a brass instrument.

15. A piano has 88 keys and is a percussion instrument.

16. A piano has 88 keys and a trumpet is a brass instrument.

17. A piano is not a percussion instrument and a piano does not have 88 keys.

18. A trumpet is not a brass instrument and a piano is a percussion instrument.

19. A flute is not a percussion instrument and a trumpet is a brass instrument.

20. It is not true that a piano is a percussion instrument and has 88 keys.

In 21–28, complete each sentence with "true" or "false" to make a correct statement.

21. When p is true and q is true, then $p \wedge q$ is _____.

22. When p is false, then $p \wedge q$ is _____.

23. If p is true, or q is true, but not both, then $p \wedge q$ is _____.

24. When $p \wedge q$ is true, then p is _____ and q is _____.

25. When $p \wedge \sim q$ is true, then p is _____ and q is _____.

26. When $\sim p \wedge q$ is true, then p is _____ and q is _____.

27. When p is false and q is true, then $\sim(p \wedge q)$ is _____.

28. If both p and q are false, then $\sim p \wedge \sim q$ is _____.

Applying Skills

In 29–36, three sentences are written. The truth values are given for the first two sentences. Determine whether the third sentence is true, is false, or has an uncertain truth value.

29. It is noon and I get lunch. (True)
It is noon. (True)
I get lunch. (?)

30. I have the hiccups and I drink some water. (False)
I have the hiccups. (True)
I drink some water. (?)

31. I have the hiccups and I drink some water. (False)
I have the hiccups. (False)
I drink some water. (?)

32. I play tennis and Anna plays golf. (False)
I play tennis. (True)
Anna plays golf. (?)

33. Pam sees a movie and Pam loves going to the theater. (True)
Pam sees a movie. (True)
Pam loves going to the theater. (?)

34. Jon and Edith like to eat ice cream. (False)
Jon likes to eat ice cream. (True)
Edith likes to eat ice cream. (?)

35. Jordan builds model trains and model planes. (False)
Jordan builds model trains. (False)
Jordan builds model planes. (?)

36. Bethany likes to chat and surf on the internet. (False)
Bethany likes to surf on the internet. (True)
Bethany likes to chat on the internet. (?)

In 37 and 38, a compound sentence is given using a conjunction. Use the truth value of the compound sentence to determine whether each sentence that follows is true or false.

37. In winter I wear a hat and scarf. (True)

   a. In winter I wear a hat.

   b. In winter I wear a scarf.

   c. In winter I do not wear a hat.

38. I do not practice and I know that I should. (True)

   a. I do not practice.

   b. I know that I should practice.

   c. I practice.

2-3 DISJUNCTIONS

In logic, a **disjunction** is a compound statement formed by combining two simple statements using the word *or*. Each of the simple statements is called a **disjunct**. When p and q represent simple statements, the disjunction p **or** q is written in symbols as $p \vee q$. For example:

   p: Andy rides his bicycle to school.
   q: Andy walks to school.
   $p \vee q$: Andy rides his bicycle to school or Andy walks to school.

In this example, when is the disjunction $p \vee q$ true and when is it false?

1. On Monday, Andy rode his bicycle part of the way to school when he met a friend. Then he walked the rest of the way to school with his friend. Here p is true and q is true. The disjunction $p \vee q$ is true.

2. On Tuesday, Andy rode his bicycle to school and did not walk to school. Here p is true and q is false. The disjunction $p \vee q$ is true.

3. On Wednesday, Andy did not ride his bicycle to school and walked to school. Here p is false and q is true. The disjunction $p \vee q$ is true.

4. On Thursday, it rained so Andy's father drove him to school. Andy did not ride his bicycle to school and did not walk to school. Here p is false and q is false. The disjunction $p \vee q$ is false.

▶ **The disjunction p or q is true when any part of the compound sentence is true: p is true, q is true, or both p and q are true.**

The only case in which the disjunction p or q is false is when both p and q are false. The truth values of the disjunction $p \vee q$ are summarized in the truth table to the right. The possible combinations of the truth values of p and of q, shown in the first two columns, are the same as those used in the truth table of the conjunction. The third column gives the truth values for the disjunction, $p \vee q$.

p	q	$p \vee q$
T	T	T
T	F	T
F	T	T
F	F	F

EXAMPLE I

Use the following statements:

> Let k represent "Kurt plays baseball."
> Let a represent "Alicia plays baseball."
> Let n represent "Nathan plays soccer."

Write each given sentence in symbolic form.

Answers

a. Kurt or Alicia play baseball. $k \vee a$

b. Kurt plays baseball or Nathan plays soccer. $k \vee n$

c. Alicia plays baseball or Alicia does not play baseball. $a \vee \sim a$

d. It is not true that Kurt or Alicia play baseball. $\sim(k \vee a)$

e. Either Kurt does not play baseball or Alicia does not play baseball. $\sim k \vee \sim a$

f. It's not the case that Alicia or Kurt play baseball. $\sim(a \vee k)$

EXAMPLE 2

Symbols are used to represent three statements. For each statement, the truth value is noted.

> k: "Every line segment has a midpoint." (True)
> m: "A line has a midpoint." (False)
> q: "A ray has one endpoint." (True)

For each sentence given in symbolic form:

a. Write a complete sentence in words to show what the symbols represent.

b. Tell whether the statement is true or false.

Answers

(1) $k \lor q$

    a. Every line segment has a midpoint or every ray has one endpoint.

    b. $T \lor T$ is a true disjunction.

(2) $k \lor m$

    a. Every line segment has a midpoint or a line has a midpoint.

    b. $T \lor F$ is a true disjunction.

(3) $m \lor {\sim}q$

    a. A line has a midpoint or a ray does not have one endpoint.

    b. $F \lor {\sim}T = F \lor F$, a false disjunction.

(4) ${\sim}(m \lor q)$

    a. It is not the case that a line has a midpoint or a ray has one endpoint.

    b. ${\sim}(F \lor T) = {\sim}T$, a false disjunction.

EXAMPLE 3

Find the solution set of each of the following if the domain is the set of positive integers less than 8.

    a. $(x < 4) \lor (x > 3)$ **b.** $(x > 3) \lor (x \text{ is odd})$ **c.** $(x > 5) \land (x < 3)$

Solution   The domain is $\{1, 2, 3, 4, 5, 6, 7\}$.

    a. The solution set of $x < 4$ is $\{1, 2, 3\}$ and the solution set of $x > 3$ is $\{4, 5, 6, 7\}$. The solution set of the disjunction $(x < 4) \lor (x > 3)$ includes all the numbers that make $x < 4$ true together with all the numbers that make $x > 3$ true.

Answer $\{1, 2, 3, 4, 5, 6, 7\}$

    Note: The solution set of the disjunction $(x < 4) \lor (x > 3)$ is the union of the solution sets of the individual parts: $\{1, 2, 3\} \cup \{4, 5, 6, 7\} = \{1, 2, 3, 4, 5, 6, 7\}$.

    b. The solution set of $(x > 3)$ is $\{4, 5, 6, 7\}$ and the solution set of $(x \text{ is odd})$ is $\{1, 3, 5, 7\}$. The solution set of the disjunction $(x > 3) \lor (x \text{ is odd})$ includes all the numbers that make either $x > 3$ true or x is odd true.

Answer $\{1, 3, 4, 5, 6, 7\}$

    Note: The solution set of the disjunction $(x > 3) \lor (x \text{ is odd})$ is the union of the solution sets: $\{4, 5, 6, 7\} \cup \{1, 3, 5, 7\} = \{1, 3, 4, 5, 6, 7\}$.

    c. The solution set of $(x > 5)$ is $\{6, 7\}$ and the solution set of $(x < 3)$ is $\{1, 2\}$. The solution set of $(x > 5) \land (x < 3)$ is $\{6, 7\} \cap \{1, 2\}$ or the empty set, $\varnothing$.

Answer $\varnothing$

Two Uses of the Word *Or*

When we use the word *or* to mean that *one* or *both* of the simple sentences are true, we call this the **inclusive *or*.** The truth table we have just shown uses truth values for the *inclusive or*.

Sometimes, however, the word *or* is used in a different way, as in "He is in grade 9 or he is in grade 10." Here it is not possible for both simple sentences to be true at the same time. When we use the word *or* to mean that *one and only one* of the simple sentences is true, we call this the **exclusive *or*.** The truth table for the *exclusive or* will be different from the table shown for disjunction. In the *exclusive or*, the disjunction *p or q* will be true when *p* is true, or when *q* is true, but not both.

In everyday conversation, it is often evident from the context which of these uses of *or* is intended. In legal documents or when ambiguity can cause difficulties, the *inclusive or* is sometimes written as *and/or*.

▶ **We will use only the *inclusive or* in this book. Whenever we speak of disjunction, *p or q* will be true when *p* is true, when *q* is true, when both *p* and *q* are true.**

Exercises

Writing About Mathematics

1. Explain the relationship between the truth set of the negation of a statement and the complement of a set.

2. Explain the difference between the *inclusive or* and the *exclusive or*.

Developing Skills

In 3–12, for each given statement: **a.** Write the statement in symbolic form, using the symbols given below. **b.** Tell whether the statement is true or false.

Let *c* represent "A gram is 100 centigrams."	(True)
Let *m* represent "A gram is 1,000 milligrams."	(True)
Let *k* represent "A kilogram is 1,000 grams."	(True)
Let *l* represent "A gram is a measure of length."	(False)

3. A gram is 1,000 milligrams or a kilogram is 1,000 grams.

4. A gram is 100 centigrams or a gram is a measure of length.

5. A gram is 100 centigrams or 1,000 milligrams.

6. A kilogram is not 1,000 grams or a gram is not 100 centigrams.

7. A gram is a measure of length or a kilogram is 1,000 grams.

8. A gram is a measure of length and a gram is 100 centigrams.

9. It is not the case that a gram is 100 centigrams or 1,000 milligrams.

10. It is false that a kilogram is not 1,000 grams or a gram is a measure of length.

11. A gram is 100 centigrams and a kilogram is 1,000 grams.

12. A gram is not 100 centigrams or is not 1,000 milligrams, and a gram is a measure of length.

In 13–20, symbols are assigned to represent sentences.

> Let *b* represent "Breakfast is a meal."
> Let *s* represent "Spring is a season."
> Let *h* represent "Halloween is a season."

For each sentence given in symbolic form: **a.** Write a complete sentence in words to show what the symbols represent. **b.** Tell whether the sentence is true or false.

13. $s \lor h$ **14.** $b \land s$ **15.** $\sim s \lor h$ **16.** $b \land \sim h$

17. $\sim b \lor \sim s$ **18.** $\sim(s \land h)$ **19.** $\sim(b \lor \sim s)$ **20.** $\sim b \land \sim s$

In 21–27, complete each sentence with the words "true" or "false" to make a correct statement.

21. When p is true, then $p \lor q$ is _____.

22. When q is true, then $p \lor q$ is _____.

23. When p is false and q is false, then $p \lor q$ is _____.

24. When $p \lor \sim q$ is false, then p is _____ and q is _____.

25. When $\sim p \lor q$ is false, then p is _____ and q is _____.

26. When p is false and q is true, then $\sim(p \lor q)$ is _____.

27. When p is false and q is true, then $\sim p \lor \sim q$ is _____.

Applying Skills

In 28–32, three sentences are written. The truth values are given for the first two sentences. Determine whether the third sentence is true, is false, or has an uncertain truth value.

28. May is the first month of the year. (False)
January is the first month of the year. (True)
May is the first month of the year or January is the first month of the year. (?)

29. I will study more or I will fail the course. (True)
I will fail the course. (False)
I will study more. (?)

30. Jen likes to play baseball and Mason likes to play baseball. (False)
Mason likes to play baseball. (True)
Jen likes to play baseball. (?)

31. Nicolette is my friend or Michelle is my friend. (True)
Nicolette is my friend. (True)
Michelle is my friend. (?)

32. I practice the cello on Monday or I practice the piano on Monday. (True)
I do not practice the piano on Monday. (False)
I practice the cello on Monday. (?)

2-4 CONDITIONALS

A sentence such as "If I have finished my homework, then I will go to the movies" is frequently used in daily conversation. This statement is made up of two simple statements:

p: I have finished my homework.
q: I will go to the movies.

The remaining words, *if . . . then,* are the connectives.

In English, this sentence is called a complex sentence. In mathematics, however, all sentences formed using connectives are called *compound sentences* or *compound statements*.

In logic, a **conditional** is a compound statement formed by using the words *if . . . then* to combine two simple statements. When p and q represent simple statements, the conditional **if p then q** is written in symbols as $p \rightarrow q$. The symbol $p \rightarrow q$ can also be read as "p implies q" or as "p only if q."

Here is another example:

p: It is January.
q: It is winter.
$p \rightarrow q$: If it is January, then it is winter.
or
It is January implies that it is winter.
or
It is January only if it is winter.

Certainly we would agree that the compound sentence *if p then q* is true for this example: "If it is January, then it is winter." However, if we reverse the order of the simple sentences to form the new conditional *if q then p*, we will get a sentence with a different meaning:

$q \rightarrow p$: If it is winter, then it is January.

When it is winter, it does not necessarily mean that it is January. It may be February, the last days of December, or the first days of March. Changing the order in which we connect two simple statements in conditional does not always give a conditional that has the same truth value as the original.

Parts of a Conditional

The parts of the conditional *if p then q* can be identified by name:

p is the **hypothesis**, which is sometimes referred to as the **premise** or the **antecedent**. It is an assertion or a sentence that begins an argument. The hypothesis usually follows the word *if*.

q is the **conclusion**, which is sometimes referred to as the **consequent**. It is the part of a sentence that closes an argument. The conclusion usually follows the word *then*.

There are different ways to write the conditional. When the conditional uses the word *if*, the hypothesis always follows *if*. When the conditional uses the word *implies*, the hypothesis always comes before *implies*. When the conditional uses the words *only if*, the conclusion follows the words *only if*.

$p \rightarrow q$: If <u>it is January</u>, then <u>it is winter</u>.
　　　　　　　hypothesis　　　　　conclusion

$p \rightarrow q$: <u>It is January</u> implies that <u>it is winter</u>.
　　　　　　hypothesis　　　　　　　conclusion

$p \rightarrow q$: <u>It is January</u> only if <u>it is winter</u>.
　　　　　　hypothesis　　　　　conclusion

All three sentences say the same thing. We are able to draw a conclusion about the season when we know that the month is January. Although the word order of a conditional may vary, the hypothesis is always written first when using symbols.

Truth Values for the Conditional $p \rightarrow q$

In order to determine the truth value of a conditional, we will consider the statement "If you get an A in Geometry, then I will buy you a new graphing calculator." Let *p* represent the hypothesis, and let *q* represent the conclusion.

　　　　p: You get an A in Geometry.
　　　　q: I will buy you a new graphing calculator.

Determine the truth values of the conditional by considering all possible combinations of the truth values for *p* and *q*.

CASE 1

You get an A in Geometry. (*p* is true.)
I buy you a new graphing calculator. (*q* is true.)
We both keep our ends of the agreement.
The conditional statement is true.

p	*q*	$p \rightarrow q$
T	T	T

CASE 2

You get an A in Geometry. (*p* is true.)
I do not buy you a new graphing calculator. (*q* is false.)
I broke the agreement because you got an A in
Geometry but I did *not* buy you a new graphing calculator.
The conditional statement is false.

p	*q*	*p* → *q*
T	F	F

CASE 3

You do not get an A in Geometry. (*p* is false.)
I buy you a new graphing calculator. (*q* is true.)
You did *not* get an A but I bought you a new graphing
calculator anyway. Perhaps I felt that a new calculator
would help you to get an A next time. I did not break my promise. My promise
only said what I would do if you did get an A.
The conditional statement is true.

p	*q*	*p* → *q*
F	T	T

CASE 4

You do not get an A in Geometry. (*p* is false.)
I do not buy you a new graphing calculator. (*q* is false.)
Since you did not get an A, I do not have to keep our
agreement.
The conditional statement is true.

p	*q*	*p* → *q*
F	F	T

Case 2 tells us that the conditional is false only when the hypothesis is true and the conclusion is false. If a conditional is thought of as an "agreement" or a "promise," this corresponds to the case when the agreement is broken.

Cases 3 and 4 tell us that when the hypothesis is false, the conclusion may or may not be true. In other words, if you do not get an A in Geometry, I may or may not buy you a new graphing calculator.

These four cases can be summarized as follows:

▶ **A conditional is false when a true hypothesis leads to a false conclusion. In all other cases, the conditional is true.**

p	*q*	*p* → *q*
T	T	T
T	F	F
F	T	T
F	F	T

Hidden Conditionals

Often the words "if . . . then" may not appear in a statement that does suggest a conditional. Instead, the expressions "when" or "in order that" may suggest that the statement is a conditional. For example:

1. "When I finish my homework I will go to the movies."
   p → *q*: *If* I finish my homework, *then* I will go to the movies.

2. "In order to succeed, you must work hard" becomes
$p \rightarrow q$: *If* you want to succeed, *then* you must work hard.

3. "$2x = 10$; therefore $x = 5$" becomes
$p \rightarrow q$: *If* $2x = 10$, *then* $x = 5$.

EXAMPLE 1

For each given sentence:

a. Identify the hypothesis p.   **b.** Identify the conclusion q.

(1) If Mrs. Shusda teaches our class, then we will learn.

(2) The assignment will be completed if I work at it every day.

(3) The task is easy when we all work together and do our best.

Solution (1) If Mrs. Shusda teaches our class, then we will learn.
                                  p q

   a. p: Mrs. Shusda teaches our class.

   b. q: We will learn.

(2) The assignment will be completed if I work at it every day.
              q p

   a. p: I work at it every day.

   b. q: The assignment will be completed.

(3) *Hidden Conditional:*

   If we all work together and do our best, then the task is easy.
              p q

   a. p: We all work together and we do our best.

   b. q: the task is easy.

Note: In (3), the hypothesis is a conjunction. If we let r represent "We all work together" and s represent "We do our best," then the conditional "If we all work together and we do our best, then the task is easy" can be symbolized as $(r \wedge s) \rightarrow q$.

EXAMPLE 2

Identify the truth value to be assigned to each conditional statement.

(1) If $4 + 4 = 8$, then $2(4) = 8$.

(2) If 2 is a prime number, then 2 is odd.

(3) If 12 is a multiple of 9, then 12 is a multiple of 3.

(4) If $2 > 3$ then $2 - 3$ is a positive integer.

Solution (1) The hypothesis p is "$4 + 4 = 8$," which is true.
The conclusion q is "$2(4) = 8$," which is true.
The conditional $p \rightarrow q$ is true. *Answer*

(2) The hypothesis p is "2 is a prime number," which is true.
The conclusion q is "2 is odd," which is false.
The conditional $p \rightarrow q$ is false. *Answer*

(3) The hypothesis p is "12 is a multiple of 9," which is false.
The conclusion q is "12 is a multiple of 3," which is true.
The conditional $p \rightarrow q$ is true. *Answer*

(4) The hypothesis p is "$2 > 3$," which is false.
The conclusion q is "$2 - 3$ is a positive integer," which is false.
The conditional $p \rightarrow q$ is true. *Answer*

EXAMPLE 3

For each given statement:

a. Write the statement in symbolic form using the symbols given below.

b. Tell whether the statement is true or false.

Let m represent "Monday is the first day of the week." (True)
Let w represent "There are 52 weeks in a year." (True)
Let h represent "An hour has 75 minutes." (False)

Answers

(1) If Monday is the first day of the week, then there are 52 weeks in a year.

a. $m \rightarrow w$

b. $T \rightarrow T$ is true.

(2) If there are 52 weeks in a year, then an hour has 75 minutes.

a. $w \rightarrow h$

b. $T \rightarrow F$ is false.

(3) If there are not 52 weeks in a year then Monday is the first day of the week.

a. $\sim w \rightarrow m$

b. $F \rightarrow T$ is true.

(4) If Monday is the first day of the week and there are 52 weeks in a year, then an hour has 75 minutes.

a. $(m \wedge w) \rightarrow h$

b. $(T \wedge T) \rightarrow F$

$T \rightarrow F$ is false.

Exercises

Writing About Mathematics

1. **a.** Show that the conditional "If x is divisible by 4, then x is divisible by 2" is true in each of the following cases:

   (1) $x = 8$ (2) $x = 6$ (3) $x = 7$

   b. Is it possible to find a value of x for which the hypothesis is true and the conclusion is false? Explain your answer.

2. For what truth values of p and q is the truth value of $p \rightarrow q$ the same as the truth value of $q \rightarrow p$?

Developing Skills

In 3–10, for each given sentence: **a.** Identify the hypothesis p. **b.** Identify the conclusion q.

3. If a polygon is a square, then it has four right angles.

4. If it is noon, then it is time for lunch.

5. When you want help, ask a friend.

6. You will finish more quickly if you are not interrupted.

7. The perimeter of a square is $4s$ if the length of one side is s.

8. If many people work at a task, it will be completed quickly.

9. $2x + 7 = 11$ implies that $x = 2$.

10. If you do not get enough sleep, you will not be alert.

In 11–16, write each sentence in symbolic form, using the given symbols.

> p: The car has a flat tire.
> q: Danny has a spare tire.
> r: Danny will change the tire.

11. If the car has a flat tire, then Danny will change the tire.

12. If Danny has a spare tire, then Danny will change the tire.

13. If the car does not have a flat tire, then Danny will not change the tire.

14. Danny will not change the tire if Danny doesn't have a spare tire.

15. The car has a flat tire if Danny has a spare tire.

16. Danny will change the tire if the car has a flat tire.

In 17–24, for each given statement: **a.** Write the statement in symbolic form, using the symbols given below. **b.** Tell whether the conditional statement is true or false, based upon the truth values given.

> b: The barbell is heavy.　　(True)
> t: Kylie trains.　　(False)
> l: Kylie lifts the barbell.　　(True)

17. If Kylie trains, then Kylie will lift the barbell.

18. If Kylie lifts the barbell, then Kylie has trained.

19. If Kylie lifts the barbell, the barbell is heavy.

20. Kylie lifts the barbell if the barbell is not heavy.

21. Kylie will not lift the barbell if Kylie does not train.

22. Kylie trains if the barbell is heavy.

23. If the barbell is not heavy and Kylie trains, then Kylie will lift the barbell.

24. If the barbell is heavy and Kylie does not train, then Kylie will not lift the barbell.

In 25–31, find the truth value to be assigned to each conditional statement.

25. If $4 + 8 = 12$, then $8 + 4 = 12$.

26. If $9 < 15$, then $19 < 25$.

27. If $1 \times 1 = 1$, then $1 \times 1 \times 1 = 1$.

28. $24 \div 3 = 8$ if $24 \div 8 = 3$.

29. $6 + 6 = 66$ if $7 + 7 = 76$.

30. $48 = 84$ if $13 = 31$.

31. If every rhombus is a polygon, then every polygon is a rhombus.

In 32–39, symbols are assigned to represent sentences, and truth values are assigned to these sentences.

> Let j represent "July is a warm month."　　(True)
> Let d represent "I am busy every day."　　(False)
> Let g represent "I work in my garden."　　(True)
> Let f represent "I like flowers."　　(True)

For each compound statement in symbolic form: **a.** Write a complete sentence in words to show what the symbols represent. **b.** Tell whether the compound statement is true or false.

32. $j \to g$　　　　**33.** $d \to \sim g$　　　　**34.** $f \to g$　　　　**35.** $\sim g \to \sim j$

36. $(j \wedge f) \to d$　　　**37.** $(j \wedge g) \to f$　　　**38.** $\sim j \to (d \wedge f)$　　　**39.** $g \to (j \vee \sim d)$

In 40–45 supply the word, phrase, or symbol that can be placed in the blank to make each resulting sentence true.

40. When p and q represent two simple sentences, the conditional *if p then q* is written symbolically as _____.

41. The conditional *if q then p* is written symbolically as _____.

42. The conditional $p \rightarrow q$ is false only when p is _____ and q is _____.

43. When the conclusion q is true, then $p \rightarrow q$ must be _____.

44. When the hypothesis p is false, then $p \rightarrow q$ must be _____.

45. If the hypothesis p is true and conditional $p \rightarrow q$ is true, then the conclusion q must be _____.

Applying Skills

In 46–50, three sentences are written in each case. The truth values are given for the first two sentences. Determine whether the third sentence is true, is false, or has an uncertain truth value.

46. If you read in dim light, then you can strain your eyes. (True)
You read in dim light. (True)
You can strain your eyes. (?)

47. If the quadrilateral has four right angles, then the quadrilateral must be a square. (False)
The quadrilateral has four right angles. (True)
The quadrilateral must be a square. (?)

48. If n is an odd number, then $2n$ is an even number. (True)
$2n$ is an even number. (True)
n is an odd number. (?)

49. If the report is late, then you will not get an A. (True)
The report is late. (False)
You will not get an A. (?)

50. Area $= \frac{1}{2}bh$ if the polygon is a triangle. (True)
The polygon is a triangle. (True)
Area $= \frac{1}{2}bh$ (?)

2-5 INVERSES, CONVERSES, AND CONTRAPOSITIVES

The conditional is the most frequently used statement in the construction of an argument or in the study of mathematics. We will use the conditional frequently in our study of geometry. In order to use the conditional statements correctly, we must understand their different forms and how their truth values are related.

There are four conditionals that can be formed from two simple statements, p and q, and their negations.

The conditional:	$p \rightarrow q$	The inverse:	$\sim p \rightarrow \sim q$
The converse:	$q \rightarrow p$	The contrapositive:	$\sim q \rightarrow \sim p$

The Inverse

The **inverse** of a conditional statement is formed by negating the hypothesis and the conclusion. For example, the inverse of the statement "If today is Monday, then I have soccer practice" is "If today is *not* Monday, then I do *not* have soccer practice." In symbols, the inverse of $(p \rightarrow q)$ is $(\sim p \rightarrow \sim q)$.

The following examples compare the truth values of given conditionals and their inverses.

1. *A true conditional can have a false inverse.*

> Let *p* represent "A number is divisible by ten."
> Let *q* represent "The number is divisible by five."

Conditional
$(p \rightarrow q)$: If a number is divisible by ten, then it is divisible by five.
           p   *q*

Inverse
$(\sim p \rightarrow \sim q)$: If a number is *not* divisible by ten, then it is *not* divisible by five.
           ~*p*   ~*q*

We can find the truth value of these two statements when the number is 15.

p:	15 is divisible by 10.	False
q:	15 is divisible by 5.	True
$p \rightarrow q$:	If 15 is divisible by 10, then 15 is divisible by 5.	F → T is T
~*p*:	15 is not divisible by 10.	True
~*q*:	15 is not divisible by 5.	False
$\sim p \rightarrow \sim q$:	If 15 is not divisible by 10, then 15 is not divisible by 5.	T → F is F

In this case, the conditional and the inverse have opposite truth values.

2. *A false conditional can have a true inverse.*

> Let *p* represent "Two angles are congruent."
> Let *q* represent "Two angles are both right angles."

Conditional $(p \rightarrow q)$: If two angles are congruent,
                p

    then the two angles are both right angles.
               q

Inverse $(\sim p \rightarrow \sim q)$: If two angles are not congruent,
               ~*p*

    then the two angles are not both right angles.
               ~*q*

We can find the truth value of these two statements with two angles *A* and *B* when m$\angle A = 60$ and m$\angle B = 60$.

p:	Angle A and $\angle B$ are congruent.	True
q:	Angle A and $\angle B$ are both right angles.	False
$p \rightarrow q$:	If $\angle A$ and $\angle B$ are congruent, then $\angle A$ and $\angle B$ are both right angles.	$T \rightarrow F$ is F
$\sim p$:	Angle A and $\angle B$ are not congruent.	False
$\sim q$:	Angle A and $\angle B$ are not both right angles.	True
$\sim p \rightarrow \sim q$:	If $\angle A$ and $\angle B$ are not congruent, then $\angle A$ and $\angle B$ are not both right angles.	$F \rightarrow T$ is T

Again, the conditional and the inverse have opposite truth values.

3. *A conditional and its inverse can have the same truth value.*

Let r represent "Twice Talia's age is 10."
Let q represent "Talia is 5 years old."

Conditional If twice Talia's age is 10, then Talia is 5 years old.
$(r \rightarrow s)$: r s

Inverse If twice Talia's age is *not* 10, then Talia is *not* 5 years old.
$(\sim r \rightarrow \sim s)$: $\sim r$ $\sim s$

When Talia is 5, r is true and s is true. The conditional is true.
When Talia is 5, $\sim r$ is false and $\sim s$ is false. The inverse is true.
Both the conditional and its inverse have the same truth value.

When Talia is 6, r is false and s is false. The conditional is true.
When Talia is 6, $\sim r$ is true and $\sim s$ is true. The conditional is true.
Again, the conditional and its inverse have the same truth value.

These three illustrations allow us to make the following conclusion:

▶ **A conditional ($p \rightarrow q$) and its inverse ($\sim p \rightarrow \sim q$) may or may not have the same truth value.**

This conclusion can be shown in the truth table. Note that the conditional ($p \rightarrow q$) and its inverse ($\sim p \rightarrow \sim q$) have the same truth value when p and q have the same truth value. The conditional ($p \rightarrow q$) and its inverse ($\sim p \rightarrow \sim q$) have opposite truth values when p and q have opposite truth values.

				Conditional	Inverse
p	q	$\sim p$	$\sim q$	$p \rightarrow q$	$\sim p \rightarrow \sim q$
T	T	F	F	T	T
T	F	F	T	F	T
F	T	T	F	T	F
F	F	T	T	T	T

The Converse

The **converse** of a conditional statement is formed by interchanging the hypothesis and conclusion. For example, the converse of the statement "If today is Monday, then I have soccer practice" is "If I have soccer practice, then today is Monday." In symbols, the converse of $(p \rightarrow q)$ is $(q \rightarrow p)$.

To compare the truth values of a conditional and its converse, we will consider some examples.

1. *A true conditional can have a false converse.*

Let p represent "x is a prime."
Let q represent "x is odd."

Conditional $(p \rightarrow q)$: If $\underbrace{x \text{ is a prime}}_{p}$, then $\underbrace{x \text{ is odd}}_{q}$.

Converse $(q \rightarrow p)$: If $\underbrace{x \text{ is odd}}_{q}$, then $\underbrace{x \text{ is a prime}}_{p}$.

When $x = 9$, p is false and q is true. Therefore, for this value of x, the conditional $(p \rightarrow q)$ is true and its converse $(q \rightarrow p)$ is false.

In this example, the conditional is true and the converse is false. The conditional and its converse do not have the same truth value.

2. *A false conditional can have a true converse.*

Let p represent "x is divisible by 2."
Let q represent "x is divisible by 6."

Conditional $(p \rightarrow q)$: If $\underbrace{x \text{ is divisible by 2}}_{p}$, then $\underbrace{x \text{ is divisible by 6}}_{q}$.

Converse $(q \rightarrow p)$: If $\underbrace{x \text{ is divisible by 6}}_{q}$, then $\underbrace{x \text{ is divisible by 2}}_{p}$.

When $x = 8$, p is true and q is false. Therefore, for this value of x, the conditional $(p \rightarrow q)$ is false and its converse $(q \rightarrow p)$ is true.

In this example, the conditional is false and the converse is true. Again, the conditional and its converse do not have the same truth value.

3. *A conditional and its converse can have the same truth value.*

Let p represent "Today is Friday."
Let q represent "Tomorrow is Saturday."

Conditional $(p \rightarrow q)$: If $\underbrace{\text{today is Friday}}_{p}$, then $\underbrace{\text{tomorrow is Saturday}}_{q}$.

Converse $(q \rightarrow p)$: If $\underbrace{\text{tomorrow is Saturday}}_{q}$, then $\underbrace{\text{today is Friday}}_{p}$.

On Friday, p is true and q is true. Therefore, both $(p \rightarrow q)$ and $(q \rightarrow p)$ are true.

On any other day of the week, p is false and q is false. Therefore, both $(p \rightarrow q)$ and $(q \rightarrow p)$ are true.

A conditional and its converse may have the same truth value.

These three examples allow us to make the following conclusion:

▶ **A conditional $(p \rightarrow q)$ and its converse $(q \rightarrow p)$ may or may not have the same truth value.**

This conclusion can be shown in the truth table. Note that the conditional $(p \rightarrow q)$ and its converse $(q \rightarrow p)$ have the same truth value when p and q have the same truth value. The conditional $(p \rightarrow q)$ and its converse $(q \rightarrow p)$ have different truth values when p and q have different truth values.

		Conditional	Converse
p	q	$p \rightarrow q$	$q \rightarrow p$
T	T	T	T
T	F	F	T
F	T	T	F
F	F	T	T

The Contrapositive

We form the inverse of a conditional by negating both the hypothesis and the conclusion. We form the converse of a conditional by interchanging the hypothesis and the conclusion. We form the **contrapositive** of a conditional by doing both of these operations: we negate and interchange the hypothesis and conclusion. In symbols, the contrapositive of $(p \rightarrow q)$ is $(\sim q \rightarrow \sim p)$.

1. *A true conditional can have a true contrapositive.*

Let p represent "Gary arrives late to class."
Let q represent "Gary is marked tardy."

Conditional If Gary arrives late to class, then Gary is marked tardy.
$(p \rightarrow q)$:
$\qquad\qquad\qquad\quad p \qquad\qquad\qquad\qquad\qquad q$

Contrapositive If Gary is *not* marked tardy,
$(\sim q \rightarrow \sim p)$:
$\qquad\qquad\qquad\qquad \sim q$

then Gary does *not* arrive late to class.
$\qquad\qquad\qquad\qquad \sim p$

If Gary arrives late to class, p is true and q is true. Therefore, $(p \rightarrow q)$ is true. Also, if p and q are true, $\sim p$ is false and $\sim q$ is false, so $(\sim q \rightarrow \sim p)$ is true.

If p is false, that is, if Gary does not arrive late to class, q is false and $(p \rightarrow q)$ is true. Similarly, if p and q are false, $\sim p$ is true and $\sim q$ is true, so $(\sim q \rightarrow \sim p)$ is true.

2. *A false conditional can have a false contrapositive.*

Conditional If $\underbrace{x \text{ is an odd number}}_{p}$, then $\underbrace{x \text{ is a prime number}}_{q}$.
($p \rightarrow q$):

Contrapositive If $\underbrace{x \text{ is not a prime number}}_{\sim q}$, then $\underbrace{x \text{ is not an odd number}}_{\sim p}$.
($\sim q \rightarrow \sim p$):

	$p \rightarrow q$	$\sim q \rightarrow \sim p$
Let $x = 2$	F → T is true	F → T is true
Let $x = 9$	T → F is false	T → F is false
Let $x = 11$	T → T is true	F → F is true
Let $x = 8$	F → F is true	T → T is true

For each value of x, the conditional and its contrapositive have the same truth value.

These illustrations allow us to make the following conclusion:

▶ **A conditional ($p \rightarrow q$) and its contrapositive ($\sim q \rightarrow \sim p$) always have the same truth value:**

When a conditional is true, its contrapositive must be true.

When a conditional is false, its contrapositive must be false.

This conclusion can be shown in the following truth table.

				Conditional	Contrapositive
p	q	$\sim p$	$\sim q$	$p \rightarrow q$	$\sim q \rightarrow \sim p$
T	T	F	F	T	T
T	F	F	T	F	F
F	T	T	F	T	T
F	F	T	T	T	T

Logical Equivalents

A conditional and its contrapositive are **logical equivalents** because they always have the same truth value.

The inverse of ($p \rightarrow q$) is ($\sim p \rightarrow \sim q$). The contrapositive of ($\sim p \rightarrow \sim q$) is formed by negating and interchanging the hypothesis and conclusion of ($\sim p \rightarrow \sim q$). The negation of $\sim p$ is p and the negation of $\sim q$ is q. Therefore, the contrapositive of ($\sim p \rightarrow \sim q$) is ($q \rightarrow p$). For instance:

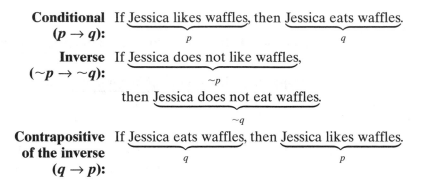

Conditional
$(p \to q)$:   If Jessica likes waffles, then Jessica eats waffles.
                    p q

Inverse
$(\sim p \to \sim q)$:   If Jessica does not like waffles,
                              $\sim p$

then Jessica does not eat waffles.
              $\sim q$

Contrapositive
of the inverse
$(q \to p)$:   If Jessica eats waffles, then Jessica likes waffles.
                    q p

Notice, however, that the contrapositive of the inverse is the same as the converse of the original conditional. Thus, the inverse and the converse of $(p \to q)$ are contrapositives of each other. Since a conditional and its contrapositive always have the same truth value, the converse and the inverse always have the same truth value. This can be verified by constructing the following truth table.

p	q	$\sim p$	$\sim q$	$\sim p \to \sim q$	$q \to p$
T	T	F	F	T	T
T	F	F	T	T	T
F	T	T	F	F	F
F	F	T	T	T	T

EXAMPLE 1

Write the inverse, converse, and contrapositive of the given conditional:
                    If today is Tuesday, then I play basketball.

Solution   Inverse: If today is not Tuesday, then I do not play basketball.
                    Converse: If I play basketball, then today is Tuesday.
          Contrapositive: If I do not play basketball, then today is not Tuesday.

EXAMPLE 2

Write the inverse, converse, and contrapositive of the given conditional:
                    If a polygon is a square then it has four right angles.

Solution   Inverse: If a polygon is not a square, then it does not have four right
                              angles.
                    Converse: If a polygon has four right angles, then it is a square.
          Contrapositive: If a polygon does not have four right angles, then it is
                              not a square.

EXAMPLE 3

Write the inverse, converse, and contrapositive of the given conditional:

If M is the midpoint of $\overline{AB}$, then $AM = MB$.

Solution   Inverse: If M is not the midpoint of $\overline{AB}$, then $AM \neq MB$.
                     Converse: If $AM = MB$, then M is the midpoint of $\overline{AB}$.
               Contrapositive: If $AM \neq MB$, then M is not the midpoint of $\overline{AB}$.

EXAMPLE 4

Given the true statement, "If the polygon is a rectangle, then it has four sides," which statement must also be true?

(1) If the polygon has four sides, then it is a rectangle.

(2) If the polygon is not a rectangle, then it does not have four sides.

(3) If the polygon does not have four sides, then it is not a rectangle.

(4) If the polygon has four sides, then it is not a rectangle.

Solution A conditional and its contrapositive always have the same truth value. The contrapositive of the given statement is "If a polygon does not have four sides then it is not a rectangle."

Answer (3)

Exercises

Writing About Mathematics

1. Samuel said that if you know that a conditional is true then you know that the converse of the conditional is true. Do you agree with Samuel? Explain why or why not.

2. Kate said that if you know the truth value of a conditional and of its converse then you know the truth value of the inverse and the contrapositive. Do you agree with Kate? Explain why or why not.

Developing Skills

In 3–6, for each statement, write in symbolic form:

a. the inverse   **b.** the converse   **c.** the contrapositive.

   3. $p \rightarrow q$ **4.** $t \rightarrow {\sim}w$ **5.** ${\sim}m \rightarrow p$ **6.** ${\sim}p \rightarrow {\sim}q$

In 7–10: **a.** Write the inverse of each conditional statement in words. **b.** Give the truth value of the conditional. **c.** Give the truth value of the inverse.

   7. If $6 > 3$, then $-6 > -3$.

   8. If a polygon is a parallelogram, then the polygon has two pairs of parallel sides.

   9. If $3(3) = 9$, then $3(4) = 12$.   **10.** 1f $2^2 = 4$, then $3^2 = 6$.

In 11–14, write the converse of each statement in words.

   11. If you lower your cholesterol, then you eat Quirky oatmeal.

   12. If you enter the Grand Prize drawing, then you will get rich.

   13. If you use Shiny's hair cream, then your hair will curl.

   14. If you feed your pet Krazy Kibble, he will grow three inches.

In 15–18: **a.** Write the converse of each conditional statement in words. **b.** Give the truth value of the conditional. **c.** Give the truth value of the converse.

   15. If a number is even, then the number is exactly divisible by 2.

   16. If 0.75 is an integer, then it is rational.

   17. If $8 = 1 + 7$, then $8^2 = 1^2 + 7^2$.

   18. If $4(5) - 6 = 20 - 6$, then $4(5) - 6 = 14$.

In 19–23: **a.** Write the contrapositive of each statement in words. **b.** Give the truth value of the conditional. **c.** Give the truth value of the contrapositive.

   19. If Rochester is a city, then Rochester is the capital of New York.

   20. If two angles form a linear pair, then they are supplementary.

   21. If $3 - 2 = 1$, then $4 - 3 = 2$.

   22. If all angles of a triangle are equal in measure, then the triangle is equiangular.

   23. If $\frac{1}{2} > 0$, then $\frac{1}{2}$ is a counting number.

In 24–28, write the numeral preceding the expression that best answers the question.

   24. When $p \rightarrow q$ is true, which related conditional must be true?
      (1) $q \rightarrow p$ (2) $\sim p \rightarrow \sim q$ (3) $p \rightarrow \sim q$ (4) $\sim q \rightarrow \sim p$

   25. Which is the contrapositive of "If March comes in like a lion, it goes out like a lamb"?
      (1) If March goes out like a lamb, then it comes in like a lion.
      (2) If March does not go out like a lamb, then it comes in like a lion.
      (3) If March does not go out like a lamb, then it does not come in like a lion.
      (4) March goes out like a lion if it comes in like a lamb.

26. Which is the converse of "If a rectangular prism is a cube, then its surface area is $6s^2$"?
(1) If a rectangular prism is not a cube, then its surface area is not $6s^2$.
(2) If the surface area of a rectangular prism is $6s^2$, then it is a cube.
(3) If the surface area of a rectangular prism is not $6s^2$, then it is not a cube.
(4) If the surface area of a cube is $6s^2$, then it is a rectangular prism.

27. Which is the inverse of "If $z = 4$, then $2z \neq 9$"?
(1) If $2z \neq 9$, then $z = 4$.   (3) If $z \neq 4$, then $2z = 9$.
(2) If $2z = 9$, then $z \neq 4$.   (4) If $z \neq 4$, then $2z \neq 9$.

28. Which is the contrapositive of "If y is greater than 3, then $2y + 10y$ is not equal to 36"?
(1) If $2y + 10y$ is not equal to 36, then y is greater than 3.
(2) If $2y + 10y$ equals 36, then y is not greater than 3.
(3) If $2y + 10y$ is not equal to 36, then y is not greater than 3.
(4) If $2y + 10y$ equals 36, then y is greater than 3.

Applying Skills

In 29–34, assume that each conditional statement is true. Then:

a. Write its converse in words and state whether the converse is always true, sometimes true, or never true.

b. Write its inverse in words and state whether the inverse is always true, sometimes true, or never true.

c. Write its contrapositive in words and state whether the contrapositive is always true, sometimes true, or never true.

29. If Derek lives in Las Vegas, then he lives in Nevada.

30. If a bin contains 3 red marbles and 3 blue marbles, then the probability of picking a red marble from the bin is $\frac{1}{2}$.

31. If a polygon has eight sides, then it is an octagon.

32. If a garden grows carrots, then it grows vegetables.

33. If the dimensions of a rectangle are 8 feet by 6 feet, then the area of the rectangle is 48 square feet.

34. If a number has 7 as a factor, then it is divisible by 7.

2-6 BICONDITIONALS

A **biconditional** is the conjunction of a conditional and its converse. For the conditional $(p \rightarrow q)$, the converse is $(q \rightarrow p)$. The biconditional can be written as $(p \rightarrow q) \wedge (q \rightarrow p)$ or in the shorter form $p \leftrightarrow q$, which is read *p if and only if q*.

Recall that a conjunction is true only when both parts of the compound statement are true. Therefore, $(p \rightarrow q) \wedge (q \rightarrow p)$ is true only when $(p \rightarrow q)$ is true and its converse $(q \rightarrow p)$ is true. In the last section you learned that a conditional $(p \rightarrow q)$ and its converse $(q \rightarrow p)$ are both true when p and q are both true or both false. This is shown in the table below.

p	q	$p \rightarrow q$	$q \rightarrow p$	$(p \rightarrow q) \wedge (q \rightarrow p)$	$p \leftrightarrow q$
T	T	T	T	T	T
T	F	F	T	F	F
F	T	T	F	F	F
F	F	T	T	T	T

▶ **The biconditional p *if and only if* q is true when p and q are both true or both false.**

In other words, $p \leftrightarrow q$ is true when p and q have the same truth value. When p and q have different truth values, the biconditional is false.

Applications of the Biconditional

There are many examples in which the biconditional is always true. Consider the following:

1. *Every definition is a true biconditional.*

Every definition can be written in reverse order. Both of the following statements are true:

- Congruent segments are segments that have the same measure.
- Line segments that have the same measure are congruent.

We can restate the definition as two true conditionals:

- If two line segments are congruent, then they have the same measure.
- If two line segments have the same measure, then they are congruent.

Therefore, this definition can be restated as a true biconditional:

- Two line segments are congruent if and only if they have the same measure.

2. *Biconditionals are used to solve equations.*

We know that when we add the same number to both sides of an equation or when we multiply both sides of an equation by the same number, the derived equation has the same solution set as the given equation. That is, any number that makes the first equation true will make the derived equation true.

For example:

p: $3x + 7 = 19$
q: $3x = 12$
$p \rightarrow q$: If $3x + 7 = 19$, then $3x = 12$. (−7 was added to both sides of the equation.)
$q \rightarrow p$: If $3x = 12$, then $3x + 7 = 19$. (7 was added to both sides of the equation.)

When $x = 4$, both p and q are true and both $p \rightarrow q$ and $q \rightarrow p$ are true.

When $x = 1$ or when x equals any number other than 3, both p and q are false and both $p \rightarrow q$ and $q \rightarrow p$ are true.

Therefore, the biconditional "$3x + 7 = 19$ if and only if $3x = 12$" is true.

The solution of an equation is a series of biconditionals:

$3x + 7 = 19$

$3x \quad = 12$ $\qquad$ $3x + 7 = 19$ if and only if $3x = 12$.

$x \quad = 4$ $\qquad$ $3x = 12$ if and only if $x = 4$.

3. A biconditional states that two logical forms are equivalent.

Two logical forms that always have the same truth values are said to be equivalent.

We have seen that a conditional and its contrapositive are logically equivalent and that the converse and inverse of a conditional are logically equivalent. There are many other statements that are logically equivalent.

The table below shows that $\sim(p \wedge q)$ and $\sim p \vee \sim q$ are logically equivalent. We can write the true biconditional $\sim(p \wedge q) \leftrightarrow (\sim p \vee \sim q)$.

p	q	$\sim p$	$\sim q$	$p \wedge q$	$\sim(p \wedge q)$	$\sim p \vee \sim q$
T	T	F	F	T	F	F
T	F	F	T	F	T	T
F	T	T	F	F	T	T
F	F	T	T	F	T	T

EXAMPLE 1

Determine the truth value to be assigned to the biconditional.

Germany is a country in Europe if and only if Berlin is the capital of Germany.

Solution "Germany is a country in Europe" is true.
"Berlin is the capital of Germany" is true.

Therefore, the biconditional "Germany is a country in Europe if and only if Berlin is the capital of Germany" is also true. *Answer*

EXAMPLE 2

The statement "I go to basketball practice on Monday and Thursday" is true. Determine the truth value to be assigned to each statement.

a. If today is Monday, then I go to basketball practice.

b. If I go to basketball practice, then today is Monday.

c. Today is Monday if and only if I go to basketball practice.

Solution Let p: "Today is Monday,"
and q: "I go to basketball practice."

a. We are asked to find the truth value of the following conditional:

$p \rightarrow q$: If today is Monday, then I go to basketball practice.

On Monday, p is true and q is true. Therefore, $p \rightarrow q$ is true.

On Thursday, p is false and q is true. Therefore, $p \rightarrow q$ is true.

On every other day, p is false and q is false. Therefore, $p \rightarrow q$ is true.

"If today is Monday, then I go to basketball practice" is always true. *Answer*

b. We are asked to find the truth value of the following conditional:

$q \rightarrow p$: If I go to basketball practice then today is Monday.

On Monday, p is true and q is true. Therefore, $q \rightarrow p$ is true.

On Thursday, p is false and q is true. Therefore, $q \rightarrow p$ is false.

On every other day, p is false and q is false. Therefore, $q \rightarrow p$ is true.

"If I go to basketball practice, then Today is Monday" is sometimes true and sometimes false. *Answer*

c. We are asked to find the truth value of the following biconditional:

$p \leftrightarrow q$: Today is Monday if and only if I go to basketball practice.

The conditionals $p \rightarrow q$ and $q \rightarrow p$ do not always have the same truth value. Therefore, the biconditional "Today is Monday if and only if I go to basketball practice" is not always true. We usually say that a statement that is not always true is false. *Answer*

EXAMPLE 3

Determine the truth value of the biconditional.

$$3y + 1 = 28 \text{ if and only if } y = 9.$$

Solution   When $y = 9$, $3y + 1 = 28$ is true and $y = 9$ is true.
Therefore, $3y + 1 = 28$ if and only if $y = 9$ is true.
When $y \neq 9$, $3y + 1 = 28$ is false and $y = 9$ is false.
Therefore, $3y + 1 = 28$ if and only if $y = 9$ is true.
$3y + 1 = 28$ if and only if $y = 9$ is always true. *Answer*

Exercises

Writing About Mathematics

1. Write the definition "A prime number is a whole number greater than 1 and has exactly two factors" as a biconditional.

2. Tiffany said that if the biconditional $p \leftrightarrow q$ is false, then either $p \rightarrow q$ is true or $q \rightarrow p$ is true but both cannot be true. Do you agree with Tiffany? Explain why or why not.

Developing Skills

In 3–16, give the truth value of each biconditional.

3. $y + 7 = 30$ if and only if $y = 23$.

4. B is between A and C if and only if $AB + AC = BC$.

5. $z + 9 = 13$ if and only if $z + 2 = 6$.

6. A parallelogram is a rhombus if and only if the parallelogram has four sides of equal length.

7. A real number is positive if and only if it is greater than zero.

8. An angle is an acute angle if and only if its degree measure is less than 90.

9. An element belongs to the intersection of sets F and G if and only if it belongs to both F and G.

10. An integer is odd if and only if it is not divisible by 2.

11. Two angles have the same measure if and only if they are right angles.

12. I live in the United States if and only if I live in New York State.

13. A rational number has a multiplicative inverse if and only if it is not zero.

14. An angle is an acute angle if and only if it has a degree measure of 50.

15. $x = 5$ if and only if $x > 3$.

16. Today is Friday if and only if tomorrow is Saturday.

Applying Skills

17. Let p represent "x is divisible by 2."
    Let q represent "x is divisible by 3."
    Let r represent "x is divisible by 6."

    a. Write the biconditional $(p \wedge q) \leftrightarrow r$ in words.

    b. Show that the biconditional is always true for the domain $\{2, 3, 5, 6, 8, 9, 11, 12\}$.

    c. Do you think that the biconditional is true for all counting numbers? Explain your answer.

18. A gasoline station displays a sign that reads "Open 24 hours a day, Monday through Friday."

    a. On the basis of the information on the sign, is the conditional "The gasoline station is closed if it is Saturday or Sunday" true?

    b. On the basis of the information on the sign, is the conditional "If the gasoline station is closed, it is Saturday or Sunday" true?

    c. On the basis of the information on the sign, is the biconditional "The gasoline station is closed if and only if it is Saturday or Sunday" true?

    d. Marsha arrives at the gasoline station on Monday and finds the station closed. Does this contradict the information on the sign?

    e. Marsha arrives at the gasoline station on Saturday and finds the station open. Does this contradict the information on the sign?

In 19–22, write a biconditional using the given conditionals and tell whether each biconditional is true or false.

19. If a triangle is isosceles, then it has two congruent sides.
    If a triangle has two congruent sides then it is isosceles.

20. If two angles are both right angles, then they are congruent.
    If two angles are congruent, then they are both right angles.

21. If today is Thursday, then tomorrow is not Saturday.
    If tomorrow is not Saturday, then today is Thursday.

22. If today is not Friday, then tomorrow is not Saturday.
    If tomorrow is not Saturday, then today is not Friday.

2-7 THE LAWS OF LOGIC

We frequently want to combine known facts in order to establish the truth of related facts. To do this, we can look for patterns that are frequently used in drawing conclusions. These patterns are called the **laws of logic**.

The Law of Detachment

A **valid argument** uses a series of statements called **premises** that have known truth values to arrive at a conclusion.

For example, Cynthia makes the following true statements to her parents:

> I want to play baseball.
> If I want to play baseball, then I need a glove.

These are the premises of Cynthia's argument. The conclusion that Cynthia wants her parents to make is that she needs a glove. Is this conclusion valid?

Let p represent "I want to play baseball."
Let q represent "I need a glove."
Then $p \rightarrow q$ represents "If I want to play baseball, then I need a glove."

p	q	$p \rightarrow q$
T	T	T
T	F	F
F	T	T
F	F	T

We know that the premises are true, that is, p is true and $p \rightarrow q$ is true. The only line of the truth table that satisfies both of these conditions is the first in which q is also true. Therefore, "I need a glove" is a true conclusion.

The example just given does not depend on the statement represented by p and q. The first line of the truth table tells us that whenever $p \rightarrow q$ is true and p is true, then q must be a true conclusion. This logical pattern is called the **Law of Detachment**:

▶ **If a conditional ($p \rightarrow q$) is true and the hypothesis (p) is true, then the conclusion (q) is true.**

EXAMPLE I

If the measure of an angle is greater than 0° and less than 90°, then the angle is an acute angle. Let "m$\angle A$ = 40°, which is greater than 0° and less than 90°" be a true statement. Prove that $\angle A$ is an acute angle.

Solution   Let p represent "m$\angle A$ = 40°, which is greater than 0° and less than 90°."

Let q represent "the angle is an acute angle."

Then $p \rightarrow q$ is true because it is a definition of an acute angle.

Also, p is true because it is given.

Then by the Law of Detachment, q is true.

Answer   "$\angle A$ is an acute angle" is true.

The Law of Disjunctive Inference

We know that a disjunction is true when one or both statements that make up the disjunction are true. The disjunction is false when both statements that make up the disjunction are false.

For example, let p represent "A real number is rational" and q represent "A real number is irrational." Then $p \lor q$ represents "A real number is rational or a real number is irrational," a true statement.

When the real number is π, then "A real number is rational" is false. Therefore "A real number is irrational" must be true.

When the real number is 7, then "A real number is irrational" is false. Therefore "A real number is rational" must be true.

The truth table shows us that when p is false and $p \lor q$ is true, only the third line of the table is satisfied. This line tells us that q is true. Also, when q is false and $p \lor q$ is true, only the second line of the table is satisfied. This line tells us that p is true.

p	q	$p \lor q$
T	T	T
T	F	T
F	T	T
F	F	F

The example just given illustrates a logical pattern that does not depend on the statements represented by p and q. When a disjunction is true and one of the disjuncts is false, then the other disjunct must be true. This logical pattern is called the **Law of Disjunctive Inference**:

▶ **If a disjunction $(p \lor q)$ is true and the disjunct (p) is false, then the other disjunct (q) is true.**

▶ **If a disjunction $(p \lor q)$ is true and the disjunct (q) is false, then the other disjunct (p) is true.**

EXAMPLE 2

What conclusion can be drawn when the following statements are true?

I will walk to school or I will ride to school with my friend.
I do not walk to school.

Solution   Since "I do not walk to school" is true, "I walk to school" is false. The disjunction "I will walk to school or I will ride to school with my friend" is true and one of the disjuncts, "I walk to school," is false. By the Law of Disjunctive Inference, the other disjunct, "I ride to school with my friend," must be true.

Alternative Solution   Let p represent "I walk to school."
Let q represent "I ride to school with my friend."

Make a truth table for the disjunction $p \lor q$ and eliminate the rows that do not apply.

We know that $p \lor q$ is true. We also know that since $\sim p$ is true, p is false.

(1) Eliminate the last row of truth values in which the disjunction is false.

(2) Eliminate the first two rows of truth values in which p is true.

(3) Only one case remains: q is true.

p	q	$p \lor q$
T	T	T
T	F	T
F	T	T
F	F	F

Answer I ride to school with my friend.

Note: In most cases, more than one possible statement can be shown to be true. For example, the following are also true statements when the given statements are true:

I do not walk to school and I ride to school with my friend.
If I do not walk to school, then I ride to school with my friend.
If I do not ride to school with my friend, then I walk to school.

EXAMPLE 3

From the following true statements, is it possible to determine the truth value of the statement "I will go to the library"?

If I have not finished my essay for English class, then I will go to the library.
I have finished my essay for English class.

Solution When "I have finished my essay for English class" is true, its negation, "I have not finished my essay for English class" is false. Therefore, the true conditional "If I have not finished my essay for English class, then I will go to the library" has a false hypothesis and the conclusion, "I will go to the library" can be either true or false.

Alternative Solution Let p represent "I have not finished my essay for English class."
Let q represent "I will go to the library."

Make a truth table for the disjunction $p \to q$ and eliminate the rows that do not apply.

We know that $p \to q$ is true. We also know that since $\sim p$ is true, p is false.

(1) Eliminate the second row of truth values in which the conditional is false.

(2) Eliminate the first row of truth values in which p is true. The second row in which p is true has already been eliminated.

(3) Two rows remain, one in which q is true and the other in which q is false.

p	q	$p \to q$
T	T	T
T	F	F
F	T	T
F	F	T

Answer "I will go to the library" could be either true or false.

EXAMPLE 4

Draw a conclusion or conclusions base on the following true statements.

> If I am tired, then I will rest.
> I do not rest.

Solution A conditional and its contrapositive are logically equivalent. Therefore, "If I do not rest, then I am not tired" is true.

By the Law of Detachment, when the hypothesis of a true conditional is true, the conclusion must be true. Therefore, since "I do not rest" is true, "I am not tired" must be true.

Alternative Solution Let p represent "I am tired."
Let q represent "I will rest."

Make a truth table for the disjunction $p \to q$ and eliminate the rows that do not apply.

We know that $p \to q$ is true. We also know that since $\sim q$ is true, q is false.

(1) Eliminate the second row of truth values in which the conditional is false.

(2) Eliminate the first and third rows of truth values in which q is true.

(3) Only one case remains: p is false.

p	q	$p \to q$
T	T	T
T	F	F
F	T	T
F	F	T

Since p is false, $\sim p$ is true. "I am not tired" is true.

Answer I am not tired.

Exercises

Writing About Mathematics

1. Clovis said that when $p \rightarrow q$ is false and $q \vee r$ is true, r must be true. Do you agree with Clovis? Explain why or why not.

2. Regina said when $p \vee q$ is true and $\sim q$ is true, then $p \wedge \sim q$ must be true. Do you agree with Regina? Explain why or why not.

Developing Skills

In 3–14, assume that the first two sentences in each group are true. Determine whether the third sentence is true, is false, or cannot be found to be true or false. Justify your answer.

3. I save up money or I do not go on the trip.
I go on the trip.
I save up money.

4. If I speed, then I get a ticket.
I speed.
I get a ticket.

5. I like swimming or kayaking.
I like kayaking.
I like swimming.

6. I like swimming or kayaking.
I do not like swimming.
I like kayaking.

7. $x \leq 18$ if $x = 14$.
$x = 14$
$x \leq 18$

8. I live in Pennsylvania if I live in Philadelphia.
I do not live in Philadelphia.
I live in Pennsylvania.

9. If I am late for dinner, then my dinner will be cold.
I am late for dinner.
My dinner is cold.

10. If I am late for dinner, then my dinner will be cold.
I am not late for dinner.
My dinner is not cold.

11. I will go to college if and only if I work this summer.
I do not work this summer.
I will go to college.

12. The average of two numbers is 20 if the numbers are 17 and 23.
The average of two numbers is 20.
The two numbers are 17 and 23.

13. If I am late for dinner, then my dinner will be cold.
My dinner is not cold.
I am not late for dinner.

14. If I do not do well in school, then I will not receive a good report card.
I do well in school.
I receive a good report card.

Applying Skills

In 15–27, assume that each given sentence is true. Write a conclusion using both premises, if possible. If no conclusion is possible, write "No conclusion." Justify your answer.

15. If I play the trumpet, I take band.
I play the trumpet.

16. $\sqrt{6}$ is rational or irrational.
$\sqrt{6}$ is not rational.

17. If $2b + 6 = 14$, then $2b = 8$.
If $2b = 8$, then $b = 4$.
$2b + 6 = 14$

18. If it is 8:15 A.M., then it is morning.
It is not morning.

19. If k is a prime, then $k \neq 8$.
$k = 8$

20. x is even and a prime if and only if $x = 2$.
$x = 2$

21. It is February or March, and it is not summer.
It is not March.

22. If x is divisible by 4, then x is divisible by 2.
x is divisible by 2.

23. On Saturdays, we go bowling or we fly kites.
Last Saturday, we did not go bowling.

24. I study computer science, and wood shop or welding.
I do not take woodshop.

25. Five is a prime if and only if five has exactly two factors.
Five is a prime.

26. If x is an integer greater than 2 and x is a prime, then x is odd.
x is not an odd integer.

27. If a ray bisects an angle, the ray divides the angle into two congruent angles.
Ray DF does not divide angle CDE into two congruent angles.

2-8 DRAWING CONCLUSIONS

Many important decisions as well as everyday choices are made by applying the principles of logic. Games and riddles also often depend on logic for their solution.

EXAMPLE 1

The three statements given below are each true. What conclusion can be found to be true?

1. If Rachel joins the choir then Rachel likes to sing.

2. Rachel will join the choir or Rachel will play basketball.

3. Rachel does not like to sing.

Solution A conditional and its contrapositive are logically equivalent. Therefore, "If Rachel does not like to sing, then Rachel will not join the choir" is true.

By the Law of Detachment, when the hypothesis of a true conditional is true, the conclusion must be true. Therefore, since "Rachel does not like to sing" is true, "Rachel will not join the choir" must be true.

Since "Rachel will not join the choir" is true, then its negation, "Rachel will join the choir," must be false.

Since "Rachel will join the choir or Rachel will play basketball" is true and "Rachel will join the choir" is false, then by the Law of Disjunctive Inference, "Rachel will play basketball" must be true.

Answer   Rachel does not join the choir. Rachel will play basketball.

Alternative   Let *c* represent "Rachel joins the choir,"
Solution   *s* represent "Rachel likes to sing,"
and *b* represent "Rachel will play basketball."

Write statements 1, 2, and 3 in symbols:

  1. $c \rightarrow s$ **2.** $c \vee b$ **3.** $\sim s$

Using Statement 1

$c \rightarrow s$ is true, so $\sim s \rightarrow \sim c$ is true. (A conditional and its contrapositive always have the same truth value.)

Using Statement 3

$\sim s$ is true and $\sim s \rightarrow \sim c$ is true, so $\sim c$ is true. (Law of Detachment)

Also, *c* is false.

Using Statement 2

$c \vee b$ is true and *c* is false, so *b* must be true. (Law of Disjunctive Inference)

Answer   Rachel does not join the choir. Rachel will play basketball.   ◾

EXAMPLE 2

If Alice goes through the looking glass, then she will see Tweedledee. If Alice sees Tweedledee, then she will see the Cheshire Cat. Alice does not see the Cheshire Cat.

Show that Alice does *not* go through the looking glass.

Solution   A conditional and its contrapositive are logically equivalent. Therefore, "If Alice does not see the Cheshire Cat, then Alice does not see Tweedledee" is true.

By the Law of Detachment, when the hypothesis of a true conditional is true, the conclusion must be true. Therefore, since "Alice does not see the Cheshire Cat" is true, "Alice does not see Tweedledee" must be true.

Again, since a conditional and its contrapositive are logically equivalent, "If Alice does not see Tweedledee, then Alice does not go through the looking glass" is also true. Applying the Law of Detachment, since "Alice does not see Tweedledee" is true, "Alice does not go through the looking glass" must be true.

Alternative Solution

Let a represent "Alice goes through the looking glass."

t represent "Alice sees Tweedledee."

c represent "Alice sees the Cheshire cat."

Then, in symbols, the given statements are:

1. $a \rightarrow t$ **2.** $t \rightarrow c$ **3.** $\sim c$

We would like to conclude that $\sim a$ is true.

Using Statement 2

$t \rightarrow c$ is true, so $\sim c \rightarrow \sim t$ is true. (A conditional and its contrapositive always have the same truth value.)

Using Statement 3

$\sim c$ is true and $\sim c \rightarrow \sim t$ is true, so $\sim t$ is true. (Law of Detachment)

Using Statement 1

$a \rightarrow t$ is true, so $\sim t \rightarrow \sim a$ is true. (A conditional and its contrapositive always have the same truth value.)

Therefore, by the Law of Detachment, since $\sim t$ is true and $\sim t \rightarrow \sim a$ is true, $\sim a$ is true.   ■

EXAMPLE 3

Three siblings, Ted, Bill, and Mary, each take a different course in one of three areas for their senior year: mathematics, art, and thermodynamics. The following statements about the siblings are known to be true.

- Ted tutors his sibling taking the mathematics course.
- The art student and Ted have an argument over last night's basketball game.
- Mary loves the drawing made by her sibling taking the art course.

What course is each sibling taking?

Solution Make a table listing each sibling and each subject. Use the statements to fill in the information about each sibling.

(1) Since Ted tutors his sibling taking the math course, Ted cannot be taking math. Place an ✗ in the table to show this.

	Math	Art	Thermodynamics
Ted	✗		
Bill			
Mary			

(2) Similarly, the second statement shows that Ted cannot be taking art. Place an ✗ to indicate this. The only possibility left is for Ted to be taking thermodynamics. Place a ✓ to indicate this, and add ✗'s in the thermodynamics column to show that no one else is taking this course.

	Math	Art	Thermodynamics
Ted	✗	✗	✓
Bill			✗
Mary			✗

(3) The third statement shows that Mary is not taking art. Therefore, Mary is taking math. Since Ted is taking thermodynamics and Mary is taking math, Bill must be taking art.

	Math	Art	Thermodynamics
Ted	✗	✗	✓
Bill	✗	✓	✗
Mary	✓	✗	✗

Answer Ted takes thermodynamics, Bill takes art, and Mary takes mathematics. ■

Exercises

Writing About Mathematics

1. Can $(p \vee q) \wedge r$ be true when p is false? If so, what are the truth values of q and of r? Justify your answer.

2. Can $(p \vee q) \wedge r$ be true when r is false? If so, what are the truth values of p and of q? Justify your answer.

Developing Skills

3. When $\sim p \wedge \sim q$ is true and $q \vee r$ is true, what is the truth value of r?

4. When $p \rightarrow q$ is false and $q \vee r$ is true, what is the truth value of r?

5. When $p \rightarrow q$ is false, what is the truth value of $q \rightarrow r$?

6. When $p \rightarrow q$ and $p \wedge q$ are both true, what are the truth values of p and of q?

7. When $p \rightarrow q$ and $p \wedge q$ are both false, what are the truth values of p and of q?

8. When $p \rightarrow q$ is true and $p \wedge q$ is false, what are the truth values of p and of q?

9. When $p \rightarrow q$ is true, $p \vee r$ is true and q is false, what is the truth value of r?

Applying Skills

10. Laura, Marta, and Shanti are a lawyer, a doctor, and an investment manager.
   - The lawyer is Marta's sister.
   - Laura is not a doctor.
   - Either Marta or Shanti is a lawyer.

   What is the profession of each woman?

11. Alex, Tony, and Kevin each have a different job: a plumber, a bookkeeper, and a teacher.
   - Alex is a plumber or a bookkeeper.
   - Tony is a bookkeeper or a teacher.
   - Kevin is a teacher.

   What is the profession of each person?

12. Victoria owns stock in three companies: Alpha, Beta, and Gamma.
   - Yesterday, Victoria sold her shares of Alpha or Gamma.
   - If she sold Alpha, then she bought more shares of Beta.
   - Victoria did not buy more shares of Beta.

   Which stock did Victoria sell yesterday?

13. Ren, Logan, and Kadoogan each had a different lunch. The possible lunches are: a ham sandwich, pizza, and chicken pot pie.
   - Ren or Logan had chicken pot pie.
   - Kadoogan did not have pizza.
   - If Logan did not have pizza, then Kadoogan had pizza.

   Which lunch did each person have?

14. Zach, Steve, and David each play a different sport: basketball, soccer, or baseball. Zach made each of the following true statements.
   - I do not play basketball.
   - If Steve does not play soccer, then David plays baseball.
   - David does not play baseball.

   What sport does each person play?

15. Taylor, Melissa, and Lauren each study one language: French, Spanish, and Latin.
  - If Melissa does not study French, then Lauren studies Latin.
  - If Lauren studies Latin, then Taylor studies Spanish.
  - Taylor does not study Spanish.

  What language does each person study?

16. Three friends, Augustus, Brutus, and Caesar, play a game in which each decides to be either a liar or a truthteller. A liar must always lie and a truthteller must always tell the truth. When you met these friends, you asked Augustus which he had chosen to be. You didn't hear his answer but Brutus volunteered, "Augustus said that he is a liar." Caesar added, "If one of us is a liar, then we are all liars." Can you determine, for each person, whether he is a liar or a truthteller?

CHAPTER SUMMARY

Definitions to Know

- **Logic** is the study of reasoning.
- In logic, a **mathematical sentence** is a sentence that contains a complete thought and can be judged to be true or false.
- A **phrase** is an expression that is only part of a sentence.
- An **open sentence** is any sentence that contains a variable.
- The **domain** or **replacement set** is the set of numbers that can replace a variable.
- The **solution set** or **truth set** is the set of all replacements that will change an open sentence to true sentences.
- A **statement** or a **closed sentence** is a sentence that can be judged to be true or false.
- A closed sentence is said to have a **truth value**, either true **(T)** or false **(F)**.
- The **negation** of a statement has the opposite truth value of a given statement.
- In logic, a **compound sentence** is a combination of two or more mathematical sentences formed by using the connectives *not*, *and*, *or*, *if . . . then*, or *if and only if*.
- A **conjunction** is a compound statement formed by combining two simple statements, called **conjuncts**, with the word *and*. The conjunction *p and q* is written symbolically as $p \wedge q$.
- A **disjunction** is a compound statement formed by combining two simple statements, called **disjuncts**, with *or*. The disjunction *p or q* is written symbolically as $p \vee q$.
- A **truth table** is a summary of all possible truth values of a logic statement.
- A **conditional** is a compound statement formed by using the words *if . . . then* to combine two simple statements. The conditional **if *p* then *q*** is written symbolically as $p \rightarrow q$.

- A **hypothesis**, also called a **premise** or **antecedent**, is an assertion that begins an argument. The hypothesis usually follows the word *if*.
- A **conclusion**, also called a **consequent**, is an ending or a sentence that closes an argument. The conclusion usually follows the word *then*.
- Beginning with a statement $(p \rightarrow q)$, the **inverse** $(\sim p \rightarrow \sim q)$ is formed by negating the hypothesis and negating the conclusion.
- Beginning with a statement $(p \rightarrow q)$, the **converse** $(q \rightarrow p)$ is formed by interchanging the hypothesis and the conclusion.
- Beginning with a conditional $(p \rightarrow q)$, the **contrapositive** $(\sim q \rightarrow \sim p)$ is formed by negating both the hypothesis and the conclusion, and then interchanging the resulting negation.
- Two statements are **logically equivalent**—or **logical equivalents**—if they always have the same truth value.
- A **biconditional** $(p \leftrightarrow q)$ is a compound statement formed by the conjunction of the conditional $p \rightarrow q$ and its converse $q \rightarrow p$.
- A **valid argument** uses a series of statements called **premises** that have known truth values to arrive at a conclusion.

Logic Statements

Negation:	$\sim p$	not p
Conjunction:	$p \wedge q$	p and q
Disjunction:	$p \vee q$	p or q
Conditional:	$p \rightarrow q$	if p then q
Inverse:	$\sim p \rightarrow \sim q$	if $\sim p$ then $\sim q$
Converse:	$q \rightarrow p$	if q then p
Contrapositive:	$\sim q \rightarrow \sim p$	if $\sim q$ then $\sim p$
Biconditional:	$p \leftrightarrow q$	p if and only if q

The truth values of the logic connectives can be summarized as follows:

p	q	Negation $\sim p$	$\sim q$	Conjunction $p \wedge q$	Disjunction $p \vee q$	Conditional $p \rightarrow q$	Inverse $\sim p \rightarrow \sim q$	Converse $q \rightarrow p$	Contrapositive $\sim q \rightarrow \sim p$	Biconditional $p \leftrightarrow q$
T	T	F	F	T	T	T	T	T	T	T
T	F	F	T	F	T	F	T	T	F	F
F	T	T	F	F	T	T	F	F	T	F
F	F	T	T	F	F	T	T	T	T	T

Laws of Logic
- The **Law of Detachment** states that when $p \rightarrow q$ is true and p is true, then q must be true.
- The **Law of Disjunctive Inference** states that when $p \vee q$ is true and p is false, then q must be true.

VOCABULARY

2-1 Logic • Truth value • Mathematical sentence • Phrase • Open sentence • Domain • Replacement Set • Solution set • Truth set • Statement • Closed sentence • Truth value (T and F) • Negation

2-2 Compound sentence • Compound statement • Conjunction • Conjunct • p and q • Tree diagram • Truth table

2-3 Disjunction • Disjunct • p or q • Inclusive *or* • Exclusive *or*

2-4 Conditional • If p then q • Hypothesis • Premise • Antecedent • Conclusion • Consequent

2-5 Inverse • Converse • Contrapositive • Logical equivalents

2-6 Biconditional

2-7 Laws of logic • Valid argument • Premises • Law of Detachment • Law of Disjunctive Inference

REVIEW EXERCISES

1. The statement "If I go to school, then I do not play basketball" is false. Using one or both of the statements "I go to school" and "I do not play basketball" or their negations, write five *true* statements.

2. The statement "If I go to school, then I do not play basketball" is false. Using one or both of the statements "I go to school" and "I do not play basketball" or their negations, write five *false* statements.

3. Mia said that the biconditional $p \leftrightarrow q$ and the biconditional $\sim p \leftrightarrow \sim q$ always have the opposite truth values. Do you agree with Mia? Explain why or why not.

In 4 and 5: **a.** Identify the hypothesis p. **b.** Identify the conclusion q.

4. If at first you don't succeed, then you should try again.

5. You will get a detention if you are late one more time.

In 6–12, tell whether each given statement is true or false.

6. If July follows June, then August follows July.

7. July follows June and July is a winter month in the northern hemisphere.

8. July is a winter month in the northern hemisphere or July follows June.

9. If August follows July, then July does not follow June.

10. July is a winter month if August is a winter month.

11. August does not follow July and July is not a winter month.

12. July follows June if and only if August follows July.

13. Which whole number, when substituted for y, will make the following sentence true?

$$(y + 5 > 9) \wedge (y < 6)$$

In 14–17, supply the word, phrase, or symbol that can be placed in each blank to make the resulting statement true.

14. $\sim(\sim p)$ has the same truth value as _____.

15. When p is true and q is false, then $p \wedge \sim q$ is _____.

16. When $p \vee \sim q$ is false, then p is _____ and q is _____.

17. If the conclusion q is true, then $p \rightarrow q$ must be _____.

In 18–22, find the truth value of each sentence when a, b, and c are all true.

18. $\sim a$ **19.** $\sim b \wedge c$ **20.** $b \rightarrow \sim c$ **21.** $a \vee \sim b$ **22.** $\sim a \leftrightarrow \sim b$

In 23–32, let p represent "$x > 5$," and let q represent "x is prime." Use the domain $\{1, 2, 3, 4, \ldots, 10\}$ to find the solution set for each of the following.

23. p **24.** $\sim p$ **25.** q **26.** $\sim q$ **27.** $p \vee q$

28. $p \wedge q$ **29.** $\sim p \wedge q$ **30.** $p \rightarrow \sim q$ **31.** $p \rightarrow q$ **32.** $p \leftrightarrow q$

33. For the conditional "If I live in Oregon, then I live in the Northwest," write: **a.** the inverse, **b.** the converse, **c.** the contrapositive, **d.** the biconditional.

34. Assume that the given sentences are true. Write a simple sentence that could be a conclusion.

- If $\angle A$ is the vertex angle of isosceles $\triangle ABC$, then $AB = AC$.
- $AB \neq AC$

35. Elmer Megabucks does not believe that girls should marry before the age of 21, and he disapproves of smoking. Therefore, he put the following provision in his will: I leave $100,000 to each of my nieces who, at the time of my death, is over 21 or unmarried, and does not smoke.

Each of his nieces is described below at the time of Elmer's death. Which nieces will inherit $100,000?

- Judy is 24, married, and smokes.
- Diane is 20, married, and does not smoke.

- Janice is 26, unmarried, and does not smoke.
- Peg is 19, unmarried, and smokes.
- Sue is 30, unmarried, and smokes.
- Sarah is 18, unmarried, and does not smoke.
- Laurie is 28, married, and does not smoke.
- Pam is 19, married, and smokes.

36. Some years after Elmer Megabucks prepared his will, he amended the conditions, by moving a comma, to read: I leave $100,000 to each of my nieces who, at the time of my death, is over 21, or unmarried and does not smoke. Which nieces described in Exercise 35 will now inherit $100,000?

37. At a swim meet, Janice, Kay, and Virginia were the first three finishers of a 200-meter backstroke competition. Virginia did not come in second. Kay did not come in third. Virginia came in ahead of Janice. In what order did they finish the competition?

38. Peter, Carlos, and Ralph play different musical instruments and different sports. The instruments that the boys play are violin, cello, and flute. The sports that the boys play are baseball, tennis, and soccer. From the clues given below, determine what instrument and what sport each boy plays.

- The violinist plays tennis.
- Peter does not play the cello.
- The boy who plays the flute does not play soccer.
- Ralph plays baseball.

39. Let p represent "x is divisible by 6."

Let q represent "x is divisible by 2."

a. If possible, find a value of x that will:

(1) make p true and q true.
(2) make p true and q false.
(3) make p false and q true.
(4) make p false and q false.

b. What conclusion can be drawn about the truth value of $p \rightarrow q$?

40. Each of the following statements is true.

- Either Peter, Jim, or Tom is Maria's brother.
- If Jim is Maria's brother, then Peter is Alice's brother.
- Alice has no brothers.
- Tom has no sisters.

Who is Maria's brother?

Exploration

1. A **tautology** is a statement that is always true. For instance, the disjunction $p \vee \sim p$ is a tautology because if p is true, the disjunction is true, and if p is false, $\sim p$ is true and the disjunction is true.

   a. Which of the following statements are tautologies?

   (1) $p \rightarrow (p \vee q)$

   (2) Either it will rain or it will snow.

   (3) $(p \wedge q) \rightarrow p$

   b. Construct two tautologies, one using symbols and the other using words.

2. A **contradiction** is a statement that is always false, that is, it cannot be true under any circumstances. For instance, the conjunction $p \wedge \sim p$ is a contradiction because if p is true, $\sim p$ is false and the conjunction is false, and if p is false, the conjunction is false.

   a. Which of the following statements are contradictions?

   (1) $(p \vee q) \wedge \sim p$

   (2) It is March or it is Tuesday, and it is not March and it is not Tuesday.

   (3) $(p \wedge \sim p) \wedge q$

   b. Construct two contradictions, one using symbols and the other using words.

3. One way to construct a tautology is to use a contradiction as the hypothesis of a conditional. For instance, since we know that $p \wedge \sim p$ is a contradiction, the conditional $(p \wedge \sim p) \rightarrow q$ is a tautology for any conclusion q.

   a. Construct two conditionals using any two contradictions as the premises.

   b. Show that the two conditionals from part **a** are tautologies.

   c. Explain why the method of using a contradiction as the hypothesis of a conditional always results in a tautology.

CUMULATIVE REVIEW   CHAPTERS 1–2

Part I

Answer all questions in this part. Each correct answer will receive 2 credits. No partial credit will be allowed.

1. Which of the following is an undefined term?

   (1) ray   (2) angle   (3) line   (4) line segment

2. The statement "If x is a prime then x is odd" is false when x equals

   (1) 1   (2) 2   (3) 3   (4) 4

3. Points J, K, and L lie on a line. The coordinate of J is -17, the coordinate of K is -8, and the coordinate of L is 13. What is the coordinate of M, the midpoint of $\overline{JL}$?

(1) -8 (2) -4 (3) -2 (4) 2

4. When "Today is Saturday" is false, which of the following statements could be either true or false?

(1) If today is Saturday, then I do not have to go to school.

(2) Today is Saturday and I do not have to go to school.

(3) Today is Saturday or I have to go to school.

(4) Today is not Saturday.

5. Which of the following is *not* a requirement in order for point H to be between points G and I?

(1) $GH = HI$ (3) $GH + HI = GI$

(2) G, H, and I are collinear. (4) G, H, and I are distinct points.

6. $\overrightarrow{UW}$ bisects $\angle TUV$. If $m\angle TUV = 34x$ and $m\angle VUW = 5x + 30$, what is $m\angle TUV$?

(1) $2.5°$ (2) $32.5°$ (3) $42.5°$ (4) $85°$

7. Which of the following must be true when $AB + BC = AC$?

(1) B is the midpoint of $\overline{AC}$ (3) C is a point on $\overline{AB}$

(2) $AB = BC$ (4) B is a point on $\overline{AC}$

8. Which of the following equalities is an example of the use of the commutative property?

(1) $3(2 + x) = 6 + 3x$ (3) $3 + (0 + x) = 3 + x$

(2) $3 + (2 + x) = (3 + 2) + x$ (4) $3(2 + x) = 3(x + 2)$

9. Which of the following must be true when p is true?

(1) $p \wedge q$ (2) $p \rightarrow q$ (3) $p \vee q$ (4) $\sim p \vee q$

10. The solution set of $x + 1 = x$ is

(1) $\varnothing$ (2) $\{0\}$ (3) $\{-1\}$ (4) {all real numbers}

Part II

Answer all questions in this part. Each correct answer will receive 2 credits. Clearly indicate the necessary steps, including appropriate formula substitutions, diagrams, graphs, charts, etc. For all questions in this part, a correct numerical answer with no work shown will receive only 1 credit.

11. The first two sentences below are true. Determine whether the third sentence is true, is false, or cannot be found to be true or false. Justify your answer.

> I win the ring toss game.
> If I win the ring toss game, then I get a goldfish.
> I get a goldfish.

12. On the number line, the coordinate of R is -5 and the coordinate of S is -1. What is the coordinate of T if S is the midpoint of $\overline{RT}$?

Part III

Answer all questions in this part. Each correct answer will receive 4 credits. Clearly indicate the necessary steps, including appropriate formula substitutions, diagrams, graphs, charts, etc. For all questions in this part, a correct numerical answer with no work shown will receive only 1 credit.

13. Let A and B be two points in a plane. Explain the meanings of the symbols $\overleftrightarrow{AB}$, $\overrightarrow{AB}$, $\overline{AB}$, and AB.

14. The ray $\overrightarrow{BD}$ is the bisector of $\angle ABC$, a straight angle. Explain why $\overrightarrow{BD}$ is perpendicular to $\overleftrightarrow{ABC}$. Use definitions to justify your explanation.

Part IV

Answer all questions in this part. Each correct answer will receive 6 credits. Clearly indicate the necessary steps, including appropriate formula substitutions, diagrams, graphs, charts, etc. For all questions in this part, a correct numerical answer with no work shown will receive only 1 credit.

15. The steps used to simplify an algebraic expression are shown below. Name the property that justifies each of the steps.

$$\begin{aligned} a(2b + 1) &= a(2b) + a(1) \\ &= (a \cdot 2) \cdot b + a(1) \\ &= (2 \cdot a) \cdot b + a(1) \\ &= 2ab + a \end{aligned}$$

16. The statement "If x is divisible by 12, then x is divisible by 4" is always true.

   a. Write the converse of the statement.

   b. Write the inverse of the statement.

   c. Are the converse and inverse true for all positive integers x? Justify your answer.

   d. Write another statement using "x is divisible by 12" and "x is divisible by 4" or the negations of these statements that is true for all positive integers x.

PROVING STATEMENTS IN GEOMETRY

After proposing 23 definitions, Euclid listed five postulates and five "common notions." These definitions, postulates, and common notions provided the foundation for the propositions or theorems for which Euclid presented proof. Modern mathematicians have recognized the need for additional postulates to establish a more rigorous foundation for these proofs.

David Hilbert (1862–1943) believed that mathematics should have a logical foundation based on two principles:

1. All mathematics follows from a correctly chosen finite set of assumptions or *axioms*.
2. This set of axioms is not contradictory.

Although mathematicians later discovered that it is not possible to formalize all of mathematics, Hilbert did succeed in putting geometry on a firm logical foundation. In 1899, Hilbert published a text, *Foundations of Geometry*, in which he presented a set of axioms that avoided the limitations of Euclid.

3-1 INDUCTIVE REASONING

Gina was doing a homework assignment on factoring positive integers. She made a chart showing the number of factors for each of the numbers from 1 to 10. Her chart is shown below.

Number	1	2	3	4	5	6	7	8	9	10
Number of factors	1	2	2	3	2	4	2	4	3	4

She noticed that in her chart, only the perfect square numbers, 1, 4, and 9 had an odd number of factors. The other numbers had an even number of factors. Gina wanted to investigate this observation further so she continued her chart to 25.

Number	11	12	13	14	15	16	17	18	19	20	21	22	23	24	25
Number of factors	2	6	2	4	4	5	2	6	2	6	4	4	2	8	3

Again her chart showed that only the perfect square numbers, 16 and 25, had an odd number of factors and the other numbers had an even number of factors. Gina concluded that this is true for all positive integers. Gina went from a few specific cases to a **generalization.**

Was Gina's conclusion valid? Can you find a counterexample to prove that a perfect square does not always have an odd number of factors? Can you find a non-perfect square that has an odd number of factors?

Scientists perform experiments once, twice, or many times in order to come to a conclusion. A scientist who is searching for a vaccine to prevent a disease will test the vaccine repeatedly to see if it is effective. This method of reasoning, in which a series of particular examples leads to a conclusion, is called **inductive reasoning.**

In geometry, we also perform experiments to discover properties of lines, angles, and polygons and to determine geometric relationships. Most of these experiments involve measurements. Because direct measurements depend on the type of instrument used to measure and the care with which the measurement is made, results can be only approximate. This is the first weakness in attempting to reach conclusions by inductive reasoning.

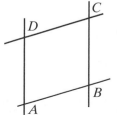

Use a ruler to draw a pair of parallel line segments by drawing a line segment along opposite edges of the ruler. Turn the ruler and draw another pair of parallel line segments that intersect the first pair. Label the intersections of the line segments A, B, C, and D. The figure, $ABCD$, is a parallelogram. It appears that the opposite sides of the parallelogram have equal measures. Use the ruler to show that this is true.

To convince yourself that this relationship is true in other parallelograms, draw other parallelograms and measure their opposite sides. In each experiment you will find that the opposite sides have the same measure. From these experiments, you can arrive at a general conclusion: The opposite sides of a parallelogram have equal measures. This is an example of inductive reasoning in geometry.

Suppose a student, after examining several parallelograms, made the generalization "All parallelograms have two acute angles and two obtuse angles." Here, a single **counterexample**, such as a parallelogram which is a rectangle and has right angles, is sufficient to show that the general conclusion that the student reached is false.

When we use inductive reasoning, we must use extreme care because we are arriving at a generalization before we have examined every possible example. This is the second weakness in attempting to reach conclusions by inductive reasoning.

When we conduct an experiment we do not give explanations for why things are true. This is the third weakness of inductive reasoning. In light of these weaknesses, when a general conclusion is reached by inductive reasoning alone, it can at best be called probably true. Such statements, that are likely to be true but not yet been proven true by a deductive proof, are called **conjectures**.

Then why study inductive reasoning? Simply because it mimics the way we naturally make new discoveries. Most new knowledge first starts with specific cases, then, through careful study, to a generalization. Only afterwards, is a *proof* or explanation usually sought. Inductive reasoning is therefore a powerful tool in discovering new mathematical facts.

SUMMARY

1. Inductive reasoning is a powerful tool in discovering and making conjectures.

2. Generalizations arising from direct measurements of specific cases are only approximate.

3. Care must be taken when applying inductive reasoning to ensure that all relevant examples are examined (no counterexamples exist).

4. Inductive reasoning does not prove or explain conjectures.

Exercises

Writing About Mathematics

1. After examining several triangles, Mindy concluded that the angles of all triangles are acute. Is Mindy correct? If not, explain to Mindy why she is incorrect.

2. Use a counterexample to show that the whole numbers are not closed under subtraction.

Developing Skills

3. a. Using geometry software or a pencil, ruler, and protractor, draw three right triangles that have different sizes and shapes.

   b. In each right triangle, measure the two acute angles and find the sum of their measures.

   c. Using inductive reasoning based on the experiments just done, make a conjecture about the sum of the measures of the acute angles of a right triangle.

4. a. Using geometry software or a pencil, ruler, and protractor, draw three quadrilaterals that have different sizes and shapes.

   b. For each quadrilateral, find the midpoint of each side of the quadrilateral, and draw a new quadrilateral using the midpoints as vertices. What appears to be true about the quadrilateral that is formed?

   c. Using inductive reasoning based on the experiments just done, make a conjecture about a quadrilateral with vertices at the midpoints of a quadrilateral.

5. a. Using geometry software or a pencil, ruler, and protractor, draw three equilateral triangles of different sizes.

   b. For each triangle, find the midpoint of each side of the triangle, and draw line segments joining each midpoint. What appears to be true about the four triangles that are formed?

   c. Using inductive reasoning based on the experiments just done, make a conjecture about the four triangles formed by joining the midpoints of an equilateral triangle.

In 6–11, describe and perform a series of experiments to investigate whether each statement is probably true or false.

6. If two lines intersect, at least one pair of congruent angles is formed.

7. The sum of the degree measures of the angles of a quadrilateral is 360.

8. If $a^2 < b^2$, then $a < b$.

9. In any parallelogram $ABCD$, $AC = BD$.

10. In any quadrilateral $DEFG$, $\overline{DF}$ bisects $\angle D$ and $\angle F$.

11. The ray that bisects an angle of a triangle intersects a side of the triangle at its midpoint.

12. Adam made the following statement: "For any counting number n, the expression $n^2 + n + 41$ will always be equal to some prime number." He reasoned:

   • When $n = 1$, then $n^2 + n + 41 = 1 + 1 + 41 = 43$, a prime number.

   • When $n = 2$, then $n^2 + n + 41 = 4 + 2 + 41 = 47$, a prime number.

Use inductive reasoning, by letting n be many different counting numbers, to show that Adam's generalization is probably true, or find a counterexample to show that Adam's generalization is false.

Applying Skills

In 13–16, state in each case whether the conclusion drawn was justified.

13. One day, Joe drove home on Route 110 and found traffic very heavy. He decided never again to drive on this highway on his way home.

14. Julia compared the prices of twenty items listed in the advertising flyers of store A and store B. She found that the prices in store B were consistently higher than those of store A. Julia decided that she will save money by shopping in store A.

15. Tim read a book that was recommended by a friend and found it interesting. He decided that he would enjoy *any* book recommended by that friend.

16. Jill fished in Lake George one day and caught nothing. She decided that there are no fish in Lake George.

17. Nathan filled up his moped's gas tank after driving 92 miles. He concluded that his moped could go at least 92 miles on one tank of gas.

Hands-On Activity

STEP 1. Out of a regular sheet of paper, construct ten cards numbered 1 to 10.

STEP 2. Place the cards face down and in order.

STEP 3. Go through the cards and turn over every card.

STEP 4. Go through the cards and turn over every second card starting with the second card.

STEP 5. Go through the cards and turn over every third card starting with the third card.

STEP 6. Continue this process until you turn over the tenth (last) card.

If you played this game correctly, the cards that are face up when you finish should be 1, 4, and 9.

a. Play this same game with cards numbered 1 to 20. What cards are face up when you finish? What property do the numbers on the cards all have in common?

b. Play this same game with cards numbered 1 to 30. What cards are face up when you finish? What property do the numbers on the cards all have in common?

c. Make a conjecture regarding the numbers on the cards that remain facing up if you play this game with any number of cards.

3-2 DEFINITIONS AS BICONDITIONALS

In mathematics, we often use inductive reasoning to make a conjecture, a statement that appears to be true. Then we use **deductive reasoning** to prove the conjecture. Deductive reasoning uses the laws of logic to combine definitions and general statements that we know to be true to reach a valid conclusion.

Before we discuss this type of reasoning, it will be helpful to review the list of definitions in Chapter 1 given on page 29 that are used in the study of Euclidean geometry.

Definitions and Logic

Each of the definitions given in Chapter 1 can be written in the form of a conditional. For example:

▶ A *scalene triangle* is a triangle that has no congruent sides.

Using the definition of a scalene triangle, we know that:

1. The definition contains a hidden conditional statement and can be rewritten using the words *If ... then* as follows:

   t: A triangle is scalene.
   p: A triangle has no congruent sides.
   t → *p*: If a triangle is scalene, then the triangle has no congruent sides.

2. In general, the converse of a true statement is not necessarily true. However, the converse of the conditional form of a definition is always true. For example, the following converse is a true statement:

   p → *t*: If a triangle has no congruent sides, then the triangle is scalene.

3. When a conditional and its converse are both true, the conjunction of these statements can be written as a true biconditional statement. Thus, $(t \rightarrow p) \wedge (p \rightarrow t)$ is equivalent to the biconditional $(t \leftrightarrow p)$.

Therefore, since both the conditional statement and its converse are true, we can rewrite the above definition as a biconditional statement, using the words *if and only if*, as follows:

▶ A triangle is scalene if and only if the triangle has no congruent sides.

Every good definition can be written as a true biconditional. Definitions will often be used to prove statements in geometry.

EXAMPLE 1

A collinear set of points is a set of points all of which lie on the same straight line.

a. Write the definition in conditional form.

b. Write the converse of the statement given in part **a.**

c. Write the biconditional form of the definition.

Solution **a.** *Conditional:* If a set of points is collinear, then all the points lie on the same straight line.

b. *Converse:* If a set of points all lie on the same straight line, then the set of points is collinear.

c. *Biconditional:* A set of points is collinear if and only if all the points lie on the same straight line.

Exercises

Writing About Mathematics

1. Doug said that "A container is a lunchbox if and only if it can be used to carry food" is not a definition because one part of the biconditional is false. Is Doug correct? If so, give a counterexample to show that Doug is correct.

2. **a.** Give a counterexample to show that the statement "If $\frac{a}{b} < 1$, then $a < b$" is not always true.

   b. For what set of numbers is the statement "If $\frac{a}{b} < 1$, then $a < b$" always true?

Developing Skills

In 3–8: **a.** Write each definition in conditional form. **b.** Write the converse of the conditional given in part **a**. **c.** Write the biconditional form of the definition.

3. An equiangular triangle is a triangle that has three congruent angles.

4. The bisector of a line segment is any line, or subset of a line that intersects the segment at its midpoint.

5. An acute angle is an angle whose degree measure is greater than 0 and less than 90.

6. An obtuse triangle is a triangle that has one obtuse angle.

7. A noncollinear set of points is a set of three or more points that do not all lie on the same straight line.

8. A ray is a part of a line that consists of a point on the line, called an endpoint, and all the points on one side of the endpoint.

In 9–14, write the biconditional form of each given definition.

9. A point B is between A and C if A, B, and C are distinct collinear points and $AB + BC = AC$.

10. Congruent segments are segments that have the same length.

11. The midpoint of a line segment is the point of that line segment that divides the segment into two congruent segments.

12. A right triangle is a triangle that has a right angle.

13. A straight angle is an angle that is the union of opposite rays and whose degree measure is 180.

14. Opposite rays are two rays of the same line with a common endpoint and no other point in common.

Applying Skills

In 15–17, write the biconditional form of the definition of each given term.

15. equilateral triangle   **16.** congruent angles   **17.** perpendicular lines

3-3 DEDUCTIVE REASONING

A **proof** in geometry is a valid argument that establishes the truth of a statement. Most proofs in geometry rely on logic. That is, they are based on a series of statements that are assumed to be true. *Deductive reasoning* uses the laws of logic to link together true statements to arrive at a true conclusion. Since definitions are true statements, they are used in a geometric proof. In the examples that follow, notice how the laws of logic are used in the proofs of geometric statements.

Using Logic to Form a Geometry Proof

Let $\triangle ABC$ be a triangle in which $\overline{AB} \perp \overline{BC}$. We can prove that $\angle ABC$ is a right angle. In this proof, use the following definition:

- Perpendicular lines are two lines that intersect to form right angles.

This definition contains a hidden conditional and can be rewritten as follows:

- *If* two lines are perpendicular, *then* they intersect to form right angles.

Recall that this definition is true for perpendicular line segments as well as for perpendicular lines. Using the specific information cited above, let p represent "$\overline{AB} \perp \overline{BC}$," and let r represent "$\angle ABC$ is a right angle."

The proof is shown by the reasoning that follows:

p: $\overline{AB} \perp \overline{BC}$
     p is true because it is given.

$p \rightarrow r$: If $\overline{AB} \perp \overline{BC}$, then $\angle ABC$ is a right angle.
     $p \rightarrow r$ is true because it is the definition of perpendicular lines.

r: $\angle ABC$ is a right angle.
     r is true by the Law of Detachment.

In the logic-based proof above, notice that the Law of Detachment is cited as a reason for reaching our conclusion. In a typical geometry proof, however, the laws of logic are used to deduce the conclusion but the laws are not listed among the reasons.

Let us restate this proof in the format used often in Euclidean geometry. We write the information known to be true as the "**given**" statements and the conclusion to be proved as the "**prove**". Then we construct a **two-column proof**. In the left column, we write *statements* that we know to be true, and in the right column, we write the *reasons* why each statement is true.

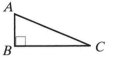

Given: In $\triangle ABC$, $\overline{AB} \perp \overline{BC}$.

Prove: $\angle ABC$ is a right angle.

Proof:

Statements	Reasons
1. $\overline{AB} \perp \overline{BC}$	**1.** Given.
2. $\angle ABC$ is a right angle.	**2.** If two lines are perpendicular, then they intersect to form right angles.

Notice how the Law of Detachment was used in this geometry proof. By combining statement 1 with reason 2, the conclusion is written as statement 2, just as p and $p \rightarrow r$ led us to the conclusion r using logic. In reason 2, we used the conditional form of the definition of perpendicular lines.

Often in a proof, we find one conclusion in order to use that statement to find another conclusion. For instance, in the following proof, we will begin by proving that $\angle ABC$ is a right angle and then use that fact with the definition of a right triangle to prove that $\triangle ABC$ is a right triangle.

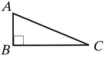

Given: In $\triangle ABC$, $\overline{AB} \perp \overline{BC}$.

Prove: $\triangle ABC$ is a right triangle.

Proof:

Statements	Reasons
1. $\overline{AB} \perp \overline{BC}$	**1.** Given.
2. $\angle ABC$ is a right angle.	**2.** If two lines are perpendicular, then they intersect to form right angles.
3. $\triangle ABC$ is a right triangle.	**3.** If a triangle has a right angle, then it is a right triangle.

Note that in this proof, we used the conditional form of the definition of perpendicular lines in reason 2 and the converse form of the definition of a right triangle in reason 3.

The proof can also be written in paragraph form, also called a **paragraph proof**. Each statement must be justified by stating a definition or another statement that has been accepted or proved to be true. The proof given on page 101 can be written as follows:

Proof: We are given that $\overline{AB} \perp \overline{BC}$. If two lines are perpendicular, then they intersect to form right angles. Therefore, $\angle ABC$ is a right angle. A right triangle is a triangle that has a right angle. Since $\angle ABC$ is an angle of $\triangle ABC$, $\triangle ABC$ is a right triangle.

EXAMPLE 1

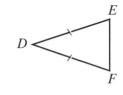

Given: $\triangle DEF$ with $DE = DF$

Prove: $\triangle DEF$ is an isosceles triangle.

Proof We will need the following definitions:

- An isosceles triangle is a triangle that has two congruent sides.
- Congruent segments are segments that have the same measure.

We can use these two definitions to first prove that the two segments with equal measure are congruent and then to prove that since the two segments, the sides, are congruent, the triangle is isosceles.

Statements	Reasons
1. $DE = DF$	**1.** Given.
2. $\overline{DE} \cong \overline{DF}$	**2.** Congruent segments are segments that have the same measure.
3. $\triangle DEF$ is isosceles.	**3.** An isosceles triangle is a triangle that has two congruent sides.

Alternative Proof The converse of the definition of congruent segments states that if two segments have the same measure, then they are congruent. The given statement, $DE = DF$, means that $\overline{DE}$ and $\overline{DF}$ have the same measure. Therefore, these two sides of $\triangle DEF$ are congruent. The converse of the definition of an isosceles triangle states that if a triangle has two congruent sides, then it is isosceles. Therefore, since $\overline{DE}$ and $\overline{DF}$ are congruent sides of $\triangle DEF$, $\triangle DEF$ is isosceles.

EXAMPLE 2

Given: $\overrightarrow{BD}$ is the bisector of $\angle ABC$.

Prove: $m\angle ABD = m\angle DBC$

Proof We will need the following definitions:

- The bisector of an angle is a ray whose endpoint is the vertex of the angle and that divides the angle into two congruent angles.
- Congruent angles are angles that have the same measure.

First use the definition of an angle bisector to prove that the angles are congruent. Then use the definition of congruent angles to prove that the angles have equal measures.

Statements	Reasons
1. $\overrightarrow{BD}$ is the bisector of $\angle ABC$.	**1.** Given.
2. $\angle ABD \cong \angle DBC$	**2.** The bisector of an angle is a ray whose endpoint is the vertex of the angle and that divides the angle into two congruent angles.
3. $m\angle ABD = m\angle DBC$	**3.** Congruent angles are angles that have the same measure.

Exercises

Writing About Mathematics

1. Is an equilateral triangle an isosceles triangle? Justify your answer.

2. Is it possible that the points A, B, and C are collinear but $AB + BC \neq AC$? Justify your answer.

Developing Skills

In 3–6, in each case: **a.** Draw a diagram to illustrate the given statement. **b.** Write a definition or definitions from geometry, in conditional form, that can be used with the *given* statement to justify the conclusion.

3. *Given:* $\overrightarrow{SP}$ bisects $\angle RST$.
  Conclusion: $\angle RSP \cong \angle PST$

4. *Given:* $\triangle ABC$ is a scalene triangle.
  Conclusion: $AB \neq BC$

5. *Given:* $\overleftrightarrow{BCD} \perp \overleftrightarrow{ACE}$
  Conclusion: $m\angle ACD = 90$

6. *Given:* $AB + BC = AC$ with $\overline{ABC}$
  Conclusion: B is between A and C.

In 7–12, in each case: **a.** Draw a diagram to illustrate the given statement. **b.** Write a definition from geometry, in conditional form, that can be used with the given statement to make a conclusion. **c.** From the given statement and the definition that you chose, draw a conclusion.

7. *Given:* $\triangle LMN$ with $\overline{LM} \perp \overline{MN}$

8. *Given:* $\overleftrightarrow{AB}$ bisects $\overline{DE}$ at F.

9. *Given:* $PQ + QR = PR$ with $\overline{PQR}$

10. *Given:* $\overrightarrow{ST}$ and $\overrightarrow{SR}$ are opposite rays.

11. *Given:* M is the midpoint of $\overline{LN}$.

12. *Given:* $0 < m\angle A < 90$

Applying Skills

In 13–15: **a.** Give the reason for each statement of the proof. **b.** Write the proof in paragraph form.

13. *Given:* M is the midpoint of $\overline{AMB}$.

  Prove: $AM = MB$

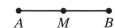

Statements	Reasons
1. M is the midpoint of $\overline{AMB}$.	**1.** _____
2. $\overline{AM} \cong \overline{MB}$	**2.** _____
3. $AM = MB$	**3.** _____ ◾

14. *Given:* $\triangle RST$ with $RS = ST$.

  Prove: $\triangle RST$ is an isosceles triangle.

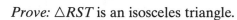

Statements	Reasons
1. $RS = ST$	**1.** _____
2. $\overline{RS} \cong \overline{ST}$	**2.** _____
3. $\triangle RST$ is isosceles.	**3.** _____ ◾

15. *Given:* In $\triangle ABC$, $\overrightarrow{CE}$ bisects $\angle ACB$.

  Prove: $m\angle ACE = m\angle BCE$

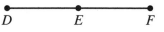

Statements	Reasons
1. $\overrightarrow{CE}$ bisects $\angle ACB$.	**1.** _____
2. $\angle ACE \cong \angle ECB$	**2.** _____
3. $m\angle ACE = m\angle ECB$	**3.** _____ ◾

16. Complete the following proof by writing the statement for each step.

  Given: $\overline{DEF}$ with $DE = EF$.

  Prove: E is the midpoint of $\overline{DEF}$.

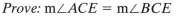

Statements	Reasons
1. _____	**1.** Given.
2. _____	**2.** Congruent segments are segments that have the same measure.
3. _____	**3.** The midpoint of a line segment is the point of that line segment that divides the segment into congruent segments. ◾

17. In $\triangle ABC$, m$\angle A < 90$ and m$\angle B < 90$. If $\triangle ABC$ is an obtuse triangle, why is m$\angle C > 90$? Justify your answer with a definition.

18. Explain why the following proof is incorrect.

Given: Isosceles $\triangle ABC$ with $\angle A$ as the vertex angle.

Prove: $BC = AC$

Statements	Reasons
1. $\triangle ABC$ is isosceles.	**1.** Given.
2. $\overline{BC} \cong \overline{AC}$	**2.** An isosceles triangle has two congruent sides.
3. $BC = AC$	**3.** Congruent segments are segments that have the same measure.

3-4 DIRECT AND INDIRECT PROOFS

A proof that starts with the given statements and uses the laws of logic to arrive at the statement to be proved is called a **direct proof**. A proof that starts with the *negation* of the statement to be proved and uses the laws of logic to show that it is false is called an **indirect proof** or a **proof by contradiction**.

An indirect proof works because when the negation of a statement is false, the statement must be true. Therefore, if we can show that the negation of the statement to be proved is false, then we can conclude that the statement is true.

Direct Proof

All of the proofs in Section 3-3 are direct proofs. In most direct proofs we use definitions together with the Law of Detachment to arrive at the desired conclusion. Example 1 uses direct proof.

EXAMPLE 1

Given: $\triangle ABC$ is an acute triangle.

Prove: m$\angle A < 90$

In this proof, we will use the following definitions:

- An acute triangle is a triangle that has three acute angles.
- An acute angle is an angle whose degree measure is greater than 0 and less than 90.

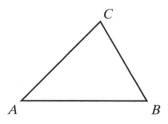

In the proof, we will use the conditional form of these definitions.

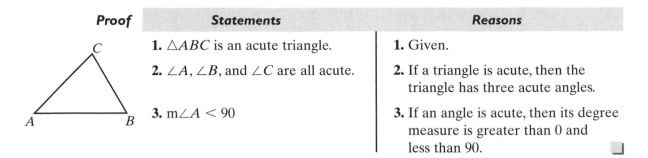

Proof	**Statements**	**Reasons**
	1. $\triangle ABC$ is an acute triangle.	**1.** Given.
	2. $\angle A$, $\angle B$, and $\angle C$ are all acute.	**2.** If a triangle is acute, then the triangle has three acute angles.
	3. $m\angle A < 90$	**3.** If an angle is acute, then its degree measure is greater than 0 and less than 90.

Note: In this proof, the "and" statement is important for the conclusion. In statement 2, the conjunction can be rewritten as "$\angle A$ is acute, and $\angle B$ is acute, and $\angle C$ is acute." We know from logic that when a conjunction is true, each conjunct is true. Also, in Reason 3, the conclusion of the conditional is a conjunction: "The degree measure is greater than 0 *and* the degree measure is less than 90." Again, since this conjunction is true, each conjunct is true.

Indirect Proof

In an indirect proof, let p be the given and q be the conclusion. Take the following steps to show that the conclusion is true:

1. Assume that the negation of the conclusion is true.

2. Use this assumption to arrive at a statement that contradicts the given statement or a true statement derived from the given statement.

3. Since the assumption leads to a contradiction, it must be false. The negation of the assumption, the desired conclusion, must be true.

Let us use an indirect proof to prove the following statement: If the measures of two segments are unequal, then the segments are not congruent.

EXAMPLE 2

Given: $\overline{AB}$ and $\overline{CD}$ such that $AB \neq CD$.

Prove: $\overline{AB}$ and $\overline{CD}$ are not congruent segments

Proof Start with an assumption that is the negation of the conclusion.

Statements	Reasons
1. $\overline{AB}$ and $\overline{CD}$ are congruent segments.	**1.** Assumption.
2. $AB = CD$	**2.** Congruent segments are segments that have the same measure.
3. $AB \neq CD$	**3.** Given.
4. $\overline{AB}$ and $\overline{CD}$ are not congruent segments.	**4.** Contradiction in 2 and 3. Therefore, the assumption is false and its negation is true.

In this proof, and most indirect proofs, our reasoning reflects the contrapositive of a definition. For example:

Definition: Congruent segments are segments that have the same measure.

Conditional: If segments are congruent, then they have the same measure.

Contrapositive: If segments do not have the same measure, then they are not congruent.

Note: To learn how the different methods of proof work, you will be asked to prove some simple statements both directly and indirectly in this section. Thereafter, you should use the method that seems more efficient, which is usually a direct proof. However, in some cases, only an indirect proof is possible.

EXAMPLE 3

Write a direct and an indirect proof of the following:

Given: m∠*CDE* ≠ 90

Prove: $\overline{CD}$ is not perpendicular to $\overline{DE}$.

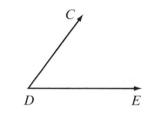

In this proof, we will use two definitions:

• If an angle is a right angle, then its degree measure is 90.

• If two intersecting lines are perpendicular, then they form right angles.

Direct Proof We will use the contrapositive form of the two definitions:

• If the degree measure of an angle is not 90, then it is not a right angle.

• If two intersecting lines do not form right angles, then they are not perpendicular.

Statements	Reasons
1. m∠CDE ≠ 90	**1.** Given.
2. ∠CDE is not a right angle.	**2.** If the degree measure of an angle is not 90, then it is not a right angle.
3. $\overline{CD}$ is not perpendicular to $\overline{DE}$.	**3.** If two intersecting lines do not form right angles, then they are not perpendicular.

Indirect Proof We will use the negation of the statement that is to be proved as an assumption, and then arrive at a contradiction using the conditional form of the two definitions.

Statements	Reasons
1. $\overline{CD}$ is perpendicular to $\overline{DE}$.	**1.** Assumption.
2. ∠CDE is a right angle.	**2.** If two intersecting lines are perpendicular, then they form right angles.
3. m∠CDE = 90	**3.** If an angle is a right angle, then its degree measure is 90.
4. m∠CDE ≠ 90.	**4.** Given.
5. $\overline{CD}$ is not perpendicular to $\overline{DE}$.	**5.** Contradiction in 3 and 4. Therefore, the assumption is false and its negation is true.

Exercises

Writing About Mathematics

1. If we are given ∠ABC, is it true that $\overleftrightarrow{AB}$ intersects $\overleftrightarrow{BC}$? Explain why or why not.

2. Glen said that if we are given line $\overleftrightarrow{ABC}$, then we know that A, B, and C are collinear and B is between A and C. Do you agree with Glen? Justify your answer.

Developing Skills

In 3–8: **a.** Draw a figure that represents the statement to be proved. **b.** Write a direct proof in two-column form. **c.** Write an indirect proof in two-column form.

3. *Given: LM = MN*

   Prove: $\overline{LM} \cong \overline{MN}$

4. *Given:* $\angle PQR$ is a straight angle.

   Prove: m$\angle PQR = 180$

5. *Given:* $\angle PQR$ is a straight angle.

   Prove: $\overrightarrow{QP}$ and $\overrightarrow{QR}$ are opposite rays.

6. *Given:* $\overrightarrow{QP}$ and $\overrightarrow{QR}$ are opposite rays.

   Prove: P, Q, and *R* are on the same line.

7. *Given:* $\angle PQR$ is a straight angle.

   Prove: P, Q, and *R* are on the same line.

8. *Given:* $\overrightarrow{EG}$ bisects $\angle DEF$.

   Prove: m$\angle DEG =$ m$\angle GEF$

9. Compare the direct proofs to the indirect proofs in problems 3–8. In these examples, which proofs were longer? Why do you think this is the case?

10. Draw a figure that represents the statement to be proved and write an indirect proof:

    Given: $\overrightarrow{EG}$ does not bisect $\angle DEF$.

    Prove: $\angle DEG$ is not congruent to $\angle GEF$.

Applying Skills

11. In order to prove a conditional statement, we let the hypothesis be the *given* and the conclusion be the *prove*.

    If $\overrightarrow{BD}$ is perpendicular to $\overleftrightarrow{ABC}$, then $\overrightarrow{BD}$ is the bisector of $\angle ABC$.

    a. Write the hypothesis of the conditional as the *given*.

    b. Write the conclusion of the conditional as the *prove*.

    c. Write a direct proof for the conditional.

12. If m$\angle EFG \neq 180$, then $\overrightarrow{FE}$ and $\overrightarrow{FG}$ are not opposite rays.

    a. Write the hypothesis of the conditional as the *given*.

    b. Write the conclusion of the conditional as the *prove*.

    c. Write an indirect proof for the conditional.

3-5 POSTULATES, THEOREMS, AND PROOF

A valid argument that leads to a true conclusion begins with true statements. In Chapter 1, we listed undefined terms and definitions that we accept as being true. We have used the undefined terms and definitions to draw conclusions.

At times, statements are made in geometry that are neither undefined terms nor definitions, and yet we know these are true statements. Some of these statements seem so "obvious" that we accept them without proof. Such a statement is called a **postulate** or an **axiom**.

Some mathematicians use the term "axiom" for a general statement whose truth is assumed without proof, and the term "postulate" for a geometric statement whose truth is assumed without proof. We will use the term "postulate" for both types of assumptions.

DEFINITION _____

A **postulate** is a statement whose truth is accepted without proof.

When we apply the laws of logic to definitions and postulates, we are able to prove other statements. These statements are called **theorems**.

DEFINITION _____

A **theorem** is a statement that is proved by deductive reasoning.

The entire body of knowledge that we know as geometry consists of undefined terms, defined terms, postulates, and theorems which we use to prove other theorems and to justify applications of these theorems.

The First Postulates Used in Proving Statements

Geometry is often concerned with measurement. In Chapter 1 we listed the properties of the number system. These properties include closure, the commutative, associative, inverse, identity, and distributive properties of addition and multiplication and the multiplication property of zero. We will use these properties as postulates in geometric proof. Other properties that we will use as postulates are the properties of equality.

When we state the relation "x is equal to y," symbolized as "$x = y$," we mean that x and y are two different names for the same element of a set, usually a number. For example:

1. When we write $AB = CD$, we mean that length of $\overline{AB}$ and the length of $\overline{CD}$ are the same number.

2. When we write $m\angle P = m\angle N$, we mean that $\angle P$ and $\angle N$ contain the same number of degrees.

Many of our definitions, for example, congruence, midpoint, and bisector, state that two measures are equal. There are three basic properties of equality.

The Reflexive Property of Equality: $a = a$

The **reflexive property of equality** is stated in words as follows:

Postulate 3.1

A quantity is equal to itself.

For example, in $\triangle CDE$, observe that:

1. The length of a segment is equal to itself:

$$CD = CD \qquad DE = DE \qquad EC = EC$$

2. The measure of an angle is equal to itself:

$$m\angle C = m\angle C \qquad m\angle D = m\angle D \qquad m\angle E = m\angle E$$

The Symmetric Property of Equality: If $a = b$, then $b = a$.

The **symmetric property of equality** is stated in words as follows:

Postulate 3.2

> An equality may be expressed in either order.

For example:

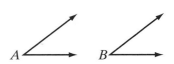

1. If $LM = NP$, then $NP = LM$.

2. If $m\angle A = m\angle B$, then $m\angle B = m\angle A$.

The Transitive Property of Equality: If $a = b$ and $b = c$, then $a = c$.

This property states that, if a and b have the same value, and b and c have the same value, it follows that a and c have the same value. The **transitive property of equality** is stated in words as follows:

Postulate 3.3

> Quantities equal to the same quantity are equal to each other.

The lengths or measures of segments and angles are numbers. In the set of real numbers, the relation "is equal to" is said to be reflexive, symmetric, and transitive. A relation for which these postulates are true is said to be an **equivalence relation**.

Congruent segments are segments with equal measures and congruent angles are angles with equal measures. This suggests that "is congruent to" is also an equivalence relation for the set of line segments. For example:

1. $\overline{AB} \cong \overline{AB}$

   A line segment is congruent to itself.

2. If $\overline{AB} \cong \overline{CD}$, then $\overline{CD} \cong \overline{AB}$.

   Congruence can be stated in either order.

3. If $\overline{AB} \cong \overline{CD}$ and $\overline{CD} \cong \overline{EF}$, then $\overline{AB} \cong \overline{EF}$.

   Segments congruent to the same segment are congruent to each other.

Therefore, we can say that "is congruent to" is an equivalence relation on the set of line segments.

We will use these postulates of equality in deductive reasoning. In constructing a valid proof, we follow these steps:

1. A diagram is used to visualize what is known and how it relates to what is to be proved.

2. State the hypothesis or premise as the *given*, in terms of the points and lines in the diagram. The premises are the given facts.

3. The conclusion contains what is to be proved. State the conclusion as the *prove*, in terms of the points and lines in the diagram.

4. We present the *proof*, the deductive reasoning, as a series of statements. Each statement in the proof should be justified by the given, a definition, a postulate, or a previously proven theorem.

EXAMPLE I

If $\overline{AB} \cong \overline{BC}$ and $\overline{BC} \cong \overline{CD}$, then $AB = CD$.

Given: $\overline{AB} \cong \overline{BC}$ and $\overline{BC} \cong \overline{CD}$

Prove: $AB = CD$

Proof

Statements	Reasons
1. $\overline{AB} \cong \overline{BC}$	1. Given.
2. $AB = BC$	2. Congruent segments are segments that have the same measure.
3. $\overline{BC} \cong \overline{CD}$	3. Given.
4. $BC = CD$	4. Congruent segments are segments that have the same measure.
5. $AB = CD$	5. Transitive property of equality (steps 2 and 4).

Alternative Proof

Statements	Reasons
1. $\overline{AB} \cong \overline{BC}$	1. Given.
2. $\overline{BC} \cong \overline{CD}$	2. Given.
3. $\overline{AB} \cong \overline{CD}$	3. Transitive property of congruence.
4. $AB = CD$	4. Congruent segments are segments that have the same measure.

EXAMPLE 2

If $\overline{AB} \perp \overline{BC}$ and $\overline{LM} \perp \overline{MN}$, then $m\angle ABC = m\angle LMN$.

Given: $\overline{AB} \perp \overline{BC}$ and $\overline{LM} \perp \overline{MN}$

Prove: $m\angle ABC = m\angle LMN$

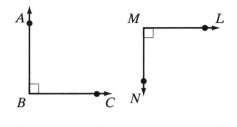

Proof	*Statements*	*Reasons*
	1. $\overline{AB} \perp \overline{BC}$	**1.** Given.
	2. $\angle ABC$ is a right angle.	**2.** Perpendicular lines are two lines that intersect to form right angles.
	3. $m\angle ABC = 90$	**3.** A right angle is an angle whose degree measure is 90.
	4. $\overline{LM} \perp \overline{MN}$	**4.** Given.
	5. $\angle LMN$ is a right angle.	**5.** Perpendicular lines are two lines that intersect to form right angles.
	6. $m\angle LMN = 90$	**6.** A right angle is an angle whose degree measure is 90.
	7. $90 = m\angle LMN$	**7.** Symmetric property of equality.
	8. $m\angle ABC = m\angle LMN$	**8.** Transitive property of equality (steps 3 and 7).

Exercises

Writing About Mathematics

1. Is "is congruent to" an equivalence relation for the set of angles? Justify your answer.

2. Is "is perpendicular to" an equivalence relation for the set of lines? Justify your answer.

Developing Skills

In 3–6, in each case: state the postulate that can be used to show that each conclusion is valid.

3. $CD = CD$

4. $2 + 3 = 5$ and $5 = 1 + 4$. Therefore, $2 + 3 = 4 + 1$.

5. $10 = a + 7$. Therefore $a + 7 = 10$.

6. $m\angle A = 30$ and $m\angle B = 30$. Therefore, $m\angle A = m\angle B$.

Applying Skills

In 7–10, write the reason of each step of the proof.

7. *Given: y = x + 4 and y = 7*

Prove: x + 4 = 7

Statements	Reasons
1. $y = x + 4$	1. _____
2. $x + 4 = y$	2. _____
3. $y = 7$	3. _____
4. $x + 4 = 7$	4. _____ ∎

8. *Given: AB + BC = AC and AB + BC = 12*

Prove: AC = 12

Statements	Reasons
1. $AB + BC = AC$	1. _____
2. $AC = AB + BC$	2. _____
3. $AB + BC = 12$	3. _____
4. $AC = 12$	4. _____ ∎

9. *Given: M is the midpoint of $\overline{LN}$ and N is the midpoint of $\overline{MP}$.*

Prove: LM = NP

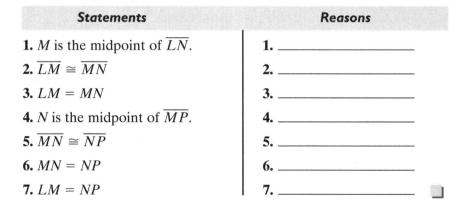

Statements	Reasons
1. M is the midpoint of $\overline{LN}$.	1. _____
2. $\overline{LM} \cong \overline{MN}$	2. _____
3. $LM = MN$	3. _____
4. N is the midpoint of $\overline{MP}$.	4. _____
5. $\overline{MN} \cong \overline{NP}$	5. _____
6. $MN = NP$	6. _____
7. $LM = NP$	7. _____ ∎

10. *Given: m∠FGH = m∠JGK and m∠HGJ = m∠JGK*

Prove: $\overrightarrow{GH}$ is the bisector of ∠FGJ.

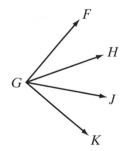

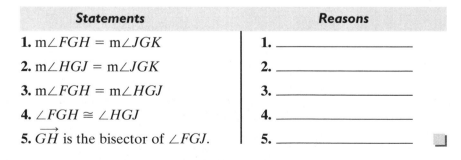

Statements	Reasons
1. m∠FGH = m∠JGK	1. _____
2. m∠HGJ = m∠JGK	2. _____
3. m∠FGH = m∠HGJ	3. _____
4. ∠FGH ≅ ∠HGJ	4. _____
5. $\overrightarrow{GH}$ is the bisector of ∠FGJ.	5. _____ ∎

11. Explain why the following proof is incorrect.

Given: $\triangle ABC$ with D a point on $\overline{BC}$.

Prove: $\angle ADB \cong \angle ADC$

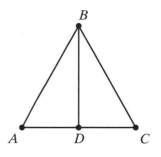

Statements	Reasons
1. $\angle ADB$ and $\angle ADC$ are right angles.	**1.** Given (from the diagram).
2. m$\angle ADB = 90$ and m$\angle ADC = 90$.	**2.** A right angle is an angle whose degree measure is 90.
3. m$\angle ADB = $ m$\angle ADC$	**3.** Transitive property of equality.
4. $\angle ADB \cong \angle ADC$	**4.** If two angles have the same measure, then they are congruent.

Hands-On Activity

Working with a partner: **a.** Determine the definitions and postulates that can be used with the given statement to write a proof. **b.** Write a proof in two-column form.

Given: $\triangle ABC$ and $\triangle BCD$ are equilateral.

Prove: $AB = CD$

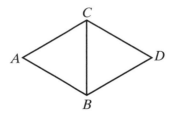

3-6 THE SUBSTITUTION POSTULATE

The substitution postulate allows us to replace one quantity, number, or measure with its equal. The **substitution postulate** is stated in words as follows:

Postulate 3.4

A quantity may be substituted for its equal in any statement of equality.

From $x = y$ and $y = 8$, we can conclude that $x = 8$. This is an example of the transitive property of equality, but we can also say that we have used the substitution property by substituting 8 for y in the equation $x = y$.

From $y = x + 7$ and $x = 3$, we can conclude that $y = 3 + 7$. This is not an example of the transitive property of equality. We have used the substitution property to replace x with its equal, 3, in the equation $y = x + 7$.

We use this postulate frequently in algebra. For example, if we find the solution of an equation, we can substitute that solution to show that we have a true statement.

$$4x - 1 = 3x + 7$$
$$x = 8$$

Check
$$4x - 1 = 3x + 7$$
$$4(8) - 1 \overset{?}{=} 3(8) + 7$$
$$31 = 31 \checkmark$$

In a system of two equations in two variables, x and y, we can solve one equation for y and substitute in the other equation. For example, we are using the substitution postulate in the third line of this solution:

$$3x + 2y = 13$$
$$y = x - 1$$

$$3x + 2(x - 1) = 13$$
$$3x + 2x - 2 = 13$$
$$5x = 15$$
$$x = 3$$
$$y = 3 - 1 = 2$$

EXAMPLE 1

Given: $CE = 2CD$ and $CD = DE$

Prove: $CE = 2DE$

Proof	**Statements**	**Reasons**
	1. $CE = 2CD$	**1.** Given.
	2. $CD = DE$	**2.** Given.
	3. $CE = 2DE$	**3.** Substitution postulate. (*Or:* A quantity may be substituted for its equal in any expression of equality.)

EXAMPLE 2

Given: m∠*ABD* + m∠*DBC* = 90 and
m∠*ABD* = m∠*CBE*

Prove: m∠*CBE* + m∠*DBC* = 90

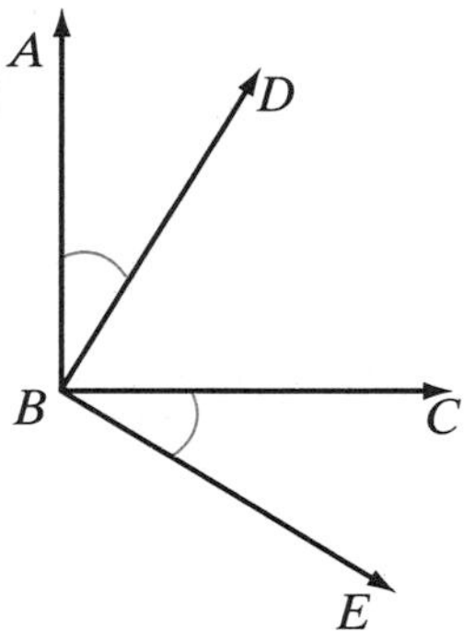

Proof

Statements	Reasons
1. m∠*ABD* + m∠*DBC* = 90	**1.** Given.
2. m∠*ABD* = m∠*CBE*	**2.** Given.
3. m∠*CBE* + m∠*DBC* = 90	**3.** Substitution postulate.

Exercises

Writing About Mathematics

1. If we know that $\overline{PQ} \cong \overline{RS}$ and that $\overline{RS} \cong \overline{ST}$, can we conclude that $\overline{PQ} \cong \overline{ST}$? Justify your answer.

2. If we know that $\overline{PQ} \cong \overline{RS}$, and that $\overline{RS} \perp \overline{ST}$, can we use the substitution postulate to conclude that $\overline{PQ} \perp \overline{ST}$? Justify your answer.

Developing Skills

In 3–10, in each case write a proof giving the reason for each statement in your proof.

3. *Given:* $MT = \frac{1}{2}RT$ and $RM = MT$

 Prove: $RM = \frac{1}{2}RT$

4. *Given:* $AD + DE = AE$ and $AD = EB$

 Prove: $EB + DE = AE$

5. *Given:* m∠*a* + m∠*b* = 180 and
 m∠*a* = m∠*c*

 Prove: m∠*c* + m∠*b* = 180

6. *Given:* $y = x + 5$ and $y + 7 = 2x$

 Prove: $x + 5 + 7 = 2x$

7. *Given:* $12 = x + y$ and $x = 8$

 Prove: $12 = 8 + y$

8. *Given:* $BC^2 = AB^2 + AC^2$ and
 $AB = DE$

 Prove: $BC^2 = DE^2 + AC^2$

9. *Given:* $AB = \sqrt{CD}, \frac{1}{2}GH = EF,$ and
 $\sqrt{CD} = EF$

 Prove: $AB = \frac{1}{2}GH$

10. *Given:* m∠*Q* + m∠*R* + m∠*S* = 75,
 m∠*Q* + m∠*S* = m∠*T*, and
 m∠*R* + m∠*T* = m∠*U*

 Prove: m∠*U* = 75

3-7 THE ADDITION AND SUBTRACTION POSTULATES

The Partition Postulate

When three points, A, B, and C, lie on the same line, the symbol $\overleftrightarrow{ABC}$ is a way of indicating the following equivalent facts about these points:

- B is on the line segment $\overline{AC}$.
- B is between A and C.
- $\overline{AB} + \overline{BC} = \overline{AC}$

Since A, B, and C lie on the same line, we can also conclude that $AB + BC = AC$. In other words, B separates $\overline{AC}$ into two segments whose sum is $\overline{AC}$. This fact is expressed in the following postulate called the **partition postulate**.

Postulate 3.5

A whole is equal to the sum of all its parts.

This postulate applies to any number of segments or to their lengths.

$$
\begin{array}{cccc}
\bullet & \bullet & \bullet & \bullet \\
A & B & C & D
\end{array}
$$

- If B is between A and C, then A, B, and C are collinear.

$$AB + BC = AC$$
$$\overline{AB} + \overline{BC} = \overline{AC}$$

- If C is between B and D, then B, C, and D are collinear.

$$BC + CD = BD$$
$$\overline{BC} + \overline{CD} = \overline{BD}$$

- If A, B, and C are collinear and B, C, and D are collinear, then A, B, C, and D are collinear, that is, $\overline{ABCD}$. We may conclude:

$$AB + BC + CD = AD$$
$$\overline{AB} + \overline{BC} + \overline{CD} = \overline{AD}$$

This postulate also applies to any number of angles or their measures.

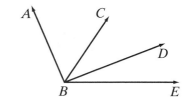

- If $\overrightarrow{BC}$ is a ray in the interior of $\angle ABD$:

$$m\angle ABC + m\angle CBD = m\angle ABD$$
$$\angle ABC + \angle CBD = \angle ABD$$

- If $\overrightarrow{BD}$ is a ray in the interior of $\angle CBE$:

$$m\angle CBD + m\angle DBE = m\angle CBE$$
$$\angle CBD + \angle DBE = \angle CBE$$

- We can also conclude that:

$$m\angle ABC + m\angle CBD + m\angle DBE = m\angle ABE$$
$$\angle ABC + \angle CBD + \angle DBE = \angle ABE$$

Note: We write $\angle x + \angle y$ to represent the sum of the angles, $\angle x$ and $\angle y$, only if $\angle x$ and $\angle y$ are adjacent angles.

Since $AB = DE$ indicates that $\overline{AB} \cong \overline{DE}$ and $m\angle ABC = m\angle DEF$ indicates that $\angle ABC \cong \angle DEF$, that is, since equality implies congruency, we can restate the partition postulate in terms of congruent segments and angles.

Postulate 3.5.1

A segment is congruent to the sum of all its parts.

Postulate 3.5.2

An angle is congruent to the sum of all its parts.

The Addition Postulate

The **addition postulate** may be stated in symbols or in words as follows:

Postulate 3.6

If $a = b$ and $c = d$, then $a + c = b + d$. If equal quantities are added to equal quantities, the sums are equal.

The following example proof uses the addition postulate.

EXAMPLE 1

Given: $\overleftrightarrow{ABC}$ and $\overleftrightarrow{DEF}$ with $AB = DE$ and $BC = EF$.

Prove: $AC = DF$

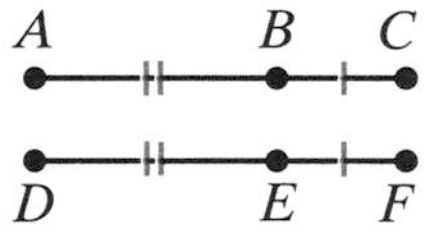

Proof

Statements	Reasons
1. $AB = DE$	**1.** Given.
2. $BC = EF$	**2.** Given.
3. $AB + BC = DE + EF$	**3.** Addition postulate. (*Or:* If equal quantities are added to equal quantities, the sums are equal.)
4. $AB + BC = AC$ $DE + EF = DF$	**4.** Partition postulate. (*Or:* A whole is equal to the sum of all its parts.)
5. $AC = DF$	**5.** Substitution postulate.

Just as the partition postulate was restated for congruent segments and congruent angles, so too can the addition postulate be restated in terms of congruent segments and congruent angles. Recall that we can add line segments $\overline{AB}$ and $\overline{BC}$ if and only if B is between A and C.

Postulate 3.6.1

> If congruent segments are added to congruent segments, the sums are congruent.

The example proof just demonstrated can be rewritten in terms of the congruence of segments.

EXAMPLE 2

Given: $\overleftrightarrow{ABC}$ and $\overleftrightarrow{DEF}$ with $\overline{AB} \cong \overline{DE}$ and $\overline{BC} \cong \overline{EF}$.

Prove: $\overline{AC} \cong \overline{DF}$

Proof

Statements	Reasons
1. $\overline{AB} \cong \overline{DE}$	**1.** Given.
2. $\overline{BC} \cong \overline{EF}$	**2.** Given.
3. $\overline{AB} + \overline{BC} \cong \overline{DE} + \overline{EF}$	**3.** Addition postulate. (*Or:* If congruent segments are added to congruent segments, the sums are congruent segments.)
4. $\overline{AB} + \overline{BC} = \overline{AC}$	**4.** Partition postulate.
5. $\overline{DE} + \overline{EF} = \overline{DF}$	**5.** Partition postulate.
6. $\overline{AC} \cong \overline{DF}$	**6.** Substitution postulate (steps 3, 4, 5).

When the addition postulate is stated for congruent angles, it is called the **angle addition postulate**:

Postulate 3.6.2

> If congruent angles are added to congruent angles, the sums are congruent.

Recall that to add angles, the angles must have a common endpoint, a common side between them, and no common interior points.

In the diagram, $\angle ABC + \angle CBD \cong \angle ABD$.

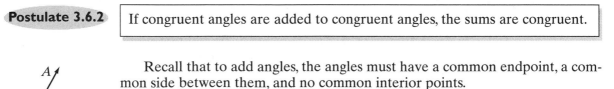

The Subtraction Postulate

The **subtraction postulate** may also be stated in symbols or in words.

Postulate 3.7

> If $a = b$, and $c = d$, then $a - c = b - d$.
>
> If equal quantities are subtracted from equal quantities, the differences are equal.

Just as the addition postulate was restated for congruent segments and congruent angles, so too may we restate the subtraction postulate in terms of congruent segments and congruent angles.

Postulate 3.7.1

> If congruent segments are subtracted from congruent segments, the differences are congruent.

Postulate 3.7.2

> If congruent angles are subtracted from congruent angles, the differences are congruent.

In Example 3, equal numbers are subtracted. In Example 4, congruent lengths are subtracted.

EXAMPLE 3

Given: $x + 6 = 14$

Prove: $x = 8$

Proof

Statements	Reasons
1. $x + 6 = 14$	**1.** Given.
2. $6 = 6$	**2.** Reflexive property.
3. $x = 8$	**3.** Subtraction postulate.

EXAMPLE 4

Given: $\overline{DEF}$, E is between D and F

Prove: $\overline{DE} \cong \overline{DF} - \overline{EF}$

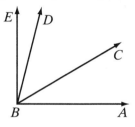

Proof	**Statements**	**Reasons**
	1. E is between D and F.	**1.** Given.
	2. $\overline{DE} + \overline{EF} \cong \overline{DF}$	**2.** Partition postulate.
	3. $\overline{EF} \cong \overline{EF}$	**3.** Reflexive property.
	4. $\overline{DE} \cong \overline{DF} - \overline{EF}$	**4.** Subtraction postulate.

Exercises

Writing About Mathematics

1. Cassie said that we do not need the subtraction postulate because subtraction can be expressed in terms of addition. Do you agree with Cassie? Explain why or why not.

2. In the diagram, $m\angle ABC = 30$, $m\angle CBD = 45$, and $m\angle DBE = 15$.

    a. Does $m\angle CBD = m\angle ABC + m\angle DBE$? Justify your answer.

    b. Is $\angle CBD \cong \angle ABC + \angle DBE$? Justify your answer.

Developing Skills

In 3 and 4, in each case fill in the missing *statement* or *reason* in the proof to show that the conclusion is valid.

3. *Given:* $\overline{AED}$ and $\overline{BFC}$, $AE = BF$, and $ED = FC$
    Prove: $AD = BC$

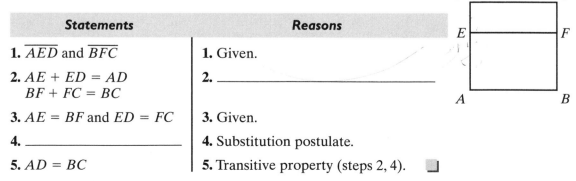

Statements	**Reasons**
1. $\overline{AED}$ and $\overline{BFC}$	**1.** Given.
2. $AE + ED = AD$ $BF + FC = BC$	**2.** _____
3. $AE = BF$ and $ED = FC$	**3.** Given.
4. _____	**4.** Substitution postulate.
5. $AD = BC$	**5.** Transitive property (steps 2, 4).

4. *Given:* $\angle SPR \cong \angle QRP$ and $\angle RPQ \cong \angle PRS$
Prove: $\angle SPQ \cong \angle QRS$

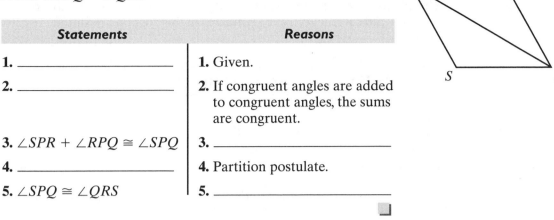

Statements	Reasons
1. _____	**1.** Given.
2. _____	**2.** If congruent angles are added to congruent angles, the sums are congruent.
3. $\angle SPR + \angle RPQ \cong \angle SPQ$	**3.** _____
4. _____	**4.** Partition postulate.
5. $\angle SPQ \cong \angle QRS$	**5.** _____

In 5–8, in each case write a proof to show that the conclusion is valid.

5. *Given:* $\overline{AC} \cong \overline{BC}$ and $\overline{MC} \cong \overline{NC}$
  Prove: $\overline{AM} \cong \overline{BN}$

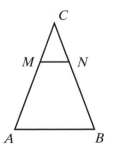

6. *Given:* $\overline{ABCD}$ with $\overline{AB} \cong \overline{CD}$
  Prove: $\overline{AC} \cong \overline{BD}$

7. *Given:* $\angle LMN \cong \angle PMQ$
  Prove: $\angle LMQ \cong \angle NMP$

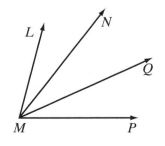

8. *Given:* $m\angle AEB = 180$ and $m\angle CED = 180$
  Prove: $m\angle AEC = m\angle BED$

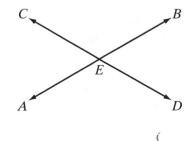

3-8 THE MULTIPLICATION AND DIVISION POSTULATES

The postulates of multiplication and division are similar to the postulates of addition and subtraction. The postulates in this section are stated in symbols and in words.

The Multiplication Postulate

Postulate 3.8

> If $a = b$, and $c = d$, then $ac = bd$.
>
> If equals are multiplied by equals, the products are equal.

When each of two equal quantities is multiplied by 2, we have a special case of this postulate, which is stated as follows:

Postulate 3.9

> Doubles of equal quantities are equal.

The Division Postulate

Postulate 3.10

> If $a = b$, and $c = d$, then $\frac{a}{c} = \frac{b}{d}$ ($c \neq 0$ and $d \neq 0$).
>
> If equals are divided by nonzero equals, the quotients are equal.

When each of two equal quantities is divided by 2, we have a special case of this postulate, which is stated as follows:

Postulate 3.11

> Halves of equal quantities are equal.

Note: Doubles and halves of congruent segments and angles will be handled in Exercise 10.

Powers Postulate

Postulate 3.12

> If $a = b$, then $a^2 = b^2$.
>
> The squares of equal quantities are equal.

If $AB = 7$, then $(AB)^2 = (7)^2$, or $(AB)^2 = 49$.

Roots Postulate

Postulate 3.13   If $a = b$ and $a > 0$, then $\sqrt{a} = \sqrt{b}$.

Recall that $\sqrt{a}$ and $\sqrt{b}$ are the positive square roots of a and of b, and so the postulate can be rewritten as:

Postulate 3.13   Positive square roots of positive equal quantities are equal.

If $(AB)^2 = 49$, then $\sqrt{(AB)^2} = \sqrt{49}$, or $AB = 7$.

EXAMPLE 1

Given: $AB = CD$, $RS = 3AB$, $LM = 3CD$

Prove: $RS = LM$

Proof

Statements	Reasons
1. $AB = CD$	**1.** Given.
2. $3AB = 3CD$	**2.** Multiplication postulate.
3. $RS = 3AB$	**3.** Given.
4. $RS = 3CD$	**4.** Transitive property of equality (steps 2 and 3).
5. $LM = 3CD$	**5.** Given.
6. $RS = LM$	**6.** Substitution postulate (steps 4 and 5).

EXAMPLE 2

Given: $5x + 3 = 38$

Prove: $x = 7$

Proof

Statements	Reasons
1. $5x + 3 = 38$	**1.** Given.
2. $3 = 3$	**2.** Reflexive property of equality.
3. $5x = 35$	**3.** Subtraction postulate.
4. $x = 7$	**4.** Division postulate.

EXAMPLE 3

Given: $m\angle ABM = \frac{1}{2}m\angle ABC, m\angle ABC = 2m\angle MBC$

Prove: $\overrightarrow{BM}$ bisects $\angle ABC$.

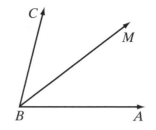

Proof	Statements	Reasons
	1. $m\angle ABM = \frac{1}{2}m\angle ABC$	**1.** Given.
	2. $m\angle ABC = 2m\angle MBC$	**2.** Given.
	3. $\frac{1}{2}m\angle ABC = m\angle MBC$	**3.** Division postulate (or halves of equal quantities are equal).
	4. $m\angle ABM = m\angle MBC$	**4.** Transitive property of equality.
	5. $\angle ABM \cong \angle MBC$	**5.** Congruent angles are angles that have the same measure.
	6. $\overrightarrow{BM}$ bisects $\angle ABC$	**6.** The bisector of an angle is a ray whose endpoint is the vertex of the angle and that divides the angle into two congruent angles.

Exercises

Writing About Mathematics

1. Explain why the word "positive" is needed in the postulate "Positive square roots of positive equal quantities are equal."

2. Barry said that "$c \neq 0$" or "$d \neq 0$" but not both could be eliminated from the division postulate. Do you agree with Barry? Explain why or why not.

Developing Skills

In 3 and 4, in each case fill in the missing *statement* or *reason* in the proof to show that the conclusion is valid.

3. *Given:* $AB = \frac{1}{4}BC$ and $BC = CD$

Prove: $CD = 4AB$

Statements	Reasons
1. $AB = \frac{1}{4}BC$	**1.** _____
2. $4 = 4$	**2.** _____
3. $4AB = BC$	**3.** _____
4. $BC = CD$	**4.** _____
5. $4AB = CD$	**5.** _____
6. $CD = 4AB$	**6.** _____ ◼

4. *Given:* $m\angle a = 3m\angle b$ and $m\angle b = 20$

Prove: $m\angle a = 60$

Statements	Reasons
1. _____	**1.** Given.
2. _____	**2.** Given.
3. $3m\angle b = 60$	**3.** _____
4. _____	**4.** Transitive property of equality. ◼

In 5–7, in each case write a proof to show that the conclusion is valid.

5. *Given:* $LM = 2MN$ and $MN = \frac{1}{2}NP$

Prove: $\overline{LM} \cong \overline{NP}$

6. *Given:* $2(3a - 4) = 16$

Prove: $a = 4$

7. *Given:* $\overline{PQRS}$, $PQ = 3QR$, and $QR = \frac{1}{3}RS$

Prove: $\overline{PQ} \cong \overline{RS}$

Applying Skills

8. On Monday, Melanie walked twice as far as on Tuesday. On Wednesday, she walked one-third as far as on Monday and two-thirds as far as on Friday. Prove that Melanie walked as far on Friday as she did on Tuesday.

9. The library, the post office, the grocery store, and the bank are located in that order on the same side of Main Street. The distance from the library to the post office is four times the distance from the post office to the grocery store. The distance from the grocery store to the bank is three times the distance from the post office to the grocery store. Prove that the distance from the library to the post office is equal to the distance from the post office to the bank. (Think of Main Street as the line segment $\overline{LPGB}$.)

10. Explain why the following versions of Postulates 3.9 and 3.11 are valid:
Doubles of congruent segments are congruent.   Halves of congruent segments are congruent.
Doubles of congruent angles are congruent.   Halves of congruent angles are congruent.

CHAPTER SUMMARY

Geometric statements can be proved by using **deductive reasoning**. Deductive reasoning applies the laws of logic to a series of true statements to arrive at a conclusion. The true statements used in deductive reasoning may be the given, definitions, postulates, or theorems that have been previously proved.

Inductive reasoning uses a series of particular examples to lead to a general conclusion. Inductive reasoning is a powerful tool in discovering and making **conjectures**. However, inductive reasoning does not prove or explain conjectures; generalizations arising from direct measurements of specific cases are only approximate; and care must be taken to ensure that all relevant examples are examined.

A **proof** using deductive reasoning may be either **direct** or **indirect**. In direct reasoning, a series of statements that include the given statement lead to the desired conclusion. In indirect reasoning, the negation of the desired conclusion leads to a statement that contradicts a given statement.

Postulates
3.1 The Reflexive Property of Equality: $a = a$
3.2 The Symmetric Property of Equality: If $a = b$, then $b = a$.
3.3 The Transitive Property of Equality: If $a = b$ and $b = c$, then $a = c$.
3.4 A quantity may be substituted for its equal in any statement of equality.
3.5 A whole is equal to the sum of all its parts.
3.5.1 A segment is congruent to the sum of all its parts.
3.5.2 An angle is congruent to the sum of all its parts.
3.6 If equal quantities are added to equal quantities, the sums are equal.
3.6.1 If congruent segments are added to congruent segments, the sums are congruent.
3.6.2 If congruent angles are added to congruent angles, the sums are congruent.
3.7 If equal quantities are subtracted from equal quantities, the differences are equal.
3.7.1 If congruent segments are subtracted from congruent segments, the differences are congruent.
3.7.2 If congruent angles are subtracted from congruent angles, the differences are congruent.
3.8 If equals are multiplied by equals, the products are equal.
3.9 Doubles of equal quantities are equal.
3.10 If equals are divided by nonzero equals, the quotients are equal.
3.11 Halves of equal quantities are equal.
3.12 The squares of equal quantities are equal.
3.13 Positive square roots of equal quantities are equal.

VOCABULARY

3-1 Generalization • Inductive reasoning • Counterexample • Conjecture

3-3 Proof • Deductive reasoning • Given • Prove • Two-column proof • Paragraph proof

3-4 Direct proof • Indirect proof • Proof by contradiction

3-5 Postulate • Axiom • Theorem • Reflexive property of equality • Symmetric property of equality • Transitive property of equality • Equivalence relation

3-6 Substitution postulate

3-7 Partition postulate • Addition postulate • Angle addition postulate • Subtraction postulate

3-8 Multiplication postulate • Division postulate • Powers postulate • Roots postulate

REVIEW EXERCISES

In 1–3, in each case: **a.** Write the given definition in a conditional form. **b.** Write the converse of the statement given as an answer to part **a. c.** Write the biconditional form of the definition.

1. An obtuse triangle is a triangle that has one obtuse angle.

2. Congruent angles are angles that have the same measure.

3. Perpendicular lines are two lines that intersect to form right angles.

4. Explain the difference between a postulate and a theorem.

5. Name the property illustrated by the following statement:

If $AB = CD$, then $CD = AB$.

6. Is the relation "is greater than" an equivalence relation for the set of real numbers? Explain your answer by demonstrating which (if any) of the properties of an equivalence relation are true and which are false.

In 7–12, in each case draw a figure that illustrates the given information and write a proof to show that the conclusion is valid.

7. *Given:* $\overleftrightarrow{AB}$ bisects $\overline{CD}$ at M.

Prove: $CM = MD$

8. *Given:* $\overline{RMST}$ is a line segment, $\overline{RM} \cong \overline{MS}$, and $\overline{MS} \cong \overline{ST}$.

Prove: $\overline{RM} \cong \overline{ST}$

9. *Given:* $\overline{ABCD}$ is a line segment and $\overline{AC} \cong \overline{BD}$.

Prove $\overline{AB} \cong \overline{CD}$

10. *Given:* $\overline{SQRP}$ is a line segment and $SQ = RP$.

Prove: $SR = QP$

11. *Given:* $\overrightarrow{BC}$ bisects $\angle ABD$ and $m\angle CBD = m\angle PQR$.

Prove: $m\angle ABC = m\angle PQR$

12. *Given:* $\overline{CD}$ and $\overline{AB}$ bisect each other at E and $CE = BE$.

Prove: $CD = AB$

13. A student wrote the following proof:

Given: $\overline{AB} \cong \overline{CD}$ and $\overline{AB} \perp \overline{BC}$

Prove: $\overline{CD} \perp \overline{BC}$

Statements	Reasons
1. $\overline{AB} \cong \overline{CD}$	**1.** Given.
2. $\overline{AB} \perp \overline{BC}$	**2.** Given.
3. $\overline{CD} \perp \overline{BC}$	**3.** Substitution postulate.

What is the error in this proof?

14. A **palindrome** is a sequence of numbers or letters that reads the same from left to right as from right to left.

a. Write the definition of a palindrome as a conditional statement.

b. Write the converse of the conditional statement in **a**.

c. Write the definition as a biconditional.

Exploration

The following "proof" leads to the statement that twice a number is equal to the number. This would mean, for example, that if $b = 1$, then $2(1) = 1$, which is obviously incorrect. What is the error in the proof?

Given: $a = b$

Prove: $b = 2b$

Statements	Reasons
1. $a = b$	**1.** Given.
2. $a^2 = ab$	**2.** Multiplication postulate.
3. $a^2 - b^2 = ab - b^2$	**3.** Subtraction postulate.
4. $(a + b)(a - b) = b(a - b)$	**4.** Substitution postulate.
5. $a + b = b$	**5.** Division postulate.
6. $b + b = b$	**6.** Substitution postulate.
7. $2b = b$	**7.** Substitution postulate.

Part I

Answer all questions in this part. Each correct answer will receive 2 credits. No partial credit will be allowed.

1. If $y = 2x - 7$ and $y = 3$, then x is equal to

(1) 1   (2) −1   (3) 5   (4) −5

2. The property illustrated in the equality $2(a + 4) = 2(4 + a)$ is

(1) the distributive property.   (3) the identity property.

(2) the associative property.   (4) the commutative property.

3. In biconditional form, the definition of the midpoint of a line segment can be written as

(1) A point on a line segment is the midpoint of that segment if it divides the segment into two congruent segments.

(2) A point on a line segment is the midpoint of that segment if it divides the segment into two congruent segments.

(3) A point on a line segment is the midpoint of that segment only if it divides the segment into two congruent segments.

(4) A point on a line segment is the midpoint of that segment if and only if it divides the segment into two congruent segments.

4. The multiplicative identity element is

(1) 1   (2) −1   (3) 0   (4) not a real number

5. The angle formed by two opposite rays is

(1) an acute angle.   (3) an obtuse angle.

(2) a right angle.   (4) a straight angle.

6. The inverse of the statement "If two angles are right angles, then they are congruent" is

(1) If two angles are not congruent, then they are not right angles.

(2) If two angles are not right angles, then they are not congruent.

(3) If two angles are congruent, then they are right angles.

(4) Two angles are not right angles if they are not congruent.

7. The statements "Today is Saturday or I go to school" and "Today is not Saturday" are both true statements. Which of the following statements is also true?

(1) Today is Saturday.   (3) I go to school.

(2) I do not go to school.   (4) Today is Saturday and I go to school.

8. $\overline{ABCD}$ is a line segment and B is the midpoint of $\overline{AC}$. Which of the following must be true?
(1) C is the midpoint of $\overline{BD}$
(2) $AB = BC$
(3) $AC = CD$
(4) $AC + BD = AD$

9. The statements "$AB = BC$" and "$DC = BC$" are true statements. Which of the following must also be true?
(1) $AB + BC = AC$
(2) A, B, and C are collinear
(3) B, C, and D are collinear
(4) $AB = DC$

10. Triangle LMN has exactly two congruent sides. Triangle LMN is
(1) a right triangle.
(2) a scalene triangle.
(3) an isosceles triangle.
(4) an equilateral triangle.

Part II

Answer all questions in this part. Each correct answer will receive 2 credits. Clearly indicate the necessary steps, including appropriate formula substitutions, diagrams, graphs, charts, etc. For all questions in this part, a correct numerical answer with no work shown will receive only 1 credit.

11. Give a reason for each step used in the solution of the equation.

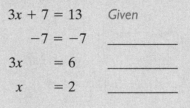

$$3x + 7 = 13 \qquad \textit{Given}$$
$$-7 = -7 \qquad \underline{\hspace{2cm}}$$
$$3x \quad = 6 \qquad \underline{\hspace{2cm}}$$
$$x \quad = 2 \qquad \underline{\hspace{2cm}}$$

12. *Given:* $\overline{DE} \perp \overline{EF}$

Prove: $\triangle DEF$ is a right triangle.

Part III

Answer all questions in this part. Each correct answer will receive 4 credits. Clearly indicate the necessary steps, including appropriate formula substitutions, diagrams, graphs, charts, etc. For all questions in this part, a correct numerical answer with no work shown will receive only 1 credit.

13. *Given:* $\triangle ABC$ with D a point on $\overline{AB}$ and $AC = AD + DB$.

Prove: $\triangle ABC$ is isosceles.

14. $\overline{PQR}$ is a line segment. $PQ = 4a - 3$, $QR = 3a + 2$ and $PR = 8a - 6$. Is Q the midpoint of $\overline{PQR}$? Justify your answer.

Part IV

Answer all questions in this part. Each correct answer will receive 6 credits. Clearly indicate the necessary steps, including appropriate formula substitutions, diagrams, graphs, charts, etc. For all questions in this part, a correct numerical answer with no work shown will receive only 1 credit.

15. $\overrightarrow{BD}$ bisects $\angle ABC$, $m\angle ABD = 3x + 18$ and $m\angle DBC = 5x - 30$. If $m\angle ABC = 7x + 12$, is $\angle ABC$ a straight angle? Justify your answer.

16. For each statement, the hypothesis is true. Write the postulate that justifies the conclusion.

   a. If $x = 5$, then $x + 7 = 5 + 7$.

   b. If $2y + 3$ represents a real number, then $2y + 3 = 2y + 3$.

   c. If $\overline{RST}$ is a line segment, then $\overline{RS} + \overline{ST} = \overline{RT}$.

   d. If $y = 2x + 1$ and $y = 15$, then $2x + 1 = 15$.

   e. If $a = 3$, then $\frac{a}{5} = \frac{3}{5}$.

CHAPTER

4

CONGRUENCE OF LINE SEGMENTS, ANGLES, AND TRIANGLES

One of the common notions stated by Euclid was the following: "Things which coincide with one another are equal to one another." Euclid used this common notion to prove the congruence of triangles. For example, Euclid's Proposition 4 states, "If two triangles have the two sides equal to two sides respectively, and have the angles contained by the equal straight lines equal, they will also have the base equal to the base, the triangle will be equal to the triangle, and the remaining angles will be equal to the remaining angles respectively, namely those which the equal sides subtend." In other words, Euclid showed that the equal sides and angle of the first triangle can be made to coincide with the sides and angle of the second triangle so that the two triangles will coincide. We will expand on Euclid's approach in our development of congruent triangles presented in this chapter.

4-1 POSTULATES OF LINES, LINE SEGMENTS, AND ANGLES

Melissa planted a new azalea bush in the fall and wants to protect it from the cold and snow this winter. She drove four parallel stakes into the ground around the bush and covered the structure with burlap fabric. During the first winter storm, this protective barrier was pushed out of shape. Her neighbor suggested that she make a tripod of three stakes fastened together at the top, forming three triangles. Melissa found that this arrangement was able to stand up to the storms. Why was this change an improvement? What geometric figure occurs most frequently in weight-bearing structures? In this chapter we will study the properties of triangles to discover why triangles keep their shape.

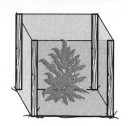

Recall that a line, $\overleftrightarrow{AB}$, is an infinite set of points that extends endlessly in both directions, but a line segment, $\overline{AB}$, is a part of $\overleftrightarrow{AB}$ and has a finite length. We can choose some point of $\overleftrightarrow{AB}$ that is not a point of $\overline{AB}$ to form a line segment of any length. When we do this, we say that we are extending the line segment.

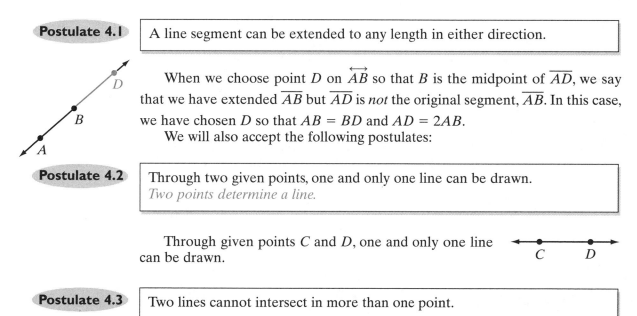

Postulate 4.1	A line segment can be extended to any length in either direction.

When we choose point D on $\overleftrightarrow{AB}$ so that B is the midpoint of $\overline{AD}$, we say that we have extended $\overline{AB}$ but $\overline{AD}$ is *not* the original segment, $\overline{AB}$. In this case, we have chosen D so that $AB = BD$ and $AD = 2AB$.

We will also accept the following postulates:

Postulate 4.2	Through two given points, one and only one line can be drawn. *Two points determine a line.*

Through given points C and D, one and only one line can be drawn.

Postulate 4.3	Two lines cannot intersect in more than one point.

$\overleftrightarrow{AEB}$ and $\overleftrightarrow{CED}$ intersect at E and cannot intersect at any other point.

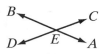

Postulate 4.4

One and only one circle can be drawn with any given point as center and the length of any given line segment as a radius.

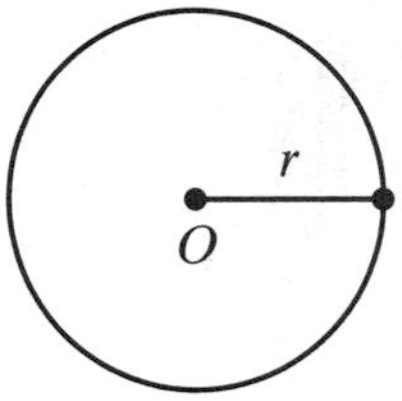

Only one circle can be drawn that has point O as its center and a radius equal in length to segment r.

We make use of this postulate in constructions when we use a compass to locate points at a given distance from a given point.

Postulate 4.5

At a given point on a given line, one and only one perpendicular can be drawn to the line.

At point P on $\overleftrightarrow{APB}$, exactly one line, $\overrightarrow{PD}$, can be drawn perpendicular to $\overleftrightarrow{APB}$ and no other line through P is perpendicular to $\overleftrightarrow{APB}$.

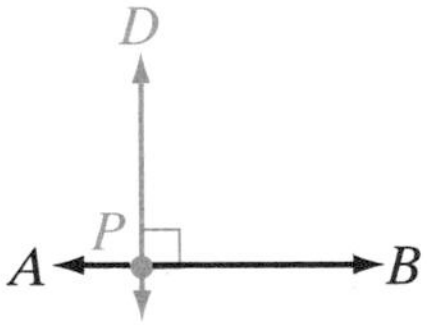

Postulate 4.6

From a given point not on a given line, one and only one perpendicular can be drawn to the line.

From point P not on $\overleftrightarrow{CD}$, exactly one line, $\overrightarrow{PE}$, can be drawn perpendicular to $\overleftrightarrow{CD}$ and no other line from P is perpendicular to $\overleftrightarrow{CD}$.

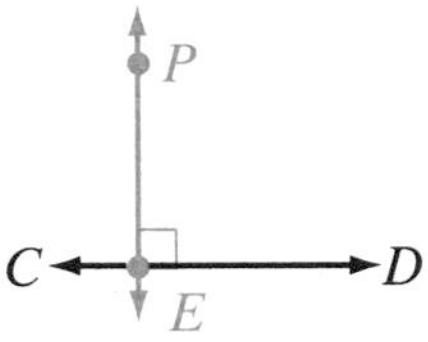

Postulate 4.7

For any two distinct points, there is only one positive real number that is the length of the line segment joining the two points.

For the distinct points A and B, there is only one positive real number, represented by AB, which is the length of $\overline{AB}$.

Since AB is also called the distance from A to B, we refer to Postulate 4.7 as the **distance postulate**.

Postulate 4.8

The shortest distance between two points is the length of the line segment joining these two points.

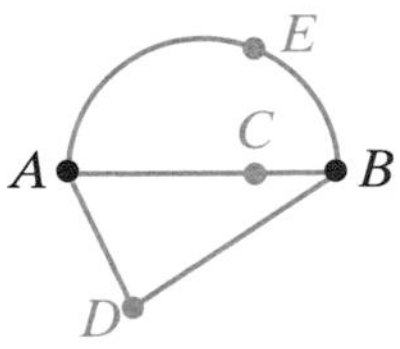

The figure shows three paths that can be taken in going from A to B.

The length of $\overline{AB}$ (the path through C, a point collinear with A and B) is less than the length of the path through D or the path through E. The measure of the shortest path from A to B is the distance AB.

Postulate 4.9 | A line segment has one and only one midpoint.

$\overline{AB}$ has a midpoint, point M, and no other point is a midpoint of $\overline{AB}$.

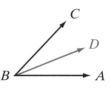

Postulate 4.10 | An angle has one and only one bisector.

Angle ABC has one bisector, $\overrightarrow{BD}$, and no other ray bisects $\angle ABC$.

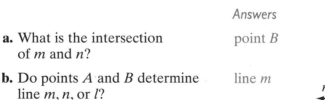

EXAMPLE 1

Use the figure to answer the following questions:

Answers

a. What is the intersection of m and n?

point B

b. Do points A and B determine line m, n, or l?

line m

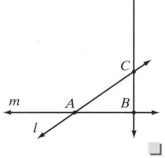

EXAMPLE 2

Lines p and n are two distinct lines that intersect line m at A. If line n is perpendicular to line m, can line p be perpendicular to line m? Explain.

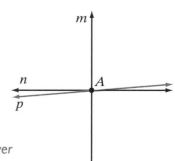

Solution   No. Only one perpendicular can be drawn to a line at a given point on the line. Since line n is perpendicular to m and lines n and p are distinct, line p cannot be perpendicular to m. *Answer*

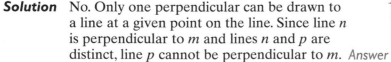

EXAMPLE 3

If $\overrightarrow{BD}$ bisects $\angle ABC$ and point E is not a point on $\overrightarrow{BD}$, can $\overrightarrow{BE}$ be the bisector of $\angle ABC$?

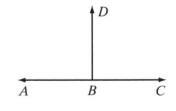

Solution   No. An angle has one and only bisector. Since point E is not a point on $\overrightarrow{BD}$, $\overrightarrow{BD}$ is not the same ray as $\overrightarrow{BE}$.

Therefore, $\overrightarrow{BE}$ cannot be the bisector of $\angle ABC$. *Answer*

Conditional Statements as They Relate to Proof

To prove a statement in geometry, we start with what is known to be true and use definitions, postulates, and previously proven theorems to arrive at the truth of what is to be proved. As we have seen in the text so far, the information that is known to be true is often stated as *given* and what is to be proved as *prove*. When the information needed for a proof is presented in a conditional statement, we use the information in the hypothesis to form a *given* statement, and the information in the conclusion to form a *prove* statement.

Numerical and Algebraic Applications

EXAMPLE 4

Rewrite the conditional statement in the *given* and *prove* format:
   If a ray bisects a straight angle, it is perpendicular to the line determined by the straight angle.

Solution   Draw and label a diagram.

Use the hypothesis, "a ray bisects a straight angle," as the *given*. Name a straight angle using the three letters from the diagram and state in the *given* that this angle is a straight angle. Name the ray that bisects the angle, using the vertex of the angle as the endpoint of the ray that is the bisector. State in the *given* that the ray bisects the angle:

   Given: $\angle ABC$ is a straight angle and $\overrightarrow{BD}$ bisects $\angle ABC$.

Use the conclusion, "if (the bisector) is perpendicular to the line determined by the straight angle," to write the *prove*. We have called the bisector $\overrightarrow{BD}$, and the line determined by the straight angle is $\overleftrightarrow{ABC}$.

   Prove: $\overrightarrow{BD} \perp \overleftrightarrow{AC}$

Answer   *Given:* $\angle ABC$ is a straight angle and $\overrightarrow{BD}$ bisects $\angle ABC$.

Prove: $\overrightarrow{BD} \perp \overleftrightarrow{AC}$

In geometry, we are interested in proving that statements are true. These true statements can then be used to help solve numerical and algebraic problems, as in Example 5.

EXAMPLE 5

$\overrightarrow{SQ}$ bisects $\angle RST$, $m\angle RSQ = 4x$, and $m\angle QST = 3x + 20$. Find the measures of $\angle RSQ$ and $\angle QST$.

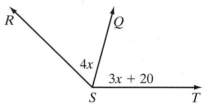

Solution   The bisector of an angle separates the angle into two congruent angles. Therefore, $\angle RSQ \cong \angle QST$. Then since congruent angles have equal measures, we may write an equation that states that $m\angle RSQ = m\angle QST$.

$$4x = 3x + 20$$
$$x = 20$$

$$m\angle RSQ = 4x \qquad\qquad m\angle QST = 3x + 20$$
$$= 4(20) \qquad\qquad\qquad = 3(20) + 20$$
$$= 80 \qquad\qquad\qquad\qquad = 60 + 20$$
$$\qquad\qquad\qquad\qquad\qquad = 80$$

Answer $m\angle RSQ = m\angle QST = 80$

Exercises

Writing About Mathematics

1. If two distinct lines $\overleftrightarrow{AEB}$ and $\overleftrightarrow{CED}$ intersect at a point F, what must be true about points E and F? Use a postulate to justify your answer.

2. If $LM = 10$, can $\overline{LM}$ be extended so that $LM = 15$? Explain why or why not.

Developing Skills

In 3–12, in each case: **a.** Rewrite the conditional statement in the *Given/Prove* format. **b.** Write a formal proof.

3. If $AB = AD$ and $DC = AD$, then $AB = DC$.

4. If $\overline{AD} \cong \overline{CD}$ and $\overline{BD} \cong \overline{CD}$, then $\overline{AD} \cong \overline{BD}$.

5. If m$\angle 1$ + m$\angle 2$ = 90 and m$\angle A$ = m$\angle 2$, then m$\angle 1$ + m$\angle A$ = 90.

6. If m$\angle A$ = m$\angle B$, m$\angle 1$ = m$\angle B$, and m$\angle 2$ = m$\angle A$, then m$\angle 1$ = m$\angle 2$.

7. If $\overline{AB} \cong \overline{CD}$, and $\overline{EF} \cong \overline{CD}$, then $\overline{AB} \cong \overline{EF}$.

8. If $2EF = DB$ and $GH = \frac{1}{2}DB$, then $EF = GH$.

9. If $CE = CF$, $CD = 2CE$, and $CB = 2CF$, then $CD = CB$.

10. If $RT = RS$, $RD = \frac{1}{2}RT$, and $RE = \frac{1}{2}RS$, then $RD = RE$.

11. If $AD = BE$ and $BC = CD$, then $AC = CE$.

12. If $\angle CDB \cong \angle CBD$ and $\angle ADB \cong \angle ABD$, then $\angle CDA \cong \angle CBA$.

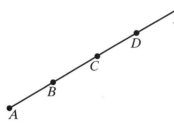

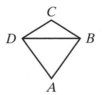

In 13–15, *Given:* $\overleftrightarrow{RST}$, $\overrightarrow{SQ}$, and $\overrightarrow{SP}$.

13. If $\angle QSR \cong \angle PST$ and m$\angle QSP$ = 96, find m$\angle QSR$.

14. If m$\angle QSR$ = 40 and $\overrightarrow{SQ} \perp \overrightarrow{SP}$, find m$\angle PST$.

15. If m$\angle PSQ$ is twice m$\angle QSR$ and $\overrightarrow{SQ} \perp \overrightarrow{SP}$, find m$\angle PST$.

Applying Skills

In 16–19, use the given conditional to **a.** draw a diagram with geometry software or pencil and paper, **b.** write a *given* and a *prove*, **c.** write a proof.

16. If a triangle is equilateral, then the measures of the sides are equal.

17. If E and F are distinct points and two lines intersect at E, then they do not intersect at F.

18. If a line through a vertex of a triangle is perpendicular to the opposite side, then it separates the triangle into two right triangles.

19. If two points on a circle are the endpoints of a line segment, then the length of the line segment is less than the length of the portion of the circle (the arc) with the same endpoints.

20. Points F and G are both on line l and on line m. If F and G are distinct points, do l and m name the same line? Justify your answer.

4-2 USING POSTULATES AND DEFINITIONS IN PROOFS

A *theorem* was defined in Chapter 3 as a statement proved by deductive reasoning. We use the laws of logic to combine definitions and postulates to prove a theorem. A carefully drawn figure is helpful in deciding upon the steps to use in the proof. However, recall from Chapter 1 that we cannot use statements that appear to be true in the figure drawn for a particular problem. For example, we may not assume that two line segments are congruent or are perpendicular because they appear to be so in the figure.

On the other hand, unless otherwise stated, we will assume that lines that appear to be straight lines in a figure actually are straight lines and that points that appear to be on a given line actually are on that line in the order shown.

EXAMPLE I

Given: $\overleftrightarrow{ABCD}$ with $\overline{AB} \cong \overline{CD}$

Prove: $\overline{AC} \cong \overline{BD}$

Proof	**Statements**	**Reasons**
	1. A, B, C, and D are collinear with B between A and C and C between B and D.	**1.** Given.
	2. $\overline{AB} + \overline{BC} = \overline{AC}$ $\overline{BC} + \overline{CD} = \overline{BD}$	**2.** Partition postulate.
	3. $\overline{AB} \cong \overline{CD}$	**3.** Given.
	4. $\overline{BC} \cong \overline{BC}$	**4.** Reflexive property.
	5. $\overline{AB} + \overline{BC} \cong \overline{CD} + \overline{BC}$	**5.** Addition postulate.
	6. $\overline{AC} \cong \overline{BD}$	**6.** Substitution postulate.

EXAMPLE 2

Given: M is the midpoint of $\overline{AB}$.

Prove: $AM = \frac{1}{2}AB$ and $MB = \frac{1}{2}AB$

Proof	Statements	Reasons
	1. M is the midpoint of $\overline{AB}$.	**1.** Given.
	2. $\overline{AM} \cong \overline{MB}$	**2.** Definition of midpoint.
	3. $AM = MB$	**3.** Definition of congruent segments.
	4. $AM + MB = AB$	**4.** Partition postulate.
	5. $AM + AM = AB$ or $2AM = AB$	**5.** Substitution postulate.
	6. $AM = \frac{1}{2}AB$	**6.** Halves of equal quantities are equal.
	7. $MB + MB = AB$ or $2MB = AB$	**7.** Substitution postulate.
	8. $MB = \frac{1}{2}AB$	**8.** Halves of equal quantities are equal.

Note: We used definitions and postulates to prove statements about length. We could *not* use information from the diagram that appears to be true.

Exercises

Writing About Mathematics

1. Explain the difference between the symbols $\overline{ABC}$ and $\overline{ACB}$. Could both of these describe segments of the same line?

2. Two lines, $\overleftrightarrow{ABC}$ and $\overleftrightarrow{DBE}$, intersect and m$\angle ABD$ is 90. Are the measures of $\angle DBC$, $\angle CBE$, and $\angle EBA$ also 90? Justify your answer.

Developing Skills

In 3–12, in each case: **a.** Rewrite the conditional statement in the *Given/Prove* format. **b.** Write a proof that demonstrates that the conclusion is valid.

3. If $\overline{AB} \cong \overline{CB}$, $\overline{FD}$ bisects $\overline{AB}$, and $\overline{FE}$ bisects $\overline{CB}$, then $\overline{AD} \cong \overline{CE}$.

4. If $\overrightarrow{CA}$ bisects $\angle DCB$, $\overrightarrow{AC}$ bisects $\angle DAB$, and $\angle DCB \cong \angle DAB$, then $\angle CAB \cong \angle DCA$.

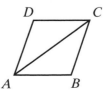

5. If $\overline{AD} \cong \overline{BE}$, then $\overline{AE} \cong \overline{BD}$.

6. If $\overline{ABCD}$ is a segment and $AB = CD$, then $AC = BD$.

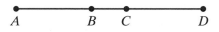

7. If $\overline{ABCD}$ is a segment, B is the midpoint of $\overline{AC}$, and C is the midpoint of $\overline{BD}$, then $AB = BC = CD$.

8. If P and T are distinct points and P is the midpoint of $\overline{RS}$, then T is not the midpoint of $\overline{RS}$.

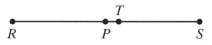

9. If $\overline{DF} \cong \overline{BE}$, then $\overline{DE} \cong \overline{BF}$.

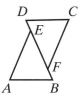

10. If $\overline{AD} \cong \overline{BC}$, E is the midpoint of $\overline{AD}$, and F is the midpoint of $\overline{BC}$, then $\overline{AE} \cong \overline{FC}$.

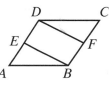

11. If $\overline{AC} \cong \overline{DB}$ and $\overline{AC}$ and $\overline{DB}$ bisect each other at E, then $\overline{AE} \cong \overline{EB}$.

12. If $\overrightarrow{DR}$ bisects $\angle CDA$, $\angle 3 \cong \angle 1$, and $\angle 4 \cong \angle 2$, then $\angle 3 \cong \angle 4$.

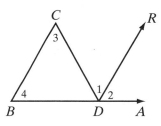

Applying Skills

13. The rays $\overrightarrow{DF}$ and $\overrightarrow{DG}$ separate $\angle CDE$ into three congruent angles, $\angle CDF$, $\angle FDG$, and $\angle GDE$. If m$\angle CDF = 7a + 10$ and m$\angle GDE = 10a - 2$, find:

a. m$\angle CDG$ **b.** m$\angle FDE$ **c.** m$\angle CDE$

d. Is $\angle CDE$ acute, right, or obtuse?

14. Segment $\overline{BD}$ is a bisector of $\overline{ABC}$ and is perpendicular to $\overline{ABC}$. $AB = 2x - 30$ and $BC = x - 10$.

a. Draw a diagram that shows $\overline{ABC}$ and $\overline{BD}$.

b. Find AB and BC.

c. Find the distance from A to $\overline{BD}$. Justify your answer.

15. Two line segments, $\overline{RS}$ and $\overline{LM}$, bisect each other and are perpendicular to each other at N, and $RN = LN$. $RS = 3x + 9$, and $LM = 5x - 17$.

a. Draw a diagram that shows $\overline{RS}$ and $\overline{LM}$.

b. Write the *given* using the information in the first sentence.

c. Prove that $RS = LM$.

d. Find RS and LM.

e. Find the distance from L to $\overline{RS}$. Justify your answer.

4-3 PROVING THEOREMS ABOUT ANGLES

In this section we will use definitions and postulates to prove some simple theorems about angles. Once a theorem is proved, we can use it as a *reason* in later proofs. Like the postulates, we will number the theorems for easy reference.

Theorem 4.1 | If two angles are right angles, then they are congruent.

Given $\angle ABC$ and $\angle DEF$ are right angles.

Prove $\angle ABC \cong \angle DEF$

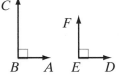

Proof

Statements	Reasons
1. $\angle ABC$ and $\angle DEF$ are right angles.	1. Given.
2. $m\angle ABC = 90$ and $m\angle DEF = 90$	2. Definition of right angle.
3. $m\angle ABC = m\angle DEF$	3. Transitive property of equality.
4. $\angle ABC \cong \angle DEF$	4. Definition of congruent angles.

We can write this proof in paragraph form as follows:

Proof: A right angle is an angle whose degree measure is 90. Therefore, $m\angle ABC$ is 90 and $m\angle DEF$ is 90. Since $m\angle ABC$ and $m\angle DEF$ are both equal to the same quantity, they are equal to each other. Since $\angle ABC$ and $\angle DEF$ have equal measures, they are congruent.

| **Theorem 4.2** | If two angles are straight angles, then they are congruent. |

Given $\angle ABC$ and $\angle DEF$ are straight angles.

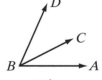

Prove $\angle ABC \cong \angle DEF$

The proof of this theorem, which is similar to the proof of Theorem 4.1, is left to the student. (See exercise 17.)

Definitions Involving Pairs of Angles

DEFINITION

Adjacent angles are two angles in the same plane that have a common vertex and a common side but do not have any interior points in common.

Angle ABC and $\angle CBD$ are adjacent angles because they have B as their common vertex, $\overrightarrow{BC}$ as their common side, and no interior points in common.

However, $\angle XWY$ and $\angle XWZ$ are not adjacent angles.

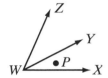

Although $\angle XWY$ and $\angle XWZ$ have W as their common vertex and $\overrightarrow{WX}$ as their common side, they have interior points in common. For example, point P is in the interior of both $\angle XWY$ and $\angle XWZ$.

DEFINITION

Complementary angles are two angles, the sum of whose degree measures is 90.

Each angle is called the **complement** of the other. If $m\angle c = 40$ and $m\angle d = 50$, then $\angle c$ and $\angle d$ are complementary angles. If $m\angle a = 35$ and $m\angle b = 55$, then $\angle a$ and $\angle b$ are complementary angles.

Complementary angles may be adjacent, as in the case of $\angle c$ and $\angle d$, or they may be nonadjacent, as in the case of $\angle a$ and $\angle b$. Note that if the two complementary angles are adjacent, their sum is a right angle: $\angle c + \angle d = \angle WZY$, a right angle.

Since $m\angle c + m\angle d = 90$, we say that $\angle c$ is the complement of $\angle d$, and that $\angle d$ is the complement of $\angle c$. When the degree measure of an angle is k, the degree measure of the complement of the angle is $(90 - k)$ because $k + (90 - k) = 90$.

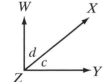

DEFINITION

Supplementary angles are two angles, the sum of whose degree measures is 180.

When two angles are supplementary, each angle is called the **supplement** of the other. If $m\angle c = 40$ and $m\angle d = 140$, then $\angle c$ and $\angle d$ are supplementary angles. If $m\angle a = 35$ and $m\angle b = 145$, then $\angle a$ and $\angle b$ are supplementary angles.

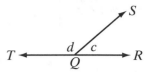

Supplementary angles may be adjacent, as in the case of $\angle c$ and $\angle d$, or they may be nonadjacent, as in the case of $\angle a$ and $\angle b$. Note that if the two supplementary angles are adjacent, their sum is a straight angle. Here $\angle c + \angle d = \angle TQR$, a straight angle.

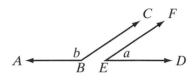

Since $m\angle c + m\angle d = 180$, we say that $\angle c$ is the supplement of $\angle d$ and that $\angle d$ is the supplement of $\angle c$. When the degree measure of an angle is k, the degree measure of the supplement of the angle is $(180 - k)$ because $k + (180 - k) = 180$.

EXAMPLE I

Find the measure of an angle if its measure is 24 degrees more than the measure of its complement.

Solution Let x = measure of complement of angle.

Then $x + 24$ = measure of angle.

The sum of the degree measures of an angle and its complement is 90.

$$x + x + 24 = 90$$
$$2x + 24 = 90$$
$$2x = 66$$
$$x = 33$$
$$x + 24 = 57$$

Answer The measure of the angle is 57 degrees.

Theorems Involving Pairs of Angles

Theorem 4.3   If two angles are complements of the same angle, then they are congruent.

Given $\angle 1$ is the complement of $\angle 2$ and $\angle 3$ is the complement of $\angle 2$.

Prove $\angle 1 \cong \angle 3$

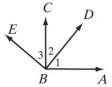

Proof	Statements	Reasons
	1. ∠1 is the complement of ∠2.	**1.** Given.
	2. m∠1 + m∠2 = 90	**2.** Complementary angles are two angles the sum of whose degree measures is 90.
	3. ∠3 is the complement of ∠2.	**3.** Given.
	4. m∠3 + m∠2 = 90	**4.** Definition of complementary angles.
	5. m∠1 + m∠2 = m∠3 + m∠2	**5.** Transitive property of equality (steps 2 and 4).
	6. m∠2 = m∠2	**6.** Reflexive property of equality.
	7. m∠1 = m∠3	**7.** Subtraction postulate.
	8. ∠1 ≅ ∠3	**8.** Congruent angles are angles that have the same measure.

Note: In a proof, there are two acceptable ways to indicate a definition as a reason. In reason 2 of the proof above, the definition of complementary angles is stated in its complete form. It is also acceptable to indicate this reason by the phrase "Definition of complementary angles," as in reason 4.

We can also give an algebraic proof for the theorem just proved.

Proof: In the figure, m∠*CBD* = *x*. Both ∠*ABD* and ∠*CBE* are complements to ∠*CBD*. Thus, m∠*ABD* = 90 − *x* and m∠*CBE* = 90 − *x*, and we conclude ∠*ABD* and ∠*CBE* have the same measure. Since angles that have the same measure are congruent, ∠*ABD* ≅ ∠*CBE*.

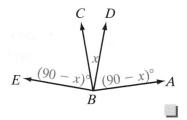

Theorem 4.4   If two angles are congruent, then their complements are congruent.

Given ∠*ABD* ≅ ∠*EFH*
∠*CBD* is the complement of ∠*ABD*.
∠*GFH* is the complement of ∠*EFH*.

Prove ∠*CBD* ≅ ∠*GHF*

This theorem can be proved in a manner similar to Theorem 4.3 but with the use of the substitution postulate. We can also use an algebraic proof.

Proof   Congruent angles have the same measure. If $\angle ABD \cong \angle EFH$, we can represent the measure of each angle by the same variable: $m\angle ABD = m\angle EFH = x$. Since $\angle CBD$ is the complement of $\angle ABD$, and $\angle GFH$ is the complement of $\angle EFH$, then $m\angle CBD = 90 - x$ and $m\angle GFH = 90 - x$. Therefore, $m\angle CBD = m\angle GFH$ and $\angle CBD \cong \angle GHF$. ∎

Theorem 4.5 | If two angles are supplements of the same angle, then they are congruent.

Given $\angle ABD$ is the supplement of $\angle DBC$, and $\angle EBC$ is the supplement of $\angle DBC$.

Prove $\angle ABD \cong \angle EBC$

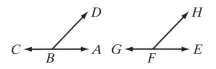

Theorem 4.6 | If two angles are congruent, then their supplements are congruent.

Given $\angle ABD \cong \angle EFH$, $\angle CBD$ is the supplement of $\angle ABD$, and $\angle GFH$ is the supplement of $\angle EFH$.

Prove $\angle CBD \cong \angle GFH$

The proofs of Theorems 4.5 and 4.6 are similar to the proofs of Theorems 4.3 and 4.4 and will be left to the student. (See exercises 18 and 19.)

More Definitions and Theorems Involving Pairs of Angles

DEFINITION

A **linear pair of angles** are two adjacent angles whose sum is a straight angle.

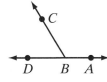

In the figure, $\angle ABD$ is a straight angle and C is not on $\overleftrightarrow{ABD}$. Therefore, $\angle ABC + \angle CBD = \angle ABD$. Note that $\angle ABC$ and $\angle CBD$ are adjacent angles whose common side is $\overrightarrow{BC}$ and whose remaining sides are opposite rays that together form a straight line, $\overleftrightarrow{AD}$.

Theorem 4.7 | If two angles form a linear pair, then they are supplementary.

Given $\angle ABC$ and $\angle CBD$ form a linear pair.

Prove $\angle ABC$ and $\angle CBD$ are supplementary.

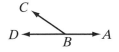

Proof In the figure, $\angle ABC$ and $\angle CBD$ form a linear pair. They share a common side, $\overrightarrow{BC}$, and their remaining sides, $\overrightarrow{BA}$ and $\overrightarrow{BD}$, are opposite rays. The sum of a linear pair of angles is a straight angle, and the degree measure of a straight angle is 180. Therefore, $m\angle ABC + m\angle CBD = 180$. Then, $\angle ABC$ and $\angle CBD$ are supplementary because supplementary angles are two angles the sum of whose degree measure is 180.

Theorem 4.8   If two lines intersect to form congruent adjacent angles, then they are perpendicular.

Given $\overleftrightarrow{ABC}$ and $\overleftrightarrow{DBE}$ with $\angle ABD \cong \angle DBC$

Prove $\overleftrightarrow{ABC} \perp \overleftrightarrow{DBE}$

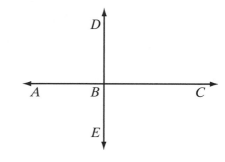

Proof The union of the opposite rays, $\overrightarrow{BA}$ and $\overrightarrow{BC}$, is the straight angle, $\angle ABC$. The measure of straight angle is 180. By the partition postulate, $\angle ABC$ is the sum of $\angle ABD$ and $\angle DBC$. Thus,

$$m\angle ABD + m\angle DBC = m\angle ABC$$
$$= 180$$

Since $\angle ABD \cong \angle DBC$, they have equal measures. Therefore,

$$m\angle ABD = m\angle DBC = \tfrac{1}{2}(180) = 90$$

The angles, $\angle ABD$ and $\angle DBC$, are right angles. Therefore, $\overleftrightarrow{ABC} \perp \overleftrightarrow{DBE}$ because perpendicular lines intersect to form right angles.

DEFINITION

Vertical angles are two angles in which the sides of one angle are opposite rays to the sides of the second angle.

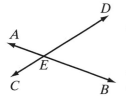

In the figure, $\angle AEC$ and $\angle DEB$ are a pair of vertical angles because $\overrightarrow{EA}$ and $\overrightarrow{EB}$ are opposite rays and $\overrightarrow{EC}$ and $\overrightarrow{ED}$ are opposite rays. Also, $\angle AED$ and $\angle CEB$ are a pair of vertical angles because $\overrightarrow{EA}$ and $\overrightarrow{EB}$ are opposite rays and $\overrightarrow{ED}$ and $\overrightarrow{EC}$ are opposite rays. In each pair of vertical angles, the opposite rays, which are the sides of the angles, form straight lines, $\overleftrightarrow{AB}$ and $\overleftrightarrow{CD}$. When two straight lines intersect, two pairs of vertical angles are formed.

Theorem 4.9 | If two lines intersect, then the vertical angles are congruent.

Given $\overleftrightarrow{AEB}$ and $\overleftrightarrow{CED}$ intersect at E.

Prove $\angle BEC \cong \angle AED$

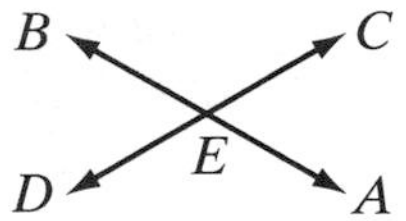

Proof

Statements	Reasons
1. $\overleftrightarrow{AEB}$ and $\overleftrightarrow{CED}$ intersect at E.	**1.** Given.
2. $\overrightarrow{EA}$ and $\overrightarrow{EB}$ are opposite rays. $\overrightarrow{ED}$ and $\overrightarrow{EC}$ are opposite rays.	**2.** Definition of opposite rays.
3. $\angle BEC$ and $\angle AED$ are vertical angles.	**3.** Definition of vertical angles.
4. $\angle BEC$ and $\angle AEC$ are a linear pair. $\angle AEC$ and $\angle AED$ are a linear pair.	**4.** Definition of a linear pair.
5. $\angle BEC$ and $\angle AEC$ are supplementary. $\angle AEC$ and $\angle AED$ are supplementary.	**5.** If two angles form a linear pair, they are supplementary. (Theorem 4.7)
6. $\angle BEC \cong \angle AED$	**6.** If two angles are supplements of the same angle, they are congruent. (Theorem 4.5)

In the proof above, reasons 5 and 6 demonstrate how previously proved theorems can be used as reasons in deducing statements in a proof. In this text, we have assigned numbers to theorems that we will use frequently in proving exercises as well as in proving other theorems. You do not need to remember the numbers of the theorems but you should memorize the statements of the theorems in order to use them as reasons when writing a proof. You may find it useful to keep a list of definitions, postulates, and theorems in a special section in your notebook or on index cards for easy reference and as a study aid.

In this chapter, we have seen the steps to be taken in presenting a proof in geometry using deductive reasoning:

1. As an aid, draw a figure that pictures the data of the theorem or the problem. Use letters to label points in the figure.

2. State the *given*, which is the hypothesis of the theorem, in terms of the figure.

3. State the *prove*, which is the conclusion of the theorem, in terms of the figure.

4. Present the *proof*, which is a series of logical arguments used in the demonstration. Each step in the proof should consist of a statement about the figure. Each statement should be justified by the given, a definition, a postulate, or a previously proved theorem. The proof may be presented in a two-column format or in paragraph form. Proofs that involve the measures of angles or of line segments can often be presented as an algebraic proof.

EXAMPLE 2

If $\overleftrightarrow{ABC}$ and $\overleftrightarrow{DBE}$ intersect at B and $\overrightarrow{BC}$ bisects $\angle EBF$, prove that $\angle CBF \cong \angle ABD$.

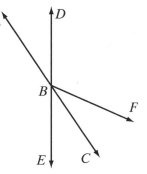

Solution *Given:* $\overleftrightarrow{ABC}$ and $\overleftrightarrow{DBE}$ intersect at B and $\overrightarrow{BC}$ bisects $\angle EBF$.

Prove: $\angle CBF \cong \angle ABD$

Proof

Statements	Reasons
1. $\overrightarrow{BC}$ bisects $\angle EBF$.	**1.** Given.
2. $\angle EBC \cong \angle CBF$	**2.** Definition of a bisector of an angle.
3. $\overleftrightarrow{ABC}$ and $\overleftrightarrow{DBE}$ intersect at B.	**3.** Given.
4. $\angle EBC$ and $\angle ABD$ are vertical angles.	**4.** Definition of vertical angles.
5. $\angle EBC \cong \angle ABD$	**5.** If two lines intersect, then the vertical angles are congruent.
6. $\angle CBF \cong \angle ABD$	**6.** Transitive property of congruence (steps 2 and 5).

Alternative Proof An algebraic proof can be given: Let $m\angle EBF = 2x$. It is given that $\overrightarrow{BC}$ bisects $\angle EBF$. The bisector of an angle separates the angle into two congruent angles: $\angle EBC$ and $\angle CBF$. Congruent angles have equal measures. Therefore, $m\angle EBC = m\angle CBF = x$.

It is also given that $\overleftrightarrow{ABC}$ and $\overleftrightarrow{DBE}$ intersect at B. If two lines intersect, the vertical angles are congruent and therefore have equal measures: $m\angle ABD = m\angle EBC = x$. Then since $m\angle CBF = x$ and $m\angle ABD = x$, $m\angle CBF = m\angle ABD$ and $\angle CBF \cong \angle ABD$.

Exercises

Writing About Mathematics

1. Josh said that Theorem 4.9 could also have been proved by showing that $\angle AEC \cong \angle BED$. Do you agree with Josh? Explain.

2. The statement of Theorem 4.7 is "If two angles form a linear pair then they are supplementary." Is the converse of this theorem true? Justify your answer.

Developing Skills

In 3–11, in each case write a proof, using the hypothesis as the given and the conclusion as the statement to be proved.

3. If m$\angle ACD$ + m$\angle DCB$ = 90, $\angle B \cong \angle DCA$, and $\angle A \cong \angle DCB$, then $\angle A$ and $\angle B$ are complements.

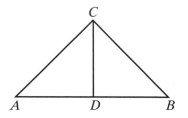

4. If $\overrightarrow{ACFG}$ and $\overrightarrow{BCDE}$ intersect at C and $\angle ADC \cong \angle BFC$, then $\angle ADE \cong \angle BFG$.

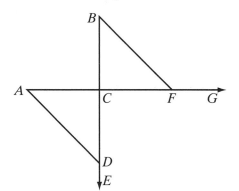

5. If $\angle ADB$ is a right angle and $\overline{CE} \perp \overline{DBE}$, then $\angle ADB \cong \angle CEB$.

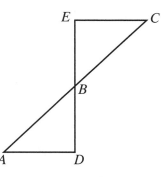

6. If $\angle ABC$ and $\angle BCD$ are right angles, and $\angle EBC \cong \angle ECB$, then $\angle EBA \cong \angle ECD$.

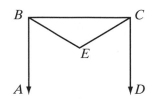

7. If $\angle ABC$ is a right angle and $\angle DBF$ is a right angle, then $\angle ABD \cong \angle CBF$.

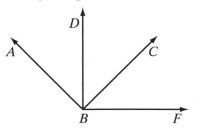

8. If $\overleftrightarrow{EF}$ intersects $\overleftrightarrow{AB}$ at H and $\overleftrightarrow{DC}$ at G, and m$\angle BHG$ = m$\angle CGH$, then $\angle BHG \cong \angle DGE$.

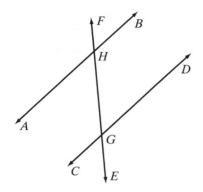

9. If $\overrightarrow{ABD}$ and $\overrightarrow{ACE}$ intersect at A, and $\angle ABC \cong \angle ACB$, then $\angle DBC \cong \angle ECB$.

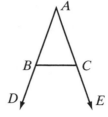

10. If $\overleftrightarrow{AEB}$ and $\overleftrightarrow{CED}$ intersect at E, and $\angle AEC \cong \angle CEB$, then $\overleftrightarrow{AEB} \perp \overleftrightarrow{CED}$.

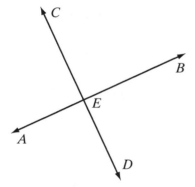

11. If $\angle ABC$ is a right angle, and $\angle BAC$ is complementary to $\angle DBA$, then $\angle BAC \cong \angle CBD$.

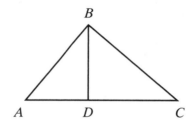

In 12–15, $\overleftrightarrow{AEB}$ and $\overleftrightarrow{CED}$ intersect at E.

12. If m$\angle BEC$ = 70, find m$\angle AED$, m$\angle DEB$, and m$\angle AEC$.

13. If m$\angle DEB$ = $2x + 20$ and m$\angle AEC$ = $3x - 30$, find m$\angle DEB$, m$\angle AEC$, m$\angle AED$, and m$\angle CEB$.

14. If m$\angle BEC$ = $5x - 25$ and m$\angle DEA$ = $7x - 65$, find m$\angle BEC$, m$\angle DEA$, m$\angle DEB$, and m$\angle AEC$.

15. If m$\angle BEC$ = y, m$\angle DEB$ = $3x$, and m$\angle DEA$ = $2x - y$, find m$\angle CEB$, m$\angle BED$, m$\angle DEA$, and m$\angle AEC$.

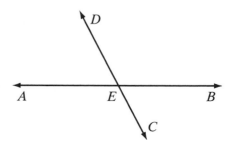

16. $\overleftrightarrow{RS}$ intersects $\overleftrightarrow{LM}$ at P, m$\angle RPL = x + y$, m$\angle LPS = 3x + 2y$, m$\angle MPS = 3x - 2y$.

   a. Solve for x and y.

   b. Find m$\angle RPL$, m$\angle LPS$, and m$\angle MPS$.

17. Prove Theorem 4.2, "If two angles are straight angles, then they are congruent."

18. Prove Theorem 4.5, "If two angles are supplements of the same angle, then they are congruent."

19. Prove Theorem 4.6, "If two angles are congruent, then their supplements are congruent."

Applying Skills

20. Two angles form a linear pair. The measure of the smaller angle is one-half the measure of the larger angle. Find the degree measure of the larger angle.

21. The measure of the supplement of an angle is 60 degrees more than twice the measure of the angle. Find the degree measure of the angle.

22. The difference between the degree measures of two supplementary angles is 80. Find the degree measure of the larger angle.

23. Two angles are complementary. The measure of the larger angle is 5 times the measure of the smaller angle. Find the degree measure of the larger angle.

24. Two angles are complementary. The degree measure of the smaller angle is 50 less than the degree measure of the larger. Find the degree measure of the larger angle.

25. The measure of the complement of an angle exceeds the measure of the angle by 24 degrees. Find the degree measure of the angle.

4-4 CONGRUENT POLYGONS AND CORRESPONDING PARTS

Fold a rectangular sheet of paper in half by placing the opposite edges together. If you tear the paper along the fold, you will have two rectangles that fit exactly on one another. We call these rectangles *congruent polygons*. **Congruent polygons** are polygons that have the same size and shape. Each angle of one polygon is congruent to an angle of the other and each edge of one polygon is congruent to an edge of the other.

In the diagram at the top of page 155, polygon *ABCD* is congruent to polygon *EFGH*. Note that the congruent polygons are named in such a way that each vertex of *ABCD* corresponds to exactly one vertex of *EFGH* and each vertex of *EFGH* corresponds to exactly one vertex of *ABCD*. This relationship is called a **one-to-one correspondence**. The order in which the vertices are named shows this one-to-one correspondence of points.

$ABCD \cong EFGH$ indicates that:

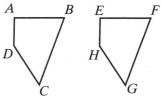

- A corresponds to E; E corresponds to A.
- B corresponds to F; F corresponds to B.
- C corresponds to G; G corresponds to C.
- D corresponds to H; H corresponds to D.

Congruent polygons should always be named so as to indicate the correspondences between the vertices of the polygons.

Corresponding Parts of Congruent Polygons

In congruent polygons $ABCD$ and $EFGH$ shown above, vertex A corresponds to vertex E. Angles A and E are called **corresponding angles**, and $\angle A \cong \angle E$.

In this example, there are four pairs of such corresponding angles:

$$\angle A \cong \angle E \quad \angle B \cong \angle F \quad \angle C \cong \angle G \quad \angle D \cong \angle H$$

In congruent polygons, corresponding angles are congruent.

In congruent polygons $ABCD$ and $EFGH$, since A corresponds to E and B corresponds to F, $\overline{AB}$ and $\overline{EF}$ are **corresponding sides**, and $\overline{AB} \cong \overline{EF}$.

In this example, there are four pairs of such corresponding sides:

$$\overline{AB} \cong \overline{EF} \quad \overline{BC} \cong \overline{FG} \quad \overline{CD} \cong \overline{GH} \quad \overline{DA} \cong \overline{HE}$$

In congruent polygons, corresponding sides are congruent.

The pairs of congruent angles and the pairs of congruent sides are called the corresponding parts of congruent polygons. We can now present the formal definition for congruent polygons.

DEFINITION _____

Two **polygons are congruent** if and only if there is a one-to-one correspondence between their vertices such that corresponding angles are congruent and corresponding sides are congruent.

This definition can be stated more simply as follows:

▶ **Corresponding parts of congruent polygons are congruent.**

Congruent Triangles

The smallest number of sides that a polygon can have is three. A *triangle* is a polygon with exactly three sides. In the figure, $\triangle ABC$ and $\triangle DEF$ are congruent triangles.

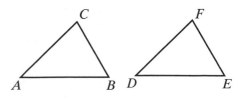

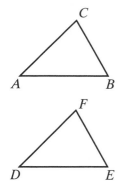

The correspondence establishes six facts about these triangles: three facts about corresponding sides and three facts about corresponding angles. In the table at the right, these six facts are stated as equalities. Since each congruence statement is equivalent to an equality statement, we will use whichever notation serves our purpose better in a particular situation.

Congruences	Equalities
$\overline{AB} \cong \overline{DE}$	$AB = DE$
$\overline{BC} \cong \overline{EF}$	$BC = EF$
$\overline{AC} \cong \overline{DF}$	$AC = DF$
$\angle A \cong \angle D$	$m\angle A = m\angle D$
$\angle B \cong \angle E$	$m\angle B = m\angle E$
$\angle C \cong \angle F$	$m\angle C = m\angle F$

For example, in one proof, we may prefer to write $\overline{AC} \cong \overline{DF}$ and in another proof to write $AC = DF$. In the same way, we might write $\angle C \cong \angle F$ or we might write $m\angle C = m\angle F$. From the definition, we may now say:

▶ **Corresponding parts of congruent triangles are equal in measure.**

In two congruent triangles, pairs of corresponding sides are always opposite pairs of corresponding angles. In the preceding figure, $\triangle ABC \cong \triangle DEF$. The order in which we write the names of the vertices of the triangles indicates the one-to-one correspondence.

1. $\angle A$ and $\angle D$ are corresponding congruent angles.

2. $\overline{BC}$ is opposite $\angle A$, and $\overline{EF}$ is opposite $\angle D$.

3. $\overline{BC}$ and $\overline{EF}$ are corresponding congruent sides.

Equivalence Relation of Congruence

In Section 3-5 we saw that the relation "is congruent to" is an equivalence relation for the set of line segments and the set of angles. Therefore, "is congruent to" must be an equivalence relation for the set of triangles or the set of polygons with a given number of sides.

1. Reflexive property: $\triangle ABC \cong \triangle ABC$.

2. Symmetric property: If $\triangle ABC \cong \triangle DEF$, then $\triangle DEF \cong \triangle ABC$.

3. Transitive property: If $\triangle ABC \cong \triangle DEF$ and $\triangle DEF \cong \triangle RST$, then $\triangle ABC \cong \triangle RST$.

Therefore, we state these properties of congruence as three postulates:

Postulate 4.11 | Any geometric figure is congruent to itself. (Reflexive Property)

Postulate 4.12 | A congruence may be expressed in either order. (Symmetric Property)

Postulate 4.13	Two geometric figures congruent to the same geometric figure are congruent to each other. (Transitive Property)

Exercises

Writing About Mathematics

1. If $\triangle ABC \cong \triangle DEF$, then $\overline{AB} \cong \overline{DE}$. Is the converse of this statement true? Justify your answer.

2. Jesse said that since $\triangle RST$ and $\triangle STR$ name the same triangle, it is correct to say $\triangle RST \cong \triangle STR$. Do you agree with Jesse? Justify your answer.

Developing Skills

In 3–5, in each case name three pairs of corresponding angles and three pairs of corresponding sides in the given congruent triangles. Use the symbol $\cong$ to indicate that the angles named and also the sides named in your answers are congruent.

3. $\triangle ABD \cong \triangle CBD$ **4.** $\triangle ADB \cong \triangle CBD$ **5.** $\triangle ABD \cong \triangle EBC$

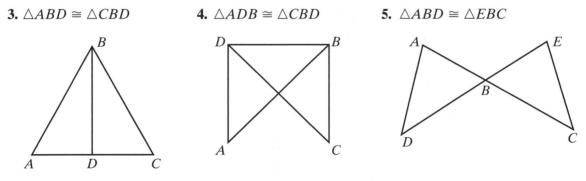

In 6–10, $LMNP$ is a square and $\overline{LSN}$ and $\overline{PSM}$ bisect each other. Name the property that justifies each statement.

6. $\triangle LSP \cong \triangle LSP$

7. If $\triangle LSP \cong \triangle NSM$, then $\triangle NSM \cong \triangle LSP$.

8. If $\triangle LSP \cong \triangle NSM$ and $\triangle NSM \cong \triangle NSP$, then $\triangle LSP \cong \triangle NSP$.

9. If $LS = SN$, then $SN = LS$.

10. If $\angle PLM \cong \angle PNM$, then $\angle PNM \cong \angle PLM$.

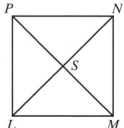

4-5 PROVING TRIANGLES CONGRUENT USING SIDE, ANGLE, SIDE

The definition of congruent polygons states that two polygons are congruent if and only if each pair of corresponding sides and each pair of corresponding angles are congruent. However, it is possible to prove two triangles congruent by proving that fewer than three pairs of sides and three pairs of angles are congruent.

Hands-On Activity

In this activity, we will use a protractor and ruler, or geometry software.
 Use the procedure below to draw a triangle given the measures of two sides and of the included angle.

STEP 1. Use the protractor or geometry software to draw an angle with the given measure.

STEP 2. Draw two segments along the rays of the angle with the given lengths. The two segments should share the vertex of the angle as a common endpoint.

STEP 3. Join the endpoints of the two segments to form a triangle.

STEP 4. Repeat steps 1 through 3 to draw a second triangle using the same angle measure and segment lengths.

 a. Follow the steps to draw two different triangles with each of the given side-angle-side measures.

  (1) 3 in., 90°, 4 in.   (3) 5 cm, 115°, 8 cm
  (2) 5 in., 40°, 5 in.   (4) 10 cm, 30°, 8 cm

 b. For each pair of triangles, measure the side and angles that were not given. Do they have equal measures?

 c. Are the triangles of each pair congruent? Does it appear that when two sides and the included angle of one triangle are congruent to the corresponding sides and angle of another, that the triangles are congruent?

 This activity leads to the following statement of side-angle-side or **SAS triangle congruence**, whose truth will be assumed without proof:

Postulate 4.14

> Two triangles are congruent if two sides and the included angle of one triangle are congruent, respectively, to two sides and the included angle of the other. (SAS)

 In $\triangle ABC$ and $\triangle DEF$, $\overline{BA} \cong \overline{ED}$, $\angle ABC \cong \angle DEF$ and $\overline{BC} \cong \overline{EF}$. It follows that $\triangle ABC \cong \triangle DEF$. The postulate used here is abbreviated SAS.

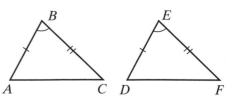

Note: When congruent sides and angles are listed, a correspondence is established. Since the vertices of congruent angles, $\angle ABC$ and $\angle DEF$, are B and E, B corresponds to E. Since $BA = ED$ and B corresponds to E, then A corresponds to D, and since $BC = EF$ and B corresponds to E, then C corresponds to F. We can write $\triangle ABC \cong \triangle DEF$. But, when naming the triangle, if we change the order of the vertices in one triangle we must change the order in the other. For example, $\triangle ABC$, $\triangle BAC$, and $\triangle CBA$ name the same triangle. If $\triangle ABC \cong \triangle DEF$, we may write $\triangle BAC \cong \triangle EDF$ or $\triangle CBA \cong \triangle FED$, but we may not write $\triangle ABC \cong \triangle EFD$.

EXAMPLE I

Given: $\triangle ABC$, $\overline{CD}$ is the bisector of $\overline{AB}$, and $\overline{CD} \perp \overline{AB}$.

Prove: $\triangle ACD \cong \triangle BCD$

Prove the triangles congruent by SAS.

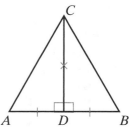

Proof	Statements	Reasons
	1. $\overline{CD}$ bisects $\overline{AB}$.	1. Given.
	2. D is the midpoint of $\overline{AB}$.	2. The bisector of a line segment intersects the segment at its midpoint.
S	**3.** $\overline{AD} \cong \overline{DB}$	3. The midpoint of a line segment divides the segment into two congruent segments.
	4. $\overline{CD} \perp \overline{AB}$	4. Given.
	5. $\angle ADC$ and $\angle BDC$ are right angles.	5. Perpendicular lines intersect to form right angles.
A	**6.** $\angle ADC \cong \angle BDC$	6. If two angles are right angles, then they are congruent.
S	**7.** $\overline{CD} \cong \overline{CD}$	7. Reflexive property of congruence.
	8. $\triangle ACD \cong \triangle BCD$	8. SAS (steps 3, 6, 7).

Note that it is often helpful to mark segments and angles that are congruent with the same number of strokes or arcs. For example, in the diagram for this proof, $\overline{AD}$ and $\overline{DB}$ are marked with a single stroke, $\angle ADC$ and $\angle BDC$ are marked with the symbol for right angles, and $\overline{CD}$ is marked with an "$\times$" to indicate a side common to both triangles.

Exercises

Writing About Mathematics

1. Each of two telephone poles is perpendicular to the ground and braced by a wire that extends from the top of the pole to a point on the level ground 5 feet from the foot of the pole. The wires used to brace the poles are of unequal lengths. Is it possible for the telephone poles to be of equal height? Explain your answer.

2. Is the following statement true? If two triangles are not congruent, then each pair of corresponding sides and each pair of corresponding angles are not congruent. Justify your answer.

Developing Skills

In 3–8, pairs of line segments marked with the same number of strokes are congruent. Pairs of angles marked with the same number of arcs are congruent. A line segment or an angle marked with an "×" is congruent to itself by the reflexive property of congruence.

In each case, is the given information sufficient to prove congruent triangles using SAS?

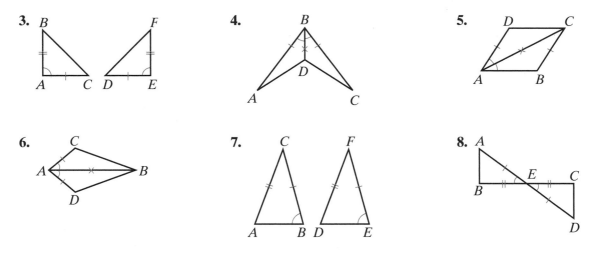

In 9–11, two sides or a side and an angle are marked to indicate that they are congruent. Name the pair of corresponding sides or corresponding angles that would have to be proved congruent in order to prove the triangles congruent by SAS.

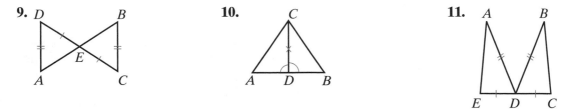

Applying Skills

In 12–14:

a. Draw a diagram with geometry software or pencil and paper and write a *given* statement using the information in the first sentence.

b. Use the information in the second sentence to write a *prove* statement.

c. Write a proof.

12. $\overline{ABC}$ and $\overline{DBE}$ bisect each other. Prove that $\triangle ABE \cong \triangle CBD$.

13. $ABCD$ is a quadrilateral; $AB = CD$; $\overline{BC} = DA$; and $\angle DAB$, $\angle ABC$, $\angle BCD$, and $\angle CDA$ are right angles. Prove that the diagonal $\overline{AC}$ separates the quadrilateral into two congruent triangles.

14. $\angle PQR$ and $\angle RQS$ are a linear pair of angles that are congruent and $PQ = QS$. Prove that $\triangle PQR \cong \triangle RQS$.

4-6 PROVING TRIANGLES CONGRUENT USING ANGLE, SIDE, ANGLE

In the last section we saw that it is possible to prove two triangles congruent by proving that fewer than three pairs of sides and three pairs of angles are congruent, that is, by proving that two sides and the included angle of one triangle are congruent to the corresponding parts of another. There are also other ways of proving two triangles congruent.

Hands-On Activity

In this activity, we will use a protractor and ruler, or geometry software.

Use the procedure below to draw a triangle given the measures of two angles and of the included side.

STEP 1. Use the protractor or geometry software to draw an angle with the first given angle measure. Call the vertex of that angle A.

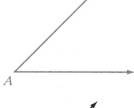

STEP 2. Draw a segment with the given length along one of the rays of $\angle A$. One of the endpoints of the segment should be A. Call the other endpoint B.

STEP 3. Draw a second angle of the triangle using the other given angle measure. Let B be the vertex of this second angle and let $\overrightarrow{BA}$ be one of the sides of this angle.

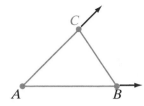

STEP 4. Let C be the intersection of the rays of $\angle A$ and $\angle B$ that are not on $\overleftrightarrow{AB}$.

STEP 5. Repeat steps 1 through 4. Let D the vertex of the first angle, E the vertex of the second angle, and F the intersection of the rays of $\angle D$ and $\angle E$.

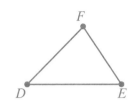

a. Follow the steps to draw *two* different triangles with each of the given angle-side-angle measures.
(1) 65°, 4 in., 35°
(2) 60°, 4 in., 60°
(3) 120°, 9 cm, 30°
(4) 30°, 9 cm, 30°

b. For each pair of triangles, measure the angle and sides that were not given. Do they have equal measures?

c. For each pair of triangles, you can conclude that the triangles are congruent.

The triangles formed, $\triangle ABC$ and $\triangle DEF$, can be placed on top of one another so that A and D coincide, B and E coincide, and C and F coincide. Therefore, $\triangle ABC \cong \triangle DEF$.

This activity leads to the following statement of angle-side-angle or **ASA triangle congruence**, whose truth will be assumed without proof:

Postulate 4.15

> Two triangles are congruent if two angles and the included side of one triangle are congruent, respectively, to two angles and the included side of the other. (ASA)

Thus, in $\triangle ABC$ and $\triangle DEF$, if $\angle B \cong \angle E$, $\overline{BA} \cong \overline{ED}$, and $\angle A \cong \angle D$, it follows that $\triangle ABC \cong \triangle DEF$. The postulate used here is abbreviated ASA. We will now use this postulate to prove two triangles congruent.

EXAMPLE 1

Given: $\overleftrightarrow{AEB}$ and $\overleftrightarrow{CED}$ intersect at E, E is the midpoint of $\overline{AEB}$, $\overline{AC} \perp \overline{AE}$, and $\overline{BD} \perp \overline{BE}$.

Prove: $\triangle AEC \cong \triangle BDE$

Prove the triangles congruent by ASA.

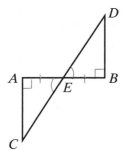

Proof	**Statements**	**Reasons**
	1. $\overleftrightarrow{AEB}$ and $\overleftrightarrow{CED}$ intersect at E.	**1.** Given.
A	**2.** $\angle AEC \cong \angle BED$	**2.** If two lines intersect, the vertical angles are congruent.
	3. E is the midpoint of $\overline{AEB}$.	**3.** Given.
S	**4.** $\overline{AE} \cong \overline{BE}$	**4.** The midpoint of a line segment divides the segment into two congruent segments.
	5. $\overline{AC} \perp \overline{AE}$ and $\overline{BD} \perp \overline{BE}$	**5.** Given.
	6. $\angle A$ and $\angle B$ are right angles.	**6.** Perpendicular lines intersect to form right angles.
A	**7.** $\angle A \cong \angle B$	**7.** If two angles are right angles, they are congruent.
	8. $\triangle AEC \cong \triangle BED$	**8.** ASA (steps 2, 4, 7). ∎

Exercises

Writing About Mathematics

1. Marty said that if two triangles are congruent and one of the triangles is a right triangle, then the other triangle must be a right triangle. Do you agree with Marty? Explain why or why not.

2. In $\triangle ABC$ and $\triangle DEF$, $\overline{AB} \cong \overline{DE}$, and $\angle B \cong \angle E$.

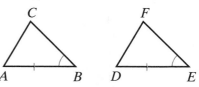

    a. Dora said that if $\overline{AC}$ is congruent to $\overline{DF}$, then it follows that $\triangle ABC \cong \triangle DEF$. Do you agree with Dora? Justify your answer.

    b. Using only the SAS and ASA postulates, if $\triangle ABC$ is not congruent to $\triangle DEF$, what sides and what angles cannot be congruent?

Developing Skills

In 3–5, tell whether the triangles in each pair can be proved congruent by ASA, using only the marked congruent parts in establishing the congruence. Give the reason for your answer.

3.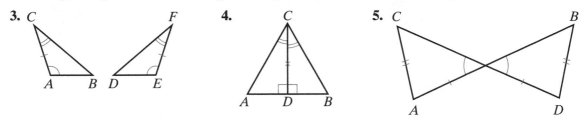

4.

5. C

In 6–8, in each case name the pair of corresponding sides or the pair of corresponding angles that would have to be proved congruent (in addition to those pairs marked congruent) in order to prove that the triangles are congruent by ASA.

6.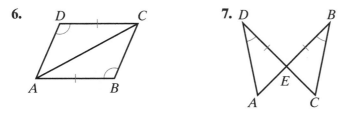

7. D

8. D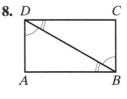

Applying Skills

9. *Given:* $\angle E \cong \angle C$, $\angle EDA \cong \angle CDB$, and *D* is the midpoint of $\overline{EC}$.
 Prove: $\triangle DAE \cong \triangle DBC$

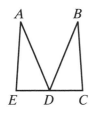

10. *Given:* $\overrightarrow{DB}$ bisects $\angle ADC$ and $\overrightarrow{BD}$ bisects $\angle ABC$.
 Prove: $\triangle ABD \cong \triangle CBD$

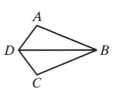

11. *Given:* $\overline{AD} \perp \overline{BC}$ and $\overrightarrow{AD}$ bisects $\angle BAC$.
 Prove: $\triangle ADC \cong \triangle ADB$

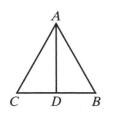

12. *Given:* $\angle DBC \cong \angle GFD$ and $\overline{AE}$ bisects $\overline{FB}$ at *D*.
 Prove: $\triangle DFE \cong \triangle DBA$

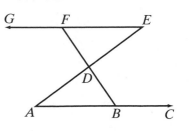

4-7 PROVING TRIANGLES CONGRUENT USING SIDE, SIDE, SIDE

Cut three straws to any lengths and put their ends together to form a triangle. Now cut a second set of straws to the same lengths and try to form a *different* triangle. Repeated experiments lead to the conclusion that it cannot be done.

As shown in $\triangle ABC$ and $\triangle DEF$, when all three pairs of corresponding sides of a triangle are congruent, the triangles must be congruent. The truth of this statement of side-side-side or **SSS triangle congruence** is assumed without proof.

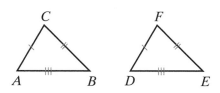

Postulate 4.16

Two triangles are congruent if the three sides of one triangle are congruent, respectively, to the three sides of the other. (SSS)

Thus, in $\triangle ABC$ and $\triangle DEF$ above, $\overline{AB} \cong \overline{DE}$, $\overline{AC} \cong \overline{DF}$, and $\overline{BC} \cong \overline{EF}$. It follows that $\triangle ABC \cong \triangle DEF$. The postulate used here is abbreviated SSS.

EXAMPLE 1

Given: Isosceles $\triangle JKL$ with $\overline{JK} \cong \overline{KL}$ and M the midpoint of $\overline{JL}$.

Prove: $\triangle JKM \cong \triangle LKM$

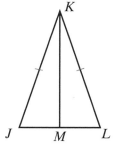

Proof Prove the triangles congruent by using SSS.

Statements	Reasons
S **1.** $\overline{JK} \cong \overline{KL}$	**1.** Given.
2. M is the midpoint of $\overline{JL}$.	**2.** Given.
S **3.** $\overline{JM} \cong \overline{LM}$	**3.** Definition of midpoint.
S **4.** $\overline{KM} \cong \overline{KM}$	**4.** Reflexive property of congruence.
5. $\triangle JKM \cong \triangle LKM$	**5.** SSS (steps 1, 3, 4).

Exercises

Writing About Mathematics

1. Josh said that if two triangles are congruent and one of the triangles is isosceles, then the other must be isosceles. Do you agree with Josh? Explain why or why not.

2. Alvan said that if two triangles are not congruent, then at least one of the three sides of one triangle is not congruent to the corresponding side of the other triangle. Do you agree with Alvan? Justify your answer.

Developing Skills

In 3–5, pairs of line segments marked with the same number of strokes are congruent. A line segment marked with "×" is congruent to itself by the reflexive property of congruence.

In each case, is the given information sufficient to prove congruent triangles?

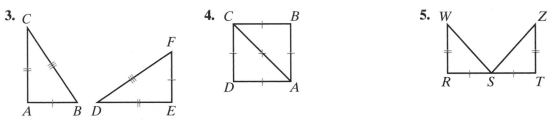

In 6–8, two sides are marked to indicate that they are congruent. Name the pair of corresponding sides that would have to be proved congruent in order to prove the triangles congruent by SSS.

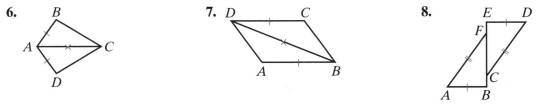

In 9–14, pairs of line segments marked with the same number of strokes are congruent. Pairs of angles marked with the same number of arcs are congruent. A line segment or an angle marked with "×" is congruent to itself by the reflexive property of congruence.

In each case, is the given information sufficient to prove congruent triangles? If so, write the abbreviation for the postulate that proves the triangles congruent.

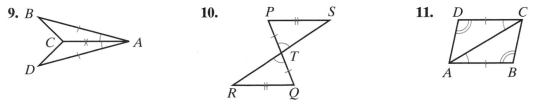

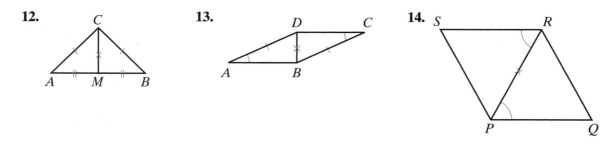

12. 13. 14.

15. If two sides and the angle opposite one of those sides in a triangle are congruent to the corresponding sides and angle of another triangle, are the two triangles congruent? Justify your answer or draw a counterexample proving that they are not.
(*Hint:* Could one triangle be an acute triangle and the other an obtuse triangle?)

Applying Skills

16. *Given*: $\overline{AEB}$ bisects $\overline{CED}$, $\overline{AC} \perp \overline{CED}$, and $\overline{BD} \perp \overline{CED}$.
  Prove: $\triangle EAC \cong \triangle EBD$

17. *Given*: $\triangle ABC$ is equilateral, D is the midpoint of $\overline{AB}$.
  Prove: $\triangle ACD \cong \triangle BCD$

18. *Given:* Triangle PQR with S on $\overline{PQ}$ and $\overline{RS} \perp \overline{PQ}$; $\triangle PSR$ is not congruent to $\triangle QSR$.
  Prove: $PS \neq QS$

19. Gina is drawing a pattern for a kite. She wants it to consist of two congruent triangles that share a common side. She draws an angle with its vertex at A and marks two points, B and C, one on each of the rays of the angle. Each point, B and C, is 15 inches from the vertex of the angle. Then she draws the bisector of $\angle BAC$, marks a point D on the angle bisector and draws $\overline{BD}$ and $\overline{CD}$. Prove that the triangles that she drew are congruent.

CHAPTER SUMMARY

Definitions to Know

- **Adjacent angles** are two angles in the same plane that have a common vertex and a common side but do not have any interior points in common.

- **Complementary angles** are two angles the sum of whose degree measures is 90.

- **Supplementary angles** are two angles the sum of whose degree measures is 180.

- **A linear pair of angles** are two adjacent angles whose sum is a straight angle.

- **Vertical angles** are two angles in which the sides of one angle are opposite rays to the sides of the second angle.

- Two **polygons are congruent** if and only if there is a one-to-one correspondence between their vertices such that corresponding angles are congruent and corresponding sides are congruent.
  - Corresponding parts of congruent polygons are congruent.
  - Corresponding parts of congruent polygons are equal in measure.

Postulates
4.1 A line segment can be extended to any length in either direction.
4.2 Through two given points, one and only one line can be drawn. (Two points determine a line.)
4.3 Two lines cannot intersect in more than one point.
4.4 One and only one circle can be drawn with any given point as center and the length of any given line segment as a radius.
4.5 At a given point on a given line, one and only one perpendicular can be drawn to the line.
4.6 From a given point not on a given line, one and only one perpendicular can be drawn to the line.
4.7 For any two distinct points, there is only one positive real number that is the length of the line segment joining the two points. (Distance Postulate)
4.8 The shortest distance between two points is the length of the line segment joining these two points.
4.9 A line segment has one and only one midpoint.
4.10 An angle has one and only one bisector.
4.11 Any geometric figure is congruent to itself. (Reflexive Property)
4.12 A congruence may be expressed in either order. (Symmetric Property)
4.13 Two geometric figures congruent to the same geometric figure are congruent to each other. (Transitive Property)
4.14 Two triangles are congruent if two sides and the included angle of one triangle are congruent, respectively, to two sides and the included angle of the other. (SAS)
4.15 Two triangles are congruent if two angles and the included side of one triangle are congruent, respectively, to two angles and the included side of the other. (ASA)
4.16 Two triangles are congruent if the three sides of one triangle are congruent, respectively, to the three sides of the other. (SSS)

Theorems
4.1 If two angles are right angles, then they are congruent.
4.2 If two angles are straight angles, then they are congruent.
4.3 If two angles are complements of the same angle, then they are congruent.
4.4 If two angles are congruent, then their complements are congruent.
4.5 If two angles are supplements of the same angle, then they are congruent.
4.6 If two angles are congruent, then their supplements are congruent.
4.7 If two angles form a linear pair, then they are supplementary.
4.8 If two lines intersect to form congruent adjacent angles, then they are perpendicular.
4.9 If two lines intersect, then the vertical angles are congruent.

VOCABULARY

4-1 Distance postulate

4-3 Adjacent angles • Complementary angles • Complement • Supplementary angles • Supplement • Linear pair of angles • Vertical angles

4-4 One-to-one correspondence • Corresponding angles • Corresponding sides • Congruent polygons

4-5 SAS triangle congruence

4-6 ASA triangle congruence

4-7 SSS triangle congruence

REVIEW EXERCISES

1. The degree measure of an angle is 15 more than twice the measure of its complement. Find the measure of the angle and its complement.

2. Two angles, $\angle LMP$ and $\angle PMN$, are a linear pair of angles. If the degree measure of $\angle LMP$ is 12 less than three times that of $\angle PMN$, find the measure of each angle.

3. Triangle JKL is congruent to triangle PQR and m$\angle K = 3a + 18$ and m$\angle Q = 5a - 12$. Find the measure of $\angle K$ and of $\angle Q$.

4. If $\overline{ABC}$ and $\overline{DBE}$ intersect at F, what is true about B and F? State a postulate that justifies your answer.

5. If $\overleftrightarrow{LM} \perp \overline{MN}$ and $\overleftrightarrow{KM} \perp \overline{MN}$, what is true about $\overleftrightarrow{LM}$ and $\overleftrightarrow{KM}$? State a postulate that justifies your answer.

6. Point R is not on $\overline{LMN}$. Is $LM + MN$ less than, equal to, or greater than $LR + RN$? State a postulate that justifies your answer.

7. If $\overrightarrow{BD}$ and $\overrightarrow{BE}$ are bisectors of $\angle ABC$, does E lie on $\overleftrightarrow{BD}$? State a postulate that justifies your answer.

8. The midpoint of $\overline{AB}$ is M. If $\overleftrightarrow{MN}$ and $\overleftrightarrow{PM}$ are bisectors of $\overline{AB}$, does P lie on $\overleftrightarrow{MN}$? Justify your answer.

9. The midpoint of $\overline{AB}$ is M. If $\overleftrightarrow{MN}$ and $\overleftrightarrow{PM}$ are perpendicular to $\overline{AB}$, does P lie on $\overleftrightarrow{MN}$? Justify your answer.

10. *Given:* m∠*A* = 50, m∠*B* = 45,
AB = 10 cm, m∠*D* = 50,
m∠*E* = 45, and *DE* = 10 cm.

Prove: △*ABC* ≅ △*DEF*

11. *Given:* $\overline{GEH}$ bisects $\overline{DEF}$ and
m∠*D* = m∠*F*.

Prove: △*GFE* ≅ △*HDE*

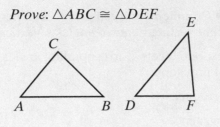

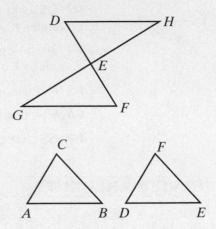

12. *Given:* $\overline{AB} \cong \overline{DE}, \overline{BC} \cong \overline{EF}$, △*ABC* is
not congruent to △*DEF*.

Prove: ∠*B* is not congruent to ∠*E*.

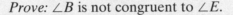

Exploration

1. If three angles of one triangle are congruent to the corresponding angles of another triangle, the triangles may or may not be congruent. Draw diagrams to show that this is true.

2. *STUVWXYZ* is a cube. Write a paragraph proof that would convince someone that △*STX*, △*UTX*, and △*STU* are all congruent to one another.

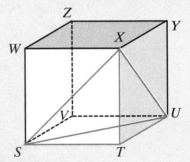

CUMULATIVE REVIEW   CHAPTERS 1–4

Part I

Answer all questions in this part. Each correct answer will receive 2 credits. No partial credit will be allowed.

1. Which of the following is an illustration of the associative property of addition?

(1) 3(4 + 7) = 3(7 + 4)

(2) 3(4 + 7) = 3(4) + 3(7)

(3) 3 + (4 + 7) = 3 + (7 + 4)

(4) 3 + (4 + 7) = (3 + 4) + 7

2. If the sum of the measures of two angles is 90, the angles are
(1) supplementary.
(3) a linear pair.
(2) complementary.
(4) adjacent angles.

3. If $AB + BC = AC$, which of the following may be false?
(1) B is the midpoint of $\overline{AC}$.
(3) B is between A and C.
(2) B is a point of $\overline{AC}$.
(4) $\overrightarrow{BA}$ and $\overrightarrow{BC}$ are opposite rays.

4. If b is a real number, then b has a multiplicative inverse only if
(1) $b = 1$
(2) $b = 0$
(3) $b \geq 0$
(4) $b \neq 0$

5. The contrapositive of "Two angles are congruent if they have the same measures" is
(1) Two angles are not congruent if they do not have the same measures.
(2) If two angles have the same measures, then they are congruent.
(3) If two angles are not congruent, then they do not have the same measures.
(4) If two angles do not have the same measures, then they are not congruent.

6. The statement "Today is Saturday and I am going to the movies" is true. Which of the following statements is false?
(1) Today is Saturday or I am not going to the movies.
(2) Today is not Saturday or I am not going to the movies.
(3) If today is not Saturday, then I am not going to the movies.
(4) If today is not Saturday, then I am going to the movies.

7. If $\triangle ABC \cong \triangle BCD$, then $\triangle ABC$ and $\triangle BCD$ must be
(1) obtuse.
(2) scalene.
(3) isosceles.
(4) equilateral.

8. If $\overleftrightarrow{ABC}$ and $\overleftrightarrow{DBE}$ intersect at B, $\angle ABD$ and $\angle CBE$ are
(1) congruent vertical angles.
(3) congruent adjacent angles.
(2) supplementary vertical angles.
(4) supplementary adjacent angles.

9. $\angle LMN$ and $\angle NMP$ form a linear pair of angles. Which of the following statements is false?
(1) $m\angle LMN + m\angle NMP = 180$
(2) $\angle LMN$ and $\angle NMP$ are supplementary angles.
(3) $\overrightarrow{ML}$ and $\overrightarrow{MP}$ are opposite rays.
(4) $\overrightarrow{ML}$ and $\overrightarrow{MN}$ are opposite rays.

10. The solution set of the equation $3(x + 2) < 5x$ is
(1) $\{x \mid x < 3\}$
(3) $\{x \mid x < 1\}$
(2) $\{x \mid x > 3\}$
(4) $\{x \mid x > 1\}$

Part II

Answer all questions in this part. Each correct answer will receive 2 credits. Clearly indicate the necessary steps, including appropriate formula substitutions, diagrams, graphs, charts, etc. For all questions in this part, a correct numerical answer with no work shown will receive only 1 credit.

11. *Given:* $\overline{PQ}$ bisects $\overline{RS}$ at M and $\angle R \cong \angle S$.

Prove: $\triangle RMQ \cong \triangle SMP$

12. *Given:* Quadrilateral *DEFG* with $DE = DG$ and $EF = GF$.

Prove: $\triangle DEF \cong \triangle DGF$

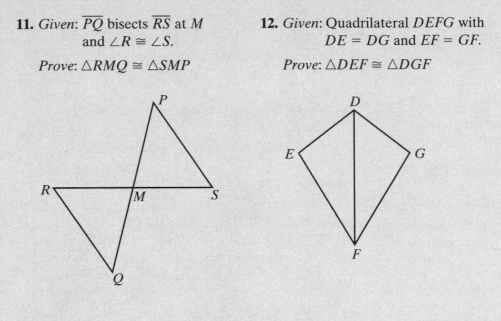

Part III

Answer all questions in this part. Each correct answer will receive 4 credits. Clearly indicate the necessary steps, including appropriate formula substitutions, diagrams, graphs, charts, etc. For all questions in this part, a correct numerical answer with no work shown will receive only 1 credit.

13. The following statements are true:

If our team does not win, we will not celebrate.

We will celebrate or we will practice.

We do not practice.

Did our team win? Justify your answer.

14. The two angles of a linear pair of angles are congruent. If the measure of one angle is represented by $2x - y$ and the measure of the other angle by $x + 4y$, find the values of x and of y.

Part IV

Answer all questions in this part. Each correct answer will receive 6 credits. Clearly indicate the necessary steps, including appropriate formula substitutions, diagrams, graphs, charts, etc. For all questions in this part, a correct numerical answer with no work shown will receive only 1 credit.

15. Josie is making a pattern for quilt pieces. One pattern is a right triangle with two acute angles that are complementary. The measure of one of the acute angles is to be 12 degrees more than half the measure of the other acute angle. Find the measure of each angle of the triangle.

16. Triangle DEF is equilateral and equiangular. The midpoint of $\overline{DE}$ is M, of $\overline{EF}$ is N, and of $\overline{FD}$ is L. Line segments $\overline{MN}$, $\overline{ML}$, and $\overline{NL}$ are drawn.

   a. Name three congruent triangles.

   b. Prove that the triangles named in **a** are congruent.

   c. Prove that $\triangle NLM$ is equilateral.

   d. Prove that $\triangle NLM$ is equiangular.

CHAPTER

5

CONGRUENCE BASED ON TRIANGLES

CHAPTER TABLE OF CONTENTS

The SSS postulate tells us that a triangle with sides of given lengths can have only one size and shape. Therefore, the area of the triangle is determined. We know that the area of a triangle is one-half the product of the lengths of one side and the altitude to that side. But can the area of a triangle be found using only the lengths of the sides? A formula to do this was known by mathematicians of India about 3200 B.C. In the Western world, Heron of Alexandria, who lived around 75 B.C., provided in his book *Metrica* a formula that we now call Heron's formula:

If A is the area of the triangle with sides of length a, b, and c, and the *semiperimeter*, s, is one-half the perimeter, that is, $s = \frac{1}{2}(a + b + c)$, then

$$A = \sqrt{s(s - a)(s - b)(s - c)}$$

In *Metrica*, Heron also provided a method of finding the approximate value of the square root of a number. This method, often called the *divide and average method*, continued to be used until calculators made the pencil and paper computation of a square root unnecessary.

5-1 LINE SEGMENTS ASSOCIATED WITH TRIANGLES

Natalie is planting a small tree. Before filling in the soil around the tree, she places stakes on opposite sides of the tree at equal distances from the base of the tree. Then she fastens cords from the same point on the trunk of the tree to the stakes. The cords are not of equal length. Natalie reasons that the tree is not perpendicular to the ground and straightens the tree until the cords are of equal lengths. Natalie used her knowledge of geometry to help her plant the tree. What was the reasoning that helped Natalie to plant the tree?

Geometric shapes are all around us. Frequently we use our knowledge of geometry to make decisions in our daily life. In this chapter you will write formal and informal proofs that will enable you to form the habit of looking for logical relationships before making a decision.

Altitude of a Triangle

DEFINITION

An **altitude of a triangle** is a line segment drawn from any vertex of the triangle, perpendicular to and ending in the line that contains the opposite side.

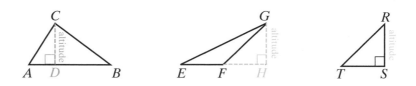

In $\triangle ABC$, if $\overline{CD}$ is perpendicular to $\overline{AB}$, then $\overline{CD}$ is the altitude from vertex C to the opposite side. In $\triangle EFG$, if $\overline{GH}$ is perpendicular to $\overleftrightarrow{EF}$, the line that contains the side $\overline{EF}$, then $\overline{GH}$ is the altitude from vertex G to the opposite side. In an obtuse triangle such as $\triangle EFG$ above, the altitude from each of the acute angles lies outside the triangle.

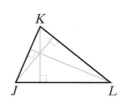

In right $\triangle TSR$, if $\overline{RS}$ is perpendicular to $\overline{TS}$, then $\overline{RS}$ is the altitude from vertex R to the opposite side $\overline{TS}$ and $\overline{TS}$ is the altitude from T to the opposite side $\overline{RS}$. In a right triangle such as $\triangle TSR$ above, the altitude from each vertex of an acute angle is a leg of the triangle. Every triangle has three altitudes as shown in $\triangle JKL$.

Median of a Triangle

DEFINITION

A **median of a triangle** is a line segment that joins any vertex of the triangle to the midpoint of the opposite side.

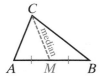

In $\triangle ABC$, if M is the midpoint of $\overline{AB}$, then $\overline{CM}$ is the median drawn from vertex C to side $\overline{AB}$. We may also draw a median from vertex A to the midpoint of side $\overline{BC}$, and a median from vertex B to the midpoint of side $\overline{AC}$. Thus, every triangle has three medians.

Angle Bisector of a Triangle

DEFINITION

An **angle bisector of a triangle** is a line segment that bisects any angle of the triangle and terminates in the side opposite that angle.

In $\triangle PQR$, if D is a point on $\overline{PQ}$ such that $\angle PRD \cong \angle QRD$, then $\overline{RD}$ is the angle bisector from R in $\triangle PQR$. We may also draw an angle bisector from the vertex P to some point on $\overline{QR}$, and an angle bisector from the vertex Q to some point on $\overline{PR}$. Thus, every triangle has three angle bisectors.

In a scalene triangle, the altitude, the median, and the angle bisector drawn from any common vertex are three distinct line segments. In $\triangle ABC$, from the common vertex B, three line segments are drawn:

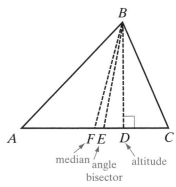

1. $\overline{BD}$ is the altitude from B because $\overline{BD} \perp \overline{AC}$.

2. $\overline{BE}$ is the angle bisector from B because $\angle ABE \cong \angle EBC$.

3. $\overline{BF}$ is the median from B because F is the midpoint of $\overline{AC}$.

In some special triangles, such as an isosceles triangle and an equilateral triangle, some of these segments coincide, that is, are the same line. We will consider these examples later.

EXAMPLE I

Given: $\overline{KM}$ is the angle bisector from K in $\triangle JKL$, and $\overline{LK} \cong \overline{JK}$.

Prove: $\triangle JKM \cong \triangle LKM$

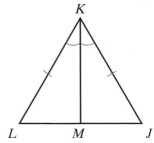

Proof	*Statements*	*Reasons*
	1. $\overline{LK} \cong \overline{JK}$	1. Given.
	2. $\overline{KM}$ is the angle bisector from K in $\triangle JKL$.	2. Given.
	3. $\overline{KM}$ bisects $\angle JKL$.	3. Definition of an angle bisector of a triangle.
	4. $\angle JKM \cong \angle LKM$	4. Definition of the bisector of an angle.
	5. $\overline{KM} \cong \overline{KM}$	5. Reflexive property of congruence.
	6. $\triangle JKM \cong \triangle LKM$	6. SAS (steps 1, 4, 5).

Exercises

Writing About Mathematics

1. Explain why the three altitudes of a right triangle intersect at the vertex of the right angle.

2. Triangle ABC is a triangle with $\angle C$ an obtuse angle. Where do the lines containing the three altitudes of the triangle intersect?

Developing Skills

3. Use a pencil, ruler, and protractor, or use geometry software, to draw $\triangle ABC$, an acute, scalene triangle with altitude $\overline{CD}$, angle bisector $\overline{CE}$, and median $\overline{CF}$.

   a. Name two congruent angles that have their vertices at C.

   b. Name two congruent line segments.

   c. Name two perpendicular line segments.

   d. Name two right angles.

4. Use a pencil, ruler, and protractor, or use geometry software, to draw several triangles. Include acute, obtuse, and right triangles.

   a. Draw three altitudes for each triangle.

   b. Make a conjecture regarding the intersection of the lines containing the three altitudes.

5. Use a pencil, ruler, and protractor, or use geometry software, to draw several triangles. Include acute, obtuse, and right triangles.

   a. Draw three angle bisectors for each triangle.

   b. Make a conjecture regarding the intersection of these three angle bisectors.

6. Use a pencil, ruler, and protractor, or use geometry software, to draw several triangles. Include acute, obtuse, and right triangles.

   a. Draw three medians to each triangle.

   b. Make a conjecture regarding the intersection of these three medians.

In 7–9, draw and label each triangle described. Complete each required proof in two-column format.

7. *Given:* In $\triangle PQR$, $\overline{PR} \cong \overline{QR}$, $\angle P \cong \angle Q$, and $\overline{RS}$ is a median.

   Prove: $\triangle PSR \cong \triangle QSR$

8. *Given:* In $\triangle DEF$, $\overline{EG}$ is both an angle bisector and an altitude.

   Prove: $\triangle DEG \cong \triangle FEG$

9. *Given:* $\overline{CD}$ is a median of $\triangle ABC$ but $\triangle ADC$ is not congruent to $\triangle BDC$.

   Prove: $\overline{CD}$ is not an altitude of $\triangle ABC$.

   (*Hint:* Use an indirect proof.)

Applying Skills

In 10–13, complete each required proof in paragraph format.

10. In a scalene triangle, $\triangle LNM$, show that an altitude, $\overline{NO}$, cannot be an angle bisector. (*Hint:* Use an indirect proof.)

11. A telephone pole is braced by two wires that are fastened to the pole at point C and to the ground at points A and B. The base of the pole is at point D, the midpoint of $\overline{AB}$. If the pole is perpendicular to the ground, are the wires of equal length? Justify your answer.

12. The formula for the area of a triangle is $A = \frac{1}{2}bh$ with b the length of one side of a triangle and h the length of the altitude to that side. In $\triangle ABC$, $\overline{CD}$ is the altitude from vertex C to side $\overline{AB}$ and M is the midpoint of $\overline{AB}$. Show that the median separates $\triangle ABC$ into two triangles of equal area, $\triangle AMC$ and $\triangle BMC$.

13. A farmer has a triangular piece of land that he wants to separate into two sections of equal area. How can the land be divided?

5-2 USING CONGRUENT TRIANGLES TO PROVE LINE SEGMENTS CONGRUENT AND ANGLES CONGRUENT

The definition of congruent triangles tells us that when two triangles are congruent, each pair of corresponding sides are congruent and each pair of corresponding angles are congruent. We use three pairs of corresponding parts, SAS, ASA, or SSS, to prove that two triangles are congruent. We can then conclude that each of the other three pairs of corresponding parts are also congruent. In this section we will prove triangles congruent in order to prove that two line segments or two angles are congruent.

EXAMPLE I

Given: $\overline{ABCD}$, $\angle A$ is a right angle, $\angle D$ is a right angle, $\overline{AE} \cong \overline{DF}$, $\overline{AB} \cong \overline{CD}$.

Prove: $\overline{EC} \cong \overline{FB}$

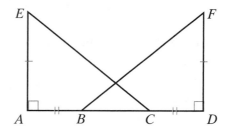

Proof The line segments that we want to prove congruent are corresponding sides of $\triangle EAC$ and $\triangle FDB$.

Therefore we will first prove that $\triangle EAC \cong \triangle FDB$. Then use that corresponding parts of congruent triangles are congruent.

Statements	Reasons
1. $\angle A$ is a right angle, $\angle D$ is a right angle.	**1.** Given.
2. $\angle A \cong \angle D$	**2.** If two angles are right angles, then they are congruent. (Theorem 4.1)
3. $\overline{AE} \cong \overline{DF}$	**3.** Given.
4. $\overline{AB} \cong \overline{CD}$	**4.** Given.
5. $\overline{AB} + \overline{BC} \cong \overline{BC} + \overline{CD}$	**5.** Addition postulate.
6. $\overline{ABCD}$	**6.** Given.
7. $\overline{AB} + \overline{BC} = \overline{AC}$ $\overline{BC} + \overline{CD} = \overline{BD}$	**7.** Partition postulate.
8. $\overline{AC} \cong \overline{BD}$	**8.** Substitution postulate (steps 5, 7).
9. $\triangle EAC \cong \triangle FDB$	**9.** SAS (steps 3, 2, 8).
10. $\overline{EC} \cong \overline{FB}$	**10.** Corresponding parts of congruent triangles are congruent.

Exercises

Writing About Mathematics

1. Triangles ABC and DEF are congruent triangles. If $\angle A$ and $\angle B$ are complementary angles, are $\angle D$ and $\angle E$ also complementary angles? Justify your answer.

2. A leg and the vertex angle of one isosceles triangle are congruent respectively to a leg and the vertex angle of another isosceles triangle. Is this sufficient information to conclude that the triangles must be congruent? Justify your answer.

Developing Skills

In 3–8, the figures have been marked to indicate pairs of congruent angles and pairs of congruent segments.

a. In each figure, name two triangles that are congruent.

b. State the reason why the triangles are congruent.

c. For each pair of triangles, name three additional pairs of parts that are congruent because they are corresponding parts of congruent triangles.

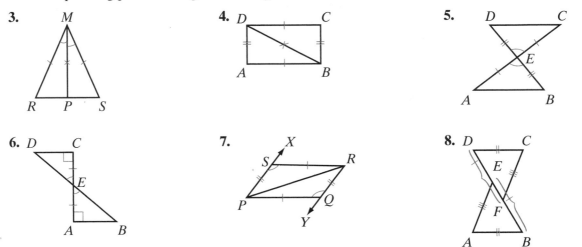

3.   **4.** D   C   **5.** D   C

6. D   C   **7.**   **8.** D   C

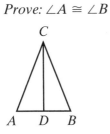

9. *Given:* $\overline{CA} \cong \overline{CB}$ and *D* is the midpoint of $\overline{AB}$.

   Prove: $\angle A \cong \angle B$

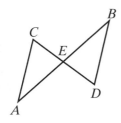

10. *Given:* $\overline{AB} \cong \overline{CD}$ and $\angle CAB \cong \angle ACD$

   Prove: $\overline{AD} \cong \overline{CB}$

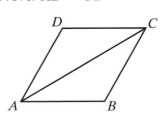

11. *Given:* $\overline{AEB}$ and $\overline{CED}$ bisect each other.

   Prove: $\angle C \cong \angle D$

12. *Given:* $\angle KLM$ and $\angle NML$ are right angles and $KL = NM$.

   Prove: $\angle K \cong \angle N$

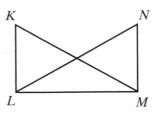

13. Triangle ABC is congruent to triangle DEF, $AB = 3x + 7$, $DE = 5x - 9$, and $BC = 4x$. Find:

   a. x **b.** AB **c.** BC **d.** EF

14. Triangle PQR is congruent to triangle LMN, $m\angle P = 7a$, $m\angle L = 4a + 15$, and $\angle P$ and $\angle Q$ are complementary. Find:

   a. a **b.** $m\angle P$ **c.** $m\angle Q$ **d.** $m\angle M$

Applying Skills

In 15 and 16, complete each required proof in paragraph format.

15. a. Prove that the median from the vertex angle of an isosceles triangle separates the triangle into two congruent triangles.

   b. Prove that the two congruent triangles in **a** are right triangles.

16. a. Prove that if each pair of opposite sides of a quadrilateral are congruent, then a diagonal of the quadrilateral separates it into two congruent triangles.

   b. Prove that a pair of opposite angles of the quadrilateral in **a** are congruent.

In 17 and 18, complete each required proof in two-column format.

17. a. Points B and C separate $\overline{ABCD}$ into three congruent segments. P is a point not on $\overleftrightarrow{AD}$ such that $\overline{PA} \cong \overline{PD}$ and $\overline{PB} \cong \overline{PC}$. Draw a diagram that shows these line segments and write the information in a *given* statement.

   b. Prove that $\angle APB \cong \angle DPC$.

   c. Prove that $\angle APC \cong \angle DPB$.

18. The line segment $\overline{PM}$ is both the altitude and the median from P to $\overline{LN}$ in $\triangle LNP$.

   a. Prove that $\triangle LNP$ is isosceles.

   b. Prove that $\overline{PM}$ is also the angle bisector from P in $\triangle LNP$.

5-3 ISOSCELES AND EQUILATERAL TRIANGLES

When working with triangles, we observed that when two sides of a triangle are congruent, the median, the altitude, and the bisector of the vertex angle separate the triangle into two congruent triangles. These congruent triangles can be used to prove that the base angles of an isosceles triangle are congruent. This observation can be proved as a theorem called the **Isosceles Triangle Theorem**.

Theorem 5.1	If two sides of a triangle are congruent, the angles opposite these sides are congruent.

Given $\triangle ABC$ with $\overline{AC} \cong \overline{BC}$

Prove $\angle A \cong \angle B$

Proof In order to prove this theorem, we will use the median to the base to separate the triangle into two congruent triangles.

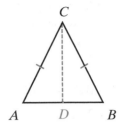

Statements	Reasons
1. Draw D, the midpoint of $\overline{AB}$.	**1.** A line segment has one and only one midpoint.
2. $\overline{CD}$ is the median from vertex C.	**2.** Definition of a median of a triangle.
3. $\overline{CD} \cong \overline{CD}$	**3.** Reflexive property of congruence.
4. $\overline{AD} \cong \overline{DB}$	**4.** Definition of a midpoint.
5. $\overline{AC} \cong \overline{BC}$	**5.** Given.
6. $\triangle ACD \cong \triangle BCD$	**6.** SSS (steps 3, 4, 5).
7. $\angle A \cong \angle B$	**7.** Corresponding parts of congruent triangles are congruent.

A **corollary** is a theorem that can easily be deduced from another theorem. We can prove two other statements that are corollaries of the isosceles triangle theorem because their proofs follow directly from the proof of the theorem.

Corollary 5.1a	The median from the vertex angle of an isosceles triangle bisects the vertex angle.

Proof: From the preceding proof that $\triangle ACD \cong \triangle BCD$, we can also conclude that $\angle ACD \cong \angle BCD$ since they, too, are corresponding parts of congruent triangles.

Corollary 5.1b	The median from the vertex angle of an isosceles triangle is perpendicular to the base.

Proof: Again, from △*ACD* ≅ △*BCD*, we can say that ∠*CDA* ≅ ∠*CDB* because they are corresponding parts of congruent triangles. If two lines intersect to form congruent adjacent angles, then they are perpendicular. Therefore, $\overline{CD} \perp \overline{AB}$.

Properties of an Equilateral Triangle

The isosceles triangle theorem has shown that in an isosceles triangle with two congruent sides, the angles opposite these sides are congruent. We may prove another corollary to this theorem for any equilateral triangle, where three sides are congruent.

Corollary 5.1c | Every equilateral triangle is equiangular.

Proof: If △*ABC* is equilateral, then $\overline{AB} \cong \overline{BC} \cong \overline{CA}$. By the isosceles triangle theorem, since $\overline{AB} \cong \overline{BC}$, ∠*A* ≅ ∠*C*, and since $\overline{BC} \cong \overline{CA}$, ∠*B* ≅ ∠*A*. Therefore, ∠*A* ≅ ∠*B* ≅ ∠*C*.

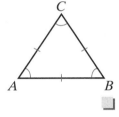

EXAMPLE I

Given: *E* not on $\overline{ABCD}$, $\overline{AB} \cong \overline{CD}$, and $\overline{EB} \cong \overline{EC}$.

Prove: $\overline{AE} \cong \overline{DE}$

Proof $\overline{AE}$ and $\overline{DE}$ are corresponding sides of △*ABE* and △*DCE*, and we will prove these triangles congruent by SAS. We are given two pairs of congruent corresponding sides and must prove that the included angles are congruent.

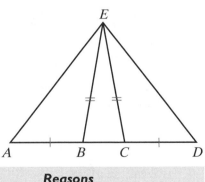

Statements	Reasons
1. $\overline{EB} \cong \overline{EC}$	**1.** Given.
2. ∠*EBC* ≅ ∠*ECB*	**2.** Isosceles triangle theorem (*Or:* If two sides of a triangle are congruent, the angles opposite these sides are congruent.).
3. $\overline{ABCD}$	**3.** Given.

Continued

(Continued)	**Statements**	**Reasons**
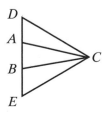	**4.** ∠*ABE* and ∠*EBC* are supplementary. ∠*DCE* and ∠*ECB* are supplementary.	**4.** If two angles form a linear pair, they are supplementary.
	5. ∠*ABE* ≅ ∠*DCE*	**5.** The supplements of congruent angles are congruent.
	6. $\overline{AB} \cong \overline{CD}$	**6.** Given.
	7. △*ABE* ≅ △*DCE*	**7.** SAS (steps 1, 5, 6).
	8. $\overline{AE} \cong \overline{DE}$	**8.** Corresponding parts of congruent triangles are congruent. ∎

Exercises

Writing About Mathematics

1. Joel said that the proof given in Example 1 could have been done by proving that △*ACE* ≅ △*DBE*. Do you agree with Joel? Justify your answer.

2. Abel said that he could prove that equiangular triangle *ABC* is equilateral by drawing median $\overline{BD}$ and showing that △*ABD* ≅ △*CBD*. What is wrong with Abel's reasoning?

Developing Skills

3. In △*ABC*, if $\overline{AB} \cong \overline{AC}$, m∠*B* = 3*x* + 15 and m∠*C* = 7*x* − 5, find m∠*B* and m∠*C*.

4. Triangle *RST* is an isosceles right triangle with *RS* = *ST* and ∠*R* and ∠*T* complementary angles. What is the measure of each angle of the triangle?

5. In equilateral △*DEF*, m∠*D* = 3*x* + *y*, m∠*E* = 2*x* + 40, and m∠*F* = 2*y*. Find *x*, *y*, m∠*D*, m∠*E*, and m∠*F*.

6. *Given:* *C* not on $\overline{DABE}$ and $\overline{CA} \cong \overline{CB}$

Prove: ∠*CAD* ≅ ∠*CBE*

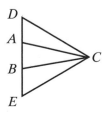

7. *Given:* Quadrilateral *ABCD* with $\overline{AB} \cong \overline{CB}$ and $\overline{AD} \cong \overline{CD}$

Prove: ∠*BAD* ≅ ∠*BCD*

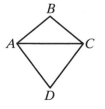

8. *Given:* $\overline{AC} \cong \overline{CB}$ and $\overline{DA} \cong \overline{EB}$

Prove: $\angle CDE \cong \angle CED$

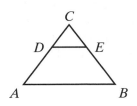

9. *Given:* $\overline{AC} \cong \overline{BC}$ and $\angle DAB \cong \angle DBA$

Prove: $\angle CAD \cong \angle CBD$

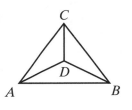

10. *Given:* Isosceles $\triangle ABC$ with $\overline{AC} \cong \overline{BC}$, D is the midpoint of $\overline{AC}$, E is the midpoint of $\overline{BC}$ and F is the midpoint of $\overline{AB}$.

  a. *Prove:* $\triangle ADF \cong \triangle BEF$

  b. *Prove:* $\triangle DEF$ is isosceles.

In 11 and 12, complete each given proof with a partner or in a small group.

11. *Given:* $\triangle ABC$ with $AB = AC$, $BG = EC$, $\overline{BE} \perp \overline{DE}$, and $\overline{CG} \perp \overline{GF}$.

Prove: $\overline{BD} \cong \overline{CF}$

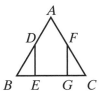

12. *Given:* E not on $\overline{ABCD}$, $\overline{AB} \cong \overline{CD}$, and $\overline{EB}$ is not congruent to $\overline{EC}$.

Prove: $\overline{AE}$ is not congruent to $\overline{DE}$.

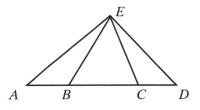

(*Hint:* Use an indirect proof.)

Applying Skills

13. Prove the isosceles triangle theorem by drawing the bisector of the vertex angle instead of the median.

14. Prove that the line segments joining the midpoints of the sides of an equilateral triangle are congruent.

15. C is a point not on $\overleftrightarrow{FBDG}$ and $BC = DC$. Prove that $\angle FBC \cong \angle GDC$.

16. In $\triangle PQR$, $m\angle R \neq m\angle Q$. Prove that $PQ \neq PR$.

5-4 USING TWO PAIRS OF CONGRUENT TRIANGLES

Often the line segments or angles that we want to prove congruent are not corresponding parts of triangles that can be proved congruent by using the given information. However, it may be possible to use the given information to prove a different pair of triangles congruent. Then the congruent corresponding parts of this second pair of triangles can be used to prove the required triangles congruent. The following is an example of this method of proof.

EXAMPLE I

Given: $\overline{AEB}$, $\overline{AC} \cong \overline{AD}$, and $\overline{CB} \cong \overline{DB}$

Prove: $\overline{CE} \cong \overline{DE}$

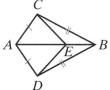

Proof Since $\overline{CE}$ and $\overline{DE}$ are corresponding parts of $\triangle ACE$ and $\triangle ADE$, we can prove these two line segments congruent if we can prove $\triangle ACE$ and $\triangle ADE$ congruent. From the given, we cannot prove immediately that $\triangle ACE$ and $\triangle ADE$ congruent. However, we can prove that $\triangle CAB \cong \triangle DAB$. Using corresponding parts of these larger congruent triangles, we can then prove that the smaller triangles are congruent.

Statements	Reasons
1. $\overline{AC} \cong \overline{AD}$	**1.** Given.
2. $\overline{CB} \cong \overline{DB}$	**2.** Given.
3. $\overline{AB} \cong \overline{AB}$	**3.** Reflexive property of congruence.
4. $\triangle CAB \cong \triangle DAB$	**4.** SSS (steps 1, 2, 3).
5. $\angle CAB \cong \angle DAB$	**5.** Corresponding parts of congruent triangles are congruent.
6. $\overline{AE} \cong \overline{AE}$	**6.** Reflexive property of congruence.
7. $\triangle CAE \cong \triangle DAE$	**7.** SAS (steps 1, 5, 6).
8. $\overline{CE} \cong \overline{DE}$	**8.** Corresponding parts of congruent triangles are congruent.

Exercises

Writing About Mathematics

1. Can Example 1 be proved by proving that $\triangle BCE \cong \triangle BDE$? Justify your answer.

2. Greg said that if it can be proved that two triangles are congruent, then it can be proved that the medians to corresponding sides of these triangles are congruent. Do you agree with Greg? Explain why or why not.

Developing Skills

3. *Given:* $\triangle ABC \cong \triangle DEF$, *M* is the midpoint of $\overline{AB}$, and *N* is the midpoint of $\overline{DE}$.

   Prove: $\triangle AMC \cong \triangle DNF$

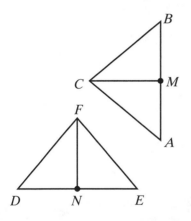

4. *Given:* $\triangle ABC \cong \triangle DEF$, $\overline{CG}$ bisects $\angle ACB$, and $\overline{FH}$ bisects $\angle DFE$.

   Prove: $\overline{CG} \cong \overline{FH}$

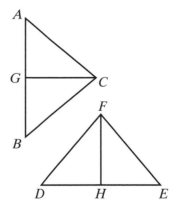

5. *Given:* $\overline{AEC}$ and $\overline{DEB}$ bisect each other, $\overline{FEG}$ intersects $\overline{AB}$ at *G* and $\overline{CD}$ at *F*.

   Prove: *E* is the midpoint of $\overline{FEG}$.

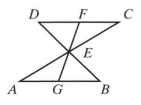

6. *Given:* $\triangle AME \cong \triangle BMF$ and $\overline{DE} \cong \overline{CF}$

   Prove: $\overline{AD} \cong \overline{BC}$

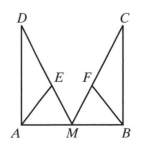

7. *Given:* $\overline{BC} \cong \overline{BA}$ and $\overrightarrow{BD}$ bisects $\angle CBA$.

    Prove: $\overrightarrow{DB}$ bisects $\angle CDA$.

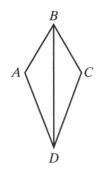

8. *Given:* $RP = RQ$ and $SP = SQ$

    Prove: $\overline{RT} \perp \overline{PQ}$

Applying Skills

9. In quadrilateral $ABCD$, $AB = CD$, $BC = DA$, and M is the midpoint of $\overline{BD}$. A line segment through M intersects $\overline{AB}$ at E and $\overline{CD}$ at F. Prove that $\overline{BMD}$ bisects $\overline{EMF}$ at M.

10. Complete the following exercise with a partner or in a small group:

Line l intersects $\overline{AB}$ at M, and P and S are any two points on l. Prove that if $PA = PB$ and $SA = SB$, then M is the midpoint of $\overline{AB}$ and l is perpendicular to $\overline{AB}$.

a. Let half the group treat the case in which P and S are on the same side of $\overline{AB}$.

b. Let half the group treat the case in which P and S are on opposite sides of $\overline{AB}$.

c. Compare and contrast the methods used to prove the cases.

5-5 PROVING OVERLAPPING TRIANGLES CONGRUENT

If we know that $\overline{AD} \cong \overline{BC}$ and $\overline{DB} \cong \overline{CA}$, can we prove that $\triangle DBA \cong \triangle CAB$? These two triangles overlap and share a common side. To make it easier to visualize the overlapping triangles that we want to prove congruent, it may be helpful to outline each with a different color as shown in the figure.

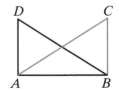

Or the triangles can be redrawn as separate triangles.

The segment $\overline{AB}$ is a side of each of the triangles $\triangle DBA$ and $\triangle CAB$. Therefore, to the given information, $\overline{AD} \cong \overline{BC}$ and $\overline{DB} \cong \overline{CA}$, we can add $\overline{AB} \cong \overline{AB}$ and prove that $\triangle DBA \cong \triangle CAB$ by SSS.

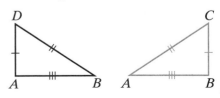

EXAMPLE 1

Given: $\overline{CD}$ and $\overline{BE}$ are medians to the legs of isosceles $\triangle ABC$.

Prove: $\overline{CD} \cong \overline{BE}$

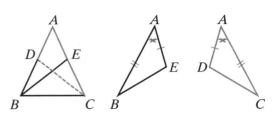

Proof Separate the triangles to see more clearly the triangles to be proved congruent. We know that the legs of an isosceles triangle are congruent. Therefore, $\overline{AB} \cong \overline{AC}$. We also know that the median is a line segment from a vertex to the midpoint of the opposite side. Therefore, D and E are midpoints of the congruent legs. The midpoint divides the line segment into two congruent segments, that is, in half, and halves of congruent segments are congruent: $\overline{AE} \cong \overline{AD}$. Now we have two pair of congruent sides: $\overline{AB} \cong \overline{AC}$ and $\overline{AE} \cong \overline{AD}$. The included angle between each of these pairs of congruent sides is $\angle A$, and $\angle A$ is congruent to itself by the reflexive property of congruence. Therefore, $\triangle ABE \cong \triangle ACD$ by SAS and $\overline{CD} \cong \overline{BE}$ because corresponding parts of congruent triangles are congruent. ■

EXAMPLE 2

Using the results of Example 1, find the length of $\overline{BE}$ if $BE = 5x - 9$ and $CD = x + 15$.

Solution
$$BE = CD$$
$$5x - 9 = x + 15$$
$$4x = 24$$
$$x = 6$$

$$BE = 5x - 9 \qquad\qquad CD = x + 15$$
$$= 5(6) - 9 \qquad\qquad\quad = 6 + 15$$
$$= 30 - 9 \qquad\qquad\qquad = 21$$
$$= 21$$

Answer 21 ■

Exercises

Writing About Mathematics

1. In Example 1, the medians to the legs of isosceles $\triangle ABC$ were proved to be congruent by proving $\triangle ABE \cong \triangle ACD$. Could the proof have been done by proving $\triangle DBC \cong \triangle ECB$? Justify your answer.

2. In Corollary 5.1b, we proved that the median to the base of an isosceles triangle is also the altitude to the base. If the median to a leg of an isosceles triangle is also the altitude to the leg of the triangle, what other type of triangle must this triangle be?

Developing Skills

3. *Given:* $\overline{AEFB}, \overline{AE} \cong \overline{FB}, \overline{DA} \cong \overline{CB}$, and $\angle A$ and $\angle B$ are right angles.

Prove: $\triangle DAF \cong \triangle CBE$ and $\overline{DF} \cong \overline{CE}$

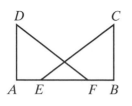

4. *Given:* $\overline{SPR} \cong \overline{SQT}, \overline{PR} \cong \overline{QT}$

Prove: $\triangle SRQ \cong \triangle STP$ and $\angle R \cong \angle T$

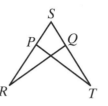

5. *Given:* $\overline{DA} \cong \overline{CB}, \overline{DA} \perp \overline{AB}$, and $\overline{CB} \perp \overline{AB}$

Prove: $\triangle DAB \cong \triangle CBA$ and $\overline{AC} \cong \overline{BD}$

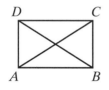

6. *Given:* $\overline{ABCD}, \angle BAE \cong \angle CBF$, $\angle BCE \cong \angle CDF, \overline{AB} \cong \overline{CD}$

Prove: $\overline{AE} \cong \overline{BF}$ and $\angle E \cong \angle F$

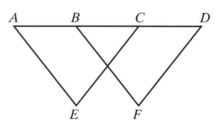

7. *Given:* $\overline{TM} \cong \overline{TN}$, M is the midpoint of $\overline{TR}$ and N is the midpoint of $\overline{TS}$.

Prove: $\overline{RN} \cong \overline{SM}$

8. *Given:* $\overline{AD} \cong \overline{CE}$ and $\overline{DB} \cong \overline{EB}$

Prove: $\angle ADC \cong \angle CEA$

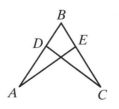

Applying Skills

In 9–11, complete each required proof in paragraph format.

9. Prove that the angle bisectors of the base angles of an isosceles triangle are congruent.

10. Prove that the triangle whose vertices are the midpoints of the sides of an isosceles triangle is an isosceles triangle.

11. Prove that the median to any side of a scalene triangle is *not* the altitude to that side.

5-6 PERPENDICULAR BISECTOR OF A LINE SEGMENT

Perpendicular lines were defined as lines that intersect to form right angles. We also proved that if two lines intersect to form congruent adjacent angles, then they are perpendicular. (Theorem 4.8)

The bisector of a line segment was defined as any line or subset of a line that intersects a line segment at its midpoint.

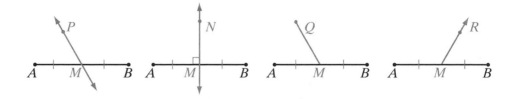

In the diagrams, $\overleftrightarrow{PM}$, $\overleftrightarrow{NM}$, $\overrightarrow{QM}$, and $\overrightarrow{MR}$ are all bisectors of $\overline{AB}$ since they each intersect $\overline{AB}$ at its midpoint, M. Only $\overleftrightarrow{NM}$ is both perpendicular to $\overline{AB}$ and the bisector of $\overline{AB}$. $\overleftrightarrow{NM}$ is the *perpendicular bisector* of $\overline{AB}$.

DEFINITION

> The **perpendicular bisector of a line segment** is any line or subset of a line that is perpendicular to the line segment at its midpoint.

In Section 3 of this chapter, we proved as a corollary to the isosceles triangle theorem that the median from the vertex angle of an isosceles triangle is perpendicular to the base.

In the diagram below, since $\overline{CM}$ is the median to the base of isosceles $\triangle ABC$, $\overline{CM} \perp \overline{AB}$. Therefore, $\overleftrightarrow{CM}$ is the perpendicular bisector of $\overline{AB}$.

(1) M is the midpoint of $\overline{AB}$: $AM = MB$.

> M is **equidistant**, or is at an equal distance, from the endpoints of $\overline{AB}$.

(2) $\overline{AB}$ is the base of isosceles $\triangle ABC$: $AC = BC$.

> C is equidistant from the endpoints of $\overline{AB}$.

These two points, M and C, determine the perpendicular bisector of $\overline{AB}$. This suggests the following theorem.

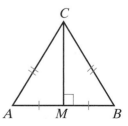

Theorem 5.2

> If two points are each equidistant from the endpoints of a line segment, then the points determine the perpendicular bisector of the line segment.

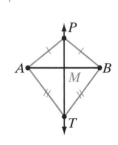

Given $\overline{AB}$ and points P and T such that $PA = PB$ and $TA = TB$.

Prove $\overleftrightarrow{PT}$ is the perpendicular bisector of $\overline{AB}$.

Strategy Let $\overleftrightarrow{PT}$ intersect $\overline{AB}$ at M. Prove $\triangle APT \cong \triangle BPT$ by SSS. Then, using the congruent corresponding angles, prove $\triangle APM \cong \triangle BPM$ by SAS. Consequently, $\overline{AM} \cong \overline{MB}$, so $\overleftrightarrow{PT}$ is a bisector. Also, $\angle AMP \cong \angle BMP$. Since two lines that intersect to form congruent adjacent angles are perpendicular, $\overline{AB} \perp \overleftrightarrow{PT}$. Therefore, $\overleftrightarrow{PT}$ is the perpendicular bisector of $\overline{AB}$.

The details of the proof of Theorem 5.2 will be left to the student. (See exercise 7.)

Theorem 5.3a

> If a point is equidistant from the endpoints of a line segment, then it is on the perpendicular bisector of the line segment.

Given Point P such that $PA = PB$.

Prove P lies on the perpendicular bisector of $\overline{AB}$.

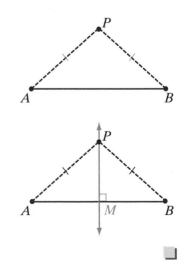

Proof Choose any other point that is equidistant from the endpoints of $\overline{AB}$, for example, M, the midpoint of $\overline{AB}$. Then $\overleftrightarrow{PM}$ is the perpendicular bisector of $\overline{AB}$ by Theorem 5.2. (If two points are each equidistant from the endpoints of a line segment, then the points determine the perpendicular bisector of the line segment.) P lies on $\overleftrightarrow{PM}$.

The converse of this theorem is also true.

Theorem 5.3b | If a point is on the perpendicular bisector of a line segment, then it is equidistant from the endpoints of the line segment.

Given Point P on the perpendicular bisector of $\overline{AB}$.

Prove $PA = PB$

Proof Let M be the midpoint of $\overline{AB}$. Then $\overline{AM} \cong \overline{BM}$ and $\overline{PM} \cong \overline{PM}$. Perpendicular lines intersect to form right angles, so $\angle PMA \cong \angle PMB$. By SAS, $\triangle PMA \cong \triangle PMB$. Since corresponding parts of congruent triangles are congruent, $\overline{PA} \cong \overline{PB}$ and $PA = PB$.

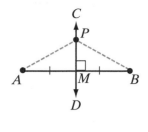

Theorems 5.3a and 5.3b can be written as a biconditional.

Theorem 5.3 | A point is on the perpendicular bisector of a line segment if and only if it is equidistant from the endpoints of the line segment.

Methods of Proving Lines or Line Segments Perpendicular

To prove that two intersecting lines or line segments are perpendicular, prove that one of the following statements is true:

1. The two lines form right angles at their point of intersection.

2. The two lines form congruent adjacent angles at their point of intersection.

3. Each of two points on one line is equidistant from the endpoints of a segment of the other.

Intersection of the Perpendicular Bisectors of the Sides of a Triangle

When we draw the three perpendicular bisectors of the sides of a triangle, it appears that the three lines are **concurrent**, that is, they intersect in one point.

Theorems 5.3a and 5.3b allow us to prove the following **perpendicular bisector concurrence theorem**.

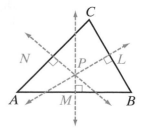

Theorem 5.4 | The perpendicular bisectors of the sides of a triangle are concurrent.

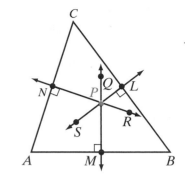

Given $\overleftrightarrow{MQ}$, the perpendicular bisector of $\overline{AB}$

$\overleftrightarrow{NR}$, the perpendicular bisector of $\overline{AC}$

$\overleftrightarrow{LS}$, the perpendicular bisector of $\overline{BC}$

Prove $\overleftrightarrow{MQ}$, $\overleftrightarrow{NR}$, and $\overleftrightarrow{LS}$ intersect in P.

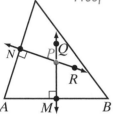

Proof (1) We can assume from the diagram that $\overleftrightarrow{MQ}$ and $\overleftrightarrow{NR}$ intersect. Let us call the point of intersection P.

(2) By theorem 5.3b, since P is a point on $\overleftrightarrow{MQ}$, the perpendicular bisector of $\overline{AB}$, P is equidistant from A and B.

(3) Similarly, since P is a point on $\overleftrightarrow{NR}$, the perpendicular bisector of $\overline{AC}$, P is equidistant from A and C.

(4) In other words, P is equidistant from A, B, and C. By theorem 5.3a, since P is equidistant from the endpoints of $\overline{BC}$, P is on the perpendicular bisector of $\overline{BC}$.

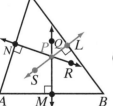

(5) Therefore, $\overleftrightarrow{MQ}$, $\overleftrightarrow{NR}$, and $\overleftrightarrow{LS}$, the three perpendicular bisectors of $\triangle ABC$, intersect in a point, P. ∎

The point where the three perpendicular bisectors of the sides of a triangle intersect is called the **circumcenter**.

EXAMPLE I

Prove that if a point lies on the perpendicular bisector of a line segment, then the point and the endpoints of the line segment are the vertices of an isosceles triangle.

Given: P lies on the perpendicular bisector of $\overline{RS}$.

Prove: $\triangle RPS$ is isosceles.

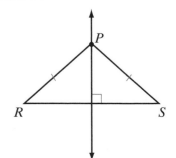

Proof	**Statements**	**Reasons**
	1. P lies on the perpendicular bisector of $\overline{RS}$.	**1.** Given.
	2. $PR = PS$	**2.** If a point is on the perpendicular bisector of a line segment, then it is equidistant from the endpoints of the line segment. (Theorem 5.3b)
	3. $\overline{PR} \cong \overline{PS}$	**3.** Segments that have the same measure are congruent.
	4. $\triangle RPS$ is isosceles.	**4.** An isosceles triangle is a triangle that has two congruent sides.

Exercises

Writing About Mathematics

1. Justify the three methods of proving that two lines are perpendicular given in this section.

2. Compare and contrast Example 1 with Corollary 5.1b, "The median from the vertex angle of an isosceles triangle is perpendicular to the base."

Developing Skills

3. If $\overline{RS}$ is the perpendicular bisector of $\overline{ASB}$, prove that $\angle ARS \cong \angle BRS$.

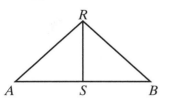

4. If $PR = PS$ and $QR = QS$, prove that $\overline{PQ} \perp \overline{RS}$ and $RT = ST$.

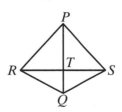

5. Polygon $ABCD$ is equilateral ($AB = BC = CD = DA$). Prove that $\overline{AC}$ and $\overline{BD}$ bisect each other and are perpendicular to each other.

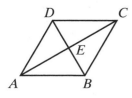

6. Given $\overline{CED}$ and $\overline{ADB}$ with $\angle ACE \cong \angle BCE$ and $\angle AED \cong \angle BED$, prove that $\overline{CED}$ is the perpendicular bisector of $\overline{ADB}$.

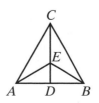

Applying Skills

7. Prove Theorem 5.2.

8. Prove that if the bisector of an angle of a triangle is perpendicular to the opposite side of the triangle, the triangle is isosceles.

9. A line through one vertex of a triangle intersects the opposite side of the triangle in adjacent angles whose measures are represented by $\frac{1}{2}a + 27$ and $\frac{3}{2}a - 15$. Is the line perpendicular to the side of the triangle? Justify your answer.

5-7 BASIC CONSTRUCTIONS

A **geometric construction** is a drawing of a geometric figure done using only a pencil, a *compass*, and a *straightedge*, or their equivalents. A **straightedge** is used to draw a line segment but is not used to measure distance or to determine equal distances. A **compass** is used to draw circles or arcs of circles to locate points at a fixed distance from given point.

The six constructions presented in this section are the basic procedures used for all other constructions. The following postulate allows us to perform these basic constructions:

Postulate 5.1 | Radii of congruent circles are congruent.

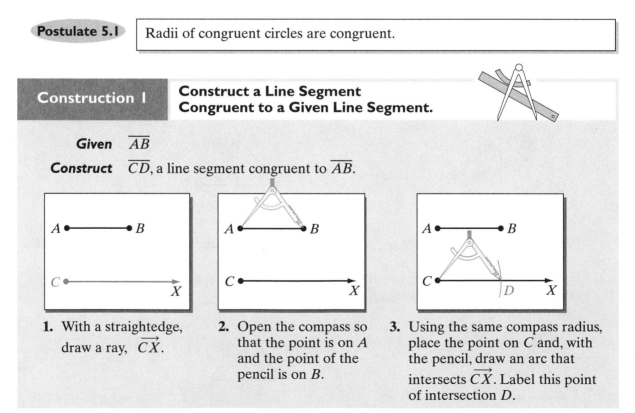

Construction 1 | **Construct a Line Segment Congruent to a Given Line Segment.**

Given $\overline{AB}$

Construct $\overline{CD}$, a line segment congruent to $\overline{AB}$.

1. With a straightedge, draw a ray, $\overrightarrow{CX}$.

2. Open the compass so that the point is on A and the point of the pencil is on B.

3. Using the same compass radius, place the point on C and, with the pencil, draw an arc that intersects $\overrightarrow{CX}$. Label this point of intersection D.

Conclusion $\overline{CD} \cong \overline{AB}$

Proof   Since $\overline{AB}$ and $\overline{CD}$ are radii of congruent circles, they are congruent.

Construction 2	**Construct an Angle Congruent to a Given Angle.**

Given $\angle A$

Construct $\angle EDF \cong \angle BAC$

1. Draw a ray with endpoint D.

2. With A as center, draw an arc that intersects each ray of $\angle A$. Label the points of intersection B and C. Using the same radius, draw an arc with D as the center that intersects the ray from D at E.

3. With E as the center, draw an arc with radius equal to BC that intersects the arc drawn in step 3. Label the intersection F.

4. Draw $\overrightarrow{DF}$.

Conclusion $\angle EDF \cong \angle BAC$

Proof   We used congruent radii to draw $\overline{AC} \cong \overline{DF}$, $\overline{AB} \cong \overline{DE}$, and $\overline{BC} \cong \overline{EF}$. Therefore, $\triangle DEF \cong \triangle ABC$ by SSS and $\angle EDF \cong \angle BAC$ because they are corresponding parts of congruent triangles.

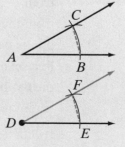

Construction 3	**Construct the Perpendicular Bisector of a Given Line Segment and the Midpoint of a Given Line Segment.**

Given $\overline{AB}$

Construct $\overleftrightarrow{CD} \perp \overline{AB}$ at M, the midpoint of $\overline{AB}$.

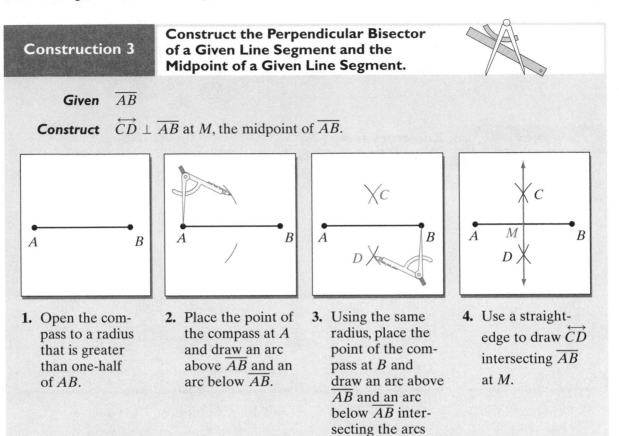

1. Open the compass to a radius that is greater than one-half of AB.

2. Place the point of the compass at A and draw an arc above $\overline{AB}$ and an arc below $\overline{AB}$.

3. Using the same radius, place the point of the compass at B and draw an arc above $\overline{AB}$ and an arc below $\overline{AB}$ intersecting the arcs drawn in step 2. Label the intersections C and D.

4. Use a straightedge to draw $\overleftrightarrow{CD}$ intersecting $\overline{AB}$ at M.

Conclusion $\overleftrightarrow{CD} \perp \overline{AB}$ at M, the midpoint of $\overline{AB}$.

Proof Since they are congruent radii, $\overline{AC} \cong \overline{BC}$ and $\overline{AD} \cong \overline{BD}$. Therefore, C and D are both equidistant from A and B. If two points are each equidistant from the endpoints of a line segment, then the points determine the perpendicular bisector of the line segment (Theorem 5.2). Thus, $\overleftrightarrow{CD}$ is the perpendicular bisector of $\overline{AB}$. Finally, M is the point on $\overline{AB}$ where the perpendicular bisector intersects $\overline{AB}$, so $AM = BM$. By definition, M is the midpoint of $\overline{AB}$.

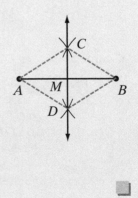

| Construction 4 | **Bisect a Given Angle.** |

Given $\angle ABC$

Construct $\overrightarrow{BF}$, the bisector of $\angle ABC$

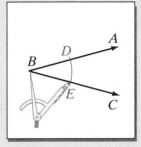 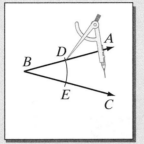 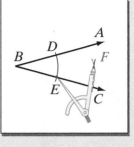 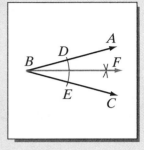

1. With B as center and any convenient radius, draw an arc that intersects $\overrightarrow{BA}$ at D and $\overrightarrow{BC}$ at E.

2. With D as center, draw an arc in the interior of $\angle ABC$.

3. Using the same radius, and with E as center, draw an arc that intersects the arc drawn in step 2. Label this intersection F.

4. Draw $\overrightarrow{BF}$.

Conclusion $\overrightarrow{BF}$ bisects $\angle ABC$; $\angle ABF \cong \angle FBC$.

Proof We used congruent radii to draw $\overline{BD} \cong \overline{BE}$ and $\overline{DF} \cong \overline{EF}$. By the reflexive property of congruence, $\overline{BF} \cong \overline{BF}$ so by SSS, $\triangle FBD \cong \triangle FBE$. Therefore, $\angle ABF \cong \angle FBC$ because they are corresponding parts of congruent triangles. Then $\overrightarrow{BF}$ bisects $\angle ABC$ because an angle bisector separates an angle into two congruent angles.

Construction 5 will be similar to Construction 4. In Construction 4, any given angle is bisected. In Construction 5, $\angle APB$ is a straight angle that is bisected by the construction. Therefore, $\angle APE$ and $\angle BPE$ are right angles and $\overleftrightarrow{PE} \perp \overleftrightarrow{AB}$.

| **Construction 5** | **Construct a Line Perpendicular to a Given Line Through a Given Point on the Line.** |

Given Point P on $\overleftrightarrow{AB}$.

Construct $\overleftrightarrow{PE} \perp \overleftrightarrow{AB}$

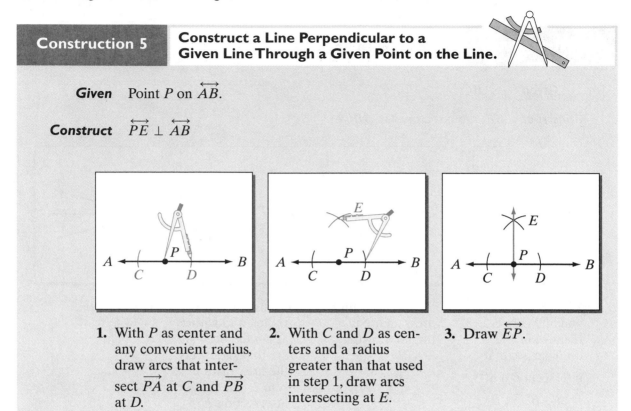

1. With P as center and any convenient radius, draw arcs that intersect $\overrightarrow{PA}$ at C and $\overrightarrow{PB}$ at D.

2. With C and D as centers and a radius greater than that used in step 1, draw arcs intersecting at E.

3. Draw $\overleftrightarrow{EP}$.

Conclusion $\overleftrightarrow{PE} \perp \overleftrightarrow{AB}$

Proof Since points C and D were constructed using congruent radii, $CP = PD$ and P is equidistant to C and D. Similarly, since E was constructed using congruent radii, $CE = ED$, and E is equidistant to C and D. If two points are each equidistant from the endpoints of a line segment, then the points determine the perpendicular bisector of the line segment (Theorem 5.2). Therefore, $\overleftrightarrow{PE}$ is the perpendicular bisector of $\overline{CD}$. Since $\overline{CD}$ is a subset of line $\overleftrightarrow{AB}$, $\overleftrightarrow{PE} \perp \overleftrightarrow{AB}$.

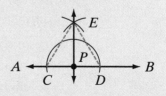

Construction 6	**Construct a Line Perpendicular to a Given Line Through a Point Not on the Given Line.**

Given   Point P not on $\overleftrightarrow{AB}$.

Construct $\overleftrightarrow{PE} \perp \overleftrightarrow{AB}$

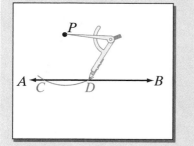

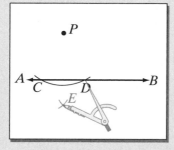

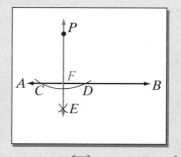

1. With P as center and any convenient radius, draw an arc that intersects $\overleftrightarrow{AB}$ in two points, C and D.

2. Open the compass to a radius greater than one-half of CD. With C and D as centers, draw intersecting arcs. Label the point of intersection E.

3. Draw $\overleftrightarrow{PE}$ intersecting $\overleftrightarrow{AB}$ at F.

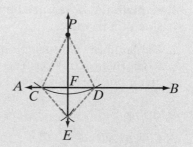

Conclusion $\overleftrightarrow{PE} \perp \overleftrightarrow{AB}$

Proof	**Statements**	**Reasons**
	1. $\overline{CP} \cong \overline{PD}, \overline{CE} \cong \overline{DE}$	1. Radii of congruent circles are congruent.
	2. $CP = PD, CE = DE$	2. Segments that are congruent have the same measure.
	3. $\overleftrightarrow{PE} \perp \overline{CD}$	3. If two points are each equidistant from the endpoints of a line segment, then the points determine the perpendicular bisector of the line segment. (Theorem 5.2)
	4. $\overleftrightarrow{PE} \perp \overleftrightarrow{AB}$	4. $\overline{CD}$ is a subset of line $\overleftrightarrow{AB}$.

EXAMPLE I

Construct the median to $\overline{AB}$ in $\triangle ABC$.

Construction A median of a triangle is a line segment that joins any vertex of the triangle to the midpoint of the opposite side. To construct the median to $\overline{AB}$, we must first find the midpoint of $\overline{AB}$.

1. Construct the perpendicular bisector of $\overline{AB}$ to locate the midpoint. Call the midpoint M.

2. Draw $\overline{CM}$.

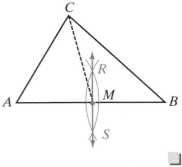

Conclusion $\overline{CM}$ is the median to $\overline{AB}$ in $\triangle ABC$.

Exercises

Writing About Mathematics

1. Explain the difference between the altitude of a triangle and the perpendicular bisector of a side of a triangle.

2. Explain how Construction 3 (Construct the perpendicular bisector of a given segment) and Construction 6 (Construct a line perpendicular to a given line through a point not on the given line) are alike and how they are different.

Developing Skills

3. *Given:* $\overline{AB}$

   Construct:

   a. A line segment congruent to $\overline{AB}$.

   b. A line segment whose measure is $2AB$.

   c. The perpendicular bisector of $\overline{AB}$.

   d. A line segment whose measure is $1\frac{1}{2}AB$.

4. *Given:* $\angle A$

   Construct:

   a. An angle congruent to $\angle A$.

   b. An angle whose measure is $2m\angle A$.

   c. The bisector of $\angle A$.

   d. An angle whose measure is $2\frac{1}{2}m\angle A$.

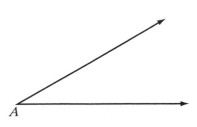

5. *Given:* Line segment $\overline{ABCD}$.

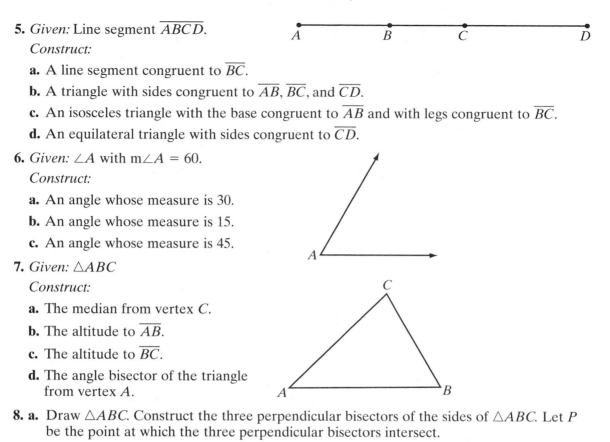

   Construct:

   a. A line segment congruent to $\overline{BC}$.

   b. A triangle with sides congruent to $\overline{AB}$, $\overline{BC}$, and $\overline{CD}$.

   c. An isosceles triangle with the base congruent to $\overline{AB}$ and with legs congruent to $\overline{BC}$.

   d. An equilateral triangle with sides congruent to $\overline{CD}$.

6. *Given:* $\angle A$ with m$\angle A = 60$.

   Construct:

   a. An angle whose measure is 30.

   b. An angle whose measure is 15.

   c. An angle whose measure is 45.

7. *Given:* $\triangle ABC$

   Construct:

   a. The median from vertex C.

   b. The altitude to $\overline{AB}$.

   c. The altitude to $\overline{BC}$.

   d. The angle bisector of the triangle from vertex A.

8. a. Draw $\triangle ABC$. Construct the three perpendicular bisectors of the sides of $\triangle ABC$. Let P be the point at which the three perpendicular bisectors intersect.

   b. Is it possible to draw a circle that passes through each of the vertices of the triangle? Explain your answer.

Hands-On Activity

Compass and straightedge constructions can also be done on the computer by using only the point, line segment, line, and circle creation tools of your geometry software and *no other software tools.*

   Working with a partner, use either a compass and straightedge, or geometry software to complete the following constructions:

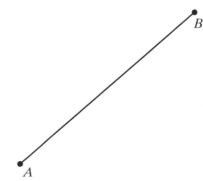

a. A square with side $\overline{AB}$.

b. An equilateral triangle with side $\overline{AB}$.

c. 45° angle ABD.

d. 30° angle ABD.

e. A circle passing through points A, B and C. (*Hint:* See the proof of Theorem 5.4 or use Theorem 5.3.)

CHAPTER SUMMARY

Definitions
to Know

- An **altitude of a triangle** is a line segment drawn from any vertex of the triangle, perpendicular to and ending in the line that contains the opposite side.

- A **median of a triangle** is a line segment that joins any vertex of the triangle to the midpoint of the opposite side.

- An **angle bisector of a triangle** is a line segment that bisects any angle of the triangle and terminates in the side opposite that angle.

- The **perpendicular bisector of a line segment** is a line, a line segment, or a ray that is perpendicular to the line segment at its midpoint.

Postulates

5.1 Radii of congruent circles are congruent.

Theorems and
Corollaries

5.1 If two sides of a triangle are congruent, the angles opposite these sides are congruent.

5.1a The median from the vertex angle of an isosceles triangle bisects the vertex angle.

5.1b The median from the vertex angle of an isosceles triangle is perpendicular to the base.

5.1c Every equilateral triangle is equiangular.

5.2 If two points are each equidistant from the endpoints of a line segment, then the points determine the perpendicular bisector of the line segment.

5.3 A point is on the perpendicular bisector of a line segment if and only if it is equidistant from the endpoints of the line segment.

5.4 The perpendicular bisectors of the sides of a triangle are concurrent.

VOCABULARY

5-1 Altitude of a triangle • Median of a triangle • Angle bisector of a triangle

5-3 Isosceles triangle theorem • Corollary

5-6 Perpendicular bisector of a line segment • Equidistant • Concurrent • Perpendicular bisector concurrence theorem • Circumcenter

5-7 Geometric construction • Straightedge • Compass

REVIEW EXERCISES

1. If $\overleftrightarrow{LMN} \perp \overline{KM}$, m$\angle LMK = x + y$ and m$\angle KMN = 2x - y$, find the value of x and of y.

2. The bisector of $\angle PQR$ in $\triangle PQR$ is $\overline{QS}$. If m$\angle PQS = x + 20$ and m$\angle SQR = 5x$, find m$\angle PQR$.

3. In $\triangle ABC$, $\overline{CD}$ is both the median and the altitude. If $AB = 5x + 3$, $AC = 2x + 8$, and $BC = 3x + 5$, what is the perimeter of $\triangle ABC$?

4. Angle PQS and angle SQR are a linear pair of angles. If $m\angle PQS = 5a + 15$ and $m\angle SQR = 8a + 35$, find $m\angle PQS$ and $m\angle SQR$.

5. Let D be the point at which the three perpendicular bisectors of the sides of equilateral $\triangle ABC$ intersect. Prove that $\triangle ADB$, $\triangle BDC$, and $\triangle CDA$ are congruent isosceles triangles.

6. Prove that if the median, $\overline{CD}$, to side $\overline{AB}$ of $\triangle ABC$ is not the altitude to side $\overline{AB}$, then $\overline{AC}$ is not congruent to $\overline{BC}$.

7. $\overline{AB}$ is the base of isosceles $\triangle ABC$ and $\overline{AB}$ is also the base of isosceles $\triangle ABD$. Prove that $\overleftrightarrow{CD}$ is the perpendicular bisector of $\overline{AB}$.

8. In $\triangle ABC$, $\overline{CD}$ is the median to $\overline{AB}$ and $\overline{CD} \cong \overline{DB}$. Prove that $m\angle A + m\angle B = m\angle ACB$. (*Hint:* Use Theorem 5.1, "If two sides of a triangle are congruent, the angles opposite these sides are congruent.")

9. **a.** Draw a line, $\overleftrightarrow{ADB}$. Construct $\overline{CD} \perp \overleftrightarrow{ADB}$.

   b. Use $\angle ADC$ to construct $\angle ADE$ such that $m\angle ADE = 45$.

   c. What is the measure of $\angle EDC$?

   d. What is the measure of $\angle EDB$?

10. **a.** Draw obtuse $\triangle PQR$ with the obtuse angle at vertex Q.

    b. Construct the altitude from vertex P.

Exploration

As you have noticed, proofs may be completed using a variety of methods. In this activity, you will explore the reasoning of others.

1. Complete a two-column proof of the following:

   Points L, M, and N separate $\overline{AB}$ into four congruent segments. Point C is not on $\overline{AB}$ and $\overline{CM}$ is an altitude of $\triangle CLM$. Prove that $\overline{CA} \cong \overline{CB}$.

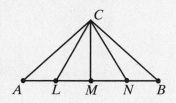

2. Cut out each *statement* and each *reason* from your proof, omitting the step numbers.

3. Trade proofs with a partner.

4. Attempt to reassemble your partner's proof.

5. Answer the following questions:

   a. Were you able to reassemble the proof?

   b. Did the reassembled proof match your partner's original proof?

   c. Did you find any components missing or have any components left-over? Why?

Part I

Answer all questions in this part. Each correct answer will receive 2 credits. No partial credit will be allowed.

1. The symbol $\overleftrightarrow{ABC}$ represents
  (1) a line segment with B between A and C.
  (2) a line with B the midpoint of $\overline{AB}$.
  (3) a line with B between A and C.
  (4) a ray with endpoint A.

2. A triangle with no two sides congruent is
  (1) a right triangle. (3) an isosceles triangle.
  (2) an equilateral triangle. (4) a scalene triangle.

3. Opposite rays have
  (1) no points in common. (3) two points in common.
  (2) one point in common. (4) all points in common.

4. The equality $a + 1 = 1 + a$ is an illustration of
  (1) the commutative property of addition.
  (2) the additive inverse property.
  (3) the multiplicative identity property.
  (4) the closure property of addition.

5. The solution set of the equation $1.5x - 7 = 0.25x + 8$ is
  (1) 120 (2) 12 (3) 11 (4) 1.2

6. What is the inverse of the statement "When spiders weave their webs by noon, fine weather is coming soon"?
  (1) When spiders do not weave their webs by noon, fine weather is not coming soon.
  (2) When fine weather is coming soon, then spiders weave their webs by noon.
  (3) When fine weather is not coming soon, spiders do not weave their webs by noon.
  (4) When spiders weave their webs by noon, fine weather is not coming soon.

7. If $\angle ABC$ and $\angle CBD$ are a linear pair of angles, then they must be
  (1) congruent angles. (3) supplementary angles.
  (2) complementary angles. (4) vertical angles.

8. Which of the following is *not* an abbreviation for a postulate that is used to prove triangles congruent?

(1) SSS   (2) SAS   (3) ASA   (4) SSA

9. If the statement "If two angles are right angles, then they are congruent" is true, which of the following statements must also be true?

(1) If two angles are not right angles, then they are not congruent.

(2) If two angles are congruent, then they are right angles.

(3) If two angles are not congruent, then they are not right angles.

(4) Two angles are congruent only if they are right angles.

10. If $\overleftrightarrow{ABC}$ is a line, which of the following may be false?

(1) B is on $\overline{AC}$.   (3) B is between A and C.

(2) $\overline{AB} + \overline{BC} = \overline{AC}$ (4) B is the midpoint of $\overline{AC}$.

Part II

Answer all questions in this part. Each correct answer will receive 2 credits. Clearly indicate the necessary steps, including appropriate formula substitutions, diagrams, graphs, charts, etc. For all questions in this part, a correct numerical answer with no work shown will receive only 1 credit.

11. Angle PQS and angle SQR are a linear pair of angles. If $m\angle PQS = 3a + 18$ and $m\angle SQR = 7a - 2$, find the measure of each angle of the linear pair.

12. Give a reason for each step in the solution of the given equation.

$$5(4 + x) = 32 - x$$
$$20 + 5x = 32 - x$$
$$20 + 5x + x = 32 - x + x$$
$$20 + 6x = 32 + 0$$
$$20 + 6x = 32$$
$$6x = 12$$
$$x = 2$$

Part III

Answer all questions in this part. Each correct answer will receive 4 credits. Clearly indicate the necessary steps, including appropriate formula substitutions, diagrams, graphs, charts, etc. For all questions in this part, a correct numerical answer with no work shown will receive only 1 credit.

13. *Given:* $\overline{AE} \cong \overline{BF}$, $\angle ABF$ is the
supplement of $\angle A$, and
$\overline{AB} \cong \overline{CD}$.

Prove: $\triangle AEC \cong \triangle BFD$

14. *Given:* $\overline{AB} \cong \overline{CB}$ and E is any
point on $\overrightarrow{BD}$, the bisector
of $\angle ABC$.

Prove: $\overline{AE} \cong \overline{CE}$

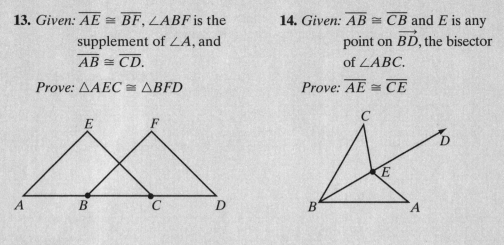

Part IV

Answer all questions in this part. Each correct answer will receive 6 credits. Clearly indicate the necessary steps, including appropriate formula substitutions, diagrams, graphs, charts, etc. For all questions in this part, a correct numerical answer with no work shown will receive only 1 credit.

15. *Given:* Point P is not on $\overline{ABCD}$ and
$PB = PC$.

Prove: $\angle ABP \cong \angle DCP$

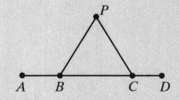

16. Prove that if $\overline{AC}$ and $\overline{BD}$ are perpendicular bisectors of each other, quadrilateral $ABCD$ is equilateral ($AB = BC = CD = DA$).

TRANSFORMATIONS AND THE COORDINATE PLANE

In our study of mathematics, we are accustomed to representing an equation as a line or a curve. This blending of geometry and algebra was not always familiar to mathematicians. In the seventeenth century, René Descartes (1596–1650), a French philosopher and mathematician, applied algebraic principles and methods to geometry. This blending of algebra and geometry is known as coordinate geometry or analytic geometry. Pierre de Fermat (1601–1665) independently developed analytic geometry before Descartes but Descartes was the first to publish his work.

Descartes showed that a curve in a plane could be completely determined if the distances of its points from two fixed perpendicular lines were known. We call these distances the Cartesian coordinates of the points; thus giving Descartes' name to the system that he developed.

6-1 THE COORDINATES OF A POINT IN A PLANE

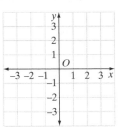

In this chapter, we will review what you know about the coordinate plane and you will study *transformations* and *symmetry*, which are common elements in nature, art, and architecture.

Two intersecting lines determine a plane. The **coordinate plane** is determined by a horizontal line, the ***x*-axis**, and a vertical line, the ***y*-axis**, which are perpendicular and intersect at a point called the **origin**. Every point on a plane can be described by two numbers, called the **coordinates** of the point, usually written as an **ordered pair**. The first number in the pair, called the ***x*-coordinate** or the **abscissa**, is the distance from the point to the *y*-axis. The second number, the ***y*-coordinate** or the **ordinate** is the distance from the point to the *x*-axis. In general, the coordinates of a point are represented as (x, y). Point O, the origin, has the coordinates $(0, 0)$.

We will accept the following postulates of the coordinate plane.

Postulate 6.1   Two points are on the same horizontal line if and only if they have the same *y*-coordinates.

Postulate 6.2   The length of a horizontal line segment is the absolute value of the difference of the *x*-coordinates.

Postulate 6.3   Two points are on the same vertical line if and only if they have the same *x*-coordinate.

Postulate 6.4   The length of a vertical line segment is the absolute value of the difference of the *y*-coordinates.

Postulate 6.5   Each vertical line is perpendicular to each horizontal line.

Locating a Point in the Coordinate Plane

An ordered pair of signed numbers uniquely determines the location of a point in the plane.

To locate a point in the coordinate plane:

1. From the origin, move along the x-axis the number of units given by the x-coordinate. Move to the right if the number is positive or to the left if the number is negative. If the x-coordinate is 0, there is no movement along the x-axis.

2. Then, from the point on the x-axis, move parallel to the y-axis the number of units given by the y-coordinate. Move up if the number is positive or down if the number is negative. If the y-coordinate is 0, there is no movement in the y direction.

For example, to locate the point $A(-3, -4)$, from O, move 3 units to the left along the x-axis, then 4 units down, parallel to the y-axis.

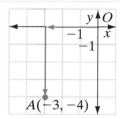

Finding the Coordinates of a Point in a Plane

The location of a point in the coordinate plane uniquely determines the coordinates of the point.

To find the coordinates of a point:

1. From the point, move along a vertical line to the x-axis. The number on the x-axis is the x-coordinate of the point.

2. From the point, move along a horizontal line to the y-axis. The number on the y-axis is the y-coordinate of the point.

For example, from point R, move in the vertical direction to 5 on the x-axis and in the horizontal direction to -6 on the y-axis. The coordinates of R are $(5, -6)$.

Note: The coordinates of a point are often referred to as **rectangular coordinates**.

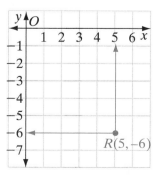

Graphing Polygons

A quadrilateral (a four-sided polygon) can be represented in the coordinate plane by locating its vertices and then drawing the sides connecting the vertices in order.

The graph shows the quadrilateral $ABCD$. The vertices are $A(3, 2)$, $B(-3, 2)$, $C(-3, -2)$ and $D(3, -2)$.

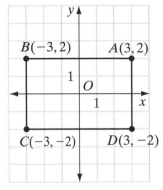

1. Points A and B have the same y-coordinate and are on the same horizontal line.

2. Points C and D have the same y-coordinate and are on the same horizontal line.

3. Points B and C have the same x-coordinate and are on the same vertical line.

4. Points A and D have the same x-coordinate and are on the same vertical line.

5. Every vertical line is perpendicular to every horizontal line.

6. Perpendicular lines are lines that intersect to form right angles.

Each angle of the quadrilateral is a right angle:

$$m\angle A = m\angle B = m\angle C = m\angle D = 90$$

From the graph, we can find the dimensions of this quadrilateral. To find AB and CD, we can count the number of units from A to B or from C to D.

$$AB = CD = 6$$

Points on the same horizontal line have the same y-coordinate. Therefore, we can also find AB and CD by subtracting their x-coordinates.

$$AB = CD = 3 - (-3)$$
$$= 3 + 3$$
$$= 6$$

To find BC and DA, we can count the number of units from B to C or from D to A.

$$BC = DA = 4$$

Points on the same vertical line have the same x-coordinate. Therefore, we can find BC and DA by subtracting their y-coordinates.

$$BC = DA = 2 - (-2)$$
$$= 2 + 2$$
$$= 4$$

EXAMPLE 1

Graph the following points: $A(4, 1)$, $B(1, 5)$, $C(-2,1)$. Then draw $\triangle ABC$ and find its area.

Solution   The graph shows $\triangle ABC$.

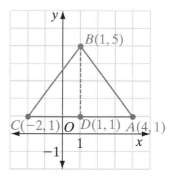

To find the area of the triangle, we need to know the lengths of the base and of the altitude drawn to that base. The base of $\triangle ABC$ is $\overline{AC}$, a horizontal line segment.

$$AC = 4 - (-2)$$
$$= 4 + 2$$
$$= 6$$

The vertical line segment drawn from B perpendicular to $\overline{AC}$ is the altitude $\overline{BD}$.

$$BD = 5 - 1$$
$$= 4$$
$$\text{Area} = \tfrac{1}{2}(AC)(BD)$$
$$= \tfrac{1}{2}(6)(4)$$
$$= 12$$

Answer   The area of $\triangle ABC$ is 12 square units.

Exercises

Writing About Mathematics

1. Mark is drawing isosceles right triangle ABC on the coordinate plane. He locates points $A(-2, -4)$ and $C(5, -4)$. He wants the right angle to be $\angle C$. What must be the coordinates of point B? Explain how you found your answer.

2. Phyllis graphed the points $D(3, 0)$, $E(0, 5)$, $F(-2, 0)$, and $G(0, -4)$ on the coordinate plane and joined the points in order. Explain how Phyllis can find the area of this polygon.

Developing Skills

In 3–12: **a.** Graph the points and connect them with straight lines in order, forming a polygon. **b.** Find the area of the polygon.

3. $A(1, 1), B(8, 1), C(1, 5)$
4. $P(0, 0), Q(5, 0), R(5, 4), S(0, 4)$
5. $C(8, -1), A(9, 3), L(4, 3), F(3, -1)$
6. $H(-4, 0), O(0, 0), M(0, 4), E(-4, 4)$

7. $H(5, -3), E(5, 3), N(-2, 0)$

8. $F(5, 1), A(5, 5), R(0, 5), M(-2, 1)$

9. $B(-3, -2), A(2, -2), R(2, 2), N(-3, 2)$

10. $P(-3, 0), O(0, 0), N(2, 2), D(-1, 2)$

11. $R(-4, 2), A(0, 2), M(0, 7)$

12. $M(-1, -1), I(3, -1), L(3, 3), K(-1, 3)$

13. Graph points $A(1, 1)$, $B(5, 1)$, and $C(5, 4)$. What must be the coordinates of point D if $ABCD$ is a quadrilateral with four right angles?

14. Graph points $P(-1, -4)$ and $Q(2, -4)$. What are the coordinates of R and S if $PQRS$ is a quadrilateral with four right angles and four congruent sides? (Two answers are possible.)

15. a. Graph points $S(3, 0)$, $T(0, 4)$, $A(-3, 0)$, and $R(0, -4)$, and draw the polygon $STAR$.

b. Find the area of $STAR$ by adding the areas of the triangles into which the axes separate the polygon.

16. a. Graph points $P(2, 0)$, $L(1, 1)$, $A(-1, 1)$, $N(-2, 0)$. $E(-1, -1)$, and $T(1, -1)$, and draw the hexagon $PLANET$.

b. Find the area of $PLANET$. (*Hint:* Use two vertical lines to separate the hexagon into parts.)

6-2 LINE REFLECTIONS

It is often possible to see the objects along the shore of a body of water reflected in the water. If a picture of such a scene is folded, the objects can be made to coincide with their images. Each point of the reflection is an image point of the corresponding point of the object.

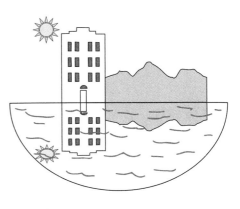

The line along which the picture is folded is the **line of reflection**, and the correspondence between the object points and the image points is called a **line reflection**. This common experience is used in mathematics to study congruent figures.

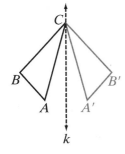

If the figure at the left were folded along line k, $\triangle ABC$ would coincide with $\triangle A'B'C$. Line k is the line of reflection, point A corresponds to point A' (in symbols, $A \rightarrow A'$) and point B corresponds to point B' ($B \rightarrow B'$). Point C is a **fixed point** because it is a point on the line of reflection. In other words, C corresponds to itself ($C \rightarrow C$). Under a reflection in line k, then, $\triangle ABC$ corresponds to $\triangle A'B'C$ ($\triangle ABC \rightarrow \triangle A'B'C$). Each of the points A, B, and C is called a *preimage* and each of the points A', B', and C is called an *image*.

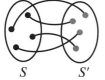

DEFINITION _____

A **transformation** is a one-to-one correspondence between two sets of points, S and S', such that every point in set S corresponds to one and only one point in set S', called its **image**, and every point in S' is the image of one and only one point in S, called its **preimage**.

The sets S and S' can be the same set and that set of points is frequently the set of points in a plane. For example, let S be the set of points in the coordinate plane. Let the image of (x, y), a point in S, be $(2 - x, y)$, a point in S'. Under this transformation:

$$(0, 1) \rightarrow (2 - 0, 1) \quad = (2, 1)$$
$$(-4, 3) \rightarrow (2 - (-4), 3) = (-2, 3)$$
$$(5, -1) \rightarrow (2 - 5, -1) = (-3, -1)$$

Every image $(2 - x, y)$ in S' is a point of the coordinate plane, that is, a point of S. Therefore, $S' = S$.

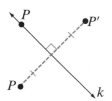

DEFINITION _____

A **reflection in line k** is a transformation in a plane such that:
1. If point P is not on k, then the image of P is P' where k is the perpendicular bisector of $\overline{PP'}$.
2. If point P is on k, the image of P is P.

From the examples that we have seen, it appears that the size and shape of the image is the same as the size and shape of the preimage. We want to prove that this is true.

Theorem 6.1	Under a line reflection, distance is preserved.

Given Under a reflection in line k, the image of A is A' and the image of B is B'.

Prove $AB = A'B'$

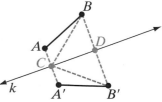

Proof We will prove this theorem by using the definition of a reflection and SAS to show that $\triangle BDC \cong \triangle B'DC$ and that $\overline{BC} \cong \overline{B'C}$. Then, using the fact that if two angles are congruent, their complements are congruent (Theorem 4.4), we can show that $\triangle ACB \cong \triangle A'CB'$. From this last statement, we can conclude that $AB = A'B'$.

(1) Let the image of A be A' and the image of B be B' under reflection in k. Let C be the midpoint of $\overline{AA'}$ and D be the midpoint of $\overline{BB'}$. Points C and D are on k, since k is the perpendicular bisector of $\overline{AA'}$ and of $\overline{BB'}$.

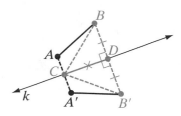

(2) $\overline{BD} \cong \overline{B'D}$ since D is the midpoint of $\overline{BB'}$, $\angle BDC \cong \angle B'DC$ since perpendicular lines intersect to form right angles, and $\overline{CD} \cong \overline{CD}$ by the reflexive property. Thus, $\triangle BDC \cong \triangle B'DC$ by SAS.

(3) From step 2, we can conclude that $\overline{BC} \cong \overline{B'C}$ and $\angle BCD \cong \angle B'CD$ since corresponding parts of congruent triangles are congruent.

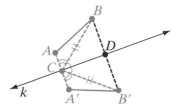

(4) Since k is the perpendicular bisector of $\overline{AA'}$, $\angle ACD$ and $\angle A'CD$ are right angles. Thus, $\angle ACB$ and $\angle BCD$ are complementary angles. Similarly, $\angle A'CB'$ and $\angle B'CD$ are complementary angles. If two angles are congruent, their complements are congruent. Since $\angle BCD$ and $\angle B'CD$ were shown to be congruent in step 3, their complements are congruent: $\angle ACB \cong \angle A'CB'$.

(5) By SAS, $\triangle ACB \cong \triangle A'CB'$. Since corresponding parts of congruent triangles are congruent, $\overline{AB} \cong \overline{A'B'}$. Therefore, $AB = A'B'$ by the definition of congruent segments.

A similar proof can be given when A and B are on opposite sides of the line k. The proof is left to the student. (See exercise 11.) ◾

Since distance is preserved under a line reflection, we can prove that the image of a triangle is a congruent triangle and that angle measure, collinearity, and midpoint are also preserved. For each of the following corollaries, the image of A is A', the image of B is B', and the image of C is C'. Therefore, in the diagram, $\triangle ABC \cong \triangle A'B'C'$ by SSS. *We will use this fact to prove the following corollaries:*

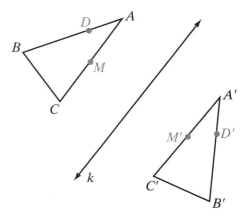

Corollary 6.1a | Under a line reflection, angle measure is preserved.

Proof: We found that $\angle ABC \cong \angle A'B'C'$. Therefore, $\angle ABC \cong \angle A'B'C'$ because the angles are corresponding parts of congruent triangles. ▪

Corollary 6.1b | Under a line reflection, collinearity is preserved.

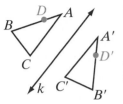

Proof: Let D be a point on $\overline{AB}$ whose image is D'. Since distance is preserved:

$$AD = A'D' \qquad DB = D'B' \qquad AB = A'B'$$

A, D, and B are collinear, D is between A and B, and $AD + DB = AB$. By substitution, $A'D' + D'B' = A'B'$.

If D' were not on $\overline{A'B'}$, $A'D' + D'B' > A'B'$ because a straight line is the shortest distance between two points. But by substitution, $A'D' + D'B' = A'B'$. Therefore, A', D' and B' are collinear and D' is between A' and B'. ▪

Corollary 6.1c | Under a line reflection, midpoint is preserved.

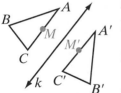

Proof: Let M be the midpoint of $\overline{AC}$ and M' the image of M. Since distance is preserved under a line reflection, $A'M' = AM$, and $M'C' = MC$.

Since M is the midpoint of $\overline{AC}$, $AM = MC$ and, by the substitution postulate, $A'M' = M'C'$.

Therefore, M' is the midpoint of $\overline{A'C'}$, that is, midpoint is preserved under a line reflection. ▪

We can summarize Theorem 6.1 and its corollaries in the following statement:

▶ **Under a line reflection, distance, angle measure, collinearity, and midpoint are preserved.**

We use r_k as a symbol for the image under a reflection in line k. For example,

$r_k(A) = B$ *means*   "The image of A under a reflection in line k is B."

$r_k(\triangle ABC) = \triangle A'B'C'$ *means*   "The image of $\triangle ABC$ under a reflection in line k is $\triangle A'B'C'$."

EXAMPLE I

If $r_k(\overline{CD}) = \overline{C'D'}$, construct $\overline{C'D'}$.

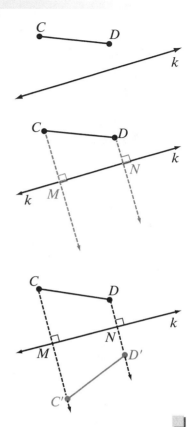

Construction

1. Construct the perpendicular line from C to k. Let the point of intersection be M.

2. Construct the perpendicular line from D to k. Let the point of intersection be N.

3. Construct point C' on $\overleftrightarrow{CM}$ such that $CM = MC'$ and point D' on $\overleftrightarrow{DN}$ such that $DN = ND'$.

4. Draw $\overline{C'D'}$.

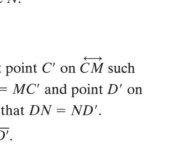

Line Symmetry

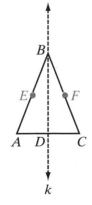

We know that the altitude to the base of an isosceles triangle is also the median and the perpendicular bisector of the base. If we imagine that isosceles triangle ABC, shown at the left, is folded along the perpendicular bisector of the base so that A falls on C, the line along which it folds, k, is a reflection line. Every point of the triangle has as its image a point of the triangle. Points B and D are fixed points because they are points of the line of reflection.

Thus, under the line reflection in k:

1. All points of $\triangle ABC$ are reflected so that

$$A \to C \quad C \to A \quad E \to F \quad F \to E \quad B \to B \quad D \to D$$

2. The sides of $\triangle ABC$ are reflected; that is, $\overline{AB} \to \overline{CB}$, a statement verifying that the legs of an isosceles triangle are congruent. Also, $\overline{AC} \to \overline{CA}$, showing that the base is its own image.

3. The angles of $\triangle ABC$ are reflected; that is, $\angle BAD \rightarrow \angle BCD$, a statement verifying that the base angles of an isosceles triangle are congruent. Also, $\angle ABC \rightarrow \angle CBA$, showing that the vertex angle is its own image.

We can note some properties of a line reflection by considering the reflection of isosceles triangle ABC in line k:

1. Distance is preserved (unchanged).

$$\overline{AB} \rightarrow \overline{CB} \text{ and } AB = CB \qquad \overline{AD} \rightarrow \overline{CD} \text{ and } AD = CD$$

2. Angle measure is preserved.

$$\angle BAD \rightarrow \angle BCD \text{ and } m\angle BAD = m\angle BCD$$
$$\angle BDA \rightarrow \angle BDC \text{ and } m\angle BDA = m\angle BDC$$

3. The line of reflection is the perpendicular bisector of every segment joining a point to its image. The line of reflection, $\overleftrightarrow{BD}$, is the perpendicular bisector of $\overline{AC}$.

4. A figure is always congruent to its image: $\triangle ABC \cong \triangle CBA$.

In nature, in art, and in industry, many forms have a pleasing, attractive appearance because of a balanced arrangement of their parts. We say that such forms have symmetry.

In each of the figures above, there is a line on which the figure could be folded so that the parts of the figure on opposite sides of the line would coincide. If we think of that line as a line of reflection, each point of the figure has as its image a point of the figure. This line of reflection is a line of symmetry, or an **axis of symmetry**, and the figure has *line symmetry*.

DEFINITION _____

A figure has **line symmetry** when the figure is its own image under a line reflection.

It is possible for a figure to have more than one axis of symmetry. In the square $PQRS$ at the right, $\overleftrightarrow{XY}$ is an axis of symmetry and $\overleftrightarrow{VW}$ is a second axis of symmetry. The diagonals, $\overline{PR}$ and $\overline{QS}$, are also segments of axes of symmetry.

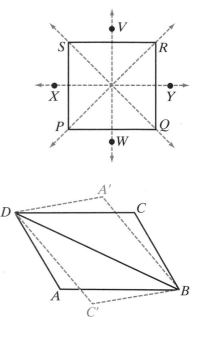

O H I O

T

Lines of symmetry may be found for some letters and for some words, as shown at the left.

Not every figure, however, has line symmetry. If the parallelogram $ABCD$ at the right is reflected in the line that contains the diagonal $\overline{BD}$, the image of A is A' and the image of C is C'. Points A' and C', however, are not points of the original parallelogram. The image of parallelogram $ABCD$ under a reflection in $\overleftrightarrow{BD}$ is $A'BC'D$. Therefore, $ABCD$ is not symmetric with respect to $\overleftrightarrow{BD}$.

We have used the line that contains the diagonal $\overline{BD}$ as a line of reflection, but note that it is not a line of symmetry. There is no line along which the parallelogram can be folded so that the image of every point of the parallelogram will be a point of the parallelogram under a reflection in that line.

EXAMPLE 2

How many lines of symmetry does the letter H have?

Solution The horizontal line through the crossbar is a line of symmetry. The vertical line midway between the vertical segments is also a line of symmetry.

Answer The letter H has two lines of symmetry.

Exercises

Writing About Mathematics

1. Explain how a number line illustrates a one-to-one correspondence between the set of points on a line and the set of real numbers.

2. Is the correspondence $(x, y) \rightarrow (2, y)$ a transformation? Explain why or why not.

Developing Skills

3. If $r_k(\triangle PQR) = \triangle P'Q'R'$ and $\triangle PQR$ is isosceles with $PQ = QR$, prove that $\triangle P'Q'R'$ is isosceles.

4. If $r_k(\triangle LMN) = \triangle L'M'N'$ and $\triangle LMN$ is a right triangle with $m\angle N = 90$, prove that $\triangle L'M'N'$ is a right triangle.

In 5–9, under a reflection in line k, the image of A is A', the image of B is B', the image of C is C', and the image of D is D.

5. Is $\triangle ABC \cong \triangle A'B'C'$? Justify your answer.

6. If $\overline{AC} \perp \overline{BC}$, is $\overline{A'C'} \perp \overline{B'C'}$? Justify your answer?

7. The midpoint of $\overline{AB}$ is M. If the image of M is M', is M' the midpoint of $\overline{A'B'}$? Justify your answer.

8. If A, M, and B lie on a line, do A', M', and B' lie on a line? Justify your answer.

9. Which point lies on the line of reflection? Justify your answer.

10. A triangle, $\triangle RST$, has line symmetry with respect to the altitude from S but does not have line symmetry with respect to the altitudes from R and T. What kind of a triangle is $\triangle RST$? Justify your answer.

Applying Skills

11. Write the proof of Theorem 6.1 for the case when points A and B are on opposite sides of line k.

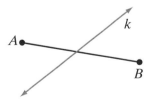

12. A baseball diamond is in the shape of a rhombus with sides 90 feet in length. Describe the lines of symmetry of a baseball diamond in terms of home plate and the three bases.

13. Print three letters that have line symmetry with respect to only one line and draw the line of symmetry.

14. Print three letters that have line symmetry with respect to exactly two lines and draw the lines of symmetry.

6-3 LINE REFLECTIONS IN THE COORDINATE PLANE

We can apply the definition of a line reflection to points in the coordinate plane.

Reflection in the y-axis

In the figure, $\triangle ABC$ is reflected in the y-axis. Its image under the reflection is $\triangle A'B'C'$. From the figure, we see that:

$$A(1, 2) \to A'(-1, 2)$$
$$B(3, 4) \to B'(-3, 4)$$
$$C(1, 5) \to C'(-1, 5)$$

For each point and its image under a reflection in the y-axis, the y-coordinate of the image is the same as the y-coordinate of the point; the x-coordinate of the image is the opposite of the x-coordinate of the point. Note that for a reflection in the y-axis, the image of $(1, 2)$ is $(-1, 2)$ and the image of $(-1, 2)$ is $(1, 2)$.

A reflection in the y-axis can be designated as $r_{y\text{-axis}}$. For example, if the image of $(1, 2)$ is $(-1, 2)$ under a reflection in the y-axis, we can write:

$$r_{y\text{-axis}}(1, 2) = (-1, 2)$$

From these examples, we form a general rule that can be proven as a theorem.

Theorem 6.2

Under a reflection in the y-axis, the image of $P(a, b)$ is $P'(-a, b)$.

Given A reflection in the y-axis.

Prove The image of $P(a, b)$ under a reflection in the y-axis is $P'(-a, b)$.

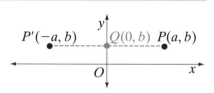

Proof By the definition of a reflection in a line, a point P' is the image of P under a reflection in a given line if and only if the line is the perpendicular bisector of $\overline{PP'}$. Therefore, we can prove that P' is the image of P under a reflection in the y-axis by showing that the y-axis is the perpendicular bisector of $\overline{PP'}$.

(1) *The y-axis is perpendicular to $\overline{PP'}$.* The line of reflection, the y-axis, is a vertical line. $P(a, b)$ and $P'(-a, b)$ have the same y-coordinates. $\overline{PP'}$ is a segment of a horizontal line because two points are on the same horizontal line if and only if they have the same y-coordinates. Every vertical line is perpendicular to every horizontal line. Therefore, the y-axis is perpendicular to $\overline{PP'}$.

(2) *The y-axis bisects $\overline{PP'}$.* Let Q be the point at which $\overline{PP'}$ intersects the y-axis. The x-coordinate of every point on the y-axis is 0. The length of a horizontal line segment is the absolute value of the difference of the x-coordinates of the endpoints.

$$PQ = |a - 0| = |a| \qquad \text{and} \qquad P'Q = |-a - 0| = |a|$$

Since $PQ = P'Q$, Q is the midpoint of $\overline{PP'}$ or the y-axis bisects $\overline{PP'}$.

Steps 1 and 2 prove that if P has the coordinates (a, b) and P' has the coordinates $(-a, b)$, the y-axis is the perpendicular bisector of $\overline{PP'}$, and therefore, the image of $P(a, b)$ is $P'(-a, b)$. ∎

Reflection in the *x*-axis

In the figure, $\triangle ABC$ is reflected in the x-axis. Its image under the reflection is $\triangle A'B'C'$. From the figure, we see that:

$$A(1, 2) \rightarrow A'(1, -2)$$
$$B(3, 4) \rightarrow B'(3, -4)$$
$$C(1, 5) \rightarrow C'(1, -5)$$

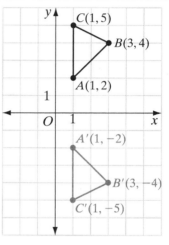

For each point and its image under a reflection in the x-axis, the x-coordinate of the image is the same as the x-coordinate of the point; the y-coordinate of the image is the opposite of the y-coordinate of the point. Note that for a reflection in the x-axis, the image of $(1, 2)$ is $(1, -2)$ and the image of $(1, -2)$ is $(1, 2)$.

A reflection in the x-axis can be designated as $r_{x\text{-axis}}$. For example, if the image of $(1, 2)$ is $(1, -2)$ under a reflection in the x-axis, we can write:

$$r_{x\text{-axis}}(1, 2) = (1, -2)$$

From these examples, we form a general rule that can be proven as a theorem.

Theorem 6.3

> Under a reflection in the x-axis, the image of $P(a, b)$ is $P'(a, -b)$.

The proof follows the same general pattern as that for a reflection in the y-axis. Prove that the x-axis is the perpendicular bisector of $\overline{PP'}$. The proof is left to the student. (See exercise 18.)

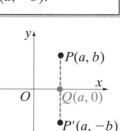

EXAMPLE 1

On graph paper:

Answers

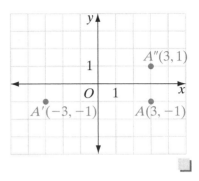

a. Locate $A(3, -1)$.

b. Locate A', the image of A under a reflection in the y-axis, and write its coordinates.

c. Locate A'', the image of A under a reflection in the x-axis, and write its coordinates.

Reflection in the Line $y = x$

In the figure, $\triangle ABC$ is reflected in the line $y = x$. Its image under the reflection is $\triangle A'B'C'$. From the figure, we see that:

$$A(1, 2) \rightarrow A'(2, 1)$$
$$B(3, 4) \rightarrow B'(4, 3)$$
$$C(1, 5) \rightarrow C'(5, 1)$$

For each point and its image under a reflection in the line $y = x$, the x-coordinate of the image is the y-coordinate of the point; the y-coordinate of the image is the x-coordinate of the point. Note that for a reflection in the line $y = x$, the image of $(1, 2)$ is $(2, 1)$ and the image of $(2, 1)$ is $(1, 2)$.

A reflection in the line $y = x$ can be designated as $r_{y=x}$. For example, if the image of $(1, 2)$ is $(2, 1)$ under a reflection in $y = x$, we can write:

$$r_{y=x}(1, 2) = (2, 1)$$

From these examples, we form a general rule that can be proven as a theorem.

Theorem 6.4 | Under a reflection in the line $y = x$, the image of $P(a, b)$ is $P'(b, a)$.

Given A reflection in the line whose equation is $y = x$.

Prove The image of $P(a, b)$ is $P'(b, a)$.

Proof For every point on the line $y = x$, the x-coordinate and the y-coordinate are equal. Locate the points $Q(a, a)$ and $R(b, b)$ on the line $y = x$.

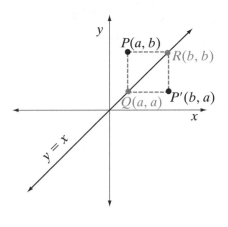

If two points have the same x-coordinate, then the distance between them is the absolute value of their y-coordinates and if two points have the same y-coordinate, then the distance between them is the absolute value of their x-coordinates.

$$PQ = |b - a| \text{ and } P'Q = |b - a|$$

Therefore, Q is equidistant from P and P'.

$$PR = |b - a| \text{ and } P'R = |b - a|$$

Therefore, R is equidistant from P and P'.

If two points are each equidistant from the endpoints of a line segment, they lie on the perpendicular bisector of the line segment (Theorem 5.3). Therefore, $\overline{RQ}$, which is a subset of the line $y = x$, is the perpendicular bisector of $\overline{PP'}$. By the definition of a line reflection across the line $y = x$, the image of $P(a, b)$ is $P'(b, a)$. ∎

EXAMPLE 2

The vertices of $\triangle DEF$ are $D(-2, -2)$, $E(0, -3)$, and $F(2, 0)$.

a. Draw $\triangle DEF$ on graph paper.

b. Draw the image of $\triangle DEF$ under a reflection in the line whose equation is $y = x$.

Answers

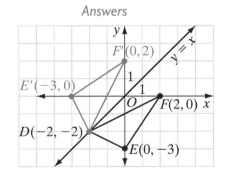

Exercises

Writing About Mathematics

1. When finding the distance from $P(a, b)$ to $Q(a, a)$, Allison wrote $PQ = |b - a|$ and Jacob wrote $PQ = |a - b|$. Explain why both are correct.

2. The image of point A is in the third quadrant under a reflection in the y-axis and in the first quadrant under a reflection in the x-axis. In what quadrant is the image of A under a reflection in the line $y = x$?

Developing Skills

In 3–7: **a.** On graph paper, locate each point and its image under $r_{x\text{-axis}}$. **b.** Write the coordinates of the image point.

   3. $(2, 5)$ **4.** $(1, 3)$ **5.** $(-2, 3)$ **6.** $(2, -4)$ **7.** $(0, 2)$

In 8–12: **a.** On graph paper, locate each point and its image under $r_{y\text{-axis}}$. **b.** Write the coordinates of the image point.

   8. $(3, 5)$ **9.** $(1, 4)$ **10.** $(2, -3)$ **11.** $(-2, 3)$ **12.** $(-1, 0)$

In 13–17: **a.** On graph paper, locate each point and its image under $r_{y=x}$. **b.** Write the coordinates of the image point.

   13. $(3, 5)$ **14.** $(-3, 5)$ **15.** $(4, -2)$ **16.** $(-1, -5)$ **17.** $(2, 2)$

Applying Skills

   18. Prove Theorem 6.3, "Under a reflection in the x-axis, the image of $P(a, b)$ is $P'(a, -b)$."

   19. When the points $A(-4, 0)$, $B(0, -4)$, $C(4, 0)$ and $D(0, 4)$ are connected in order, square $ABCD$ is drawn.

     a. Show that the line $y = x$ is a line of symmetry for the square.

     b. Show that the y-axis is a line of symmetry for the square.

   20. Show that the y-axis is not a line of symmetry for the rectangle whose vertices are $E(0, -3)$, $F(5, -3)$, $G(5, 3)$, and $H(0, 3)$.

   21. Write the equation of two lines that are lines of symmetry for the rectangle whose vertices are $E(0, -3)$, $F(6, -3)$, $G(6, 3)$, and $H(0, 3)$.

Hands-On Activity 1

In this activity, you will learn how to construct a reflection in a line using a compass and a straightedge, or geometry software. (*Note:* Compass and straightedge constructions can also be done on the computer by using only the point, line segment, line, and circle creation tools of your geometry software and no other software tools.)

STEP 1. Draw a line segment. Label the endpoints A and B. Draw a reflection line k.

STEP 2. Construct line l perpendicular to line k through point A. Let M be the point where lines l and k intersect.

STEP 3. Construct line segment $\overline{A'M}$ congruent to $\overline{AM}$ along line l. Using point M as the center, draw a circle with radius equal to AM. Let A' be the point where the circle intersects line l on the ray that is the opposite ray of $\overrightarrow{MA}$.

STEP 4. Repeat steps 2 and 3 for point B in order to construct B'.

STEP 5. Draw $\overline{A'B'}$.

Result: $\overline{A'B'}$ is the image of $\overline{AB}$ under a reflection in line k.

For each figure, construct the reflection in the given line.

a. Segment $\overline{AB}$ with vertices $A(-4, 2)$ and $B(2, 4)$, and line k through points $C(-2, 1)$ and $D(0, 5)$.

b. Angle EFG with vertices $E(4, 3)$, $F(1, 6)$, and $G(6, 6)$, and line l through points $H(1, 3)$ and $I(3, 1)$.

c. Triangle JKL with vertices $J(-2, 2)$, $K(-5, 1)$, and $L(-1, 4)$, and line p through points $M(-7, -3)$ and $N(2, 0)$.

d. Segments $\overline{PQ}$ and $\overline{RS}$ with vertices $P(-3, 4)$, $Q(3, 2)$, $R(-1, 0)$, and $S(1, 6)$, and line q through points $T(0, -3)$ and $U(3, 0)$.

Hands-On Activity 2

In each figure, one is the image of the other under a reflection in a line. For each figure, construct the reflection line using a compass and a straightedge.

a. • A

• A'

b. B B'

A A'

c. Triangle ABC with vertices $A(-1, 4)$, $B(4, 3)$, and $(-2, 8)$, and triangle $A'B'C'$ with vertices $A'(-3, 2)$, $B'(-2, -3)$, $C'(-7, 3)$.

d. Triangle ABC with vertices $A(0, 4)$, $B(7, 1)$, and $C(6, 6)$, and triangle $A'B'C'$ with vertices $A'(7, -3)$, $B'(4, 4)$, $C'(9, 3)$.

6-4 POINT REFLECTIONS IN THE COORDINATE PLANE

The figure at the right illustrates another type of reflection, a reflection in a point. In the figure, $\triangle A'B'C'$ is the image of $\triangle ABC$ under a reflection in point P. If a line segment is drawn connecting any point to its image, then the point of reflection is the midpoint of that segment. In the figure:

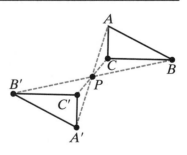

- Point A' is on $\overleftrightarrow{AP}$, $AP = PA'$, and P is the midpoint of $\overline{AA'}$.

- Point B' is on $\overleftrightarrow{BP}$, $BP = PB'$, and P is the midpoint of $\overline{BB'}$.

- Point C' is on $\overleftrightarrow{CP}$, $CP = PC'$, and P is the midpoint of $\overline{CC'}$.

DEFINITION

A **point reflection in P** is a transformation of the plane such that:

1. If point A is not point P, then the image of A is A' and P the midpoint of $\overline{AA'}$.

2. The point P is its own image.

Properties of Point Reflections

Looking at triangles ABC and $A'B'C'$ and point of reflection P, we observe some properties of a point reflection:

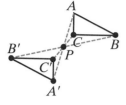

1. Distance is preserved: $AB = A'B'$.

2. Angle measure is preserved: $m\angle ACB = m\angle A'C'B'$.

3. A figure is always congruent to its image.

4. Collinearity is preserved. The image of any point on $\overline{AB}$ is a point on $\overline{A'B'}$.

5. Midpoint is preserved. The image of the midpoint of $\overline{AB}$ is the midpoint of $\overline{A'B'}$.

We can state the first property as a theorem and the remaining properties as corollaries to this theorem.

Theorem 6.5

> Under a point reflection, distance is preserved.

Given Under a reflection in point P, the image of A is A' and the image of B is B'.

Prove $AB = A'B'$

Proof Since $\angle APB$ and $\angle A'PB'$ are vertical angles, $\angle APB \cong \angle A'PB'$. In a point reflection, if a point X is not point P, then the image of X is X' and P the midpoint of $\overline{XX'}$. Therefore, P is the midpoint of both $\overline{AA'}$ and $\overline{BB'}$, and $\overline{BP} \cong \overline{BP'}$ and $\overline{AP} \cong \overline{PA'}$. By SAS, $\triangle APB \cong \triangle A'PB'$ and $AB = A'B'$.

The case when either point A or B is point P is left to the student. (See exercise 9.)

Corollary 6.5a

> Under a point reflection, angle measure is preserved.

Corollary 6.5b

> Under a point reflection, collinearity is preserved.

Corollary 6.5c

> Under a point reflection, midpoint is preserved.

We can prove that angle measure, collinearity, and midpoint are preserved using the same proofs that we used to prove the corollaries of Theorem 6.1. (See exercises 10–12.)

Theorem 6.5 and its corollaries can be summarized in the following statement.

▶ **Under a point reflection, distance, angle measure, collinearity, and midpoint are preserved.**

We use R_P as a symbol for the image under a reflection in point P. For example,

$R_P(A) = B$ *means*   "The image of A under a reflection in point P is B."

$R_{(1,2)}(A) = A'$ *means*   "The image of A under a reflection in point $(1, 2)$ is A'."

Point Symmetry

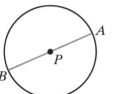

DEFINITION
A figure has **point symmetry** if the figure is its own image under a reflection in a point.

A circle is the most common example of a figure with point symmetry. Let P be the center of a circle, A be any point on the circle, and B be the other point at which $\overleftrightarrow{AP}$ intersect the circle. Since every point on a circle is equidistant from the center, $PA = PB$, P is the midpoint of $\overline{AB}$ and B, a point on the circle, is the image of A under a reflection in P.

Other examples of figures that have point symmetry are letters such as **S** and **N** and numbers such as **8**.

Point Reflection in the Coordinate Plane

In the coordinate plane, the origin is the most common point that is used to used to define a point reflection.

In the diagram, points $A(3, 5)$ and $B(-2, 4)$ are reflected in the origin. The coordinates of A', the image of A, are $(-3, -5)$ and the coordinates of B', the image of B, are $(2, -4)$. These examples suggest the following theorem.

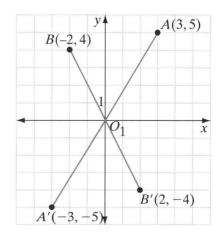

Theorem 6.6

Under a reflection in the origin, the image of $P(a, b)$ is $P'(-a, -b)$.

Given A reflection in the origin.

Prove Under a reflection in the origin, the image of $P(a, b)$ is $P'(-a, -b)$.

Proof Let P' be the image of $P(a, b)$ under a reflection in the origin, O. Then:

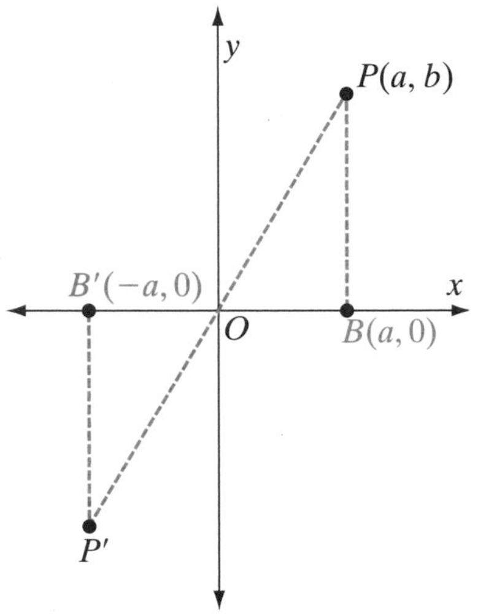

$$OP = OP' \text{ and } \overline{OP} \cong \overline{OP'}$$

Let $B(a, 0)$ be the point of intersection of the x-axis and a vertical line through P. Then:

$$OB = |0 - a| = |a|$$

Let B' be the point $(-a, 0)$. Then:

$$OB' = |0 - (-a)| = |a|$$
$$OB = OB' \text{ and } \overline{OB} \cong \overline{OB'}$$

$\angle POB \cong \angle P'OB'$ because vertical angles are congruent. Therefore, $\triangle POB \cong \triangle P'OB'$ by SAS. In particular,

$$PB = P'B' = |0 - b|$$
$$= |b|$$

Since $\angle OBP$ is a right angle, $\angle OB'P'$ is a right angle and $\overline{P'B'}$ is a vertical line. Therefore, P' has the same x-coordinate as B' and is b units in the opposite direction from the x-axis as P. The coordinates of P' are $(-a, -b)$.

Note: $-a$ and $-b$ are the opposites of a and b. The image of $(-4, 2)$ is $(4, -2)$ and the image of $(-5, -1)$ is $(5, 1)$.

We can symbolize the reflection in the origin as R_O. Therefore, we can write:

$$R_O(a, b) = (-a, -b)$$

EXAMPLE 1

a. What are the coordinates of B, the image of $A(-3, 2)$ under a reflection in the origin?

b. What are the coordinates of C, the image of $A(-3, 2)$ under a reflection in the x-axis?

c. What are the coordinates of D, the image of C under a reflection in the y-axis?

d. Does a reflection in the origin give the same result as a reflection in the x-axis followed by a reflection in the y-axis? Justify your answer.

Solution   **a.** Since $R_O(a, b) = (-a, -b)$, $R_O(-3, 2) = (3, -2)$. The coordinates of B are $(3, -2)$. *Answer*

b. Since $r_{x\text{-axis}}(a, b) = (a, -b)$, $r_{x\text{-axis}}(-3, 2) = (-3, -2)$. The coordinates of C are $(-3, -2)$. *Answer*

c. Since $r_{y\text{-axis}}(a, b) = (-a, b)$, $r_{y\text{-axis}}(-3, -2) = (3, -2)$. The coordinates of D are $(3, -2)$. *Answer*

d. For the point $A(-3, 2)$, a reflection in the origin gives the same result as a reflection in the x-axis followed by a reflection in the y-axis. In general,

$$R_O(a, b) = (-a, -b)$$

while

$$r_{x\text{-axis}}(a, b) = (a, -b)$$
$$r_{y\text{-axis}}(a, -b) = (-a, -b)$$

Therefore, a reflection in the origin gives the same result as a reflection in the x-axis followed by a reflection in the y-axis. *Answer*

Exercises

Writing About Mathematics

1. Ada said if the image of $\overline{AB}$ under a reflection in point P is $\overline{A'B'}$, then the image of $\overline{A'B'}$ under a reflection in point P is $\overline{AB}$. Do you agree with Ada? Justify your answer.

2. Lines l and m intersect at P. $\overline{A'B'}$ is the image of $\overline{AB}$ under a reflection in line l and $\overline{A''B''}$ is the image of $\overline{A'B'}$ under a reflection in line m. Is the result the same if $\overline{AB}$ is reflected in P? Justify your answer.

Developing Skills

In 3–8, give the coordinates of the image of each point under R_O.

3. $(1, 5)$ **4.** $(-2, 4)$ **5.** $(-1, 0)$ **6.** $(0, 3)$ **7.** $(6, 6)$ **8.** $(-1, -5)$

Applying Skills

9. Write a proof of Theorem 6.5 for the case when either point A or B is point P, that is, when the reflection is in one of the points.

10. Prove Corollary 6.5a, "Under a point reflection, angle measure is preserved."

11. Prove Corollary 6.5b, "Under a point reflection, collinearity is preserved."

12. Prove Corollary 6.5c, "Under a point reflection, midpoint is preserved."

13. The letters **S** and **N** have point symmetry. Print 5 other letters that have point symmetry.

14. Show that the quadrilateral with vertices $P(-5, 0)$, $Q(0, -5)$, $R(5, 0)$, and $S(0, 5)$ has point symmetry with respect to the origin.

15. **a.** What is the image of $A(2, 6)$ under a reflection in $P(2, 0)$?

    b. What is the image of $B(3, 6)$ under a reflection in $P(2, 0)$?

    c. Is the reflection in the point $P(2, 0)$ the same as the reflection in the x-axis? Justify your answer.

16. **a.** What is the image of $A(4, 4)$ under a reflection in the point $P(4, -2)$?

    b. What is the image of $C(1, -2)$ under a reflection in the point $P(4, -2)$?

    c. The point $D(2, 0)$ lies on the segment $\overline{AB}$. Does the image of D lie on the image of $\overline{AB}$? Justify your answer.

6-5 TRANSLATIONS IN THE COORDINATE PLANE

It is often useful or necessary to move objects from one place to another. If we move a table from one place in the room to another, the bottom of each leg moves the same distance in the same direction.

DEFINITION _____

A **translation** is a transformation of the plane that moves every point in the plane the same distance in the same direction.

If $\triangle A'B'C'$ is the image of $\triangle ABC$ under a translation, $AA' = BB' = CC'$. It appears that the size and shape of the figure are unchanged, so that $\triangle ABC \cong \triangle A'B'C'$. Thus, under a translation, as with a reflection, a figure is congruent to its image. The following example in the coordinate plane shows that this is true. In the coordinate 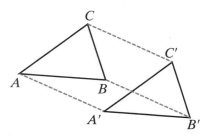 plane, the distance is given in terms of horizontal distance (change in the x-coordinates) and vertical distance (change in the y-coordinates).

In the figure, $\triangle DEF$ is translated by moving every point 4 units to the right and 5 units down. From the figure, we see that:

$$D(1,2) \rightarrow D'(5,-3)$$

$$E(4,2) \rightarrow E'(8,-3)$$

$$F(1,6) \rightarrow F'(5,1)$$

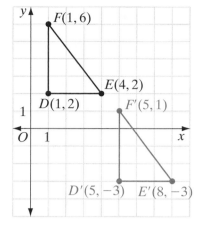

- $\overline{DE}$ and $\overline{D'E'}$ are horizontal segments.

$$DE = |1-4| \qquad D'E' = |5-8|$$
$$= 3 \qquad\qquad = 3$$

- $\overline{DF}$ and $\overline{D'F'}$ are vertical segments.

$$DF = |2-6| \qquad D'F' = |-3-1|$$
$$= 4 \qquad\qquad = 4$$

- $\angle FDE$ and $\angle F'D'E'$ are right angles.

Therefore, $\triangle DEF \cong \triangle D'E'F'$ by SAS.

This translation moves every point 4 units to the right $(+4)$ and 5 units down (-5).

1. The x-coordinate of the image is 4 more than the x-coordinate of the point:

$$x \rightarrow x + 4$$

2. The y-coordinate of the image is 5 less than the y-coordinate of the point:

$$y \rightarrow y - 5$$

From this example, we form a general rule:

DEFINITION _____

A **translation of a units in the horizontal direction and b units in the vertical direction** is a transformation of the plane such that the image of $P(x, y)$ is

$$P'(x + a, y + b).$$

Note: If the translation moves a point to the right, a is positive; if it moves a point to the left, a is negative; if the translation moves a point up, b is positive; if it moves a point down, b is negative.

Theorem 6.7 | Under a translation, distance is preserved.

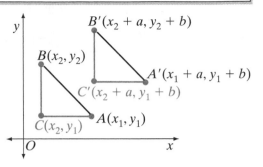

Given A translation in which the image of $A(x_1, y_1)$ is $A'(x_1 + a, y_1 + b)$ and the image of $B(x_2, y_2)$ is $B'(x_2 + a, y_2 + b)$.

Prove $AB = A'B'$

Proof Locate $C(x_2, y_1)$, a point on the same horizontal line as A and the same vertical line as B. Then by definition, the image of C is

$$C'(x_2 + a, y_1 + b).$$

Thus:

$$BC = |y_2 - y_1| \qquad\qquad AC = |x_1 - x_2|$$
$$B'C' = |(y_2 + b) - (y_1 + b)| \qquad A'C' = |(x_1 + a) - (x_2 + a)|$$
$$= |y_2 - y_1 + b - b| \qquad\qquad = |x_1 - x_2 + a - a|$$
$$= |y_2 - y_1| \qquad\qquad\qquad = |x_1 - x_2|$$

Since horizontal and vertical lines are perpendicular, $\angle BCA$ and $\angle B'C'A'$ are both right angles. By SAS, $\triangle ABC \cong \triangle A'B'C'$. Then, AB and $A'B'$ are the equal measures of the corresponding sides of congruent triangles. ∎

When we have proved that distance is preserved, we can prove that angle measure, collinearity, and midpoint are preserved. The proofs are similar to the proofs of the corollaries of Theorem 6.1 and are left to the student. (See exercise 10.)

Corollary 6.7a | Under a translation, angle measure is preserved.

Corollary 6.7b | Under a translation, collinearity is preserved.

Corollary 6.7c | Under a translation, midpoint is preserved.

We can write Theorem 6.7 and its corollaries as a single statement.

▶ **Under a translation, distance, angle measure, collinearity, and midpoint are preserved.**

Let $T_{a,b}$ be the symbol for a translation of a units in the horizontal direction and b units in the vertical direction. We can write:

$$T_{a,b}(x, y) = (x + a, y + b)$$

EXAMPLE I

The coordinates of the vertices of $\triangle ABC$ are $A(3, -3)$, $B(3, 2)$, and $C(6, 1)$.

a. Find the coordinates of the vertices of $\triangle A'B'C'$, the image of $\triangle ABC$ under $T_{-5,3}$.

b. Sketch $\triangle ABC$ and $\triangle A'B'C'$ on graph paper.

Solution **a.** Under the given translation, every point moves 5 units to the left and 3 units up.

$$A(3, -3) \rightarrow A'(3 - 5, -3 + 3) = A'(-2, 0)$$
$$B(3, 2) \rightarrow B'(3 - 5, 2 + 3) = B'(-2, 5)$$
$$C(6, 1) \rightarrow C'(6 - 5, 1 + 3) = C'(1, 4)$$

b.

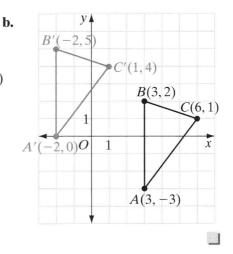

Translational Symmetry

DEFINITION

A figure has **translational symmetry** if the image of every point of the figure is a point of the figure.

Patterns used for decorative purposes such as wallpaper or borders on clothing often appear to have *translational symmetry*. True translational symmetry would be possible, however, only if the pattern could repeat without end.

Exercises

Writing About Mathematics

1. Explain why there can be no fixed points under a translation other than $T_{0,0}$.

2. Hunter said that if A' is the image of A under a reflection in the y-axis and A'' is the image of A' under a reflection in the line $x = 3$, then A'' is the image of A under the translation $(x, y) \rightarrow (x + 6, y)$.

   a. Do you agree with Hunter when A is a point with an x-coordinate greater than or equal to 3? Justify your answer.

   b. Do you agree with Hunter when A is a point with an x-coordinate greater than 0 but less than 3? Justify your answer.

   c. Do you agree with Hunter when A is a point with a negative x-coordinate? Justify your answer.

Developing Skills

3. The diagram consists of nine congruent rectangles. Under a translation, the image of A is G. Find the image of each of the given points under the same translation.

   a. J **b.** B **c.** I **d.** F **e.** E

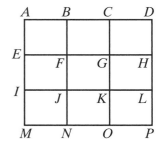

4. **a.** On graph paper, draw and label $\triangle ABC$, whose vertices have the coordinates $A(1, 2)$, $B(6, 3)$, and $C(4, 6)$.

   b. Under the translation $P(x, y) \rightarrow P'(x + 5, y - 3)$, the image of $\triangle ABC$ is $\triangle A'B'C'$. Find the coordinates of A', B', and C'.

   c. On the same graph drawn in part **a**, draw and label $\triangle A'B'C'$.

5. **a.** On graph paper, draw and label $\triangle ABC$ if the coordinates of A are $(-2, -2)$, the coordinates of B are $(2, 0)$, and the coordinates of C are $(3, -3)$.

   b. On the same graph, draw and label $\triangle A'B'C'$, the image of $\triangle ABC$ under the translation $T_{-4,7}$.

   c. Give the coordinates of the vertices of $\triangle A'B'C'$.

6. If the rule of a translation is written $(x, y) \rightarrow (x + a, y + b)$, what are the values of a and b for a translation where every point moves 6 units to the right on a graph?

7. In a translation, every point moves 4 units down. Write a rule for this translation in the form $(x, y) \rightarrow (x + a, y + b)$.

8. The coordinates of $\triangle ABC$ are $A(2, 1)$, $B(4, 1)$, and $C(5, 5)$.

a. On graph paper, draw and label $\triangle ABC$.

b. Write a rule for the translation in which the image of A is $C(5, 5)$.

c. Use the rule from part **b** to find the coordinates of B', the image of B, and C', the image of C, under this translation.

d. On the graph drawn in part **a**, draw and label $\triangle C'B'C'$, the image of $\triangle ABC$.

9. The coordinates of the vertices of $\triangle ABC$ are $A(0, -1)$, $B(2, -1)$, and $C(3, 3)$.

a. On graph paper, draw and label $\triangle ABC$.

b. Under a translation, the image of C is $B(2, -1)$. Find the coordinates of A', the image of A, and of B', the image of B, under this same translation.

c. On the graph drawn in part **a**, draw and label $\triangle A'B'B'$, the image of $\triangle ABC$.

d. How many points, if any, are fixed points under this translation?

Applying Skills

10. Prove the corollaries of Theorem 6.7.

a. Corollary 6.7a, "Under a translation, angle measure is preserved."

b. Corollary 6.7b, "Under a translation, collinearity is preserved."

c. Corollary 6.7c, "Under a translation, midpoint is preserved."

(*Hint:* See the proofs of the corollaries of Theorem 6.1 on page 217.)

11. The coordinates of the vertices of $\triangle LMN$ are $L(-6, 0)$, $M(-2, 0)$, and $N(-2, 2)$.

a. Draw $\triangle LMN$ on graph paper.

b. Find the coordinates of the vertices of $\triangle L'M'N'$, the image of $\triangle LMN$ under a reflection in the line $x = -1$, and draw $\triangle L'M'N'$ on the graph drawn in **a**.

c. Find the coordinates of the vertices of $\triangle L''M''N''$, the image of $\triangle L'M'N'$ under a reflection in the line $x = 4$, and draw $\triangle L''M''N''$ on the graph drawn in **a**.

d. Find the coordinates of the vertices of $\triangle PQR$, the image of $\triangle LMN$ under the translation $T_{10,0}$.

e. What is the relationship between $\triangle L''M''N''$ and $\triangle PQR$?

12. The coordinates of the vertices of $\triangle DEF$ are $D(-2, -3)$, $E(1, -3)$, and $F(0, 0)$.

a. Draw $\triangle DEF$ on graph paper.

b. Find the coordinates of the vertices of $\triangle D'E'F'$, the image of $\triangle DEF$ under a reflection in the line $y = 0$ (the x-axis), and draw $\triangle D'E'F'$ on the graph drawn in **a**.

c. Find the coordinates of the vertices of $\triangle D''E''F''$, the image of $\triangle D'E'F'$ under a reflection in the line $y = 3$, and draw $\triangle D''E''F''$ on the graph drawn in **a**.

d. Find the coordinates of the vertices of $\triangle RST$, the image of $\triangle DEF$ under the translation $T_{0,6}$.

e. What is the relationship between $\triangle D''E''F''$ and $\triangle RST$?

13. The coordinates of the vertices of $\triangle ABC$ are $A(1, 2)$, $B(5, 2)$, and $C(4, 5)$.

a. Draw $\triangle ABC$ on graph paper.

b. Find the coordinates of the vertices of $\triangle A'B'C'$, the image of $\triangle ABC$ under a reflection in the line $y = 0$ (the x-axis), and draw $\triangle A'B'C'$ on the graph drawn in **a**.

c. Find the coordinates of the vertices of $\triangle A''B''C''$, the image of $\triangle A'B'C'$ under a reflection in the line $y = -3$, and draw $\triangle A''B''C''$ on the graph drawn in **a**.

d. Is there a translation, $T_{a,b}$, such that $\triangle A''B''C''$ is the image of $\triangle ABC$? If so, what are the values of a and b?

14. In exercises, 11, 12, and 13, is there a relationship between the distance between the two lines of reflection and the x-values and y-values in the translation?

15. The coordinates of the vertices of $\triangle ABC$ are $A(1, 2)$, $B(5, 2)$, and $C(4, 5)$.

a. Draw $\triangle ABC$ on graph paper.

b. Find the coordinates of the vertices of $\triangle A'B'C'$, the image of $\triangle ABC$ under a reflection in any horizontal line, and draw $\triangle A'B'C'$ on the graph drawn in **a**.

c. Find the coordinates of the vertices of $\triangle A''B''C''$, the image of $\triangle A'B'C'$ under a reflection in the line that is 3 units below the line of reflection that you used in **b**, and draw $\triangle A''B''C''$ on the graph drawn in **a**.

d. What is the translation $T_{a,b}$ such that $\triangle A''B''C''$ is the image of $\triangle ABC$? Is this the same translation that you found in exercise 13?

6-6 ROTATIONS IN THE COORDINATE PLANE

Think of what happens to all of the points of a wheel as the wheel is turned. Except for the fixed point in the center, every point moves through a part of a circle, or *arc*, so that the position of each point is changed by a *rotation* of the same number of degrees.

DEFINITION _____

A **rotation** is a transformation of a plane about a fixed point P through an angle of d degrees such that:

1. For A, a point that is not the fixed point P, if the image of A is A', then $PA = PA'$ and $m\angle APA' = d$.

2. The image of the center of rotation P is P.

In the figure, P is the center of rotation. If A is rotated about P to A', and B is rotated the same number of degrees to B', then $m\angle APA' = m\angle BPB'$. Since P is the center of rotation, $PA = PA'$ and $PB = PB'$, and

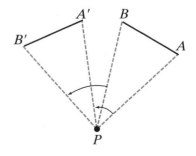

$$m\angle APA' - m\angle BPA' = m\angle APB$$
$$m\angle BPB' - m\angle BPA' = m\angle A'PB'$$

Therefore, m$\angle APB$ = m$\angle A'PB'$ and $\triangle APB \cong \triangle A'PB'$ by SAS. Because corresponding parts of congruent triangles are congruent and equal in measure, $AB = A'B'$, that is, that distance is preserved under a rotation. We have just proved the following theorem:

Theorem 6.8

> Distance is preserved under a rotation about a fixed point.

As we have seen when studying other transformations, when distance is preserved, angle measure, collinearity, and midpoint are also preserved.

▶ **Under a rotation about a fixed point, distance, angle measure, collinearity, and midpoint are preserved.**

We use $R_{P,d}$ as a symbol for the image under a rotation of d degrees about point P. For example, the statement "$R_{O,30°}(A) = B$" can be read as "the image of A under a rotation of 30° degrees about the origin is B."

A rotation in the counterclockwise direction is called a **positive rotation**. For instance, B is the image of A under a rotation of 30° about P.

A rotation in the clockwise direction is called a **negative rotation**. For instance, C is the image of A under a rotation of $-45°$ about P.

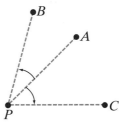

Rotational Symmetry

DEFINITION _____

A figure is said to have **rotational symmetry** if the figure is its own image under a rotation and the center of rotation is the only fixed point.

Many letters, as well as designs in the shapes of wheels, stars, and polygons, have *rotational symmetry*. When a figure has rotational symmetry under a rotation of $d°$, we can rotate the figure by $d°$ to an image that is an identical figure. Each figure shown below has rotational symmetry.

Any *regular polygon* (a polygon with all sides congruent and all angles congruent) has rotational symmetry. When regular pentagon $ABCDE$ is rotated $\frac{360°}{5}$, or 72°, about its center, the image of every point of the figure is a point of the figure. Under this rotation:

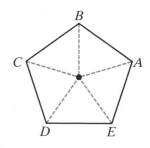

$$A \rightarrow B \quad B \rightarrow C \quad C \rightarrow D \quad D \rightarrow E \quad E \rightarrow A$$

The figure would also have rotational symmetry if it were rotated through a multiple of 72° (144°, 216°, or 288°). If it were rotated through 360°, every point would be its own image. Since this is true for every figure, we do not usually consider a 360° rotation as rotational symmetry.

Rotations in the Coordinate Plane

The most common rotation in the coordinate plane is a **quarter turn** about the origin, that is, a counterclockwise rotation of 90° about the origin.

In the diagram, the vertices of right triangle ABC are $A(0,0)$, $B(3, 4)$ and $C(3, 0)$. When rotated 90° about the origin, A remains fixed because it is the center of rotation. The image of C, which is on the x-axis and 3 units from the origin, is $C'(0, 3)$ on the y-axis and 3 units from the origin. Since $\overline{CB}$ is a vertical line 4 units long, its image is a horizontal line 4 units long and to the left of the y-axis. Therefore, the image of B is $B'(-4, 3)$. Notice that the x-coordinate of B' is the negative of the y-coordinate of B and the y-coordinate of B' is the x-coordinate of B.

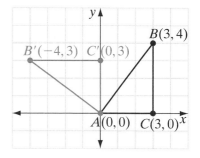

The point $B(3, 4)$ and its image $B'(-4, 3)$ in the above example suggest a rule for the coordinates of the image of any point $P(x, y)$ under a counterclockwise rotation of 90° about the origin.

Theorem 6.9

Under a counterclockwise rotation of 90° about the origin, the image of $P(a, b)$ is $P'(-b, a)$.

Proof: We will prove this theorem by using a rectangle with opposite vertices at the origin and at P. Note that in quadrants I and II, when b is positive, $-b$ is negative, and in quadrants III and IV, when b is negative, $-b$ is positive.

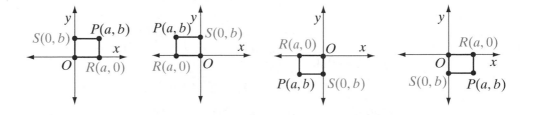

Let $P(a, b)$ be any point not on an axis and $O(0, 0)$ be the origin. Let $R(a, 0)$ be the point on the x-axis on the same vertical line as P and $S(0, b)$ be the point on the y-axis on the same horizontal line as P. Therefore, $PR = |b - 0| = |b|$ and $PS = |a - 0| = |a|$.

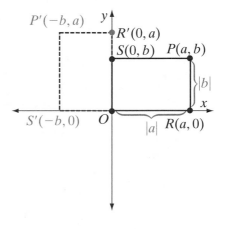

Under a rotation of $90°$ about the origin, the image of $ORPS$ is $OR'P'S'$. Then, $OR = OR'$ and $m\angle R'OR = 90$. This means that since $\overline{OR}$ is a horizontal segment, $\overline{OR'}$ is a vertical segment with the same length. Therefore, since R is on the x-axis, R' is on the y-axis and the coordinates of R' are $(0, a)$. Similarly, since S is on the y-axis, S' is on the x-axis and the coordinates of S' are $(-b, 0)$. Point P' is on the same horizontal line as R' and therefore has the same y-coordinate as R', and P' is on the same vertical line as S' and has the same x-coordinate as S'. Therefore, the coordinates of P' are $(-b, a)$.

The statement of Theorem 6.9 may be written as:

$$R_{O,90°}(x, y) = (-y, x) \quad \text{or} \quad R_{90°}(x, y) = (-y, x)$$

When the point that is the center of rotation is not named, it is understood to be O, the origin.

Note: The symbol R is used to designate both a point reflection and a rotation.

1. When the symbol R is followed by a letter that designates a point, it represents a reflection in that point.

2. When the symbol R is followed by both a letter that designates a point and the number of degrees, it represents a rotation of the given number of degrees about the given point.

3. When the symbol R is followed by the number of degrees, it represents a rotation of the given number of degrees about the origin.

EXAMPLE I

Point P is at the center of equilateral triangle ABC (the point at which the perpendicular bisectors of the sides intersect) so that:

$$PA = PB = PC \quad \text{and} \quad m\angle APB = m\angle BPC = m\angle CPA$$

Under a rotation about P for which the image of A is B, find:

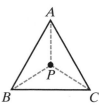

Answers

a. The number of degrees in the rotation.   **a.** $\frac{360°}{3} = 120°$

b. The image of B.   **b.** C

c. The image of $\overline{CA}$.   **c.** $\overline{AB}$

d. The image of $\angle CAB$.   **d.** $\angle ABC$

EXAMPLE 2

What are the coordinates of the image of $P(-2, 3)$ under $R_{90°}$?

Solution The image of (x, y) is $(-y, x)$. Therefore, the image of $(-2, 3)$ is $(-3, -2)$.

    Answer $(-3, -2)$

Exercises

Writing About Mathematics

1. A point in the coordinate plane is rotated $180°$ about the origin by rotating the point counterclockwise $90°$ and then rotating the image counterclockwise $90°$.

   a. Choose several points in the coordinate plane and find their images under a rotation of $180°$. What is the image of $P(x, y)$ under this rotation?

   b. For what other transformation does $P(x, y)$ have the same image?

   c. Is a $180°$ rotation about the origin equivalent to the transformation found in part **b**?

2. Let Q be the image of $P(x, y)$ under a clockwise rotation of $90°$ about the origin and R be the image of Q under a clockwise rotation of $90°$ about the origin. For what two different transformations is R the image of P?

Developing Skills

For 3 and 4, refer to the figure at the right.

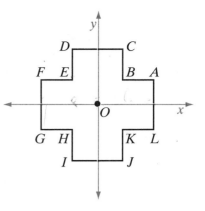

3. What is the image of each of the given points under $R_{90°}$?

    a. A **b.** B **c.** C **d.** G

    e. H **f.** J **g.** K **h.** L

4. What is the image of each of the given points under $R_{-90°}$?

    a. A **b.** B **c.** C **d.** G

    e. H **f.** J **g.** K **h.** L

5. The vertices of rectangle $ABCD$ are $A(2, -1)$, $B(5, -1)$, $C(5, 4)$, and $D(2, 4)$.

    a. What are the coordinates of the vertices of $A'B'C'D'$, the image of $ABCD$ under a counterclockwise rotation of 90° about the origin?

    b. Is $A'B'C'D'$ a rectangle?

    c. Find the coordinates of the midpoint of $\overline{AB}$ and of $\overline{A'B'}$. Is the midpoint of $\overline{A'B'}$ the image of the midpoint of $\overline{AB}$ under this rotation?

6. a. What are the coordinates of Q, the image of $P(1, 3)$ under a counterclockwise rotation of 90° about the origin?

    b. What are the coordinates of R, the image of Q under a counterclockwise rotation of 90° about the origin?

    c. What are the coordinates of S, the image of R under a counterclockwise rotation of 90° about the origin?

    d. What are the coordinates of P', the image of P under a clockwise rotation of 90° about the origin?

    e. Explain why S and P' are the same point.

6-7 GLIDE REFLECTIONS

When two transformations are performed, one following the other, we have a **composition of transformations.** The first transformation produces an image and the second transformation is performed on that image. One such composition that occurs frequently is the composition of a line reflection and a translation.

DEFINITION

A **glide reflection** is a composition of transformations of the plane that consists of a line reflection and a translation in the direction of the line of reflection performed in either order.

The vertices of $\triangle ABC$ are $A(1, 2)$, $B(5, 3)$, and $C(3, 4)$. Under a reflection in the y-axis, the image of $\triangle ABC$ is $\triangle A'B'C'$ whose vertices are $A'(-1, 2)$, $B'(-5, 3)$, and $C'(-3, 4)$. Under the translation $T_{0,-4}$, the image of $\triangle A'B'C'$ is $\triangle A''B''C''$ whose vertices are $A''(-1, -2)$, $B''(-5, -1)$, and $C''(-3, 0)$. Under a glide reflection, the image of $\triangle ABC$ is $\triangle A''B''C''$. Note that the line of reflection, the y-axis, is a vertical line. The translation is in the vertical direction because the x-coordinate of the translation is 0. Distance is preserved under a line reflection and under a translation: $\triangle ABC \cong \triangle A'B'C' \cong \triangle A''B''C''$. Distance is preserved under a glide reflection. We have just proved the following theorem:

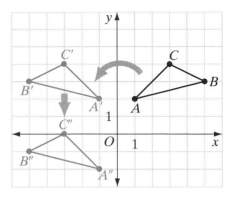

Theorem 6.10 | Under a glide reflection, distance is preserved.

As we have seen with other transformations, when distance is preserved, angle measure, collinearity, and midpoint are also preserved. Therefore, we may make the following statement:

▶ **Under a glide reflection, distance, angle measure, collinearity, and midpoint are preserved.**

In this chapter, we have studied five transformations: line reflection, point reflection, translation, rotation, and glide reflection. Each of these transformations is called an *isometry* because it preserves distance.

DEFINITION
An **isometry** is a transformation that preserves distance.

EXAMPLE I

The vertices of $\triangle PQR$ are $P(2, 1)$, $Q(4, 1)$, and $R(4, 3)$.

a. Find $\triangle P'Q'R'$, the image of $\triangle PQR$ under $r_{y=x}$ followed by $T_{-3,-3}$.

b. Find $\triangle P''Q''R''$, the image of $\triangle PQR$ under $T_{-3,-3}$ followed by $r_{y=x}$.

c. Are $\triangle P'Q'R'$ and $\triangle P''Q''R''$ the same triangle?

d. Are $r_{y=x}$ followed by $T_{-3,-3}$ and $T_{-3,-3}$ followed by $r_{y=x}$ the same glide reflection? Explain.

e. Write a rule for this glide reflection.

Solution **a.** $r_{y=x}(x, y) = (y, x)$ $T_{-3,-3}(y, x) = (y - 3, x - 3)$

$r_{y=x}(2, 1) = (1, 2)$ $T_{-3,-3}(1, 2) = (-2, -1)$

$r_{y=x}(4, 1) = (1, 4)$ $T_{-3,-3}(1, 4) = (-2, 1)$

$r_{y=x}(4, 3) = (3, 4)$ $T_{-3,-3}(3, 4) = (0, 1)$

The vertices of $\triangle P'Q'R'$ are $P'(-2, -1)$, $Q'(-2, 1)$, and $R'(0, 1)$. *Answer*

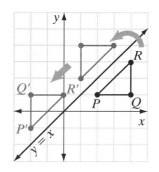

b. $T_{-3,-3}(x, y) = (x - 3, y - 3)$ $r_{y=x}(x - 3, y - 3) = (y - 3, x - 3)$

$T_{-3,-3}(2, 1) = (-1, -2)$ $r_{y=x}(-1, -2) = (-2, -1)$

$T_{-3,-3}(4, 1) = (1, -2)$ $r_{y=x}(1, -2) = (-2, 1)$

$T_{-3,-3}(4, 3) = (1, 0)$ $r_{y=x}(1, 0) = (0, 1)$

The vertices of $\triangle P''Q''R''$ are $P''(-2, -1)$, $Q''(-2, 1)$, and $R''(0, 1)$. *Answer*

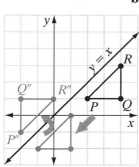

c. $\triangle P'Q'R'$ and $\triangle P''Q''R''$ are the same triangle. *Answer*

d. The image of (x, y) under $r_{y=x}$ followed by $T_{-3,-3}$ is
$$(x, y) \rightarrow (y, x) \rightarrow (y - 3, x - 3)$$

The image of (x, y) under $T_{-3,-3}$ followed by $r_{y=x}$ is
$$(x, y) \rightarrow (x - 3, y - 3) \rightarrow (y - 3, x - 3)$$

$r_{y=x}$ followed by $T_{-3,-3}$ and $T_{-3,-3}$ followed by $r_{y=x}$ are the same glide reflection. *Answer*

e. $(x, y) \rightarrow (y - 3, x - 3)$ *Answer*

EXAMPLE 2

Is a reflection in the y-axis followed by the translation $T_{5,5}$ a glide reflection?

Solution The y-axis is a vertical line. The translation $T_{5,5}$ is not a translation in the vertical direction. Therefore, a reflection in the y-axis followed by the translation $T_{5,5}$ is not a glide reflection.

Exercises

Writing About Mathematics

1. Does a glide reflection have any fixed points? Justify your answer.

2. In a glide reflection, the line reflection and the translation can be done in either order. Is the composition of any line reflection and any translation always the same in either order? Justify your answer.

Developing Skills

In 3–8, **a.** find the coordinates of the image of the triangle whose vertices are $A(1, 1)$, $B(5, 4)$, and $C(3, 5)$ under the given composition of transformations. **b.** Sketch $\triangle ABC$ and its image. **c.** Explain why the given composition of transformations is or is not a glide reflection. **d.** Write the coordinates of the image of (a, b) under the given composition of transformations.

3. A reflection in the x-axis followed by a translation of 4 units to the right.

4. A reflection in the y-axis followed by a translation of 6 units down.

5. $T_{5,0}$ followed by $r_{x\text{-axis}}$

6. $T_{3,3}$ followed by $r_{y\text{-axis}}$

7. $T_{3,3}$ followed by $r_{y=x}$

8. R_{90} followed by $T_{1,2}$

Applying Skills

9. The point $A'(-4, 5)$ is the image of A under a glide reflection that consists of a reflection in the x-axis followed by the translation $T_{-2,0}$. What are the coordinates of A?

10. Under a glide reflection, the image of $A(2, 4)$ is $A'(-2, -7)$ and the image of $B(3, 7)$ is $B'(-3, -4)$.

   a. If the line of reflection is a vertical line, write an equation for the line reflection and a rule for the translation.

   b. What are the coordinates of C', the image of $C(0, 7)$, under the same glide reflection?

11. Under a glide reflection, the image of $A(3, 4)$ is $A'(-1, -4)$ and the image of $B(5, 5)$ is $B'(1, -5)$.

   a. If the line of reflection is a horizontal line, write a rule for the line reflection and a rule for the translation.

   b. What are the coordinates of C', the image of $C(4, 2)$, under the same glide reflection?

12. **a.** For what transformation or transformations are there no fixed points?

   b. For what transformation or transformations is there exactly one fixed point?

   c. For what transformation or transformations are there infinitely many fixed points?

6-8 DILATIONS IN THE COORDINATE PLANE

In this chapter, we have learned about transformations in the plane that are isometries, that is, transformations that preserve distance. There is another transformation in the plane that preserves angle measure but not distance. This transformation is a *dilation*.

For example, in the coordinate plane, a dilation of 2 with center at the origin will stretch each ray by a factor of 2. If the image of A is A', then A' is a point on $\overrightarrow{OA}$ and $OA' = 2OA$.

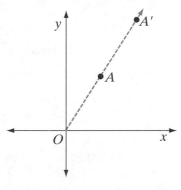

DEFINITION _____

A **dilation** of k is a transformation of the plane such that:

1. The image of point O, the center of dilation, is O.
2. When k is positive and the image of P is P', then $\overrightarrow{OP}$ and $\overrightarrow{OP'}$ are the same ray and $OP' = kOP$.
3. When k is negative and the image of P is P', then $\overrightarrow{OP}$ and $\overrightarrow{OP'}$ are opposite rays and $OP' = -kOP$.

Note: In step 3, when k is negative, $-k$ is positive.

In the coordinate plane, the center of dilation is usually the origin. If the center of dilation is not the origin, the coordinates of the center will be given.

In the coordinate plane, under a dilation of k with the center at the origin:

$$P(x, y) \rightarrow P'(kx, ky) \quad \text{or} \quad D_k(x, y) = (kx, ky)$$

For example, the image of $\triangle ABC$ is $\triangle A'B'C'$ under a dilation of $\frac{1}{2}$. The vertices of $\triangle ABC$ are $A(2, 6)$, $B(6, 4)$, and $C(4, 0)$. Under a dilation of $\frac{1}{2}$, the rule is

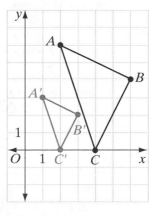

$$D_{\frac{1}{2}}(x, y) = \left(\tfrac{1}{2}x, \tfrac{1}{2}y\right)$$
$$A(2, 6) \rightarrow A'(1, 3)$$
$$B(6, 4) \rightarrow B'(3, 2)$$
$$C(4, 0) \rightarrow C'(2, 0)$$

By the definition of a dilation, distance is not preserved. We will prove in Chapter 12 that angle measure, collinearity, and midpoint are preserved.

▶ **Under a dilation about a fixed point, distance is *not* preserved.**

EXAMPLE I

The coordinates of parallelogram $EFGH$ are $E(0,0)$, $F(3,0)$, $G(4,2)$, and $H(1,2)$. Under D_3, the image of $EFGH$ is $E'F'G'H'$.

a. Find the coordinates of the vertices of $E'F'G'H'$.

b. Let M be the midpoint of $\overline{EF}$. Find the coordinates of M and of M', the image of M. Verify that the midpoint is preserved.

Solution   **a.** $D_3(x, y) = (3x, 3y)$. Therefore, $E'(0,0)$, $F'(9,0)$, $G'(12,6)$, and $H'(3,6)$. *Answer*

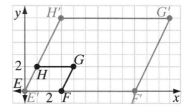

b. Since E and F lie on the x-axis, EF is the absolute value of the difference of their x-coordinates. $EF = |3 - 0| = 3$. Therefore, the midpoint, M, is $1\frac{1}{2}$ units from E or from F on the x-axis. The coordinates of M are $\left(0 + 1\frac{1}{2}, 0\right) = \left(1\frac{1}{2}, 0\right)$.

Since E' and F' lie on the x-axis, $E'F'$ is the absolute value of the difference of their x-coordinates. $E'F' = |9 - 0| = 9$. Therefore, the midpoint, M', is $4\frac{1}{2}$ units from E' or from F' on the x-axis. The coordinates of M are $\left(0 + 4\frac{1}{2}, 0\right) = \left(4\frac{1}{2}, 0\right)$.

$D_3\left(1\frac{1}{2}, 0\right) = \left(4\frac{1}{2}, 0\right)$. Therefore, the image of M is M' and midpoint is preserved. *Answer*

EXAMPLE 2

Find the coordinates of P', the image of $P(4, -5)$ under the composition of transformations: D_2 followed by $r_{x\text{-axis}}$.

Solution   The dilation is to be performed first: $D_2(4, -5) = (8, -10)$. Then perform the reflection, using the result of the dilation: $r_{x\text{-axis}}(8, -10) = (8, 10)$.

Answer $P' = (8, 10)$

Exercises

Writing About Mathematics

1. Let $\triangle A'B'C'$ be the image of $\triangle ABC$ under a dilation of k. When $k > 1$, how do the lengths of the sides of $\triangle A'B'C'$ compare with the lengths of the corresponding sides of $\triangle ABC$? Is a dilation an isometry?

2. Let $\triangle A'B'C'$ be the image of $\triangle ABC$ under a dilation of k. When $0 < k < 1$, how do the lengths of the sides of $\triangle A'B'C'$ compare with the lengths of the corresponding sides of $\triangle ABC$?

Developing Skills

In 3–6, use the rule $(x, y) \rightarrow (4x, 4y)$ to find the coordinates of the image of each given point.

3. $(3, 7)$ **4.** $(-4, 2)$

5. $(2, 0)$ **6.** $(-1, 9)$

In 7–10, find the coordinates of the image of each given point under D_5.

7. $(2, 2)$ **8.** $(1, 10)$

9. $(-3, 5)$ **10.** $(0, 4)$

In 11–14, each given point is the image under D_2. Find the coordinates of each preimage.

11. $(6, -2)$ **12.** $(4, 0)$

13. $(-6, -5)$ **14.** $(10, 7)$

In 15–20, find the coordinates of the image of each given point under the given composition of transformations.

15. $D_4(1, 5)$ followed by $r_{y\text{-axis}}$ **16.** $D_{-2}(-3, 2)$ followed by $R_{90°}$

17. $T_{2,1}(4, -2)$ followed by $D_{\frac{1}{2}}$ **18.** $D_3(5, -1)$ followed by $r_{x\text{-axis}}$

19. $T_{1,-1}(-4, 2)$ followed by D_{-2} **20.** $D_2(6, 3)$ followed by $r_{y=x}$

In 21–24, each transformation is the composition of a dilation and a reflection in either the x-axis or the y-axis. In each case, write a rule for composition of transformations for which the image of A is A'.

21. $A(2, 5) \rightarrow A'(4, -10)$ **22.** $A(3, -1) \rightarrow A'(-21, -7)$

23. $A(10, 4) \rightarrow A'(5, -2)$ **24.** $A(-20, 8) \rightarrow A'(5, 2)$

Applying Skills

25. The vertices of rectangle $ABCD$ are $A(1, -1)$, $B(3, -1)$, $C(3, 3)$, and $D(1, 3)$.

    a. Find the coordinates of the vertices of $A'B'C'D'$, the image of $ABCD$ under D_4.

    b. Show that $A'B'C'D'$ is a rectangle.

26. Show that when $k > 0$, a dilation of k with center at the origin followed by a reflection in the origin is the same as a dilation of $-k$ with center at the origin.

27. a. Draw $\triangle ABC$, whose vertices are $A(2, 3)$, $B(4, 3)$, and $C(4, 6)$.

  b. Using the same set of axes, graph $\triangle A'B'C'$, the image of $\triangle ABC$ under a dilation of 3.

  c. Using $\triangle ABC$ and its image $\triangle A'B'C'$, show that distance is *not* preserved under the dilation.

Hands-On Activity

In this activity, we will verify that angle measure, collinearity, and midpoint are preserved under a dilation.

 Using geometry software, or a pencil, ruler and a protractor, draw the pentagon $A(1, 2)$, $B(4, 2)$, $C(6, 4)$, $D(2, 8)$, $E(1, 6)$, and point $P(4, 6)$ on $\overline{CD}$. For each given dilation: **a.** Measure the corresponding angles. Do they appear to be congruent? **b.** Does the image of P appear to be on the image of $\overline{CD}$? **c.** Does the image of the midpoint of $\overline{AE}$ appear to be the midpoint of the image of $\overline{AE}$?

(1) D_3 (3) $D_{\frac{1}{2}}$

(2) D_{-3} (4) $D_{-\frac{1}{2}}$

6-9 TRANSFORMATIONS AS FUNCTIONS

A **function** is a set of ordered pairs in which no two pairs have the same first element. For example, the equation $y = x + 5$ describes the set of ordered pairs of real numbers in which the second element, y, is 5 more than the first element, x. The set of first elements is the **domain** of the function and the set of second elements is the **range**. Some pairs of this function are $(-3, 2)$, $(-2.7, 2.3)$, $\left(\frac{2}{3}, \frac{17}{3}\right)$, $(0, 5)$, and $(1, 6)$. Since the set of real numbers is infinite, the set of pairs of this function cannot be listed completely and are more correctly defined when they are described. Each of the following notations describes the function in which the second element is 5 more than the first.

$$f = \{(x, y) \mid y = x + 5\} \qquad f : x \rightarrow x + 5 \qquad y = x + 5 \qquad f(x) = x + 5$$

Note that $f(x)$ and y each name the second element of the ordered pair.

Each of the transformations defined in this chapter is a one-to-one function. It is a set of ordered pairs in which the first element of each pair is a point of the plane and the second element is the image of that point under a transformation. For each point in the plane, there is one and only one image. For example, compare the function $f(x) = x + 5$ and a line reflection.

f: A one-to-one algebraic function

$$f(x) = x + 5$$

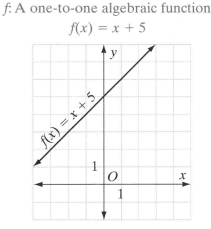

r$_m$: A reflection in line *m*

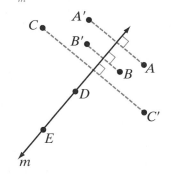

1. For every *x* in the domain there is one and only one *y* in the range. For example:

$$f(3) = 8$$
$$f(0) = 5$$
$$f(-2) = 3$$

2. Every *f*(*x*) or *y* in the range corresponds to one and only one value of *x* in the range. For example:

If *f*(*x*) = 4, then *x* = −1.

If *f*(*x*) = 9, then *x* = 4.

3. For the function *f*:

domain = {real numbers}

range = {real numbers}

1. For every point in the plane, there is one and only one image of that point in the plane. For example:

$$r_m(A) = A'$$
$$r_m(C) = C'$$
$$r_m(D) = D$$

2. Every point is the image of one and only one preimage. For example:

The point *B'* has the preimage *B*.

The point *E* has itself as the preimage.

The point *C'* has the preimage *C*.

3. For the function *r*$_m$:

domain = {points in the plane}

range = {points in the plane}

Composition of Transformations

We defined the composition of transformations as a combination of two transformations in which the first transformation produces an image and the second transformation is performed on that image.

For example, to indicate that *A'* is the image of *A* under a reflection in the line *y* = *x* followed by the translation $T_{2,0}$, we can write

$$T_{2,0}(r_{y=x}(A)) = A'.$$

If the coordinates of A are $(2, 5)$, we start with A and perform the transformations from right to left as indicated by the parentheses, evaluating what is in parentheses first.

$$r_{y=x}(2, 5) = (5, 2) \quad \text{followed by} \quad T_{2,0}(5, 2) = (7, 2)$$

or

$$T_{2,0}(r_{y=x}(2, 5)) = T_{2,0}(5, 2)$$

$$= (7, 2)$$

A small raised circle is another way of indicating a composition of transformations. $T_{2,0}(r_{y=x}(A)) = A'$ can also be written as

$$T_{2,0} \circ r_{y=x}(A) = A'.$$

Again, we start with the coordinates of A and move from right to left. Find the image of A under the reflection and then, using the coordinates of that image, find the coordinates under the translation.

Orientation

The figures below show the images of $\triangle ABC$ under a point reflection, a translation, and a rotation. In each figure, the vertices, when traced from A to B to C are in the clockwise direction, called the orientation of the points. The vertices of the images, when traced in the same order, from A' to B' to C' are also in the clockwise direction. Therefore, we say that under a point reflection, a translation, or a rotation, **orientation** is unchanged.

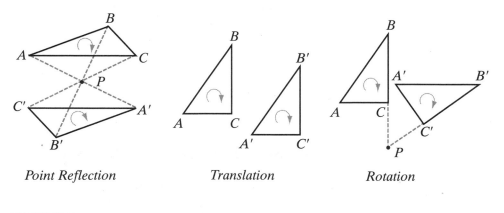

| Point Reflection | Translation | Rotation |

DEFINITION
A **direct isometry** is a transformation that preserves distance and orientation.

Point reflection, rotation, and translation are direct isometries.

The figure at the right shows $\triangle ABC$ and its image under a line reflection. In this figure, the vertices, when traced from A to B to C, are again in the clockwise direction. The vertices of the images, when traced in the same order, from A' to B' to C', are in the counterclockwise direction. Therefore, under a line reflection, orientation is changed or reversed.

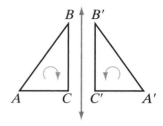

Line Reflection

DEFINITION _____

An **opposite isometry** is a transformation that preserves distance but changes the order or orientation from clockwise to counterclockwise or from counterclockwise to clockwise.

A line reflection is an opposite isometry.

EXAMPLE I

The vertices of $\triangle RST$ are $R(1,1)$, $S(6,3)$, and $T(2,5)$.

a. Sketch $\triangle RST$.

b. Find the coordinates of the vertices of $\triangle R'S'T'$ under the composition $r_{x\text{-axis}} \circ r_{y\text{-axis}}$ and sketch $\triangle R'S'T'$.

c. Is the composition a direct isometry?

d. For what single transformation is the image the same as that of $r_{x\text{-axis}} \circ r_{y\text{-axis}}$?

Solution

a.

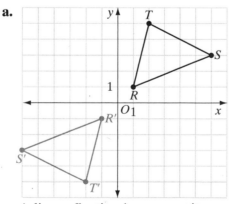

b. $r_{x\text{-axis}} \circ r_{y\text{-axis}}(1, 1) = r_{x\text{-axis}}(-1, 1)$
$$= (-1, -1)$$

$r_{x\text{-axis}} \circ r_{y\text{-axis}}(6, 3) = r_{x\text{-axis}}(-6, 3)$
$$= (-6, -3)$$

$r_{x\text{-axis}} \circ r_{y\text{-axis}}(2, 5) = r_{x\text{-axis}}(-2, 5)$
$$= (-2, -5)$$

c. A line reflection is an opposite isometry. The composition of two opposite isometries is a direct isometry. This composition is a direct isometry.

d. $r_{x\text{-axis}} \circ r_{y\text{-axis}}(x, y) = (-x, -y)$ and $R_o(x, y) = (-x, -y)$. The composition of a reflection in the y-axis followed by a reflection in the x-axis is a reflection in the origin.

EXAMPLE 2

The transformation $(x, y) \rightarrow (-y + 3, x - 2)$ is the composition of what two transformations?

Solution   Under a rotation of 90° about the origin, $(x, y) \rightarrow (-y, x)$.

When this rotation is followed by the translation
$T_{3,-2}(-y, x) = (-y + 3, x - 2)$.

Answer $T_{3,-2} \circ R_{90°}(x,y) = T_{3,-2}(-y, x) = (-y + 3, x - 2)$

This solution is not unique.

Alternative Solution   Under the translation $T_{-2,-3}(x, y) = (x - 2, y - 3)$.

When this translation is followed by the rotation
$R_{90°}(x - 2, y - 3) = (-(y - 3), x - 2) = (-y + 3, x - 2)$.

Answer $R_{90°} \circ T_{-2,-3}(x, y) = R_{90°}(x - 2, y - 3) = (-y + 3, x - 2)$

Exercises

Writing About Mathematics

1. Owen said that since a reflection in the x-axis followed by a reflection in the y-axis has the same result as a reflection in the y-axis followed by a reflection in the x-axis, the composition of line reflections is a commutative operation. Do you agree with Owen? Justify your answer.

2. Tyler said that the composition of an even number of opposite isometries is a direct isometry and that the composition of an odd number of opposite isometries is an opposite isometry. Do you agree with Tyler? Justify your answer.

Developing Skills

In 3–11, find the image of $A(3, -2)$ under the given composition of transformations.

3. $r_{x\text{-axis}} \circ r_{y\text{-axis}}$

4. $r_{y\text{-axis}} \circ r_{x\text{-axis}}$

5. $R_O \circ r_{x\text{-axis}}$

6. $R_{90°} \circ R_O$

7. $T_{-1,4} \circ r_{y=x}$

8. $R_{90°} \circ R_{90°}$

9. $r_{y\text{-axis}} \circ r_{y=x}$

10. $R_O \circ T_{3,-2}$

11. $T_{-2,4} \circ D_{-2}$

12. A reflection in the line $y = x$ followed by a rotation of 90° about the origin is equivalent to what single transformation?

13. A rotation of 90° about the origin followed by another rotation of 90° about the origin is equivalent to what single transformation?

14. The vertices of $\triangle DEF$ are $D(3,2)$, $E(5,5)$, and $F(4,-1)$.

   a. If $T_{-3,0} \circ r_{x\text{-axis}}(\triangle DEF) = \triangle D'E'F'$, find the coordinates of the vertices of $\triangle D'E'F'$.

   b. The composition $T_{-3,0} \circ r_{x\text{-axis}}$ is an example of what type of transformation?

   c. Is the composition $T_{-3,0} \circ r_{x\text{-axis}}$ a direct or an opposite isometry?

Applying Skills

15. Prove that under a reflection in the line $y = -x$, $A(a, b) \rightarrow A'(-b, -a)$.

   Suggested plan: Let $B(a, -a)$ and $C(-b, b)$ be points on the line $y = -x$. Show that the line $y = -x$ is the perpendicular bisector of $\overline{AA'}$.

16. Prove that the composition of a reflection in the line $y = -x$ followed by a reflection in the line $y = x$ is a reflection in the origin.

17. Prove that the composition of a reflection in the line $y = x$ followed by a reflection in the line $y = -x$ is a reflection in the origin.

CHAPTER SUMMARY

Definitions to Know

- A **transformation** is a one-to-one correspondence between two sets of points, S and S', when every point in S corresponds to one and only one point in S' called its **image**, and every point in S' is the image of one and only one point in S' called its **preimage**.

- A **reflection in line k** is a transformation in a plane such that:

  1. If point A is not on k, then the image of A is A' where k is the perpendicular bisector of $\overline{AA'}$.

  2. If point A is on k, the image of A is A.

- A figure has **line symmetry** when the figure is its own image under a line reflection.

- A **point reflection in P** is a transformation plane such that:

  1. If point A is not point P, then the image of A is A' where P the midpoint of $\overline{AA'}$.

  2. The point P is its own image.

- A **translation** is a transformation in a plane that moves every point in the plane the same distance in the same direction.

- A **translation of a units in the horizontal direction and b units in the vertical direction** is a transformation in a plane such that the image of $P(x, y)$ is $P'(x + a, y + b)$.

- A **rotation** is a transformation of a plane about a fixed point P through an angle of d degrees such that:

  1. For A, a point that is not the fixed point, if the image of A is A', then $PA = PA'$ and $m\angle APA' = d$.

  2. The image of the center of rotation P is P.

- A **quarter turn** is a counterclockwise rotation of $90°$.

- A **composition of transformations** is a combination of two transformations in which the first transformation produces an image and the second transformation is performed on that image.

- A **glide reflection** is a composition of transformations of the plane that consists of a line reflection and a translation in the direction of the line of reflection in either order.

- An **isometry** is a transformation that preserves distance.

- A **dilation** of k is a transformation of the plane such that:

  1. The image of point O, the center of dilation, is O.

  2. When k is positive and the image of P is P', then $\overrightarrow{OP}$ and $\overrightarrow{OP'}$ are the same ray and $OP' = kOP$.

  3. When k is negative and the image of P is P', then $\overrightarrow{OP}$ and $\overrightarrow{OP'}$ are opposite rays and $OP' = -kOP$.

- A **function** is a set of ordered pairs in which no two pairs have the same first element.

- The set of first elements is the **domain** of the function.

- The set of second elements is the **range** of the function.

- A **direct isometry** is a transformation that preserves distance and orientation.

- An **opposite isometry** is a transformation that preserves distance and reverses orientation.

Postulates

6.1 Two points are on the same horizontal line if and only if they have the same y-coordinates.

6.2 The length of a horizontal line segment is the absolute value of the difference of the x-coordinates.

6.3 Two points are on the same vertical line if and only if they have the same x-coordinate.

6.4 The length of a vertical line segment is the absolute value of the difference of the y-coordinates.

6.5 Each vertical line is perpendicular to each horizontal line.

Theorems and Corollaries

6.1 Under a line reflection, distance is preserved.

6.1a Under a reflection, angle measure is preserved.

6.1b Under a reflection, collinearity is preserved.

6.1c Under a reflection, midpoint is preserved.

6.2   Under a reflection in the y-axis, the image of $P(a, b)$ is $P'(-a, b)$.

6.3   Under a reflection in the x-axis, the image of $P(a, b)$ is $P'(a, -b)$.

6.4   Under a reflection in the line $y = x$, the image of $P(a, b)$ is $P'(b, a)$.

6.5   Under a point reflection, distance is preserved.

6.5a   Under a point reflection, angle measure is preserved.

6.5b   Under a point reflection, collinearity is preserved.

6.5c   Under a point reflection, midpoint is preserved.

6.6   Under a reflection in the origin, the image of $P(a, b)$ is $P'(-a, -b)$.

6.7   Under a translation, distance is preserved.

6.7a   Under a translation, angle measure is preserved.

6.7b   Under a translation, collinearity is preserved.

6.7c   Under a translation, midpoint is preserved.

6.8   Distance is preserved under a rotation about a fixed point.

6.9   Under a counterclockwise rotation of 90° about the origin, the image of $P(a, b)$ is $P'(-b, a)$.

6.10   Under a glide reflection, distance is preserved.

VOCABULARY

6-1 Coordinate plane • x-axis • y-axis • Origin • Coordinates • Ordered pair • x-coordinate (abscissa) • y-coordinate (ordinate) • (x, y)

6-2 Line of reflection • Line reflection • Fixed point • Transformation • Preimage • Image • Reflection in line k • r_k • Axis of symmetry • Line symmetry

6-3 $r_{y\text{-axis}}$ • $r_{x\text{-axis}}$ • $r_{y=x}$

6-4 Point reflection in P • R_P • Point symmetry • R_O

6-5 Translation • Translation of a units in the horizontal direction and b units in the vertical direction • $T_{a,b}$ • Translational symmetry

6-6 Rotation • $R_{P,d}$ • Positive rotation • Negative rotation • Rotational symmetry • Quarter turn

6-7 Composition of transformations • Glide reflection • Isometry

6-8 Dilation • D_k

6-9 Function • Domain • Range • Orientation • Direct isometry • Opposite Isometry

REVIEW EXERCISES

1. Write the equation of the vertical line that is 2 units to the left of the origin.

2. Write the equation of the horizontal line that is 4 units below the origin.

3. a. On graph paper, draw the polygon $ABCD$ whose vertices are $A(-4, 0)$, $B(0, 0)$, $C(3, 3)$, and $D(-4, 3)$.

b. Find the area of polygon $ABCD$.

In 4–11: **a.** Find the image of $P(5, -3)$ under each of the given transformations. **b.** Name a fixed point of the transformation if one exists.

4. $r_{x\text{-axis}}$ **5.** R_O **6.** $R_{90°}$ **7.** $T_{0,3}$

8. $r_{x\text{-axis}} \circ T_{2,2}$ **9.** $r_{x\text{-axis}} \circ r_{y=x}$ **10.** $r_{y=x} \circ r_{x\text{-axis}}$ **11.** $R_{(2,2)} \circ T_{5,-3}$

12. Draw a quadrilateral that has exactly one line of symmetry.

13. Draw a quadrilateral that has exactly four lines of symmetry.

14. Print a letter that has rotational symmetry.

15. The letters **S** and **N** have point symmetry. Print another letter that has point symmetry.

16. What transformations are opposite isometries?

17. What transformation is not an isometry?

18. a. On graph paper, locate the points $A(3, 2)$, $B(3, 7)$, and $C(-2, 7)$. Draw $\triangle ABC$.

b. Draw $\triangle A'B'C'$, the image of $\triangle ABC$ under a reflection in the origin, and write the coordinates of its vertices.

c. Draw $\triangle A''B''C''$, the image of $\triangle A'B'C'$ under a reflection in the y-axis, and write the coordinates of its vertices.

d. Under what single transformation is $\triangle A''B''C''$ the image of $\triangle ABC$?

19. a. On graph paper, locate the points $R(-4, -1)$, $S(-1, -1)$, and $T(-1, 2)$. Draw $\triangle RST$.

b. Draw $\triangle R'S'T'$, the image of $\triangle RST$ under a reflection in the origin, and write the coordinates of its vertices.

c. Draw $\triangle R''S''T''$, the image of $\triangle R'S'T'$ under a reflection in the line whose equation is $y = 4$, and write the coordinates of its vertices.

20. a. Write the coordinates of the image, under the correspondence $(x, y) \rightarrow (x, 0)$, of each of the following ordered pairs:

$$(3, 5), (3, 3), (-1, -1), (-1, 5).$$

b. Explain why $(x, y) \rightarrow (x, 0)$ is not a transformation.

21. The vertices of $\triangle MAT$ have coordinates $M(1, 3)$, $A(2, 2)$, and $T(-2, 2)$.

a. Find $\triangle M'A'T'$, the image of $\triangle MAT$ under the composition $r_{x\text{-axis}} \circ D_3$.

b. Find $\triangle M''A''T''$, the image of $\triangle MAT$ under the composition $D_3 \circ r_{x\text{-axis}}$.

c. Are $r_{x\text{-axis}} \circ D_3$ and $D_3 \circ r_{x\text{-axis}}$ equivalent transformations? Justify your answer.

Exploration

Designs for wallpaper, wrapping paper, or fabric often repeat the same pattern in different arrangements. Such designs are called **tessellations**. Find examples of the use of line reflections, point reflections, translations and glide reflections in these designs.

CUMULATIVE REVIEW   Chapters 1–6

Part I

Answer all questions in this part. Each correct answer will receive 2 credits. No partial credit will be allowed.

1. The product $-7a(3a - 1)$ can be written as

 (1) $-21a - 7a$ (2) $-21a + 7a$ (3) $-21a^2 + 7a$ (4) $-21a^2 - 1$

2. In the coordinate plane, the point whose coordinates are $(-2, 0)$ is

 (1) on the x-axis   (3) in the first quadrant
 (2) on the y-axis   (4) in the fourth quadrant

3. If D is not on $\overline{ABC}$, then

 (1) $AD + DC = AC$ (3) $AD + DC < AC$
 (2) $AD + DC > AC$ (4) $AB + BC > AD + DC$

4. If $\overleftrightarrow{ABC}$ is the perpendicular bisector of $\overline{DBE}$, which of the following could be *false*?

 (1) $AD = AE$ (2) $DB = BE$ (3) $CD = CE$ (4) $AB = BC$

5. $\triangle DEF$ is not congruent to $\triangle LMN$, $DE = LM$, and $EF = MN$. Which of the following must be true?

 (1) $\triangle DEF$ and $\triangle LMN$ are not both right triangles.
 (2) $m\angle D \neq m\angle L$
 (3) $m\angle F \neq m\angle N$
 (4) $m\angle E \neq m\angle M$

6. Under a reflection in the origin, the image of $(-2, 4)$ is

 (1) $(-2, -4)$ (2) $(2, 4)$ (3) $(2, -4)$ (4) $(4, -2)$

7. If $PQ = RQ$, then which of the following must be true?

 (1) Q is the midpoint of $\overline{PR}$.
 (2) Q is on the perpendicular bisector of $\overline{PR}$.
 (3) $PQ + QR = PR$
 (4) Q is between P and R.

8. Which of the following always has a line of symmetry?
(1) a right triangle   (3) an isosceles triangle
(2) a scalene triangle   (4) an acute triangle

9. In the coordinate plane, two points lie on the same vertical line. Which of the following must be true?
(1) The points have the same x-coordinates.
(2) The points have the same y-coordinates.
(3) The points lie on the y-axis.
(4) The points lie on the x-axis.

10. Angle A and angle B are complementary angles. If $m\angle A = x + 42$ and $m\angle B = 2x - 12$, the measure of the smaller angle is
(1) 20   (2) 28   (3) 62   (4) 88

Part II

Answer all questions in this part. Each correct answer will receive 2 credits. Clearly indicate the necessary steps, including appropriate formula substitutions, diagrams, graphs, charts, etc. For all questions in this part, a correct numerical answer with no work shown will receive only 1 credit.

11. The image of $\triangle ABC$ under a reflection in the line $y = x$ is $\triangle ADE$. If the coordinates of A are $(-2, b)$, what is the value of b? Explain your answer.

12. What are the coordinates of the midpoint of a line segment whose endpoints are $A(2, -5)$ and $B(2, 3)$?

Part III

Answer all questions in this part. Each correct answer will receive 4 credits. Clearly indicate the necessary steps, including appropriate formula substitutions, diagrams, graphs, charts, etc. For all questions in this part, a correct numerical answer with no work shown will receive only 1 credit.

13. In quadrilateral $ABCD$, prove that $\angle ABC \cong \angle ADC$ if $\overline{AEC}$ is the perpendicular bisector of $\overline{BED}$.

14. The measure of $\angle R$ is 12 degrees more than three times the measure of $\angle S$. If $\angle S$ and $\angle R$ are supplementary angles, find the measure of each angle.

Part IV

Answer all questions in this part. Each correct answer will receive 6 credits. Clearly indicate the necessary steps, including appropriate formula substitutions, diagrams, graphs, charts, etc. For all questions in this part, a correct numerical answer with no work shown will receive only 1 credit.

15. The measures of the angles of $\triangle ABC$ are unequal ($m\angle A \neq m\angle B$, $m\angle B \neq m\angle C$, and $m\angle A \neq m\angle C$). Prove that $\triangle ABC$ is a scalene triangle.

16. *Given:* $T_{3,-2}(\triangle ABC) = \triangle A'B'C'$ and $\triangle XYZ$ with $XY = A'B'$, $YZ = B'C'$, and $\angle Y \cong \angle B$.

Prove: $\triangle ABC \cong \triangle XYZ$

CHAPTER

7

GEOMETRIC INEQUALITIES

Euclid's Proposition 20 of Book 1 of the *Elements* states, "In any triangle, two sides taken together in any manner are greater than the remaining one."

The Epicureans, a group of early Greek philosophers, ridiculed this theorem, stating that it is evident even to a donkey since if food is placed at one vertex of a triangle and the donkey at another, the donkey will make his way along one side of the triangle rather than traverse the other two, to get to the food. But no matter how evident the truth of a statement may be, it is important that it be logically established in order that it may be used in the proof of theorems that follow. Many of the inequality theorems of this chapter depend on this statement for their proof.

7-1 BASIC INEQUALITY POSTULATES

Each time the athletes of the world assemble for the Olympic Games, they attempt to not only perform better than their competitors at the games but also to surpass previous records in their sport. News commentators are constantly comparing the winning time of a bobsled run or a 500-meter skate with the world records and with individual competitors' records.

In previous chapters, we have studied pairs of congruent lines and pairs of congruent angles that have equal measures. But pairs of lines and pairs of angles are often not congruent and have unequal measures. In this chapter, we will apply the basic inequality principles that we used in algebra to the lengths of line segments and the measures of angles. These inequalities will enable us to formulate many important geometric relationships.

Postulate Relating a Whole Quantity and Its Parts

In Chapter 3 we stated postulates of equality. Many of these postulates suggest related postulates of inequality.

Consider the partition postulate:

▶ **A whole is equal to the sum of all its parts.**

This corresponds to the following postulate of inequality:

Postulate 7.1

A whole is greater than any of its parts.

In arithmetic: Since $14 = 9 + 5$, then $14 > 9$ and $14 > 5$.

In algebra: If a, b, and c represent positive numbers and $a = b + c$, then $a > b$ and $a > c$.

In geometry: The lengths of line segments and the measures of angles are positive numbers.

Consider these two applications:

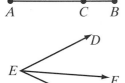

- If $\overline{ACB}$ is a line segment, then $AB = AC + CB$, $AB > AC$, and $AB > CB$.
- If $\angle DEF$ and $\angle FEG$ are adjacent angles, $m\angle DEG = m\angle DEF + m\angle FEG$, $m\angle DEG > m\angle DEF$, and $m\angle DEG > m\angle FEG$.

Transitive Property

Consider this statement of the transitive property of equality:

▶ **If a, b, and c are real numbers such that $a = b$ and $b = c$, then $a = c$.**

This corresponds to the following **transitive property of inequality**:

Postulate 7.2 | If a, b, and c are real numbers such that $a > b$ and $b > c$, then $a > c$.

In arithmetic: If $12 > 7$ and $7 > 3$, then $12 > 3$.

In algebra: If $5x + 1 > 2x$ and $2x > 16$, then $5x + 1 > 16$.

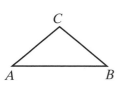

In geometry: If $BA > BD$ and $BD > BC$, then $BA > BC$.
Also, if $m\angle BCA > m\angle BCD$ and $m\angle BCD > m\angle BAC$, then $m\angle BCA > m\angle BAC$.

Substitution Postulate

Consider the substitution postulate as it relates to equality:

▶ **A quantity may be substituted for its equal in any statement of equality.**

Substitution also holds for inequality, as demonstrated in the following postulate:

Postulate 7.3 | A quantity may be substituted for its equal in any statement of inequality.

In arithmetic: If $10 > 2 + 5$ and $2 + 5 = 7$, then $10 > 7$.

In algebra: If $5x + 1 > 2y$ and $y = 4$, then $5x + 1 > 2(4)$.

In geometry: If $AB > BC$ and $BC = AC$, then $AB > AC$.
Also, if $m\angle C > m\angle A$ and $m\angle A = m\angle B$, then $m\angle C > m\angle B$.

The Trichotomy Postulate

We know that if x represents the coordinate of a point on the number line, then x can be a point to the left of 3 when $x < 3$, x can be the point whose coordinate is 3 if $x = 3$, or x can be a point to the right of 3 if $x > 3$. We can state this as a postulate that we call the **trichotomy postulate**, meaning that it is divided into three cases.

Postulate 7.4 | Given any two quantities, a and b, one and only one of the following is true:
$$a < b \quad \text{or} \quad a = b \quad \text{or} \quad a > b.$$

EXAMPLE 1

Given: m∠*DAC* = m∠*DAB* + m∠*BAC* and
   m∠*DAB* > m∠*ABC*

Prove: m∠*DAC* > m∠*ABC*

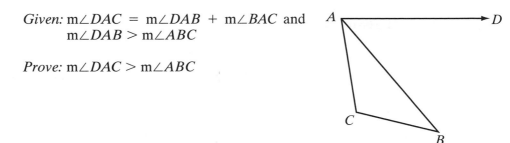

Proof

Statements	Reasons
1. m∠*DAC* = m∠*DAB* + m∠*BAC*	**1.** Given.
2. m∠*DAC* > m∠*DAB*	**2.** A whole is greater than any of its parts.
3. m∠*DAB* > m∠*ABC*	**3.** Given.
4. m∠*DAC* > m∠*ABC*	**4.** Transitive property of inequality.

EXAMPLE 2

Given: *Q* is the midpoint of $\overline{PS}$ and
   RS < *QS*.

Prove: *RS* < *PQ*

Proof

Statements	Reasons
1. *Q* is the midpoint of $\overline{PS}$.	**1.** Given.
2. $\overline{PQ} \cong \overline{QS}$	**2.** The midpoint of a line segment is the point that divides the segment into two congruent segments.
3. *PQ* = *QS*.	**3.** Congruent segments have equal measures.
4. *RS* < *QS*	**4.** Given.
5. *RS* < *PQ*	**5.** Substitution postulate.

Exercises

Writing About Mathematics

1. Is inequality an equivalence relation? Explain why or why not.

2. Monica said that when $AB > BC$ is false, $AB < BC$ must be true. Do you agree with Monica? Explain your answer.

Developing Skills

In 3–12: **a.** Draw a diagram to illustrate the hypothesis and tell whether each conclusion is true or false. **b.** State a postulate or a definition that justifies your answer.

3. If $\overline{ADB}$ is a line segment, then $DB < AB$.

4. If D is not on $\overleftrightarrow{AC}$, then $CD + DA < CA$.

5. If $\angle BCD + \angle DCA = \angle BCA$, then m$\angle BCD <$ m$\angle BCA$.

6. If $\overrightarrow{DB}$ and $\overrightarrow{DA}$ are opposite rays with point C not on $\overrightarrow{DB}$ or $\overrightarrow{DA}$, then
m$\angle BDC +$ m$\angle CDA = 180$.

7. If $\overrightarrow{DB}$ and $\overrightarrow{DA}$ are opposite rays and m$\angle BDC > 90$, then m$\angle CDA > 90$.

8. If $\overline{ADB}$ is a line segment, then $DA > BD$, or $DA = BD$, or $DA < BD$.

9. If $AT > AS$ and $AS > AR$, then $AT > AR$.

10. If m$\angle 1 >$ m$\angle 2$ and m$\angle 2 >$ m$\angle 3$, then m$\angle 1 >$ m$\angle 3$.

11. If $SR > KR$ and $SR = TR$, then $TR > KR$.

12. If m$\angle 3 <$ m$\angle 2$ and m$\angle 2 =$ m$\angle 1$, then m$\angle 3 <$ m$\angle 1$.

Applying Skills

13. *Given:* $\triangle ABC$ is isosceles, $\overline{AC} \cong \overline{BC}$,
m$\angle CBD >$ m$\angle CBA$

Prove: m$\angle CBD >$ m$\angle A$

14. *Given:* $\overline{PQRS}$ and $PQ = RS$

Prove: **a.** $PR > PQ$ **b.** $PR > RS$

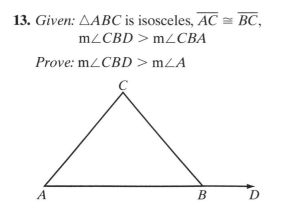

In 15 and 16, use the figure to the right.

15. If $\overline{KLM}$ and $\overline{LM} \cong \overline{NM}$, prove that $KM > NM$.

16. If $KM > KN$, $KN > NM$, and $NM = NL$, prove that $KM > NL$.

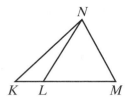

7-2 INEQUALITY POSTULATES INVOLVING ADDITION AND SUBTRACTION

Postulates of equality and examples of inequalities involving the numbers of arithmetic can help us to understand the inequality postulates presented here.
Consider the addition postulate:

▶ **If equal quantities are added to equal quantities, then the sums are equal.**

Addition of inequalities requires two cases:

Postulate 7.5

> If equal quantities are added to unequal quantities, then the sums are unequal in the same order.

Postulate 7.6

> If unequal quantities are added to unequal quantities in the same order, then the sums are unequal in the same order.

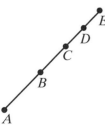

In arithmetic: Since $12 > 5$, then $12 + 3 > 5 + 3$ or $15 > 8$.
Since $12 > 5$ and $3 > 2$, then $12 + 3 > 5 + 2$ or $15 > 7$.

In algebra: If $x - 5 > 10$, then $x - 5 + 5 > 10 + 5$ or $x > 15$.
If $x - 5 > 10$ and $5 > 3$, then $x - 5 + 5 > 10 + 3$ or $x > 13$.

In geometry: If $\overline{ABCD}$ and $AB > CD$, then $AB + BC > BC + CD$ or $AC > BD$.
If $\overline{ABCDE}$, $AB > CD$, and $BC > DE$, then $AB + BC > CD + DE$ or $AC > CE$.

We can subtract equal quantities from unequal quantities without changing the order of the inequality, but the result is uncertain when we subtract unequal quantities from unequal quantities.
Consider the subtraction postulate:

▶ **If equal quantities are subtracted from equal quantities, then the differences are equal.**

Subtraction of inequalities is restricted to a single case:

Postulate 7.7

> If equal quantities are subtracted from unequal quantities, then the differences are unequal in the same order.

However, when unequal quantities are subtracted from unequal quantities, the results may or may not be unequal and the order of the inequality may or may not be the same.

For example:

- $5 > 2$ and $4 > 1$, but it is not true that $5 - 4 > 2 - 1$ since $1 = 1$.
- $12 > 10$ and $7 > 1$, but it is not true that $12 - 7 > 10 - 1$ since $5 < 9$.
- $12 > 10$ and $2 > 1$, and it is true that $12 - 2 > 10 - 1$ since $10 > 9$.

EXAMPLE 1

Given: $m\angle BDE < m\angle CDA$

Prove: $m\angle BDC < m\angle EDA$

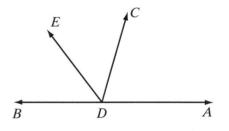

Proof	**Statements**	**Reasons**
	1. $m\angle BDE < m\angle CDA$	**1.** Given.
	2. $m\angle BDE + m\angle EDC$ $< m\angle EDC + m\angle CDA$	**2.** If equal quantities are added to unequal quantities, then the sums are unequal in the same order.
	3. $m\angle BDC = m\angle BDE + m\angle EDC$	**3.** The whole is equal to the sum of its parts.
	4. $m\angle EDA = m\angle EDC + m\angle CDA$	**4.** The whole is equal to the sum of its parts.
	5. $m\angle BDC < m\angle EDA$	**5.** Substitution postulate for inequalities.

Exercises

Writing About Mathematics

1. Dana said that $13 > 11$ and $8 > 3$. Therefore, $13 - 8 < 11 - 3$ tells us that if unequal quantities are subtracted from unequal quantities, the difference is unequal in the opposite order. Do you agree with Dana? Explain why or why not.

2. Ella said that if unequal quantities are subtracted from equal quantities, then the differences are unequal in the opposite order. Do you agree with Ella? Explain why or why not.

Developing Skills

In 3–10, in each case use an inequality postulate to prove the conclusion.

3. If $10 > 7$, then $18 > 15$.

4. If $4 < 14$, then $15 < 25$.

5. If $x + 3 > 12$, then $x > 9$.

6. If $y - 5 < 5$, then $y < 10$.

7. If $8 > 6$ and $5 > 3$, then $13 > 9$.

8. If $7 < 12$, then $5 < 10$.

9. If $y > 8$, then $y - 1 > 7$.

10. If $a = b$, then $180 - a > 90 - b$.

Applying Skills

11. *Given:* $AB = AD$, $BC < DE$

  Prove: $AC < AE$

12. *Given:* $AE > BD$, $AF = BF$

  Prove: $FE > FD$

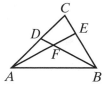

13. *Given:* $m\angle DAC > m\angle DBC$ and $AE = BE$

  Prove: **a.** $m\angle EAB = m\angle EBA$

  b. $m\angle DAB > m\angle CBA$

14. In August, Blake weighed more than Caleb. In the next two months, Blake and Caleb had each gained the same number of pounds. Does Blake still weigh more than Caleb? Justify your answer.

15. In December, Blake weighed more than Andre. In the next two months, Blake lost more than Andre lost. Does Blake still weigh more than Andre? Justify your answer.

7-3 INEQUALITY POSTULATES INVOLVING MULTIPLICATION AND DIVISION

Since there are equality postulates for multiplication and division similar to those of addition and subtraction, we would expect that there are inequality postulates for multiplication and division similar to those of addition and subtraction. Consider these examples that use both positive and negative numbers.

If $9 > 3$, then $9(4) > 3(4)$ or $36 > 12.$	If $-9 < -3$, then $-9(4) < -3(4)$ or $-36 < -12.$
If $1 < 5$, then $1(3) < 5(3)$ or $3 < 15.$	If $-1 > -5$, then $-1(3) > -5(3)$ or $-1 > -15.$
If $9 > 3$, then $9(-4) < 3(-4)$ or $-36 < -12.$	If $-9 < -3$, then $-9(-4) > -3(-4)$ or $36 > 12.$
If $1 < 5$, then $1(-3) > 5(-3)$ or $-3 > -15.$	If $-1 > -5$, then $-1(-3) < -5(-3)$ or $3 < 15.$

Notice that in the top four examples, we are multiplying by positive numbers and the order of the inequality does not change. In the bottom four examples, we are multiplying by negative numbers and the order of the inequality does change.

These examples suggest the following postulates of inequality:

Postulate 7.8

If unequal quantities are multiplied by positive equal quantities, then the products are unequal in the same order.

Postulate 7.9

If unequal quantities are multiplied by negative equal quantities, then the products are unequal in the opposite order.

Since we know that division by $a \neq 0$ is the same as multiplication by $\frac{1}{a}$ and that a and $\frac{1}{a}$ are always either both positive or both negative, we can write similar postulates for division of inequalities.

Postulate 7.10

If unequal quantities are divided by positive equal quantities, then the quotients are unequal in the same order.

Postulate 7.11 | If unequal quantities are divided by negative equal quantities, then the quotients are unequal in the opposite order.

Care must be taken when using inequality postulates involving multiplication and division because multiplying or dividing by a *negative* number will reverse the order of the inequality.

EXAMPLE 1

Given: $BA = 3BD$, $BC = 3BE$, and $BE > BD$

Prove: $BC > BA$

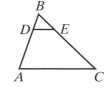

Proof	**Statements**	**Reasons**
	1. $BE > BD$	**1.** Given.
	2. $3BE > 3BD$	**2.** If unequal quantities are multiplied by positive equal quantities, then the products are unequal in the same order.
	3. $BC = 3BE$, $BA = 3BD$	**3.** Given.
	4. $BC > BA$	**4.** Substitution postulate for inequalities.

EXAMPLE 2

Given: $m\angle ABC > m\angle DEF$, $\overrightarrow{BG}$ bisects $\angle ABC$, $\overrightarrow{EH}$ bisects $\angle DEF$.

Prove: $m\angle ABG > m\angle DEH$

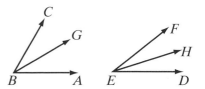

Proof An angle bisector separates the angle into two congruent parts. Therefore, the measure of each part is one-half the measure of the angle that was bisected, so $m\angle ABG = \frac{1}{2}m\angle ABC$ and $m\angle DEH = \frac{1}{2}m\angle DEF$.

Since we are given that $m\angle ABC > m\angle DEF$, $\frac{1}{2}m\angle ABC > \frac{1}{2}m\angle DEF$ because if unequal quantities are multiplied by positive equal quantities, the products are unequal in the same order. Therefore, by the substitution postulate for inequality, $m\angle ABG > m\angle DEH$.

Exercises

Writing About Mathematics

1. Since $1 < 2$, is it always true that $a < 2a$? Explain why or why not.

2. Is it always true that if $a > b$ and $c > d$, then $ac > bd$? Justify your answer.

Developing Skills

In 3–8, in each case state an inequality postulate to prove the conclusion.

3. If $8 > 7$, then $24 > 21$.

4. If $30 < 35$, then $-6 > -7$.

5. If $8 > 6$, then $4 > 3$.

6. If $3x > 15$, then $x > 5$.

7. If $\frac{x}{2} > -4$, then $-x < 8$.

8. If $\frac{y}{6} < 3$, then $y < 18$.

In 9–17: If a, b, and c are positive real numbers such that $a > b$ and $b > c$, tell whether each relationship is *always true*, *sometimes true*, or *never true*. If the statement is always true, state the postulate illustrated. If the statement is sometimes true, give one example for which it is true and one for which it is false. If the statement is never true, give one example for which it is false.

9. $ac > bc$

10. $a + c > b + c$

11. $c - a > c - b$

12. $a - c > b - c$

13. $a - b > b - c$

14. $\frac{c}{a} > \frac{c}{b}$

15. $\frac{a}{c} > \frac{b}{c}$

16. $-ac > -bc$

17. $a > c$

Applying Skills

18. *Given:* $BD < BE$, D is the midpoint of $\overline{BA}$, E is the midpoint of $\overline{BC}$.

Prove: $BA < BC$

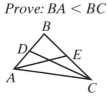

19. *Given:* $m\angle DBA > m\angle CAB$, $m\angle CBA = 2m\angle DBA$, $m\angle DAB = 2m\angle CAB$

Prove: $m\angle CBA > m\angle DAB$

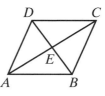

20. *Given:* $AB > AD$, $AE = \frac{1}{2}AB$, $AF = \frac{1}{2}AD$

Prove: $AE > AF$

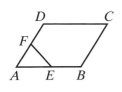

21. *Given:* $m\angle CAB < m\angle CBA$, $\overline{AD}$ bisects $\angle CAB$, $\overline{BE}$ bisects $\angle CBA$.

Prove: $m\angle DAB < m\angle EBA$

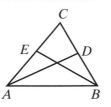

7-4 AN INEQUALITY INVOLVING THE LENGTHS OF THE SIDES OF A TRIANGLE

The two quantities to be compared are often the lengths of line segments or the distances between two points. The following postulate was stated in Chapter 4.

▶ **The shortest distance between two points is the length of the line segment joining these two points.**

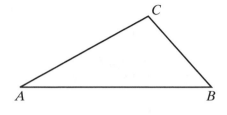

The vertices of a triangle are three noncollinear points. The length of $\overline{AB}$ is AB, the shortest distance from A to B. Therefore, $AB < AC + CB$. Similarly, $BC < BA + AC$ and $AC < AB + BC$. We have just proved the following theorem, called the **triangle inequality theorem**:

Theorem 7.1

> The length of one side of a triangle is less than the sum of the lengths of the other two sides.

In the triangle shown above, $AB > AC > BC$. To show that the lengths of three line segments can be the measures of the sides of a triangle, we must show that the length of any side is less than the sum of the other two lengths of the other two sides.

EXAMPLE I

Which of the following may be the lengths of the sides of a triangle?

(1) 4, 6, 10   (2) 8, 8, 16   (3) 6, 8, 16   (4) 10, 12, 14

Solution The length of a side of a triangle must be less than the sum of the lengths of the other two sides. If the lengths of the sides are $a < b < c$, then $a < b$ means that $a < b + c$ and $b < c$ means that $b < c + a$. Therefore, we need only test the longest side.

(1) Is $10 < 4 + 6$? No

(2) Is $16 < 8 + 8$? No

(3) Is $16 < 6 + 8$? No

(4) Is $14 < 10 + 12$? Yes *Answer*

EXAMPLE 2

Two sides of a triangle have lengths 3 and 7. Find the range of possible lengths of the third side.

Solution (1) Let s = length of third side of triangle.

(2) Of the lengths 3, 7, and s, the longest side is either 7 or s.

(3) If the length of the longest side is s, then $s < 3 + 7$ or $s < 10$.

(4) If the length of the longest side is 7, then $7 < s + 3$ or $4 < s$.

Answer $4 < s < 10$

EXAMPLE 3

Given: Isosceles triangle ABC with $AB = CB$ and M the midpoint of $\overline{AC}$.

Prove: $AM < AB$

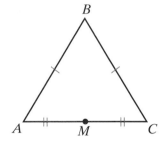

Proof

Statements	Reasons
1. $AC < AB + CB$	**1.** The length of one side of a triangle is less than the sum of the lengths of the other two sides.
2. $AB = CB$	**2.** Given.
3. $AC < AB + AB$ or $AC < 2AB$	**3.** Substitution postulate for inequalities.
4. M is the midpoint of $\overline{AC}$.	**4.** Given.
5. $AM = MC$	**5.** Definition of a midpoint.
6. $AC = AM + MC$	**6.** Partition postulate.
7. $AC = AM + AM = 2AM$	**7.** Substitution postulate.
8. $2AM < 2AB$	**8.** Substitution postulate for inequality.
9. $AM < AB$	**9.** Division postulate for inequality.

Exercises

Writing About Mathematics

1. If 7, 12, and *s* are the lengths of three sides of a triangle, and *s* is not the longest side, what are the possible values of *s*?

2. **a.** If $a < b < c$ are any real numbers, is $a < b + c$ always true? Justify your answer.

   b. If $a < b < c$ are the lengths of the sides of a triangle, is $a < b + c$ always true? Justify your answer.

Developing Skills

In 3–10, tell in each case whether the given lengths can be the measures of the sides of a triangle.

3. 3, 4, 5	**4.** 5, 8, 13	**5.** 6, 7, 10	**6.** 3, 9, 15
7. 2, 2, 3	**8.** 1, 1, 2	**9.** 3, 4, 4	**10.** 5, 8, 11

In 11–14, find values for *r* and *t* such that the inequality $r < s < t$ best describes *s*, the length of the third sides of a triangle for which the lengths of the other two sides are given.

11. 2 and 4   **12.** 12 and 31   **13.** $\frac{13}{2}$ and $\frac{13}{2}$ **14.** 9.6 and 12.5

15. Explain why $x, 2x,$ and $3x$ cannot represent the lengths of the sides of a triangle.

16. For what values of *a* can $a, a + 2, a - 2$ represent the lengths of the sides of a triangle? Justify your answer.

Applying Skills

17. *Given: ABCD* is a quadrilateral.

   Prove: AD < AB + BC + CD

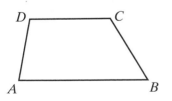

18. *Given:* $\triangle ABC$ with *D* a point on $\overline{BC}$ and $AD = DC.$

   Prove: AB < BC

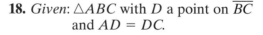

19. *Given:* Point *P* in the interior of $\triangle XYZ, \overline{YPQ}$

   Prove: PY + PZ < XY + XZ

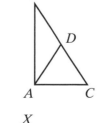

Hands-On Activity

One side of a triangle has a length of 6. The lengths of the other two sides are integers that are less than or equal to 6.

a. Cut one straw 6 inches long and two sets of straws to integral lengths of 1 inch to 6 inches. Determine which lengths can represent the sides of a triangle.

Or

Use geometry software to determine which lengths can represent the sides of a triangle.

b. List all sets of three integers that can be the lengths of the sides of the triangle. For example,

$$\{6, 3, 5\}$$

is one set of lengths.

c. List all sets of three integers less than or equal to 6 that *cannot* be the lengths of the sides of the triangle.

d. What patterns emerge in the results of parts **b** and **c**?

7-5 AN INEQUALITY INVOLVING AN EXTERIOR ANGLE OF A TRIANGLE

Exterior Angles of a Polygon

At each vertex of a polygon, an angle is formed that is the union of two sides of the polygon. Thus, for polygon $ABCD$, $\angle DAB$ is an angle of the polygon, often called an interior angle. If, at vertex A, we draw $\overrightarrow{AE}$, the opposite ray of $\overrightarrow{AD}$, we form $\angle BAE$, an exterior angle of the polygon at vertex A.

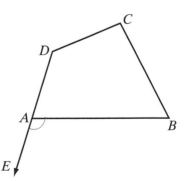

DEFINITION _____

An **exterior angle of a polygon** is an angle that forms a linear pair with one of the interior angles of the polygon.

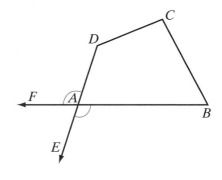

At vertex A, we can also draw $\overrightarrow{AF}$, the opposite ray of $\overrightarrow{AB}$, to form $\angle DAF$, another exterior angle of the polygon at vertex A. At each vertex of a polygon, two exterior angles can be drawn. Each of these exterior angles forms a linear pair with the interior angle at A, and the angles in a linear pair are supplementary. The two exterior angles at A are congruent angles because they are vertical angles. Either can be drawn as the exterior angle at A.

Exterior Angles of a Triangle

An exterior angle of a triangle is formed outside the triangle by extending a side of the triangle.

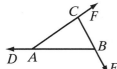

The figure to the left shows $\triangle ABC$ whose three interior angles are $\angle CAB$, $\angle ABC$, and $\angle BCA$. By extending each side of $\triangle ABC$, three exterior angles are formed, namely, $\angle DAC$, $\angle EBA$, and $\angle FCB$.

For each exterior angle, there is an **adjacent interior angle** and two **remote** or **nonadjacent interior angles**. For $\triangle ABC$, these angles are as follows:

Vertex	Exterior Angle	Adjacent Interior Angle	Nonadjacent Interior Angles
A	$\angle DAC$	$\angle CAB$	$\angle ABC$ and $\angle BCA$
B	$\angle EBA$	$\angle ABC$	$\angle CAB$ and $\angle BCA$
C	$\angle FCB$	$\angle BCA$	$\angle CAB$ and $\angle ABC$

With these facts in mind, we are now ready to prove another theorem about inequalities in geometry called the **exterior angle inequality theorem**.

Theorem 7.2

The measure of an exterior angle of a triangle is greater than the measure of either nonadjacent interior angle.

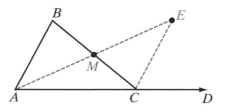

Given $\triangle ABC$ with exterior $\angle BCD$ at vertex C; $\angle A$ and $\angle B$ are nonadjacent interior angles with respect to $\angle BCD$.

Prove   m$\angle BCD >$ m$\angle B$

Proof	Statements	Reasons
	1. Let M be the midpoint of $\overline{BC}$.	**1.** Every line segment has one and only one midpoint.
	2. Draw $\overrightarrow{AM}$, extending the ray through M to point E so that $\overline{AM} \cong \overline{EM}$.	**2.** Two points determine a line. A line segment can be extended to any length.
	3. Draw $\overline{EC}$.	**3.** Two points determine a line.
	4. $m\angle BCD = m\angle BCE + m\angle ECD$	**4.** A whole is equal to the sum of its parts.
	5. $\overline{BM} \cong \overline{CM}$	**5.** Definition of midpoint.
	6. $\overline{AM} \cong \overline{EM}$	**6.** Construction (step 2).
	7. $\angle AMB \cong \angle EMC$	**7.** Vertical angles are congruent.
	8. $\triangle AMB \cong \triangle EMC$	**8.** SAS (steps 5, 7, 6).
	9. $\angle B \cong \angle MCE$	**9.** Corresponding parts of congruent triangles are congruent.
	10. $m\angle BCD > m\angle MCE$	**10.** A whole is greater than any of its parts.
	11. $m\angle BCD > m\angle B$	**11.** Substitution postulate for inequalities.

These steps prove that the measure of an exterior angle is greater than the measure of one of the nonadjacent interior angles, $\angle B$. A similar proof can be used to prove that the measure of an exterior angle is greater than the measure of the other nonadjacent interior angle, $\angle A$. This second proof uses N, the midpoint of $\overline{AC}$, a line segment $\overline{BNG}$ with $\overline{BN} \cong \overline{NG}$, and a point F extending ray $\overrightarrow{BC}$ through C.

The details of this proof will be left to the student. (See exercise 14.)

EXAMPLE 1

The point D is on $\overline{AB}$ of $\triangle ABC$.

  a. Name the exterior angle at D of $\triangle ADC$.

  b. Name two nonadjacent interior angles of the exterior angle at D of $\triangle ADC$.

  c. Why is $m\angle CDB > m\angle DCA$?

  d. Why is $AB > AD$?

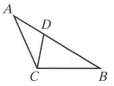

Solution   **a.** $\angle CDB$

b. $\angle DCA$ and $\angle A$

c. The measure of an exterior angle of a triangle is greater than the measure of either nonadjacent interior angle.

d. The whole is greater than any of its parts. ◻

EXAMPLE 2

Given: Right triangle ABC, m$\angle C = 90$, $\angle BAD$ is an exterior angle at A.

Prove: $\angle BAD$ is obtuse.

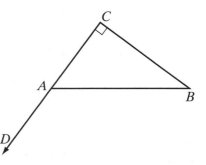

Proof

Statements	Reasons
1. $\angle BAD$ is an exterior angle.	**1.** Given.
2. m$\angle BAD > $ m$\angle C$	**2.** Exterior angle inequality theorem.
3. m$\angle C = 90$	**3.** Given.
4. m$\angle BAD > 90$	**4.** Substitution postulate for inequalities.
5. m$\angle BAD + $ m$\angle BAC = 180$	**5.** If two angles form a linear pair, then they are supplementary.
6. $180 > $ m$\angle BAD$	**6.** The whole is greater than any of its parts.
7. $180 > $ m$\angle BAD > 90$	**7.** Steps 4 and 6.
8. $\angle BAD$ is obtuse.	**8.** An obtuse angle is an angle whose degree measure is greater than 90 and less than 180. ◻

Exercises

Writing About Mathematics

1. Evan said that every right triangle has at least one exterior angle that is obtuse. Do you agree with Evan? Justify your answer.

2. Connor said that every right triangle has at least one exterior angle that is a right angle. Do you agree with Connor? Justify your answer.

Developing Skills

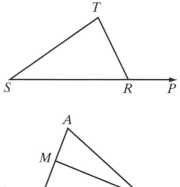

3. a. Name the exterior angle at R.

b. Name two nonadjacent interior angles of the exterior angle at R.

In 4–13, $\triangle ABC$ is scalene and $\overline{CM}$ is a median to side $\overline{AB}$.

a. Tell whether each given statement is *True* or *False*.

b. If the statement is true, state the definition, postulate, or theorem that justifies your answer.

4. $AM = MB$

5. $m\angle ACB > m\angle ACM$

6. $m\angle AMC > m\angle ABC$

7. $AB > AM$

8. $m\angle CMB > m\angle ACM$

9. $m\angle CMB > m\angle CAB$

10. $BA > MB$

11. $m\angle ACM = m\angle BCM$

12. $m\angle BCA > m\angle MCA$

13. $m\angle BMC = m\angle AMC$

Applying Skills

14. *Given:* $\triangle ABC$ with exterior $\angle BCD$ at vertex C; $\angle A$ and $\angle B$ are nonadjacent interior angles with respect to $\angle BCD$.

Prove: $m\angle BCD > m\angle A$
(Complete the proof of Theorem 7.2).

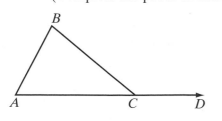

15. *Given:* $\angle ABD + \angle DBE = \angle ABE$ and $\angle ABE + \angle EBC = \angle ABC$

Prove: $m\angle ABD < m\angle ABC$

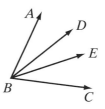

16. *Given:* Isosceles $\triangle DEF$ with
      $DE = FE$ and exterior $\angle EFG$

Prove: m$\angle EFG >$ m$\angle EFD$

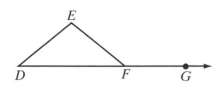

17. *Given:* Right $\triangle ABC$ with m$\angle C = 90$

Prove: $\angle A$ is acute.

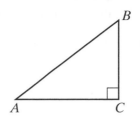

18. *Given:* $\triangle SMR$ with $\overline{STM}$ extended
      through M to P

Prove: m$\angle RMP >$ m$\angle SRT$

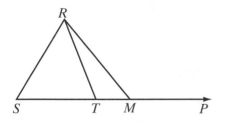

19. *Given:* Point F not on $\overleftrightarrow{ABCDE}$ and
      $FC = FD$

Prove: m$\angle ABF >$ m$\angle EDF$

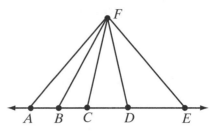

7-6 INEQUALITIES INVOLVING SIDES AND ANGLES OF A TRIANGLE

We know that if the lengths of two sides of a triangle are equal, then the measures of the angles opposite these sides are equal. Now we want to compare the measures of two angles opposite sides of unequal length.

Let the measures of the sides of $\triangle ABC$ be $AB = 12$, $BC = 5$, and $CA = 9$.

Write the lengths in order: $12 \quad > 9 \quad > 5$
Name the sides in order: $AB \quad > CA \quad > BC$
Name the angles opposite these sides in order: m$\angle C >$ m$\angle B >$ m$\angle A$

Notice how the vertex of the angle opposite a side of the triangle is always the point that is not an endpoint of that side.

Theorem 7.3

If the lengths of two sides of a triangle are unequal, then the measures of the angles opposite these sides are unequal and the larger angle lies opposite the longer side.

To prove this theorem, we will extend the shorter side of a triangle to a length equal to that of the longer side, forming an isosceles triangle. We can then use the isosceles triangle theorem and the exterior angle inequality theorem to compare angle measures.

Given $\triangle ABC$ with $AB > BC$

Prove $m\angle ACB > m\angle BAC$

Proof

Statements	Reasons
1. $\triangle ABC$ with $AB > BC$.	1. Given.
2. Extend $\overline{BC}$ through C to point D so that $BD = BA$.	2. A line segment may be extended to any length.
3. Draw $\overline{AD}$.	3. Two points determine a line.
4. $\triangle ABD$ is isosceles.	4. Definition of isosceles triangle.
5. $m\angle BAD = m\angle BDA$	5. Base angles of an isosceles triangle are equal in measure.
6. For $\triangle ACD$, $m\angle BCA > m\angle BDA$.	6. Exterior angle inequality theorem.
7. $m\angle BCA > m\angle BAD$	7. Substitution postulate for inequalities.
8. $m\angle BAD > m\angle BAC$	8. A whole is greater than any of its parts.
9. $m\angle BCA > m\angle BAC$	9. Transitive property of inequality.

The converse of this theorem is also true, as can be seen in this example:
Let the measures of the angles of $\triangle ABC$ be $m\angle A = 40$, $m\angle B = 80$, and $m\angle C = 60$.

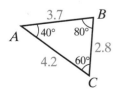

Write the angle measures in order: 80   > 60   > 40
Name the angles in order: $m\angle B > m\angle C > m\angle A$
Name the sides opposite these angles in order: AC > AB > BC

Theorem 7.4

If the measures of two angles of a triangle are unequal, then the lengths of the sides opposite these angles are unequal and the longer side lies opposite the larger angle.

We will write an indirect proof of this theorem. Recall that in an indirect proof, we assume the opposite of what is to be proved and show that the assumption leads to a contradiction.

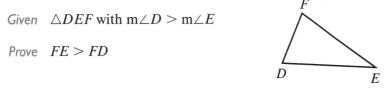

Given $\triangle DEF$ with m$\angle D >$ m$\angle E$

Prove $FE > FD$

Proof By the trichotomy postulate: $FE > FD$ or $FE = FD$ or $FE < FD$. We assume the negation of the conclusion, that is, we assume $FE \leq FD$. Therefore, either $FE = FD$ or $FE < FD$.

If $FE = FD$, then m$\angle D =$ m$\angle E$ because base angles of an isosceles triangle are equal in measure. This contradicts the given premise, m$\angle D >$ m$\angle E$. Thus, $FE = FD$ is a false assumption.

If $FE < FD$, then, by Theorem 7.3, we must conclude that m$\angle D <$ m$\angle E$. This also contradicts the given premise that m$\angle D >$ m$\angle E$. Thus, $FE < FD$ is also a false assumption.

Since $FE = FD$ and $FE < FD$ are both false, $FE > FD$ must be true and the theorem has been proved. ◼

EXAMPLE 1

One side of $\triangle ABC$ is extended to D. If m$\angle A = 45$, m$\angle B = 50$, and m$\angle BCD = 95$, which is the longest side of $\triangle ABC$?

Solution The exterior angle and the interior angle at vertex C form a linear pair and are supplementary. Therefore:

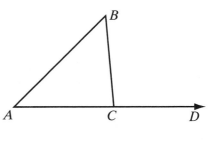

$$\text{m}\angle BCA = 180 - \text{m}\angle BCD$$
$$= 180 - 95$$
$$= 85$$

Since $85 > 50 > 45$, the longest side of the triangle is $\overline{BA}$, the side opposite $\angle BCA$. *Answer* ◼

EXAMPLE 2

In $\triangle ADC$, $\overline{CB}$ is drawn to $\overline{ABD}$ and $\overline{CA} \cong \overline{CB}$. Prove that $CD > CA$.

Proof Consider $\triangle CBD$. The measure of an exterior angle is greater than the measure of a nonadjacent interior angle, so m$\angle CBA >$ m$\angle CDA$. Since $\overline{CA} \cong \overline{CB}$, $\triangle ABC$ is isosceles. The base angles of an isosceles triangle have equal measures, so m$\angle A =$ m$\angle CBA$. A quantity may be substituted for its equal in an inequality, so m$\angle A >$ m$\angle CDA$.

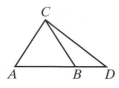

If the measures of two angles of a triangle are unequal, then the lengths of the sides opposite these angles are unequal and the longer side is opposite the larger angle. Therefore, $CD > AC$.

Exercises

Writing About Mathematics

1. **a.** Write the contrapositive of the statement "If the lengths of two sides of a triangle are unequal, then the measures of the angles opposite these sides are unequal."

   b. Is this contrapositive statement true?

2. The Isosceles Triangle Theorem states that if two sides of a triangle are congruent, then the angles opposite these sides are congruent.

   a. Write the converse of the Isosceles Triangle Theorem.

   b. How is this converse statement related to the contrapositive statement written in exercise 1?

Developing Skills

3. If $AB = 10$, $BC = 9$, and $CA = 11$, name the largest angle of $\triangle ABC$.

4. If m$\angle D = 60$, m$\angle E = 70$, and m$\angle F = 50$, name the longest side of $\triangle DEF$.

In 5 and 6, name the shortest side of $\triangle ABC$, using the given information.

5. In $\triangle ABC$, m$\angle C = 90$, m$\angle B = 35$, and m$\angle A = 55$.

6. In $\triangle ABC$, m$\angle A = 74$, m$\angle B = 58$, and m$\angle C = 48$.

In 7 and 8, name the smallest angle of $\triangle ABC$, using the given information.

7. In $\triangle ABC$, $AB = 7$, $BC = 9$, and $AC = 5$.

8. In $\triangle ABC$, $AB = 5$, $BC = 12$, and $AC = 13$.

9. In $\triangle RST$, an exterior angle at R measures 80 degrees. If m$\angle S >$ m$\angle T$, name the shortest side of the triangle.

10. If $\angle ABD$ is an exterior angle of $\triangle BCD$, m$\angle ABD = 118$, m$\angle D = 60$, and m$\angle C = 58$, list the sides of $\triangle BCD$ in order starting with the longest.

11. If $\angle EFH$ is an exterior angle of $\triangle FGH$, m$\angle EFH = 125$, m$\angle G = 65$, m$\angle H = 60$, list the sides of $\triangle FGH$ in order starting with the shortest.

12. In $\triangle RST$, $\angle S$ is obtuse and m$\angle R <$ m$\angle T$. List the lengths of the sides of the triangle in order starting with the largest.

Applying Skills

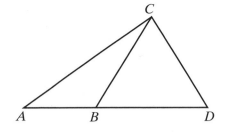

13. *Given:* C is a point that is not on $\overline{ABD}$,

$\quad\quad$ m$\angle ABC >$ m$\angle CBD$.

$\quad$ *Prove:* $AC > BC$

14. Let $\triangle ABC$ be any right triangle with the right angle at C and hypotenuse $\overline{AB}$.

$\quad$ **a.** Prove that $\angle A$ and $\angle B$ are acute angles.

$\quad$ **b.** Prove that the hypotenuse is the longest side of the right triangle.

15. Prove that every obtuse triangle has two acute angles.

CHAPTER SUMMARY

Definitions to Know	• An **exterior angle of a polygon** is an angle that forms a linear pair with one of the interior angles of the polygon.
	• Each exterior angle of a triangle has an **adjacent interior angle** and two **remote** or **nonadjacent interior angles**.

Postulates

7.1 A whole is greater than any of its parts.

7.2 If a, b, and c are real numbers such that $a > b$ and $b > c$, then $a > c$.

7.3 A quantity may be substituted for its equal in any statement of inequality.

7.4 Given any two quantities, a and b, one and only one of the following is true: $a < b$, or $a = b$, or $a > b$.

7.5 If equal quantities are added to unequal quantities, then the sums are unequal in the same order.

7.6 If unequal quantities are added to unequal quantities in the same order, then the sums are unequal in the same order.

7.7 If equal quantities are subtracted from unequal quantities, then the differences are unequal in the same order.

7.8 If unequal quantities are multiplied by positive equal quantities, then the products are unequal in the same order.

7.9 If unequal quantities are multiplied by negative equal quantities, then the products are unequal in the opposite order.

7.10 If unequal quantities are divided by positive equal quantities, then the quotients are unequal in the same order.

7.11 If unequal quantities are divided by negative equal quantities, then the quotients are unequal in the opposite order.

Theorems

7.1 The length of one side of a triangle is less than the sum of the lengths of the other two sides.

7.2 The measure of an exterior angle of a triangle is greater than the measure of either nonadjacent interior angle.

7.3 If the lengths of two sides of a triangle are unequal, then the measures of the angles opposite these sides are unequal and the larger angle lies opposite the longer side.

7.4 If the measures of two angles of a triangle are unequal, then the lengths of the sides opposite these angles are unequal and the longer side lies opposite the larger angle.

VOCABULARY

7-1 Transitive property of inequality • Trichotomy postulate

7-4 Triangle inequality theorem

7-5 Exterior angle of a polygon • Adjacent interior angle • Nonadjacent interior angle • Remote interior angle • Exterior angle inequality theorem

REVIEW EXERCISES

In 1–8, state a definition, postulate, or theorem that justifies each of the following statements about the triangles in the figure.

1. $AC > BC$

2. If $DA < DB$ and $DB < DC$, then $DA < DC$.

3. $m\angle DBC > m\angle A$

4. If $m\angle C > m\angle CDB$, then $DB > BC$.

5. If $DA < DB$, then $DA + AC < DB + AC$.

6. $DA + AC > DC$

7. If $m\angle A > m\angle C$, then $DC > DA$.

8. $m\angle ADC > m\angle ADB$

9. *Given:* $\overline{AEC}, \overline{BDC}, AE > BD$, and $EC > DC$

   Prove: $m\angle B > m\angle A$

10. *Given:* $\triangle ABC \cong \triangle CDA$, $AD > DC$

   Prove: **a.** $m\angle ACD > m\angle CAD$
   b. $\overline{AC}$ does not bisect $\angle A$.

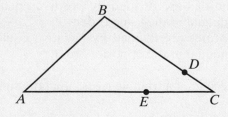

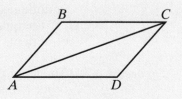

11. In isosceles triangle ABC, $\overline{CA} \cong \overline{CB}$. If D is a point on $\overline{AC}$ between A and C, prove that $DB > DA$.

12. In isosceles triangle RST, $RS = ST$. Prove that $\angle SRP$, the exterior angle at R, is congruent to $\angle STQ$, the exterior angle at T.

13. Point B is 4 blocks north and 3 blocks east of A. All streets run north and south or east and west except a street that slants from C to B. Of the three paths from A and B that are marked:

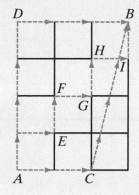

a. Which path is shortest? Justify your answer.

b. Which path is longest? Justify your answer.

Exploration

The Hinge Theorem states: If two sides of one triangle are congruent to two sides of another triangle, and the included angle of the first is larger than the included angle of the second, then the third side of the first triangle is greater than the third side of the second triangle.

1. a. With a partner or in a small group, prove the Hinge Theorem.

   b. Compare your proof with the proofs of the other groups. Were different diagrams used? Were different approaches used? Were these approaches valid?

The converse of the Hinge Theorem states: If the two sides of one triangle are congruent to two sides of another triangle, and the third side of the first triangle is greater than the third side of the second triangle, then the included angle of the first triangle is larger than the included angle of the second triangle.

2. a. With a partner or in a small group, prove the converse of the Hinge Theorem.

   b. Compare your proof with the proofs of the other groups. Were different diagrams used? Were different approaches used? Were these approaches valid?

Part I

Answer all questions in this part. Each correct answer will receive 2 credits. No partial credit will be allowed.

1. The solution set of the equation $2x - 3.5 = 5x - 18.2$ is

(1) 49   (2) −49   (3) 4.9   (4) −4.9

2. Which of the following is an example of the transitive property of inequality?

(1) If $a > b$, then $b < a$.

(2) If $a > b$, then $a + c > b + c$.

(3) If $a > b$ and $c > 0$, then $ac > bc$.

(4) If $a > b$ and $b > c$, then $a > c$.

3. Point M is the midpoint of $\overline{ABMC}$. Which of the following is *not* true?

(1) $AM = MC$ (2) $AB < MC$ (3) $AM > BC$ (4) $BM < MC$

4. The degree measure of the larger of two complementary angles is 30 more than one-half the measure of the smaller. The degree measure of the smaller is

(1) 40   (2) 50   (3) 80   (4) 100

5. Which of the following could be the measures of the sides of a triangle?

(1) 2, 2, 4   (2) 1, 3, 5   (3) 7, 12, 20   (4) 6, 7, 12

6. Which of the following statements is true for all values of x?

(1) $x = 5$ and $x \neq 5$

(2) $x < 5$ or $x > 5$

(3) If $x > 5$, then $x > 3$.

(4) If $x > 3$, then $x > 5$.

7. In $\triangle ABC$ and $\triangle DEF$, $\overline{AB} \cong \overline{DE}$, and $\angle A \cong \angle D$. In order to prove $\triangle ABC \cong \triangle DEF$ using ASA, we need to prove that

(1) $\angle B \cong \angle E$

(2) $\angle C \cong \angle F$

(3) $\overline{BC} \cong \overline{EF}$

(4) $\overline{AC} \cong \overline{DF}$

8. Under a reflection in the y-axis, the image of $(-2, 5)$ is

(1) $(2, 5)$ (2) $(2, -5)$ (3) $(-2, -5)$ (4) $(5, -2)$

9. Under an opposite isometry, the property that is changed is

(1) distance

(2) angle measure

(3) collinearity

(4) orientation

10. Points P and Q lie on the perpendicular bisector of $\overline{AB}$. Which of the following statements must be true?

(1) $\overline{AB}$ is the perpendicular bisector of $\overline{PQ}$.

(2) $PA = PB$ and $QA = QB$.

(3) $PA = QA$ and $PB = QB$.

(4) P is the midpoint of $\overline{AB}$ or Q is the midpoint of $\overline{AB}$.

Part II

Answer all questions in this part. Each correct answer will receive 2 credits. Clearly indicate the necessary steps, including appropriate formula substitutions, diagrams, graphs, charts, etc. For all questions in this part, a correct numerical answer with no work shown will receive only 1 credit.

11. Each of the following statements is true.

If the snow continues to fall, our meeting will be cancelled.

Our meeting is not cancelled.

Can you conclude that snow does not continue to fall? List the logic principles needed to justify your answer.

12. The vertices of $\triangle ABC$ are $A(0, 3)$, $B(4, 3)$, and $C(3, 5)$. Find the coordinates of the vertices of $\triangle A'B'C'$, the image of $\triangle ABC$ under the composition $r_{y=x} \circ T_{-4,-5}$.

Part III

Answer all questions in this part. Each correct answer will receive 4 credits. Clearly indicate the necessary steps, including appropriate formula substitutions, diagrams, graphs, charts, etc. For all questions in this part, a correct numerical answer with no work shown will receive only 1 credit.

13. *Given:* $\overline{PR}$ bisects $\overline{ARB}$ but $\overline{PR}$ is not perpendicular to $\overline{ARB}$.

Prove: $AP \neq BP$

14. *Given:* In quadrilateral $ABCD$, $\overrightarrow{AC}$ bisects $\angle DAB$ and $\overrightarrow{CA}$ bisects $\angle DCB$.

Prove: $\angle B \cong \angle D$

Part IV

Answer all questions in this part. Each correct answer will receive 6 credits. Clearly indicate the necessary steps, including appropriate formula substitutions, diagrams, graphs, charts, etc. For all questions in this part, a correct numerical answer with no work shown will receive only 1 credit.

15. The intersection of $\overleftrightarrow{PQ}$ and $\overleftrightarrow{RS}$ is T. If $m\angle PTR = x$, $m\angle QTS = y$, and $m\angle RTQ = 2x + y$, find the measures of $\angle PTR$, $\angle QTS$, $\angle RTQ$, and $\angle PTS$.

16. In $\triangle ABC$, $m\angle A < m\angle B$ and $\angle DCB$ is an exterior angle at C. The measure of $\angle BCA = 6x + 8$, and the measure of $\angle DCB = 4x + 12$.

a. Find $m\angle BCA$ and $m\angle DCB$.

b. List the interior angles of the triangle in order, starting with the smallest.

c. List the sides of the triangles in order starting with the smallest.

CHAPTER

8

SLOPES AND EQUATIONS OF LINES

In coordinate geometry, a straight line can be characterized as a line whose slope is a constant. Do curves have slopes and if so, can they be determined?

In the late 17th and early 18th centuries, two men independently developed methods to answer these and other questions about curves and the areas that they determine. The slope of a curve at a point is defined to be the slope of the *tangent* to that curve at that point. Descartes worked on the problem of finding the slope of a tangent to a curve by considering the slope of a tangent to a circle that intersected the curve at a given point and had its center on an axis. Gottfried Leibniz (1646–1716) in Germany and Isaac Newton (1642–1727) in England each developed methods for determining the slope of a tangent to a curve at any point as well as methods for determining the area bounded by a curve or curves. Newton acknowledged the influence of the work of mathematicians and scientists who preceded him in his statement, "If I have seen further, it is by standing on the shoulders of giants." The work of Leibniz and Newton was the basis of differential and integral calculus.

8-1 THE SLOPE OF A LINE

In the coordinate plane, horizontal and vertical lines are used as reference lines. Slant lines intersect horizontal lines at acute and obtuse angles. The ratio that measures the slant of a line in the coordinate plane is the *slope* of the line.

Finding the Slope of a Line

Through two points, one and only one line can be drawn. In the coordinate plane, if the coordinates of two points are given, it is possible to use a ratio to determine the measure of the slant of the line. This ratio is the **slope** of the line.

Through the points $A(-1, -2)$ and $B(2, 7)$, $\overleftrightarrow{AB}$ is drawn. Let $C(2, -2)$ be the point at which the vertical line through B intersects the horizontal line through A. The slope of $\overline{AB}$ is the ratio of the change in vertical distance, BC, to the change in horizontal distance, AC. Since B and C are on the same vertical line, BC is the difference in the y-coordinates of B and C. Since A and C are on the same horizontal line, AC is the difference in the x-coordinates of A and C.

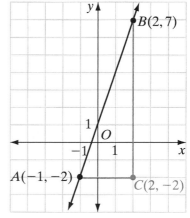

$$\text{slope of } \overline{AB} = \frac{\text{change in vertical distance}}{\text{change in horizontal distance}}$$
$$= \frac{BC}{AC}$$
$$= \frac{7 - (-2)}{2 - (-1)}$$
$$= \frac{9}{3}$$
$$= 3$$

This ratio is the same for any segment of the line $\overleftrightarrow{AB}$. Suppose we change the order of the points $(-1, -2)$ and $(2, 7)$ in performing the computation. We then have:

$$\text{slope of } \overline{BA} = \frac{\text{change in vertical distance}}{\text{change in horizontal distance}}$$
$$= \frac{CB}{CA}$$
$$= \frac{(-2) - 7}{(-1) - 2}$$
$$= \frac{-9}{-3}$$
$$= 3$$

The result of both computations is the same. When we compute the slope of a line that is determined by two points, it does not matter which point is considered as the first point and which the second.

Also, when we find the slope of a line using two points on the line, it does not matter which two points on the line we use because all segments of a line have the same slope as the line.

Procedure

To find the slope of a line:

1. Select any two points on the line.

2. Find the vertical change, that is, the change in y-values by subtracting the y-coordinates in any order.

3. Find the horizontal change, that is, the change in x-values, by subtracting the x-coordinates in the same order as the y-coordinates.

4. Write the ratio of the vertical change to the horizontal change.

In general, the slope, m, of a line that passes through any two points $A(x_1, y_1)$ and $B(x_2, y_2)$, where $x_1 \neq x_2$, is the ratio of the difference of the y-values of these points to the difference of the corresponding x-values.

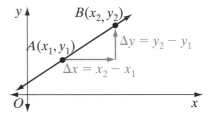

$$\text{slope of } \overleftrightarrow{AB} = m = \frac{y_2 - y_1}{x_2 - x_1}$$

The difference in x-values, $x_2 - x_1$, can be represented by Δx, read as "delta x." Similarly, the difference in y-values, $y_2 - y_1$, can be represented by Δy, read as "delta y." Therefore, we write:

$$\text{slope of a line} = m = \frac{\Delta y}{\Delta x}$$

The slope of a line is positive if the line slants upward from left to right, negative if the line slants downward from left to right, or zero if the line is horizontal. If the line is vertical, it has no slope.

Positive Slope

The points C and D are two points on $\overleftrightarrow{AB}$. Let the coordinates of C be $(1, 2)$ and the coordinates of D be $(3, 3)$. As the values of x increase, the values of y also increase. The graph of $\overleftrightarrow{AB}$ slants upward.

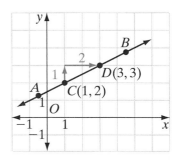

$$\text{slope of } \overleftrightarrow{AB} = \frac{3 - 2}{3 - 1} = \frac{1}{2}$$

The slope of $\overleftrightarrow{AB}$ is positive.

Negative Slope

Points C and D are two points on $\overleftrightarrow{EF}$. Let the coordinates of C be $(2, 3)$ and the coordinates of D be $(5, 1)$. As the values of x increase, the values of y decrease. The graph of $\overleftrightarrow{EF}$ slants downward.

$$\text{slope of } \overleftrightarrow{EF} = \frac{1 - 3}{5 - 2} = -\frac{2}{3}$$

The slope of $\overleftrightarrow{EF}$ is negative.

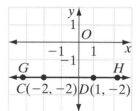

Zero Slope

Points C and D are two points on $\overleftrightarrow{GH}$. Let the coordinates of C be $(-2, -2)$ and the coordinates of D be $(1, -2)$. As the values of x increase, the values of y remain the same. The graph of $\overleftrightarrow{GH}$ is a horizontal line.

$$\text{slope of } \overleftrightarrow{GH} = \frac{-2 - (-2)}{1 - (-2)} = \frac{0}{3} = 0$$

The slope of $\overleftrightarrow{GH}$ is 0.
The slope of any horizontal line is 0.

No Slope

Points C and D are two points on $\overleftrightarrow{ML}$. Let the coordinates of C be $(2, -2)$ and the coordinates of D be $(2, 1)$. The values of x remain the same for all points as the values of y increase. The graph of $\overleftrightarrow{ML}$ is a vertical line.
The slope of $\overleftrightarrow{ML}$ is $\frac{-2 - 1}{2 - 2} = \frac{-3}{0}$, which is undefined.
$\overleftrightarrow{ML}$ has no slope.
A vertical line has no slope.

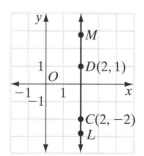

A fundamental property of a straight line is that its slope is constant. Therefore, any two points on a line may be used to compute the slope of the line.

EXAMPLE 1

Find the slope of the line that is determined by points $(-2, 4)$ and $(4, 2)$.

Solution Plot points $(-2, 4)$ and $(4, 2)$.

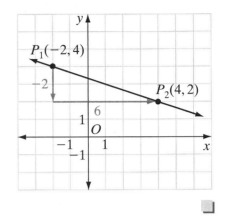

Let point $(-2, 4)$ be $P_1(x_1, y_1)$ and let point $(4, 2)$ be $P_2(x_2, y_2)$.
Then, $x_1 = -2, y_1 = 4, x_2 = 4$, and $y_2 = 2$.

$$\text{slope of } \overleftrightarrow{P_1P_2} = \frac{\Delta y}{\Delta x} = \frac{y_2 - y_1}{x_2 - x_1}$$

$$= \frac{2 - 4}{4 - (-2)}$$

$$= \frac{-2}{6}$$

$$= -\frac{1}{3} \quad \textit{Answer}$$

EXAMPLE 2

Through point $(1, 4)$, draw the line whose slope is $-\frac{3}{2}$.

Solution

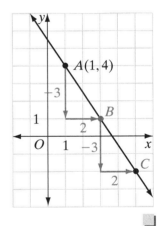

How to Proceed

(1) Graph point $A(1, 4)$.

(2) Note that, since slope $= \frac{\Delta y}{\Delta x} = -\frac{3}{2} = \frac{-3}{2}$, when y decreases by 3, x increases by 2. Start at point $A(1, 4)$ and move 3 units downward and 2 units to the right to locate point B.

(3) Start at B and repeat these movements to locate point C.

(4) Draw a line that passes through points $A, B,$ and C.

Exercises

Writing About Mathematics

1. How is the symbol Δy read and what is its meaning?

2. Brad said that since 0 is the symbol for "nothing," no slope is the same as zero slope. Do you agree with Brad? Explain why or why not.

Developing Skills

In 3–11, in each case: **a.** Plot both points and draw the line that they determine. **b.** Find the slope of this line if the line has a defined slope. **c.** State whether the line through these points would slant upward, slant downward, be horizontal, or be vertical.

3. $(0, 1)$ and $(4, 5)$ **4.** $(1, 0)$ and $(4, 9)$ **5.** $(0, 0)$ and $(-3, 6)$

6. $(-1, 5)$ and $(3, 9)$ **7.** $(5, -3)$ and $(1, -1)$ **8.** $(-2, 4)$ and $(-2, 2)$

9. $(-1, -2)$ and $(7, -8)$ **10.** $(4, 2)$ and $(8, 2)$ **11.** $(-1, 3)$ and $(2, -3)$

In 12–23, in each case, draw a line with the given slope, m, through the given point.

12. $(0, 1); m = 2$ **13.** $(-1, 3); m = 3$ **14.** $(2, 5); m = -1$

15. $(-4, 5); m = \frac{2}{3}$ **16.** $(-3, 2); m = 0$ **17.** $(-4, 7); m = -2$

18. $(1, 3); m = 1$ **19.** $(-2, 3); m = -\frac{3}{2}$ **20.** $(-1, 5); m = -\frac{1}{3}$

21. $(-1, 0); m = \frac{5}{4}$ **22.** $(0, -2); m = \frac{2}{3}$ **23.** $(-2, 0); m = \frac{1}{2}$

Applying Skills

24. a. Graph the points $A(2, 4)$ and $B(8, 4)$.

    b. From point A, draw a line that has a slope of 2.

    c. From point B, draw a line that has a slope of -2.

    d. Let the intersection of the lines drawn in **b** and **c** be C. What are the coordinates of C?

    e. Draw the altitude from vertex C to base $\overline{AB}$ of $\triangle ABC$. Prove that $\triangle ABC$ is an isosceles triangle.

25. Points $A(3, -2)$ and $B(9, -2)$ are two vertices of rectangle $ABCD$ whose area is 24 square units. Find the coordinates of C and D. (Two answers are possible.)

26. A path to the top of a hill rises 75 feet vertically in a horizontal distance of 100 feet. Find the slope of the path up the hill.

27. The doorway of a building under construction is 3 feet above the ground. A ramp to reach the doorway is to have a slope of $\frac{2}{5}$. How far from the base of the building should the ramp begin?

8-2 THE EQUATION OF A LINE

We have learned two facts that we can use to write the equation of a line.

▶ **Two points determine a line.**

▶ **The slope of a straight line is constant.**

The second statement on the bottom of page 295 can be written as a biconditional:

Postulate 8.1

> A, B, and C lie on the same line if and only if the slope of $\overline{AB}$ is equal to the slope of $\overline{BC}$.

Let $A(-3, -1)$ and $B(6, 5)$ be two points on $\overleftrightarrow{AB}$. Let $P(x, y)$ be any other point on $\overleftrightarrow{AB}$. We can write the equation of $\overleftrightarrow{AB}$ by using the following fact:

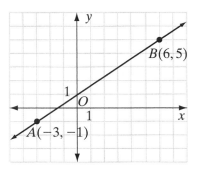

$$\text{slope of } \overline{AB} = \text{slope of } \overline{BP}$$
$$\frac{-1 - 5}{-3 - 6} = \frac{y - 5}{x - 6}$$
$$\frac{-6}{-9} = \frac{y - 5}{x - 6}$$
$$\frac{2}{3} = \frac{y - 5}{x - 6}$$
$$3(y - 5) = 2(x - 6)$$
$$3y - 15 = 2x - 12$$
$$3y = 2x + 3$$
$$y = \tfrac{2}{3}x + 1$$

Recall that when the equation is solved for y in terms of x, the coefficient of x is the slope of the line and the constant term is the **y-intercept**, the y-coordinate of the point where the line intersects the y-axis.

The **x-intercept** is the x-coordinate of the point where the line intersects the x-axis.

Procedure

To find the equation of a line given two points on the line:

1. Find the slope of the line using the coordinates of the two given points.

2. Let $P(x, y)$ be any point on the line. Write a ratio that expresses the slope of the line in terms of the coordinates of P and the coordinates of one of the given points.

3. Let the slope found in step 2 be equal to the slope found in step 1.

4. Solve the equation written in step 3 for y.

When we are given the coordinates of one point and the slope of the line, the equation of the line can be determined. For example, if (a, b) is a point on the line whose slope is m, then the equation is:

$$\frac{y - b}{x - a} = m$$

This equation is called the **point-slope form** of the equation of a line.

EXAMPLE 1

The slope of a line through the point $A(3, 0)$ is -2.

a. Use the point-slope form to write an equation of the line.

b. What is the y-intercept of the line?

c. What is the x-intercept of the line?

Solution **a.** Let $P(x, y)$ be any point on the line. The slope of $\overline{AP} = -2$.

$$\frac{y - 0}{x - 3} = -2$$
$$y = -2(x - 3)$$
$$y = -2x + 6$$

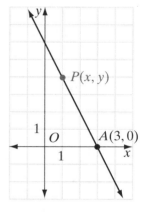

b. The y-intercept is the y-coordinate of the point at which the line intersects the y-axis, that is, the value of y when x is 0. When $x = 0$,

$$y = -2(0) + 6$$
$$= 0 + 6$$
$$= 6$$

The y-intercept is 6. When the equation is solved for y, the y-intercept is the constant term.

c. The x-intercept is the x-coordinate of the point at which the line intersects the x-axis, that is, the value of x when y is 0. Since $(3, 0)$ is a given point on the line, the x-intercept is 3.

Answers **a.** $y = -2x + 6$ **b.** 6 **c.** 3

EXAMPLE 2

a. Show that the three points $A(-2, -3)$, $B(0, 1)$, and $C(3, 7)$ lie on a line.

b. Write an equation of the line through A, B, and C.

Solution **a.** The points $A(-2, -3)$, $B(0, 1)$, and $C(3, 7)$ lie on the same line if and only if the slope of $\overline{AB}$ is equal to the slope of $\overline{BC}$.

$$\text{slope of } \overline{AB} = \frac{-3 - 1}{-2 - 0} \qquad\qquad \text{slope of } \overline{BC} = \frac{1 - 7}{0 - 3}$$
$$= \frac{-4}{-2} \qquad\qquad\qquad\qquad\quad = \frac{-6}{-3}$$
$$= 2 \qquad\qquad\qquad\qquad\qquad = 2$$

The slopes are equal. Therefore, the three points lie on the same line.

b. Use the point-slope form of the equation of a line. Let (x, y) be any other point on the line. You can use any of the points, A, B, or C, with (x, y) and the slope of the line, to write an equation. We will use $A(-2, -3)$.

$$\frac{y - (-3)}{x - (-2)} = 2$$
$$y + 3 = 2(x + 2)$$
$$y + 3 = 2x + 4$$
$$y = 2x + 1$$

Answers **a.** Since the slope of $\overline{AB}$ is equal to the slope of $\overline{BC}$, A, B, and C lie on a line.

b. $y = 2x + 1$

Alternative Solution

(1) Write the **slope-intercept form** of an equation of a line: $\qquad\qquad y = mx + b$

(2) Substitute the coordinates of A in that equation: $\qquad\qquad -3 = m(-2) + b$

(3) Substitute the coordinates of C in that equation: $\qquad\qquad 7 = m(3) + b$

(4) Write the system of two equations from (2) and (3): $\qquad\qquad -3 = -2m + b$
$\qquad\qquad\qquad\qquad\qquad\qquad\qquad\qquad 7 = 3m + b$

(5) Solve the equation $-3 = -2m + b$ for b in terms of m: $\qquad\qquad b = 2m - 3$

(6) Substitute the value of b found in (5) for b in the second equation and solve for m:
$\qquad\qquad 7 = 3m + b$
$\qquad\qquad 7 = 3m + (2m - 3)$
$\qquad\qquad 7 = 5m - 3$
$\qquad\qquad 10 = 5m$
$\qquad\qquad 2 = m$

(7) Substitute this value of m in either equation to find the value of b:
$\qquad\qquad b = 2m - 3$
$\qquad\qquad b = 2(2) - 3$
$\qquad\qquad b = 1$

The equation is $y = 2x + 1$.

We can show that each of the given points lies on the line whose equation is $y = 2x + 1$ by showing that each pair of values makes the equation true.

$(-2, -3)$ $\qquad\qquad$ $(0, 1)$ $\qquad\qquad$ $(3, 7)$

$y = 2x + 1$ $\qquad\qquad$ $y = 2x + 1$ $\qquad\qquad$ $y = 2x + 1$

$-3 \overset{?}{=} 2(-2) + 1$ $\qquad$ $1 \overset{?}{=} 2(0) + 1$ $\qquad$ $7 \overset{?}{=} 2(3) + 1$

$-3 = -3$ ✔ $\qquad\qquad$ $1 = 1$ ✔ $\qquad\qquad$ $7 = 7$ ✔

Answers **a, b:** Since the coordinates of each point make the equation $y = 2x + 1$ true, the three points lie on a line whose equation is $y = 2x + 1$.

Exercises

Writing About Mathematics

1. Jonah said that A, B, C, and D lie on the same line if the slope of $\overline{AB}$ is equal to the slope of $\overline{CD}$. Do you agree with Jonah? Explain why or why not.

2. Sandi said that the point-slope form cannot be used to find the equation of a line with no slope.
   a. Do you agree with Sandi? Justify your answer.
   b. Explain how you can find the equation of a line with no slope.

Developing Skills

In 3–14, write the equation of each line.

3. Through $(1, -2)$ and $(5, 10)$
4. Through $(0, -1)$ and $(1, 0)$
5. Through $(2, -2)$ and $(0, 6)$
6. Slope 2 and through $(-2, -4)$
7. Slope -4 through $(1, 1)$
8. Slope $\frac{1}{2}$ through $(5, 4)$
9. Slope -3 and y-intercept 5
10. Slope 1 and x-intercept 4
11. Through $(1, 5)$ and $(4, 5)$
12. Through $(1, 5)$ and $(1, -2)$
13. x-intercept 2 and y-intercept 4
14. No slope and x-intercept 2
15. **a.** Do the points $P(3, 3)$, $Q(5, 4)$, and $R(-1, 1)$ lie on the same line?

   b. If P, Q, and R lie on the same line, find the equation of the line. If P, Q, and R do not lie on the same line, find the equations of the lines $\overleftrightarrow{PQ}$, $\overleftrightarrow{QR}$, and $\overleftrightarrow{PR}$.

16. a. Do the points $L(1, 3)$, $M(5, 6)$, and $N(-4, 0)$ lie on the same line?

    b. If L, M, and N lie on the same line, find the equation of the line. If L, M, and N do not lie on the same line, find the equations of the lines $\overleftrightarrow{LM}$, $\overleftrightarrow{MN}$, and $\overleftrightarrow{LN}$.

Applying Skills

17. At a TV repair shop, there is a uniform charge for any TV brought in for repair plus an hourly fee for the work done. For a TV that needed two hours of repair, the cost was $100. For a TV that needed one and a half hours of repair, the cost was $80.

    a. Write an equation that can be used to find the cost of repair, y, when x hours of work are required. Write the given information as ordered pairs, $(2, 100)$ and $(1.5, 80)$.

    b. What would be the cost of repairing a TV that requires 3 hours of work?

    c. What does the coefficient of x in the equation that you wrote in **a** represent?

    d. What does the constant term in the equation that you wrote in **a** represent?

18. An office manager buys printer cartridges from a mail order firm. The bill for each order includes the cost of the cartridges plus a shipping cost that is the same for each order. The bill for 5 cartridges was $98 and a later bill for 3 cartridges, at the same rate, was $62.

    a. Write an equation that can be used to find y, the amount of the bill for an order, when x cartridges are ordered. Write the given information as ordered pairs, $(5, 98)$ and $(3, 62)$.

    b. What would be the amount of the bill for 8 cartridges?

    c. What does the coefficient of x in the equation that you wrote in **a** represent?

    d. What does the constant term in the equation that you wrote in **a** represent?

19. Show that if the equation of the line can be written as $\frac{x}{a} + \frac{y}{b} = 1$, then the line intersects the x-axis at $(a, 0)$ and the y-axis at $(0, b)$.

8-3 MIDPOINT OF A LINE SEGMENT

The *midpoint* of a line segment is the point of that line segment that divides the segment into two congruent segments. In the figure, $A(-1, 4)$ and $B(7, 4)$ determine a horizontal segment, $\overline{AB}$, whose midpoint, M, can be found by first finding the distance from A to B. Since $AB = 7 - (-1) = 8$ units, $AM = 4$ units, and $MB = 4$ units.

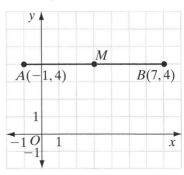

We can find the x-coordinate of M by adding AM to the x-coordinate of A or by subtracting MB from the x-coordinate of B. The x-coordinate of M is $-1 + 4 = 3$ or $7 - 4 = 3$. Since A, B, and M are all on the same horizontal line, they all have the same y-coordinate, 4. The coordinates of the midpoint M are $(3, 4)$. The x-coordinate of M is the average of the x-coordinates of A and B.

$$x\text{-coordinate of } M = \frac{-1 + 7}{2}$$
$$= \frac{6}{2}$$
$$= 3$$

Similarly, $C(3, -3)$ and $D(3, 1)$ determine a vertical segment, $\overline{CD}$, whose midpoint, N, can be found by first finding the distance from C to D. Since $CD = 1 - (-3) = 4$ units, $CN = 2$ units, and $ND = 2$ units. We can find the y-coordinate of N by adding 2 to the y-coordinate of C or by subtracting 2 from the y-coordinate of D. The y-coordinate of N is $-3 + 2 = -1$ or $1 - 2 = -1$. Since C, D, and N are all on the same vertical line,

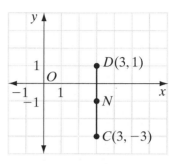

they all have the same x-coordinate, 3. The coordinates of the midpoint N are $(3, -1)$. The y-coordinate of N is the average of the y-coordinates of C and D.

$$y\text{-coordinate of } N = \frac{1 + (-3)}{2}$$
$$= \frac{-2}{2}$$
$$= -1$$

These examples suggest the following relationships:

▶ **If the endpoints of a horizontal segment are (a, c) and (b, c), then the coordinates of the midpoint are:**

$$\left(\frac{a + b}{2}, c\right)$$

▶ **If the endpoints of a vertical segment are (d, e) and (d, f), then the coordinates of the midpoint are:**

$$\left(d, \frac{e + f}{2}\right)$$

In the figure, $P(2, 1)$ and $Q(8, 5)$ are the endpoints of $\overline{PQ}$. A horizontal line through P and a vertical line through Q intersect at $R(8, 1)$.

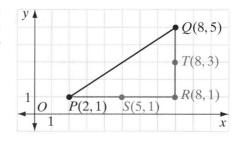

- The coordinates of S, the midpoint of $\overline{PR}$, are $\left(\frac{2 + 8}{2}, 1\right) = (5, 1)$.

- The coordinates of T, the midpoint of $\overline{QR}$, are $\left(8, \frac{5 + 1}{2}\right) = (8, 3)$.

Now draw a vertical line through S and a horizontal line through T. These lines appear to intersect at a point on $\overline{PQ}$ that we will call M. This point has the coordinates $(5, 3)$. We need to show that this point is a point on $\overline{PQ}$ and is the midpoint of $\overline{PQ}$.

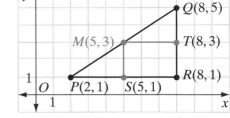

The point M is on $\overline{PQ}$ if and only if the slope of $\overline{PM}$ is equal to the slope of $\overline{MQ}$.

$$\text{slope of } \overline{PM} = \frac{3-1}{5-2} \qquad \text{slope of } \overline{MQ} = \frac{5-3}{8-5}$$
$$= \frac{2}{3} \qquad\qquad\qquad = \frac{2}{3}$$

Since these slopes are equal, P, M, and Q lie on a line.

The point M is the midpoint of $\overline{PQ}$ if $PM = MQ$. We can show that $PM = MQ$ by showing that they are corresponding parts of congruent triangles.

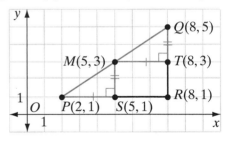

- $PS = 5 - 2 = 3$ and $MT = 8 - 5 = 3$. Therefore, $\overline{PS} \cong \overline{MT}$.

- $SM = 3 - 1 = 2$ and $TQ = 5 - 3 = 2$. Therefore, $\overline{SM} \cong \overline{TQ}$.

- Since vertical lines are perpendicular to horizontal lines, $\angle PSM$ and $\angle MTQ$ are right angles and therefore congruent.

- Therefore, $\triangle PSM \cong \triangle MTQ$ by SAS and $\overline{PM} \cong \overline{MQ}$ because corresponding parts of congruent triangles are congruent.

We can conclude that the coordinates of the midpoint of a line segment whose endpoints are $(2, 1)$ and $(8, 5)$ are $\left(\frac{2+8}{2}, \frac{1+5}{2}\right) = (5, 3)$.

This example suggests the following theorem:

Theorem 8.1 If the endpoints of a line segment are (x_1, y_1) and (x_2, y_2), then the coordinates of the midpoint of the segment are $\left(\frac{x_1 + x_2}{2}, \frac{y_1 + y_2}{2}\right)$.

Given The endpoints of $\overline{AB}$ are $A(x_1, y_1)$ and $B(x_2, y_2)$.

Prove The coordinates of the midpoint of $\overline{AB}$ are $\left(\frac{x_1 + x_2}{2}, \frac{y_1 + y_2}{2}\right)$.

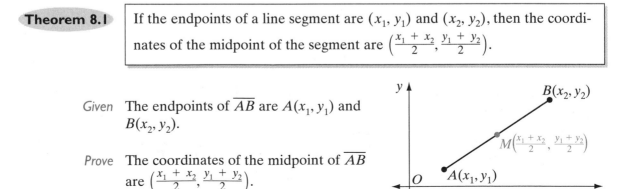

Proof In this proof we will use the following facts from previous chapters that we have shown to be true:

- Three points lie on the same line if the slope of the segment joining two of the points is equal to the slope of the segment joining one of these points to the third.

- If two points lie on the same horizontal line, they have the same y-coordinate and the length of the segment joining them is the absolute value of the difference of their x-coordinates.

- If two points lie on the same vertical line, they have the same x-coordinate and the length of the segment joining them is the absolute value of the difference of their y-coordinates.

We will follow a strategy similar to the one used in the previous example. First, we will prove that the point M with coordinates $\left(\frac{x_1 + x_2}{2}, \frac{y_1 + y_2}{2}\right)$ is on $\overline{AB}$, and then we will use congruent triangles to show that $\overline{AM} \cong \overline{MB}$. From the definition of a midpoint of a segment, this will prove that M is the midpoint of $\overline{AB}$.

(1) Show that $M\left(\frac{x_1 + x_2}{2}, \frac{y_1 + y_2}{2}\right)$ lies on $\overline{AB}$:

$$\text{slope of } \overline{AM} = \frac{\frac{y_1 + y_2}{2} - y_1}{\frac{x_1 + x_2}{2} - x_1} \qquad\qquad \text{slope of } \overline{MB} = \frac{y_2 - \frac{y_1 + y_2}{2}}{x_2 - \frac{x_1 + x_2}{2}}$$

$$= \frac{y_1 + y_2 - 2y_1}{x_1 + x_2 - 2x_1} \qquad\qquad\qquad = \frac{2y_2 - (y_1 + y_2)}{2x_2 - (x_1 + x_2)}$$

$$= \frac{y_2 - y_1}{x_2 - x_1} \qquad\qquad\qquad\qquad = \frac{y_2 - y_1}{x_2 - x_1}$$

Points A, M, and B lie on the same line because the slope of $\overline{AM}$ is equal to the slope of $\overline{MB}$.

(2) Let C be the point on the same vertical line as B and the same horizontal line as A. The coordinates of C are (x_2, y_1).

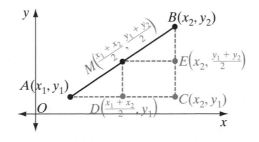

The midpoint of $\overline{AC}$ is $D\left(\frac{x_1 + x_2}{2}, y_1\right)$.

The midpoint of $\overline{BC}$ is $E\left(x_2, \frac{y_1 + y_2}{2}\right)$.

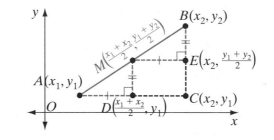

$$AD = \left| \frac{x_1 + x_2}{2} - x_1 \right| \qquad\qquad ME = \left| x_2 - \frac{x_1 + x_2}{2} \right|$$

$$= \left| \frac{x_1 + x_2 - 2x_1}{2} \right| \qquad\qquad = \left| \frac{2x_2 - x_1 - x_2}{2} \right|$$

$$= \left| \frac{x_2 - x_1}{2} \right| \qquad\qquad = \left| \frac{x_2 - x_1}{2} \right|$$

Therefore, $AD = ME$ and $\overline{AD} \cong \overline{ME}$.

$$MD = \left| y_1 - \frac{y_1 + y_2}{2} \right| \qquad\qquad BE = \left| \frac{y_1 + y_2}{2} - y_2 \right|$$

$$= \left| \frac{2y_1 - y_1 - y_2}{2} \right| \qquad\qquad = \left| \frac{y_1 + y_2 - 2y_2}{2} \right|$$

$$= \left| \frac{y_1 - y_2}{2} \right| \qquad\qquad = \left| \frac{y_1 - y_2}{2} \right|$$

Therefore, $MD = BE$ and $\overline{MD} \cong \overline{BE}$.

Vertical lines are perpendicular to horizontal lines. $\overline{AD} \perp \overline{MD}$ and $\overline{ME} \perp \overline{BE}$. Therefore, $\angle ADM$ and $\angle MEB$ are right angles and are congruent.

$\triangle ADM \cong \triangle MEB$ by SAS and $\overline{AM} \cong \overline{MB}$. Therefore, $M\left(\frac{x_1 + x_2}{2}, \frac{y_1 + y_2}{2} \right)$ is the midpoint of $\overline{AB}$. ◻

We generally refer to the formula given in this theorem, that is, $\left(\frac{x_1 + x_2}{2}, \frac{y_1 + y_2}{2} \right)$, as the **midpoint formula**.

EXAMPLE I

Find the coordinates of the midpoint of the segment $\overline{CD}$ whose endpoints are $C(-1, 5)$ and $D(4, -1)$.

Solution   Let $(x_1, y_1) = (-1, 5)$ and $(x_2, y_2) = (4, -1)$.

The coordinates of the midpoint are $\left(\frac{x_1 + x_2}{2}, \frac{y_1 + y_2}{2} \right) = \left(\frac{-1 + 4}{2}, \frac{5 + (-1)}{2} \right)$

$$= \left(\frac{3}{2}, \frac{4}{2} \right)$$

$$= \left(\frac{3}{2}, 2 \right) \; \textit{Answer} \quad ◻$$

EXAMPLE 2

$M(1, -2)$ is the midpoint of $\overline{AB}$ and the coordinates of A are $(-3, 2)$. Find the coordinates of B.

Solution Let the coordinates of $A = (x_1, y_1) = (-3, 2)$ and the coordinates of $B = (x_2, y_2)$.

The coordinates of the midpoint are $\left(\frac{x_1 + x_2}{2}, \frac{y_1 + y_2}{2}\right) = (1, -2)$.

$$\frac{-3 + x_2}{2} = 1 \qquad\qquad \frac{2 + y_2}{2} = -2$$

$$-3 + x_2 = 2 \qquad\qquad 2 + y_2 = -4$$

$$x_2 = 5 \qquad\qquad\qquad y_2 = -6$$

Answer The coordinates of B are $(5, -6)$.

EXAMPLE 3

The vertices of $\triangle ABC$ are $A(1, 1)$, $B(7, 3)$, and $C(2, 6)$. Write an equation of the line that contains the median from C to $\overline{AB}$.

Solution A median of a triangle is a line segment that joins any vertex to the midpoint of the opposite side. Let M be the midpoint of $\overline{AB}$.

(1) Find the coordinates of M. Let (x_1, y_1) be $(1, 1)$ and (x_2, y_2) be $(7, 3)$. The coordinates of M are:

$$\left(\frac{x_1 + x_2}{2}, \frac{y_1 + y_2}{2}\right) = \left(\frac{1 + 7}{2}, \frac{1 + 3}{2}\right)$$

$$= (4, 2)$$

(2) Write the equation of the line through $C(2, 6)$ and $M(4, 2)$. Let $P(x, y)$ be any other point on the line.

$$\text{slope of } \overline{PC} = \text{slope of } \overline{CM}$$

$$\frac{y - 6}{x - 2} = \frac{6 - 2}{2 - 4}$$

$$\frac{y - 6}{x - 2} = -2$$

$$y - 6 = -2(x - 2)$$

$$y - 6 = -2x + 4$$

$$y = -2x + 10$$

Answer $y = -2x + 10$

Exercises

Writing About Mathematics

1. If $P(a, c)$ and $Q(b, c)$ are two points in the coordinate plane, show that the coordinates of the midpoint are $\left(a + \frac{b - a}{2}, c\right)$. $\left(\textit{Hint:} \text{ Show that } a + \frac{b - a}{2} = \frac{a + b}{2}.\right)$

2. If $P(a, c)$ and $Q(b, c)$ are two points in the coordinate plane, show that the coordinates of the midpoint are $\left(b - \frac{b - a}{2}, c\right)$. $\left(\textit{Hint:} \text{ Show that } b - \frac{b - a}{2} = \frac{a + b}{2}.\right)$

Developing Skills

In 3–14, find the midpoint of the each segment with the given endpoints.

3. $(1, 7), (5, 1)$

4. $(-2, 5), (8, 7)$

5. $(0, 8), (10, 0)$

6. $(0, -2), (4, 6)$

7. $(-5, 1), (5, -1)$

8. $(6, 6), (2, 5)$

9. $(1, 0), (0, 8)$

10. $(-3, 8), (5, 8)$

11. $(-3, -5), (-1, -1)$

12. $(7, -2), (-1, 9)$

13. $\left(\frac{1}{2}, 3\right), \left(1, 2\frac{1}{2}\right)$

14. $\left(\frac{1}{3}, 9\right), \left(\frac{2}{3}, 3\right)$

In 15–20, M is the midpoint of $\overline{AB}$. Find the coordinates of the third point when the coordinates of two of the points are given.

15. $A(2, 7), M(1, 6)$

16. $A(3, 3), M(3, 9)$

17. $B(4, 7), M(5, 5)$

18. $B(4, -2), M\left(\frac{3}{2}, 0\right)$

19. $A(3, 3), B(1, 10)$

20. $A(0, 7), M\left(0, \frac{7}{2}\right)$

Applying Skills

21. The points $A(1, 1)$ and $C(9, 7)$ are the vertices of rectangle $ABCD$ and B is a point on the same horizontal line as A.

    a. What are the coordinates of vertices B and D?

    b. Show that the midpoint of diagonal $\overline{AC}$ is also the midpoint of diagonal $\overline{BD}$.

22. The points $P(x_1, y_1)$ and $R(x_2, y_2)$ are vertices of rectangle $PQRS$ and Q is a point on the same horizontal line as P.

    a. What are the coordinates of vertices Q and S?

    b. Show that the midpoint of diagonal $\overline{PR}$ is also the midpoint of diagonal $\overline{QS}$.

23. The vertices of $\triangle ABC$ are $A(-1, 4), B(5, 2),$ and $C(5, 6)$.

    a. What are the coordinates of M, the midpoint of $\overline{AB}$?

    b. Write an equation of $\overleftrightarrow{CM}$ that contains the median from C.

    c. What are the coordinates of N, the midpoint of $\overline{AC}$?

    d. Write an equation of $\overleftrightarrow{BN}$ that contains the median from B.

e. What are the coordinates of the intersection of $\overleftrightarrow{CM}$ and $\overleftrightarrow{BN}$?

f. What are the coordinates of P, the midpoint of $\overline{BC}$?

g. Write an equation of $\overleftrightarrow{AP}$ that contains the median from A.

h. Does the intersection of $\overleftrightarrow{CM}$ and $\overleftrightarrow{BN}$ lie on $\overleftrightarrow{AP}$?

i. Do the medians of this triangle intersect in one point?

8-4 THE SLOPES OF PERPENDICULAR LINES

Let l_1 be a line whose equation is $y = m_1 x$ where m_1 is not equal to 0. Then $O(0, 0)$ and $A(1, m_1)$ are two points on the line.

$$\begin{aligned}\text{slope of } l_1 &= \frac{m_1 - 0}{1 - 0} \\ &= \frac{m_1}{1} \\ &= m_1\end{aligned}$$

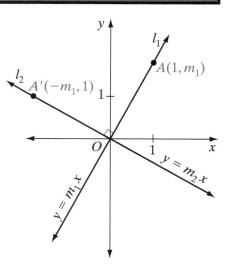

Under a counterclockwise rotation of 90° about the origin, the image of $A(1, m_1)$ is $A'(-m_1, 1)$. Since $\angle AOA'$ is a right angle, $\overleftrightarrow{OA} \perp \overleftrightarrow{OA'}$.

Let l_2 be the line $\overleftrightarrow{OA'}$ through $A'(-m_1, 1)$ and $O(0, 0)$, and let the slope of l_2 be m_2. Then:

$$\begin{aligned}m_2 &= \frac{0 - 1}{0 - (-m_1)} \\ &= \frac{-1}{0 + m_1} \\ &= -\frac{1}{m_1}\end{aligned}$$

We have shown that when two lines through the origin are perpendicular, the slope of one is the negative reciprocal of the slope of the other.

Is the rule that we found for the slopes of perpendicular lines through the origin true for perpendicular lines that do not intersect at the origin? We will show this by first establishing that translations preserve slope:

Theorem 8.2

Under a translation, slope is preserved, that is, if line l has slope m, then under a translation, the image of l also has slope m.

Proof: Let $P(x_1, y_1)$ and $Q(x_2, y_2)$ be two points on line l. Then:

$$\text{slope of } l = \frac{y_2 - y_1}{x_2 - x_1}$$

Under a translation $T_{a,b}$, the images of P and Q have coordinates $P'(x_1 + a, y_1 + b)$ and $Q'(x_2 + a, y_2 + b)$. Therefore, the slope of l', the image of l, is

$$\begin{aligned}
\text{slope of } l' &= \frac{(y_2 + b) - (y_1 + b)}{(x_2 + a) - (x_1 + a)} \\
&= \frac{y_2 - y_1 + b - b}{x_2 - x_1 + a - a} \\
&= \frac{y_2 - y_1}{x_2 - x_1}
\end{aligned}$$

As a simple application of Theorem 8.2, we can show that the slopes of any two perpendicular lines are negative reciprocals of each other:

Theorem 8.3a

> If two non-vertical lines are perpendicular, then the slope of one is the negative reciprocal of the other.

Proof: Let l_1 and l_2 be two perpendicular lines that intersect at (a, b). Under the translation $(x, y) \rightarrow (x - a, y - b)$, the image of (a, b) is $(0, 0)$.

Theorem 8.2 tells us that if the slope of l_1 is m, then the slope of its image, l_1', is m. Since l_1 and l_2 are perpendicular, their images, l_1' and l_2', are also perpendicular because translations preserve angle measure. Using what we established at the beginning of the section, since the slope of l_1' is m, the slope of l_2' is $-\frac{1}{m}$. Slope is preserved under a translation. Therefore, the slope of l_2 is $-\frac{1}{m}$.

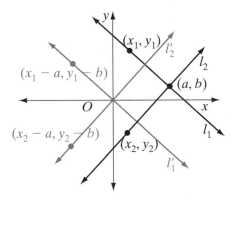

The proof of Theorem 8.3a is called a **transformational proof** since it uses transformations to prove the desired result.

Is the converse of Theorem 8.3a also true? To demonstrate that it is, we need to show that if the slope of one line is the negative reciprocal of the slope of the other, then the lines are perpendicular. We will use an indirect proof.

Theorem 8.3b

If the slopes of two lines are negative reciprocals, then the two lines are perpendicular.

Given   Two lines, $\overleftrightarrow{AB}$ and $\overleftrightarrow{AC}$, that intersect at
A. The slope of $\overleftrightarrow{AB}$ is m and the slope
of $\overleftrightarrow{AC}$ is $-\frac{1}{m}$, the negative reciprocal
of m.

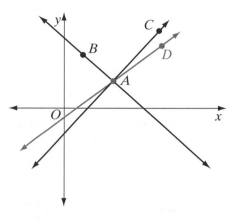

Prove $\overleftrightarrow{AB} \perp \overleftrightarrow{AC}$

Proof

Statements	Reasons
1. $\overleftrightarrow{AC}$ is not perpendicular to $\overleftrightarrow{AB}$.	**1.** Assumption.
2. Construct $\overleftrightarrow{AD}$ perpendicular to $\overleftrightarrow{AB}$ at A.	**2.** At a point on a given line, one and only one perpendicular can be drawn.
3. Slope of $\overleftrightarrow{AB}$ is m.	**3.** Given.
4. The slope of $\overleftrightarrow{AD}$ is $-\frac{1}{m}$.	**4.** If two lines are perpendicular, the slope of one is the negative reciprocal of the slope of the other.
5. The slope of $\overleftrightarrow{AC}$ is $-\frac{1}{m}$.	**5.** Given.
6. A, C, and D are on the same line, that is, $\overleftrightarrow{AC}$ and $\overleftrightarrow{AD}$ are the same line.	**6.** Three points lie on the same line if and only if the slope of the segment joining two of the points is equal to the slope of a segment joining another pair of these points.
7. $\overleftrightarrow{AC} \perp \overleftrightarrow{AB}$	**7.** Contradiction (steps 1, 6).

We can restate Theorems 8.3a and 8.3b as a biconditional.

Theorem 8.3 Two non-vertical lines are perpendicular if and only if the slope of one is the negative reciprocal of the slope of the other.

EXAMPLE 1

What is the slope of l_2, the line perpendicular to l_1, if the equation of l_1 is $x + 2y = 4$?

Solution (1) Solve the equation of l_1 for y:

$$x + 2y = 4$$
$$2y = -x + 4$$
$$y = -\tfrac{1}{2}x + 2$$

(2) Find the slope of l_1:

$$y = -\tfrac{1}{2}x + 2$$

slope

(3) Find the slope of l_2, the negative reciprocal of the slope of l_1:

slope of $l_2 = 2$

Answer The slope of l_2 is 2.

EXAMPLE 2

Show that $\triangle ABC$ is a right triangle if its vertices are $A(1,1)$, $B(4,3)$, and $C(2,6)$.

Solution The slope of $\overline{AB}$ is $\frac{3-1}{4-1} = \frac{2}{3}$.

The slope of $\overline{BC}$ is $\frac{3-6}{4-2} = \frac{-3}{2} = -\frac{3}{2}$.

The slope of $\overline{AC}$ is $\frac{1-6}{1-2} = \frac{-5}{-1} = 5$.

The slope of $\overline{BC}$ is the negative reciprocal of the slope of $\overline{AB}$. Therefore, $\overleftrightarrow{BC}$ is perpendicular to $\overleftrightarrow{AB}$, $\angle B$ is a right angle, and $\triangle ABC$ is a right triangle.

Exercises

Writing About Mathematics

1. Explain why the slope of a line perpendicular to the line whose equation is $x = 5$ cannot be found by using the negative reciprocal of the slope of that line.

2. The slope of a line l is $-\frac{1}{3}$. Kim said that the slope of a line perpendicular to l is $-\frac{1}{-\frac{1}{3}}$. Santos said that the slope of a line perpendicular to l is 3. Who is correct? Explain your answer.

Developing Skills

In 3–12: **a.** Find the slope of the given line. **b.** Find the slope of the line perpendicular to the given line.

3. $y = 4x - 7$

4. $y = x + 2$

5. $x + y = 8$

6. $2x - y = 3$

7. $3x = 5 - 2y$

8. through $(1, 1)$ and $(5, 3)$

9. through $(0, 4)$ and $(2, 0)$

10. y-intercept -2 and x-intercept 4

11. through $(4, 4)$ and $(4, -2)$

12. parallel to the x-axis through $(5, 1)$

In 13–16, find the equation of the line through the given point and perpendicular to the given line.

13. $\left(-\frac{1}{2}, -2\right)$; $2x + 7y = -15$

14. $(0, 0)$; $-2x + 4y = 12$

15. $(7, 3)$; $y = -\frac{1}{3}x - 3$

16. $(2, -2)$; $y = 1$

17. Is the line whose equation is $y = -3x + 5$ perpendicular to the line whose equation is $3x + y = 6$?

18. Two perpendicular lines have the same y-intercept. If the equation of one of these lines is $y = \frac{1}{2}x - 1$, what is the equation of the other line?

19. Two perpendicular lines intersect at $(2, -1)$. If $x - y = 3$ is the equation of one of these lines, what is the equation of the other line?

20. Write an equation of the line that intersects the y-axis at $(0, -1)$ and is perpendicular to the line whose equation is $x + 2y = 6$.

In 21–24, the coordinates of the endpoints of a line segment are given. For each segment, find the equation of the line that is the perpendicular bisector of the segment.

21. $A(2, 2)$, $B(-1, 1)$

22. $A\left(-\frac{1}{2}, 3\right)$, $B\left(\frac{3}{2}, 1\right)$

23. $A(3, -9)$, $B(3, 9)$

24. $A(-4, -1)$, $B(3, -3)$

Applying Skills

25. The vertices of DEF triangle are $D(-3, 4)$, $E(-1, -2)$, and $F(3, 2)$.

    a. Find an equation of the altitude from vertex D of $\triangle DEF$.

    b. Is the altitude from D also the median from D? Explain your answer.

    c. Prove that $\triangle DEF$ is isosceles.

26. If a four-sided polygon has four right angles, then it is a rectangle. Prove that if the vertices of a polygon are $A(3, -2)$, $B(5, 1)$, $C(-1, 5)$, and $D(-3, 2)$, then $ABCD$ is a rectangle.

27. The vertices of $\triangle ABC$ are $A(2,2)$, $B(6,6)$ and $C(6,0)$.

   a. What is the slope of $\overline{AB}$?

   b. Write an equation for the perpendicular bisector of $\overline{AB}$.

   c. What is the slope of $\overline{BC}$?

   d. Write an equation for the perpendicular bisector of $\overline{BC}$.

   e. What is the slope of $\overline{AC}$?

   f. Write an equation for the perpendicular bisector of $\overline{AC}$.

   g. Verify that the perpendicular bisectors of $\triangle ABC$ intersect in one point.

28. The coordinates of the vertices of $\triangle ABC$ are $A(-2,0)$, $B(4,0)$, and $C(0,4)$.

   a. Write the equation of each altitude of the triangle.

   b. Find the coordinates of the point of intersection of these altitudes.

29. The coordinates of $\triangle LMN$ are $L(2,5)$, $M(2,-3)$, and $N(7,-3)$.

   a. Using the theorems of Section 7-6, prove that $\angle L$ and $\angle N$ are acute angles.

   b. List the sides of the triangles in order, starting with the shortest.

   c. List the angles of the triangles in order, starting with the smallest.

Hands-On Activity

The following activity may be completed using graph paper, pencil, compass, and straight-edge, or geometry software.

In the exercises of Section 6-5, we saw how a translation in the horizontal direction can be achieved by a composition of two line reflections in vertical lines and a translation in the vertical direction can be achieved by a composition of two line reflections in horizontal lines. In this activity, we will see how any translation is a composition of two line reflections.

STEP 1. Draw any point A on a coordinate plane.

STEP 2. Translate the point A to its image A' under the given translation, $T_{a,b}$.

STEP 3. Draw line $\overleftrightarrow{AA'}$.

STEP 4. Draw any line l_1 perpendicular to $\overleftrightarrow{AA'}$.

STEP 5. Reflect the point A in line l_1. Let its image be A''.

STEP 6. Let l_2 be the line that is the perpendicular bisector of $\overline{A'A''}$.

Result: The given translation, $T_{a,b}$, is the composition of the two line reflections, r_{l_1} and r_{l_2}, in that order, that is, $T_{a,b} = r_{l_1} \circ r_{l_2}$. (Recall that r_{l_1} is performed first.)

For **a–c**, using the procedure above write the equations of two lines under which reflections in the two lines are equal to the given translation. Check your answers using the given coordinates.

a. $T_{4,4}$

   $D(0,0)$, $E(5,3)$, $F(2,2)$

b. $T_{-3,2}$

   $D(-1,2)$, $E(-5,3)$, $F(5,6)$

c. $T_{-1,-3}$

   $D(6,6)$, $E(-2,-5)$, $F(1,6)$

8-5 COORDINATE PROOF

Many of the proofs that we did in the preceding chapters can also be done using coordinates. In particular, we can use the formula for slope, the slopes of perpendicular lines, and the coordinates of the midpoint of a line segment presented in this chapter to prove theorems about triangles. In later chapters we will use coordinates to prove theorems about polygons, parallel lines, and distances.

There are two types of proofs in coordinate geometry:

1. *Proofs Involving Special Cases.* When the coordinates of the endpoints of a segment or the vertices of a polygon are given as ordered pairs of numbers, we are proving something about a specific segment or polygon. (See Example 1.)

2. *Proofs of General Theorems.* When the given information is a figure that represents a particular type of polygon, we must state the coordinates of its vertices in general terms using variables. Those coordinates can be any convenient variables. Since it is possible to use a transformation that is an isometry to move a triangle without changing its size and shape, a geometric figure can be placed so that one of its sides is a segment of the x-axis. If two line segments or adjacent sides of a polygon are perpendicular, they can be represented as segments of the x-axis and the y-axis. (See Example 2.)

To prove that line segments bisect each other, show that the coordinates of the midpoints of each segment are the same ordered pair, that is, are the same point.

To prove that two lines are perpendicular to each other, show that the slope of one line is the negative reciprocal of the slope of the other.

The vertices shown in the diagrams below can be used when working with triangles.

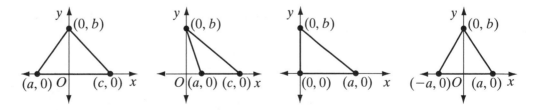

The triangle with vertices $(a, 0), (0, b), (c, 0)$ can be any triangle. It is convenient to place one side of the triangle on the x-axis and the vertex opposite that side on the y-axis. The triangle can be acute if a and c have opposite signs or obtuse if a and c have the same sign.

A triangle with vertices at $(a, 0), (0, 0), (0, b)$ is a right triangle because it has a right angle at the origin.

A triangle with vertices at $(-a, 0), (0, b), (a, 0)$ is isosceles because the altitude and the median are the same line segment.

When a general proof involves the midpoint of a segment, it is helpful to express the coordinates of the endpoints of the segment as variables divisible by 2. For example, if we had written the coordinates of the vertices of a right triangle as $(d, 0), (0, e)$ and $(f, 0)$, we could simply let $d = 2a$, $e = 2b$, and $f = 2c$ so that the coordinates would be $(2a, 0), (0, 2b)$ and $(2c, 0)$. The coordinates of midpoints of the sides of this triangle would be simpler using these coordinates.

EXAMPLE 1

Prove that $\overline{AB}$ and $\overline{CD}$ bisect each other and are perpendicular to each other if the coordinates of the endpoints of these segments are $A(-3, 5)$, $B(5, 1)$, $C(-2, -3)$, and $D(4, 9)$.

Solution   This is a proof involving a special case.

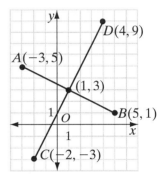

The midpoint of $\overline{AB}$ is
$$\left(\frac{-3 + 5}{2}, \frac{5 + 1}{2}\right) = \left(\frac{2}{2}, \frac{6}{2}\right) = (1, 3).$$

The midpoint of $\overline{CD}$ is
$$\left(\frac{-2 + 4}{2}, \frac{-3 + 9}{2}\right) = \left(\frac{2}{2}, \frac{6}{2}\right) = (1, 3).$$

The slope of $\overline{AB}$ is $\frac{5 - 1}{-3 - 5} = \frac{4}{-8} = -\frac{1}{2}$.

The slope of $\overline{CD}$ is $\frac{9 - (-3)}{4 - (-2)} = \frac{12}{6} = 2$.

$\overline{AB}$ and $\overline{CD}$ bisect each other because they have a common midpoint, $(1, 3)$. $\overline{AB}$ and $\overline{CD}$ are perpendicular because the slope of one is the negative reciprocal of the slope of the other.

EXAMPLE 2

Prove that the midpoint of the hypotenuse of a right triangle is equidistant from the vertices.

Given: Right triangle ABC whose vertices are $A(2a, 0)$, $B(0, 2b)$, and $C(0, 0)$. Let M be the midpoint of the hypotenuse $\overline{AB}$.

Prove: $AM = BM = CM$

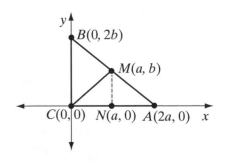

Proof This is a proof of a general theorem. Since it is a right triangle, we can place one vertex at the origin, one side of the triangle on the *x*-axis, and a second side on the *y*-axis so that these two sides form the right angle. We will use coordinates that are divisible by 2 to simplify computation of midpoints.

(1) The midpoint of a line segment is a point of that line segment that separates the line segment into two congruent segments. Therefore, $\overline{AM} \cong \overline{BM}$.

(2) The coordinates of M are $\left(\frac{2a + 0}{2}, \frac{0 + 2b}{2} \right) = (a, b)$.

(3) From M, draw a vertical segment that intersects $\overline{AC}$ at N. The *x*-coordinate of N is a because it is on the same vertical line as M. The *y*-coordinate of N is 0 because it is on the same horizontal line as A and C. The coordinates of N are $(a, 0)$.

(4) The midpoint of $\overline{AC}$ is $\left(\frac{0 + 2a}{2}, \frac{0 + 0}{2} \right) = (a, 0)$. N is the midpoint of $\overline{AC}$ and $\overline{AN} \cong \overline{NC}$.

(5) The vertical segment $\overline{MN}$ is perpendicular to the horizontal segment $\overline{AC}$. Perpendicular lines are two lines that intersect to form right angles. Therefore, $\angle ANM$ and $\angle CMN$ are right angles. All right angles are congruent, so $\angle ANM \cong \angle CMN$. Also, $\overline{MN} \cong \overline{MN}$.

(6) Then, $\triangle AMN \cong \triangle CMN$ by SAS (steps 4 and 5).

(7) $\overline{AM} \cong \overline{CM}$ because they are corresponding parts of congruent triangles.

(8) $\overline{AM} \cong \overline{BM}$ (step 1) and $\overline{AM} \cong \overline{CM}$ (step 7). Therefore, $\overline{AM} \cong \overline{BM} \cong \overline{CM}$ or $AM = BM = CM$. The midpoint of the hypotenuse of a right triangle is equidistant from the vertices.

Exercises

Writing About Mathematics

1. Ryan said that $(a, 0)$, $(0, b)$, $(c, 0)$ can be the vertices of any triangle but if $\frac{a}{b} = -\frac{b}{c}$, then the triangle is a right triangle. Do you agree with Ryan? Explain why or why not.

2. Ken said that $(a, 0)$, $(0, b)$, $(c, 0)$ can be the vertices of any triangle but if a and c have the same sign, then the triangle is obtuse. Do you agree with Ken? Explain why or why not.

Developing Skills

3. The coordinates of the endpoints of $\overline{AB}$ are $A(0, -2)$ and $B(4, 6)$. The coordinates of the endpoints of $\overline{CD}$ are $C(-4, 5)$ and $D(8, -1)$. Using the midpoint formula, show that the line segments bisect each other.

4. The vertices of polygon $ABCD$ are $A(2, 2)$, $B(5, -2)$, $C(9, 1)$, and $D(6, 5)$. Prove that the diagonals $\overline{AC}$ and $\overline{BD}$ are perpendicular and bisect each other using the midpoint formula.

5. The vertices of a triangle are $L(0, 1)$, $M(2, 5)$, and $N(6, 3)$.

   a. Find the coordinates K, the midpoint of the base, $\overline{LN}$.

   b. Show that $\overline{MK}$ is an altitude from M to $\overline{LN}$.

   c. Using parts **a** and **b**, prove that $\triangle LMN$ is isosceles.

6. The vertices of $\triangle ABC$ are $A(1, 7)$, $B(9, 3)$, and $C(3, 1)$.

   a. Prove that $\triangle ABC$ is a right triangle.

   b. Which angle is the right angle?

   c. Which side is the hypotenuse?

   d. What are the coordinates of the midpoint of the hypotenuse?

   e. What is the equation of the median from the vertex of the right angle to the hypotenuse?

   f. What is the equation of the altitude from the vertex of the right angle to the hypotenuse?

   g. Is the triangle an isosceles right triangle? Justify your answer using parts **e** and **f**.

7. The coordinates of the vertices of $\triangle ABC$ are $A(-4, 0)$, $B(0, 8)$, and $C(12, 0)$.

   a. Draw the triangle on graph paper.

   b. Find the coordinates of the midpoints of each side of the triangle.

   c. Find the slope of each side of the triangle.

   d. Write the equation of the perpendicular bisector of each side of the triangle.

   e. Find the coordinates of the circumcenter of the triangle.

8. A rhombus is a quadrilateral with four congruent sides. The vertices of rhombus $ABCD$ are $A(2, -3)$, $B(5, 1)$, $C(10, 1)$, and $D(7, -3)$.

   a. Prove that the diagonals $\overline{AC}$ and $\overline{BD}$ bisect each other.

   b. Prove that the diagonals $\overline{AC}$ and $\overline{BD}$ are perpendicular to each other.

Applying Skills

9. The vertices of rectangle $PQRS$ are $P(0, 0)$, $Q(a, 0)$, $R(a, b)$, and $S(0, b)$. Use congruent triangles to prove that the diagonals of a rectangle are congruent, that is, $\overline{PR} \cong \overline{QS}$.

10. The vertices of square $EFGH$ are $E(0, 0)$, $F(a, 0)$, $G(a, a)$, and $H(0, a)$. Prove that the diagonals of a square, $\overline{EG}$ and $\overline{FH}$, are the perpendicular bisectors of each other using the midpoint formula.

11. Use congruent triangles to prove that $(0, 0)$, $(2a, 0)$, and (a, b) are the vertices of an isosceles triangle. (*Suggestion:* Draw the altitude from (a, b).)

12. Use a translation to prove that $(-a, 0)$, $(0, b)$, and $(a, 0)$ are the vertices of an isosceles triangle. (*Hint:* A translation will let you use the results of Exercise 11.)

13. The coordinates of the vertices of $\triangle ABC$ are $A(0, 0)$, $B(2a, 2b)$, and $C(2c, 2d)$.

a. Find the coordinates of E, the midpoint of $\overline{AB}$ and of F, the midpoint of $\overline{AC}$.

b. Prove that the slope of $\overline{EF}$ is equal to the slope of $\overline{BC}$.

14. The endpoints of segment $\overline{AB}$ are $(-a, 0)$ and $(a, 0)$.

a. Use congruent triangles to show that $P(0, b)$ and $Q(0, c)$ are both equidistant from the endpoints of $\overline{AB}$.

b. Show that $\overleftrightarrow{PQ}$ is the perpendicular bisector of $\overline{AB}$.

8-6 CONCURRENCE OF THE ALTITUDES OF A TRIANGLE

The postulates of the coordinate plane and the statements that we have proved about the slopes of perpendicular lines make it possible for us to prove that the three altitudes of a triangle are *concurrent*, that is, they intersect in one point.

Theorem 8.4 | The altitudes of a triangle are concurrent.

Proof: We can place the triangle anywhere in the coordinate plane. We will place it so that $\overline{AC}$ lies on the x-axis and B lies on the y-axis. Let $A(a, 0)$, $B(0, b)$, and $C(c, 0)$ be the vertices of $\triangle ABC$. Let $\overline{AE}$ be the altitude from A to $\overline{BC}$, $\overline{BO}$ be the altitude from B to $\overline{AC}$, and $\overline{CF}$ be the altitude from C to $\overline{AB}$.

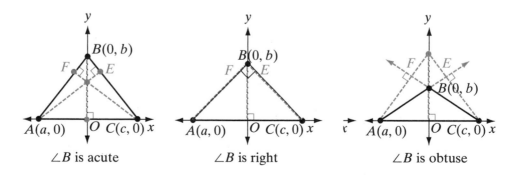

| $\angle B$ is acute | $\angle B$ is right | $\angle B$ is obtuse |

In the figures, $\angle A$ and $\angle C$ are acute angles.

We will show that altitudes $\overline{AE}$ and $\overline{BO}$ intersect in the same point as altitudes $\overline{CF}$ and $\overline{BO}$.

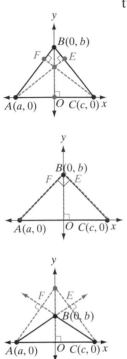

Intersection of altitudes $\overline{AE}$ and $\overline{BO}$	Intersection of altitudes $\overline{CF}$ and $\overline{BO}$
1. The slope of side $\overline{BC}$ is $$\frac{0-b}{c-0} = -\frac{b}{c}.$$	**1.** The slope of side $\overline{AB}$ is $$\frac{0-b}{a-0} = -\frac{b}{a}.$$
2. The slope of altitude $\overline{AE}$, which is perpendicular to $\overline{BC}$, is $\frac{c}{b}$.	**2.** The slope of altitude $\overline{CF}$, which is perpendicular to $\overline{AB}$, is $\frac{a}{b}$.
3. The equation of $\overline{AE}$ is $$\frac{y-0}{x-a} = \frac{c}{b} \text{ or } y = \frac{c}{b}x - \frac{ac}{b}.$$	**3.** The equation of $\overline{CF}$ is $$\frac{y-0}{x-c} = \frac{a}{b} \text{ or } y = \frac{a}{b}x - \frac{ac}{b}.$$
4. Since $\overline{AC}$ is a horizontal line, $\overline{BO}$ is a vertical line, a segment of the y-axis since B is on the y-axis.	**4.** Since $\overline{AC}$ is a horizontal line, $\overline{BO}$ is a vertical line, a segment of the y-axis since B is on the y-axis.
5. The equation of $\overleftrightarrow{BO}$ is $x = 0$.	**5.** The equation of $\overleftrightarrow{BO}$ is $x = 0$.
6. The coordinates of the intersection of $\overline{AE}$ and $\overline{BO}$ can be found by finding the common solution of their equations: $y = \frac{c}{b}x - \frac{ac}{b}$ and $x = 0$.	**6.** The coordinates of the intersection of $\overline{CF}$ and $\overline{BO}$ can be found by finding the common solution of their equations: $y = \frac{a}{b}x - \frac{ac}{b}$ and $x = 0$.
7. Since one of the equations is $x = 0$, replace x by 0 in the other equation: $$\begin{aligned} y &= \frac{c}{b}x - \frac{ac}{b} \\ &= \frac{c}{b}(0) - \frac{ac}{b} \\ &= -\frac{ac}{b} \end{aligned}$$	**7.** Since one of the equations is $x = 0$, replace x by 0 in the other equation: $$\begin{aligned} y &= \frac{a}{b}x - \frac{ac}{b} \\ &= \frac{a}{b}(0) - \frac{ac}{b} \\ &= -\frac{ac}{b} \end{aligned}$$
8. The coordinates of the intersection of $\overline{AE}$ and $\overline{BO}$ are $\left(0, -\frac{ac}{b}\right)$.	**8.** The coordinates of the intersection of $\overline{CF}$ and $\overline{BO}$ are $\left(0, -\frac{ac}{b}\right)$.

The altitudes of a triangle are concurrent at $\left(0, -\frac{ac}{b}\right)$.

Note: If B is a right angle, $\overline{CB}$ is the altitude from C to $\overline{AB}$, $\overline{AB}$ is the altitude from A to $\overline{CB}$, and $\overline{BO}$ is the altitude from B to $\overline{AC}$. The intersection of these three altitudes is $B(0, b)$.

The point where the altitudes of a triangle intersect is called the **orthocenter**.

EXAMPLE 1

The coordinates of the vertices of $\triangle PQR$ are $P(0,0)$, $Q(-2,6)$, and $R(4,0)$. Find the coordinates of the orthocenter of the triangle.

Solution Let $\overline{PL}$ be the altitude from P to $\overline{QR}$.

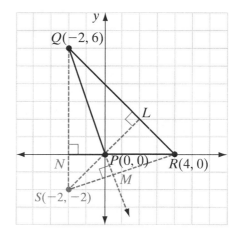

The slope of $\overline{QR}$ is $\frac{6-0}{-2-4} = \frac{6}{-6} = -1$.

The slope of $\overline{PL}$ is 1.

The equation of $\overline{PL}$ is $\frac{y-0}{x-0} = 1$ or $y = x$.

Let $\overline{QN}$ be the altitude from Q to $\overline{PR}$. The point of intersection, N, is on the line determined by P and R.

The slope of $\overline{PR}$ is 0 since $\overline{PR}$ is a horizontal line. Therefore, $\overline{QN}$ is a segment of a vertical line that has no slope.

The equation of $\overline{QN}$ is $x = -2$.

The intersection S of the altitudes $\overline{QN}$ and $\overline{PL}$ is the common solution of the equations $x = -2$ and $y = x$. Therefore, the coordinates of the intersection S are $(-2, -2)$. By Theorem 8.4, point S is the orthocenter of the triangle or the point where the altitudes are concurrent.

Answer The orthocenter of $\triangle PQR$ is $S(-2, -2)$.

Alternative Solution Use the result of the proof given in this section. The coordinates of the point of intersection of the altitudes are $\left(0, -\frac{ac}{b}\right)$. In order for this result to apply, Q must lie on the y-axis and P and R must lie on the x-axis. Since P and R already lie on the x-axis, we need only to use a translation to move Q in the horizontal direction to the y-axis. Use the translation $(x, y) \rightarrow (x + 2, y)$:

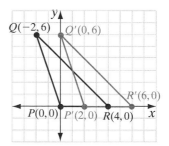

$$P(0,0) \rightarrow P'(2,0) \quad Q(-2,6) \rightarrow Q'(0,6) \quad R(4,0) \rightarrow R'(6,0)$$

Therefore,

$$A(a, 0) = P'(2, 0) \text{ or } a = 2$$
$$B(0, b) = Q'(0, 6) \text{ or } b = 6$$
$$C(c, 0) = R'(6, 0) \text{ or } c = 6.$$

The coordinates of S', the point at which the altitudes of $\triangle P'R'Q'$ intersect, are

$$\left(0, -\tfrac{ac}{b}\right) = \left(0, -\tfrac{2(6)}{6}\right) = (0, -2)$$

The intersection of the altitudes of $\triangle PQR$ is S, the preimage of $S'(0, -2)$ under the translation $(x, y) \rightarrow (x + 2, y)$. Therefore, the coordinates of S are $(-2, -2)$.

Answer   The orthocenter of $\triangle PQR$ is $(-2, -2)$.   ◻

EXAMPLE 2

The coordinates of the vertices of $\triangle ABC$ are $A(0, 0)$, $B(3, 4)$ and $C(2, 1)$. Find the coordinates of the orthocenter of the triangle.

Solution   The slope of $\overline{AC}$ is $\tfrac{1 - 0}{2 - 0} = \tfrac{1}{2}$.

Let $\overline{BD}$ be the altitude from B to $\overline{AC}$.

The slope of altitude $\overline{BD}$ is -2.

The equation of line $\overleftrightarrow{BD}$ is

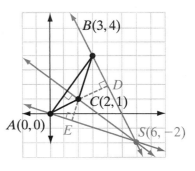

$$\tfrac{4 - y}{3 - x} = -2$$
$$4 - y = -2(3 - x)$$
$$-y = -6 + 2x - 4$$
$$y = -2x + 10$$

The slope of $\overline{BC}$ is $\tfrac{4 - 1}{3 - 2} = \tfrac{3}{1} = 3$.

Let $\overline{AE}$ be the altitude from A to $\overline{BC}$.

The slope of altitude $\overline{AE}$ is $-\tfrac{1}{3}$.

The equation of line $\overleftrightarrow{AE}$ is

$$\tfrac{y - 0}{x - 0} = -\tfrac{1}{3}$$
$$y = -\tfrac{1}{3}x$$

The orthocenter S is the common solution of the equations $y = -2x + 10$ and $y = -\tfrac{1}{3}x$.

$$-2x + 10 = -\tfrac{1}{3}x$$
$$-1\tfrac{2}{3}x = -10$$
$$-\tfrac{5}{3}x = -10$$
$$x = -10\left(\tfrac{-3}{5}\right)$$
$$x = 6$$

The x-coordinate is 6; the y-coordinate is $-\tfrac{1}{3}(6) = -2$.

Answer   The coordinates of the orthocenter are $(6, -2)$ ◻

Exercises

Writing About Mathematics

In 1–2, the vertices of $\triangle ABC$ are $A(a, 0)$, $B(0, b)$, and $C(c, 0)$, as shown in the diagrams of the proof of Theorem 8.4. Assume that $b > 0$.

1. Esther said that if A is to the left of the origin and C is to the right of the origin, then the point of intersection of the altitudes is above the origin. Do you agree with Esther? Explain why or why not.

2. Simon said that if $\angle A$ is an obtuse angle, and both A and C are to the right of the origin, then the point of intersection of the altitudes is above the origin. Do you agree with Simon? Explain why or why not.

Developing Skills

3. The coordinates of $\triangle DEF$ are $D(-9, 0)$, $E(0, 12)$, and $F(16, 0)$.

  a. Show that $\triangle DEF$ is a right triangle.

  b. Show that E is the orthocenter of the triangle.

In 4–9, find the coordinates of the orthocenter of each triangle with the given vertices.

4. $A(-2, 0)$, $B(0, 6)$, $C(3, 0)$

5. $D(-12, 0)$, $E(0, 8)$, $F(6, 0)$

6. $L(7, 2)$, $M(2, 12)$, $N(11, 2)$

7. $P(-3, 4)$, $Q(1, 8)$, $R(3, 4)$

8. $G(-5, 2)$, $H(4, 8)$, $I(5, 1)$

9. $J(0, -3)$, $K(3, 4)$, $L(2, -1)$

Applying Skills

10. Two of the vertices of $\triangle ABC$ are $A(-3, 0)$ and $C(6, 0)$. The altitudes from these vertices intersect at $P(0, 3)$.

  a. $\overleftrightarrow{AB}$ is a line through A, perpendicular to $\overline{CP}$. Write the equation of $\overleftrightarrow{AB}$.

  b. $\overleftrightarrow{CB}$ is a line through C, perpendicular to $\overline{AP}$. Write the equation of $\overleftrightarrow{CB}$.

  c. Find B, the intersection of $\overleftrightarrow{AB}$ and $\overleftrightarrow{CB}$.

  d. Write an equation of the line that contains the altitude from B to $\overline{AC}$.

  e. Show that P is a point on that line.

11. Two of the vertices of $\triangle ABC$ are $A(2, -2)$ and $C(5, 5)$. The altitudes from these vertices intersect at $P(1, 1)$.

  a. $\overleftrightarrow{AB}$ is a line through A, perpendicular to $\overline{CP}$. Write the equation of $\overleftrightarrow{AB}$.

  b. $\overleftrightarrow{CB}$ is a line through C, perpendicular to $\overline{AP}$. Write the equation of $\overleftrightarrow{CB}$.

  c. Find B, the intersection of $\overleftrightarrow{AB}$ and $\overleftrightarrow{CB}$.

d. Write an equation of the line that contains the altitude from B to $\overline{AC}$.

e. Show that P is a point on that line.

Hands-On Activity

In this activity we will use a compass and a straightedge to construct the orthocenters of various triangles.

 For each triangle: **a.** Graph the triangle on paper or using geometry software. **b.** Using a compass, straightedge, and pencil, or geometry software, construct the orthocenter. (If using the computer, you are only allowed to use the point, line segment, line, and circle creation tools of your geometry software and no other software tools.)

(1) $A(-3, 0)$, $B(0, 2)$, $C(4, 0)$

(2) $D(-4, -7)$, $E(0, 5)$, $F(3, -1)$

(3) $G(-4, 2)$, $H(6, 0)$, $I(0, -4)$

CHAPTER SUMMARY

Definitions to Know

- The **slope**, m, of a line that passes through any two points $A(x_1, y_1)$ and $B(x_2, y_2)$, where $x_1 \neq x_2$, is the ratio of the difference of the y-values of these points to the difference of the corresponding x-values.

$$\text{slope of } \overleftrightarrow{AB} = m = \frac{y_2 - y_1}{x_2 - x_1}$$

- Three or more lines are **concurrent** if they intersect in one point.

- The **orthocenter** of a triangle is the point at which the altitudes of a triangle intersect.

Postulate

8.1 A, B, and C lie on the same line if and only if the slope of $\overline{AB}$ is equal to the slope of $\overline{BC}$.

Theorems

8.1 If the endpoints of a line segment are (x_1, y_1) and (x_2, y_2), then the coordinates of the midpoint of the segment are $\left(\dfrac{x_1 + x_2}{2}, \dfrac{y_1 + y_2}{2}\right)$.

8.2 Under a translation, slope is preserved, that is, if line l has slope m, then under a translation, the image of l also has slope m.

8.3 Two non-vertical lines are perpendicular if and only if the slope of one is the negative reciprocal of the other.

8.4 The altitudes of a triangle are concurrent.

VOCABULARY

8-1 Slope • Δx • Δy

8-2 y-intercept • x-intercept • Point-slope form of an equation • Slope-intercept form of an equation

8-3 Midpoint formula

8-4 Transformational proof

8-6 Orthocenter

REVIEW EXERCISES

In 1–3, the coordinates of the endpoints of a line segment are given. **a.** Find the coordinates of each midpoint. **b.** Find the slope of each segment. **c.** Write an equation of the line that is determined by each pair of points.

1. $(0, 0)$ and $(6, -4)$ **2.** $(-3, 2)$ and $(7, -4)$ **3.** $(1, -1)$ and $(5, -5)$

4. The coordinates of P are $(-2, 5)$ and the coordinates of Q are $(6, 1)$.

  a. What is the slope of $\overleftrightarrow{PQ}$?

  b. What is the equation of $\overleftrightarrow{PQ}$?

  c. What are the coordinates of the midpoint of $\overline{PQ}$?

  d. What is the equation of the perpendicular bisector of $\overline{PQ}$?

5. The vertices of $\triangle RST$ are $R(-2, -2)$, $S(1, 4)$, and $T(7, 1)$.

  a. Show that $\triangle RST$ is a right triangle.

  b. Find the coordinates of the midpoint of $\overline{RT}$.

  c. Write the equation of the line that contains the median from S.

  d. Show that the median of the triangle from S is also the altitude from S.

  e. Prove that $\triangle RST$ is an isosceles triangle.

6. Two of the vertices of $\triangle ABC$ are $A(1, 2)$ and $B(9, 6)$. The slope of $\overline{AC}$ is 1 and the slope of $\overline{BC}$ is $-\frac{1}{3}$. What are the coordinates of B?

7. The vertices of $\triangle DEF$ are $D(1, 1)$, $E(5, 5)$, $F(-1, 5)$.

  a. Find the coordinates of the midpoint of each side of the triangle.

  b. Find the slope of each side of the triangle.

  c. Find the slope of each altitude of the triangle.

  d. Write an equation of the perpendicular bisector of each side of the triangle.

  e. Show that the three perpendicular bisectors intersect in a point and find the coordinates of that point.

8. The vertices of $\triangle ABC$ are $A(-7, 1)$, $B(5, -3)$, and $C(-3, 5)$.

 a. Prove that $\triangle ABC$ is a right triangle.

 b. Let M be the midpoint of $\overline{AB}$ and N be the midpoint of $\overline{AC}$. Prove that $\triangle AMN \cong \triangle CMN$ and use this result to show that M is equidistant from the vertices of $\triangle ABC$.

9. The vertices of $\triangle DEF$ are $D(-2, -3)$, $E(5, 0)$, and $F(-2, 3)$

 a. Find the coordinates of M, the midpoint of $\overline{DF}$.

 b. Show that $\overline{DE} \cong \overline{FE}$.

10. The coordinates of the vertices of quadrilateral $ABCD$ are $A(-5, -4)$, $B(1, -6)$, $C(-1, -2)$, and $D(-4, -1)$. Show that $ABCD$ has a right angle.

Exploration

 The following exploration may be completed using graph paper or using geometry software.

We know that the area of a triangle is equal to one-half the product of the lengths of the base and height. We can find the area of a right triangle in the coordinate plane if the base and height are segments of the x-axis and y-axis. The steps that follow will enable us to find the area of any triangle in the coordinate plane. For example, find the area of a triangle if the vertices have the coordinates $(2, 5)$, $(5, 9)$, and $(8, -2)$.

STEP 1. Plot the points and draw the triangle.

STEP 2. Through the vertex with the smallest x-coordinate, draw a vertical line.

STEP 3. Through the vertex with the largest x-coordinate, draw a vertical line.

STEP 4. Through the vertex with the smallest y-coordinate, draw a horizontal line.

STEP 5. Through the vertex with the largest y-coordinate, draw a horizontal line.

STEP 6. The triangle is now enclosed by a rectangle with horizontal and vertical sides. Find the length and width of the rectangle and the area of the rectangle.

STEP 7. The rectangle is separated into four triangles: the given triangle and three triangles that have a base and altitude that are a horizontal and a vertical line. Find the area of these three triangles.

STEP 8. Find the sum of the three areas found in step 7. Subtract this sum from the area of the rectangle. The difference is the area of the given triangle.

Repeat steps 1 through 8 for each of the triangles with the given vertices.

 a. $(-2, 0), (3, 7), (6, -1)$ **b.** $(-3, -6), (0, 4), (5, -1)$ **c.** $(0, -5), (6, 6), (2, 5)$

Can you use this procedure to find the area of the quadrilateral with vertices at $(0, 2), (5, -2), (5, 3)$, and $(2, 5)$?

CUMULATIVE REVIEW   Chapters 1–8

Part I

Answer all questions in this part. Each correct answer will receive 2 credits. No partial credit will be allowed.

1. When the coordinates of A are $(2, -3)$ and of B are $(2, 7)$, AB equals
- (1) 4
- (2) -4
- (3) 10
- (4) -10

2. The slope of the line whose equation is $2x - y = 4$ is
- (1) $\frac{1}{2}$
- (2) 2
- (3) -2
- (4) 4

3. Which of the following is an example of the associative property for multiplication?
- (1) $3(2 + 5) = 3(5 + 2)$
- (2) $3(2 \cdot 5) = (3 \cdot 2) \cdot 5$
- (3) $3(2 + 5) = 3(2) + 5$
- (4) $3 + (2 + 5) = (3 + 2) + 5$

4. The endpoints of $\overline{AB}$ are $A(0, 6)$ and $B(-4, 0)$. The coordinates of the midpoint of $\overline{AB}$ are
- (1) $(-2, 3)$
- (2) $(2, -3)$
- (3) $(-2, -3)$
- (4) $(2, 3)$

5. In isosceles triangle DEF, $DE = DF$. Which of the following is true?
- (1) $\angle D \cong \angle E$
- (2) $\angle F \cong \angle E$
- (3) $\angle D \cong \angle F$
- (4) $\angle D \cong \angle E \cong \angle F$

6. The slope of line l is $\frac{2}{3}$. The slope of a line perpendicular to l is
- (1) $\frac{2}{3}$
- (2) $-\frac{2}{3}$
- (3) $\frac{3}{2}$
- (4) $-\frac{3}{2}$

7. The coordinates of two points are $(0, 6)$ and $(3, 0)$. The equation of the line through these points is
- (1) $y = 2x + 6$
- (2) $y = -2x + 6$
- (3) $y = \frac{1}{2}x + 3$
- (4) $y = -\frac{1}{2}x + 3$

8. The converse of the statement "If two angles are right angles then they are congruent" is
- (1) If two angles are congruent then they are right angles.
- (2) If two angles are not right angles then they are not congruent.
- (3) Two angles are congruent if and only if they are right angles.
- (4) Two angles are congruent if they are right angles.

9. The measure of an angle is twice the measure of its supplement. The measure of the smaller angle is
- (1) 30
- (2) 60
- (3) 90
- (4) 120

10. Under a reflection in the y-axis, the image of $(4, -2)$ is
- (1) $(-4, 2)$
- (2) $(4, -2)$
- (3) $(-4, -2)$
- (4) $(2, 4)$

Part II

Answer all questions in this part. Each correct answer will receive 2 credits. Clearly indicate the necessary steps, including appropriate formula substitutions, diagrams, graphs, charts, etc. For all questions in this part, a correct numerical answer with no work shown will receive only 1 credit.

11. a. Draw $\triangle ABC$ in the coordinate plane if the coordinates of A are $(-3, 1)$, of B are $(1, -5)$, and of C are $(5, 2)$.

   b. Under a translation, the image of A is $A'(1, -1)$. Draw $\triangle A'B'C'$, the image of $\triangle ABC$ under this translation, and give the coordinates of B' and C'.

   c. If this translation can be written as $(x, y) \rightarrow (x + a, y + b)$, what are the values of a and b?

12. The coordinates of the vertices of $\triangle DEF$ are $D(-1, 6)$, $E(3, 3)$, and $F(1, 2)$. Is $\triangle DEF$ a right triangle? Justify your answer.

Part III

Answer all questions in this part. Each correct answer will receive 4 credits. Clearly indicate the necessary steps, including appropriate formula substitutions, diagrams, graphs, charts, etc. For all questions in this part, a correct numerical answer with no work shown will receive only 1 credit.

13. In the diagram, $\overrightarrow{TQ}$ bisects $\angle RTS$ and $\overline{TQ} \perp \overline{RS}$. Prove that $\triangle RST$ is isosceles.

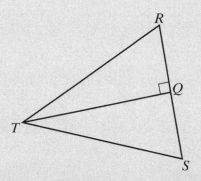

14. The following statements are true:

- If Evanston is not the capital of Illinois, then Chicago is not the capital.

- Springfield is the capital of Illinois or Chicago is the capital of Illinois.

- Evanston is not the capital of Illinois.

Use the laws of logic to prove that Springfield is the capital of Illinois.

Part IV

Answer all questions in this part. Each correct answer will receive 6 credits. Clearly indicate the necessary steps, including appropriate formula substitutions, diagrams, graphs, charts, etc. For all questions in this part, a correct numerical answer with no work shown will receive only 1 credit.

15. Find the coordinates of the point of intersection of the lines whose equations are $y = 3x - 1$ and $x + 2y = 5$.

16. *Given:* $\overline{CGF}$ and $\overline{DGE}$ bisect each other at G.

Prove: $\overline{CD} \cong \overline{FE}$

CHAPTER 9

CHAPTER TABLE OF CONTENTS

PARALLEL LINES

"If a straight line falling on two straight lines makes the interior angles on the same side less than two right angles, then the two straight lines, if produced indefinitely, meet on that side on which the angles are less than two right angles."

This statement, Euclid's fifth postulate, is called **Euclid's parallel postulate**. Throughout history this postulate has been questioned by mathematicians because many felt it was too complex to be a postulate.

Throughout the history of mathematics, attempts were made to prove this postulate or state a related postulate that would make it possible to prove Euclid's parallel postulate. Other postulates have been proposed that appear to be simpler and which could provide the basis for a proof of the parallel postulate.

The form of the parallel postulate most commonly used in the study of elementary geometry today was proposed by John Playfair (1748–1819). **Playfair's postulate** states:

- Through a point not on a given line there can be drawn one and only one line parallel to the given line.

9-1 PROVING LINES PARALLEL

You have already studied many situations involving intersecting lines that lie in the same plane. When all the points or lines in a set lie in a plane, we say that these points or these lines are **coplanar**. Let us now consider situations involving coplanar lines that do not intersect in one point.

DEFINITION _____

Parallel lines are coplanar lines that have no points in common, or have all points in common and, therefore, coincide.

The word "lines" in the definition means straight lines of unlimited extent. We say that segments and rays are parallel if the lines that contain them are parallel.

We indicate that $\overleftrightarrow{AB}$ is parallel to $\overleftrightarrow{CD}$ by writing $\overleftrightarrow{AB} \parallel \overleftrightarrow{CD}$. The parallel lines $\overleftrightarrow{AB}$ and $\overleftrightarrow{CD}$ extended indefinitely never intersect and have no points in common.

The parallel lines $\overleftrightarrow{AB}$ and $\overleftrightarrow{CD}$ may have all points in common, that is, be two different names for the same line. A line is parallel to itself. Thus, $\overleftrightarrow{AB} \parallel \overleftrightarrow{AB}$, $\overleftrightarrow{CD} \parallel \overleftrightarrow{CD}$ and $\overleftrightarrow{AB} \parallel \overleftrightarrow{CD}$.

In Chapter 4, we stated the following postulate:

▶ **Two distinct lines cannot intersect in more than one point.**

This postulate, together with the definition of parallel lines, requires that one of three possibilities exist for any two coplanar lines, $\overleftrightarrow{AB}$ and $\overleftrightarrow{CD}$:

1. $\overleftrightarrow{AB}$ and $\overleftrightarrow{CD}$ have no points in common.
$\overleftrightarrow{AB}$ and $\overleftrightarrow{CD}$ are parallel.

2. $\overleftrightarrow{AB}$ and $\overleftrightarrow{CD}$ have only one point in common.
$\overleftrightarrow{AB}$ and $\overleftrightarrow{CD}$ intersect.

3. $\overleftrightarrow{AB}$ and $\overleftrightarrow{CD}$ have all points in common.
$\overleftrightarrow{AB}$ and $\overleftrightarrow{CD}$ are the same line.

These three possibilities can also be stated in the following postulate:

Postulate 9.1

Two distinct coplanar lines are either parallel or intersecting.

EXAMPLE 1

If line *l* is not parallel to line *p*, what statements can you make about these two lines?

Solution Since *l* is not parallel to *p*, *l* and *p* cannot be the same line, and they have exactly one point in common. *Answer*

Parallel Lines and Transversals

$\overleftrightarrow{AB}$ intersects $\overleftrightarrow{CD}$

When two lines intersect, four angles are formed that have the same vertex and no common interior points. In this set of four angles, there are two pair of congruent vertical angles and four pair of supplementary adjacent angles. When two lines are intersected by a third line, two such sets of four angles are formed.

DEFINITION _____

A **transversal** is a line that intersects two other coplanar lines in two different points.

Two lines, *l* and *m*, are cut by a transversal, *t*. Two sets of angles are formed, each containing four angles. Each of these angles has one ray that is a subset of *l* or of *m* and one ray that is a subset of *t*. In earlier courses, we learned names to identify these sets of angles.

- The angles that have a part of a ray between *l* and *m* are **interior angles**.
  Angles 3, 4, 5, 6 are interior angles.

- The angles that do not have a part of a ray between *l* and *m* are **exterior angles**.
  Angles 1, 2, 7, 8 are exterior angles.

- **Alternate interior angles** are on opposite sides of the transversal and do not have a common vertex.
  Angles 3 and 6 are alternate interior angles, and angles 4 and 5 are alternate interior angles.

- **Alternate exterior angles** are on opposite sides of the transversal and do not have a common vertex.
  Angles 1 and 8 are alternate exterior angles, and angles 2 and 7 are alternate exterior angles.

- **Interior angles on the same side of the transversal** do not have a common vertex.
  Angles 3 and 5 are interior angles on the same side of the transversal, and angles 4 and 6 are interior angles on the same side of the transversal.

- **Corresponding angles** are one exterior and one interior angle that are on the same side of the transversal and do not have a common vertex.
  Angles 1 and 5, angles 2 and 6, angles 3 and 7, and angles 4 and 8 are pairs of corresponding angles.

In the diagram shown on page 330, the two lines cut by the transversal are not parallel lines. However, when two lines are parallel, many statements may be postulated and proved about these angles.

Theorem 9.1a | If two coplanar lines are cut by a transversal so that the alternate interior angles formed are congruent, then the two lines are parallel.

Given $\overleftrightarrow{AB}$ and $\overleftrightarrow{CD}$ are cut by transversal $\overleftrightarrow{EF}$ at points E and F, respectively; $\angle 1 \cong \angle 2$.

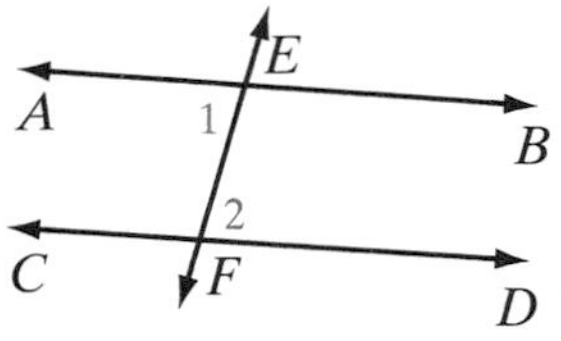

Prove $\overleftrightarrow{AB} \parallel \overleftrightarrow{CD}$

Proof To prove this theorem, we will use an indirect proof.

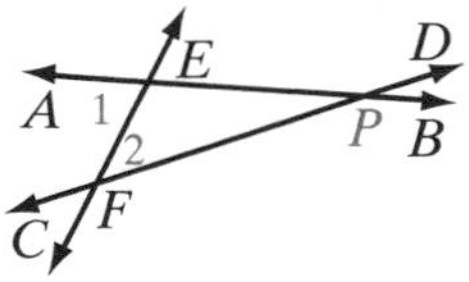

Statements	**Reasons**
1. $\overleftrightarrow{AB}$ is not parallel to $\overleftrightarrow{CD}$.	**1.** Assumption.
2. $\overleftrightarrow{AB}$ and $\overleftrightarrow{CD}$ are cut by transversal $\overleftrightarrow{EF}$ at points E and F, respectively.	**2.** Given.
3. $\overleftrightarrow{AB}$ and $\overleftrightarrow{CD}$ intersect at some point P, forming $\triangle EFP$.	**3.** Two distinct coplanar lines are either parallel or intersecting.
4. $m\angle 1 > m\angle 2$	**4.** The measure of an exterior angle of a triangle is greater than the measure of either nonadjacent interior angle.
5. But $\angle 1 \cong \angle 2$.	**5.** Given.
6. $m\angle 1 = m\angle 2$	**6.** Congruent angles are equal in measure.
7. $\overleftrightarrow{AB} \parallel \overleftrightarrow{CD}$	**7.** Contradiction in steps 4 and 6.

Now that we have proved Theorem 9.1, we can use it in other theorems that also prove that two lines are parallel.

Theorem 9.2a | If two coplanar lines are cut by a transversal so that the corresponding angles are congruent, then the two lines are parallel.

Given $\overleftrightarrow{EF}$ intersects $\overleftrightarrow{AB}$ and $\overleftrightarrow{CD}$; $\angle 1 \cong \angle 5$.

Prove $\overleftrightarrow{AB} \parallel \overleftrightarrow{CD}$

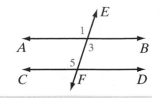

Proof	**Statements**	**Reasons**
	1. $\overleftrightarrow{EF}$ intersects $\overleftrightarrow{AB}$ and $\overleftrightarrow{CD}$; $\angle 1 \cong \angle 5$	1. Given.
	2. $\angle 1 \cong \angle 3$	2. Vertical angles are congruent.
	3. $\angle 3 \cong \angle 5$	3. Transitive property of congruence.
	4. $\overleftrightarrow{AB} \parallel \overleftrightarrow{CD}$	4. If two coplanar lines are cut by a transversal so that the alternate interior angles formed are congruent, then the two lines are parallel.

Theorem 9.3a   If two coplanar lines are cut by a transversal so that the interior angles on the same side of the transversal are supplementary, then the lines are parallel.

Given $\overleftrightarrow{EF}$ intersects $\overleftrightarrow{AB}$ and $\overleftrightarrow{CD}$, and $\angle 5$ is the supplement of $\angle 4$.

Prove $\overleftrightarrow{AB} \parallel \overleftrightarrow{CD}$

Proof Angle 4 and angle 3 are supplementary since they form a linear pair. If two angles are supplements of the same angle, then they are congruent. Therefore, $\angle 3 \cong \angle 5$. Angles 3 and 5 are a pair of congruent alternate interior angles. If two coplanar lines are cut by a transversal so that the alternate interior angles formed are congruent, then the lines are parallel. Therefore, $\overleftrightarrow{AB} \parallel \overleftrightarrow{CD}$.

Theorem 9.4   If two coplanar lines are each perpendicular to the same line, then they are parallel.

Given $\overleftrightarrow{AB} \perp \overleftrightarrow{EF}$ and $\overleftrightarrow{CD} \perp \overleftrightarrow{EF}$.

Prove $\overleftrightarrow{AB} \parallel \overleftrightarrow{CD}$

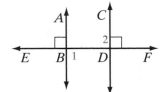

Strategy Show that a pair of alternate interior angles are congruent.

The proof of Theorem 9.4 is left to the student. (See exercise 10.)

Methods of Proving Lines Parallel

To prove that two coplanar lines that are cut by a transversal are parallel, prove that any one of the following statements is true:

1. A pair of alternate interior angles are congruent.

2. A pair of corresponding angles are congruent.

3. A pair of interior angles on the same side of the transversal are supplementary.

4. Both lines are perpendicular to the same line.

EXAMPLE 2

If $m\angle A = 100 + 3x$ and $m\angle B = 80 - 3x$, explain why $\overline{AD} \parallel \overline{BC}$.

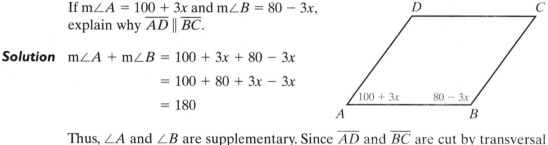

Solution

$$m\angle A + m\angle B = 100 + 3x + 80 - 3x$$
$$= 100 + 80 + 3x - 3x$$
$$= 180$$

Thus, $\angle A$ and $\angle B$ are supplementary. Since $\overline{AD}$ and $\overline{BC}$ are cut by transversal $\overline{AB}$ to form supplementary interior angles on the same side of the transversal, the segments are parallel, namely, $\overline{AD} \parallel \overline{BC}$. ∎

EXAMPLE 3

If $\overline{BD}$ bisects $\angle ABC$, and $\overline{BC} \cong \overline{CD}$, prove $\overline{CD} \parallel \overrightarrow{BA}$.

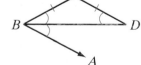

Proof

(1) Since $\overline{BC} \cong \overline{CD}$, $\angle CBD \cong \angle D$ because the base angles of an isosceles triangle are congruent.

(2) Since $\overline{BD}$ bisects $\angle ABC$, $\angle CBD \cong \angle DBA$ because the bisector of an angle divides the angle into two congruent angles.

(3) Therefore, by the transitive property of congruence, $\angle DBA \cong \angle D$.

(4) Then, $\angle DBA$ and $\angle D$ are congruent alternate interior angles when $\overline{CD}$ and $\overrightarrow{BA}$ are intersected by transversal $\overline{BD}$. Therefore, $\overline{CD} \parallel \overrightarrow{BA}$ because if two coplanar lines are cut by a transversal so that the alternate interior angles formed are congruent, then the two lines are parallel. ∎

Exercises

Writing About Mathematics

1. Two lines are cut by a transversal. If $\angle 1$ and $\angle 2$ are vertical angles and $\angle 1$ and $\angle 3$ are alternate interior angles, what type of angles do $\angle 2$ and $\angle 3$ form?

2. Is it true that if two lines that are not parallel are cut by a transversal, then the alternate interior angles are not congruent? Justify your answer.

Developing Skills

In 3–8, the figure shows eight angles formed when $\overleftrightarrow{AB}$ and $\overleftrightarrow{CD}$ are cut by transversal $\overleftrightarrow{EF}$. For each of the following, state the theorem or theorems that prove $\overleftrightarrow{AB} \parallel \overleftrightarrow{CD}$.

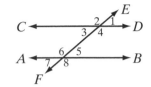

3. $m\angle 3 = 70$ and $m\angle 5 = 70$

4. $m\angle 2 = 140$ and $m\angle 6 = 140$

5. $m\angle 3 = 60$ and $m\angle 6 = 120$

6. $m\angle 2 = 150$ and $m\angle 5 = 30$

7. $m\angle 2 = 160$ and $m\angle 8 = 160$

8. $m\angle 4 = 110$ and $m\angle 7 = 70$

Applying Skills

9. Write an *indirect* proof of Theorem 9.2a, "If two coplanar lines are cut by a transversal so that the corresponding angles are congruent, then the two lines are parallel."

10. Prove Theorem 9.4, "If two coplanar lines are each perpendicular to the same line, then they are parallel."

In 11 and 12, $ABCD$ is a quadrilateral.

11. If $m\angle A = 3x$ and $m\angle B = 180 - 3x$. Show that $\overline{AD} \parallel \overline{BC}$.

12. If $\overline{DC} \perp \overline{BC}$ and $m\angle ADC = 90$, prove $\overline{AD} \parallel \overline{BC}$.

13. If $\overline{AB}$ and $\overline{CD}$ bisect each other at point E, prove:

a. $\triangle CEA \cong \triangle DEB$

b. $\angle ECA \cong \angle EDB$

c. $\overline{CA} \parallel \overline{DB}$

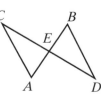

14. Prove that if two coplanar lines are cut by a transversal, forming a pair of alternate exterior angles that are congruent, then the two lines are parallel.

9-2 PROPERTIES OF PARALLEL LINES

In the study of logic, we learned that a conditional and its converse do not always have the same truth value. Once a conditional statement has been proved to be true, it may be possible to prove that its converse is also true. In this section, we will prove converse statements of some of the theorems proved in the previous section. The proof of these converse statements requires the following postulate and theorem:

Postulate 9.2

> Through a given point not on a given line, there exists one and only one line parallel to the given line.

Theorem 9.5

> If, in a plane, a line intersects one of two parallel lines, it intersects the other.

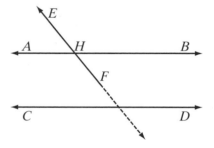

Given $\overleftrightarrow{AB} \parallel \overleftrightarrow{CD}$ and $\overleftrightarrow{EF}$ intersects $\overleftrightarrow{AB}$ at H.

Prove $\overleftrightarrow{EF}$ intersects $\overleftrightarrow{CD}$.

Proof Assume $\overleftrightarrow{EF}$ does not intersect $\overleftrightarrow{CD}$. Then $\overleftrightarrow{EF} \parallel \overleftrightarrow{CD}$. Therefore, through H, a point not on $\overleftrightarrow{CD}$, two lines, $\overleftrightarrow{AB}$ and $\overleftrightarrow{EF}$ are each parallel to $\overleftrightarrow{CD}$. This contradicts the postulate that states that through a given point not on a given line, one and only one line can be drawn parallel to a given line. Since our assumption leads to a contradiction, the assumption must be false and its negation, $\overleftrightarrow{EF}$ intersects $\overleftrightarrow{CD}$ must be true.

Now we are ready to prove the converse of Theorem 9.1a.

Theorem 9.1b

> If two parallel lines are cut by a transversal, then the alternate interior angles formed are congruent.

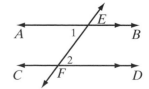

Given $\overleftrightarrow{AB} \parallel \overleftrightarrow{CD}$, transversal $\overleftrightarrow{EF}$ intersects $\overleftrightarrow{AB}$ at E and $\overleftrightarrow{CD}$ at F.

Prove $\angle 1 \cong \angle 2$

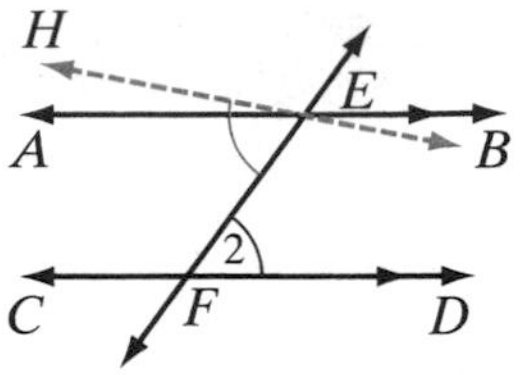

Proof   We can use an indirect proof. Assume $\angle 1$ is not congruent to $\angle 2$. Construct $\overrightarrow{EH}$ so that $\angle HEF \cong \angle 2$. Since $\angle HEF$ and $\angle 2$ are congruent alternate interior angles, $\overleftrightarrow{HE} \parallel \overleftrightarrow{CD}$. But $\overleftrightarrow{AB}$ is a line through E, and we are given $\overleftrightarrow{AB} \parallel \overleftrightarrow{CD}$. This contradicts the postulate that states that through a given point not on a given line, there exists one and only one line parallel to the given line. Thus, the assumption is false and $\angle 1 \cong \angle 2$.

Note that Theorem 9.1b is the converse of Theorem 9.1a. We may state the two theorems in biconditional form:

Theorem 9.1

> Two coplanar lines cut by a transversal are parallel if and only if the alternate interior angles formed are congruent.

Each of the next two theorems is also a converse of a theorem stated in Section 9-1.

Theorem 9.2b

> If two parallel lines are cut by a transversal, then the corresponding angles are congruent. (Converse of Theorem 9.2a)

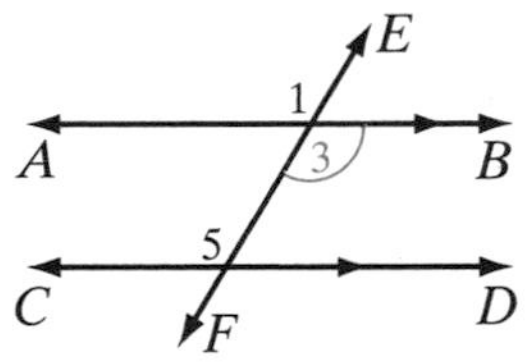

Given $\overleftrightarrow{AB} \parallel \overleftrightarrow{CD}$ and transversal $\overleftrightarrow{EF}$

Prove $\angle 1 \cong \angle 5$

Proof

Statements	Reasons
1. $\overleftrightarrow{AB} \parallel \overleftrightarrow{CD}$ and transversal $\overleftrightarrow{EF}$	1. Given.
2. $\angle 3 \cong \angle 5$	2. If two parallel lines are cut by a transversal, then the alternate interior angles formed are congruent.
3. $\angle 1 \cong \angle 3$	3. Vertical angles are congruent.
4. $\angle 1 \cong \angle 5$	4. Transitive property of congruence.

Theorem 9.3b | If two parallel lines are cut by a transversal, then two interior angles on the same side of the transversal are supplementary. (Converse of Theorem 9.3a)

Given $\overleftrightarrow{AB} \parallel \overleftrightarrow{CD}$ and transversal $\overleftrightarrow{EF}$

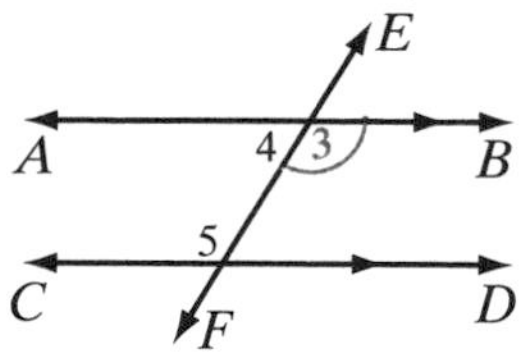

Prove $\angle 4$ is the supplement of $\angle 5$.

Strategy Show that $\angle 3 \cong \angle 5$ and that $\angle 4$ is the supplement of $\angle 3$. If two angles are congruent, then their supplements are congruent. Therefore, $\angle 4$ is also the supplement of $\angle 5$.

The proof of this theorem is left to the student. (See exercise 18.) Since Theorems 9.2b and 9.3b are converses of Theorems 9.2a and 9.3a, we may state the theorems in biconditional form:

Theorem 9.2 | Two coplanar lines cut by a transversal are parallel if and only if corresponding angles are congruent.

Theorem 9.3 | Two coplanar lines cut by a transversal are parallel if and only if interior angles on the same side of the transversal are supplementary.

EXAMPLE I

Transversal $\overleftrightarrow{EF}$ intersects $\overleftrightarrow{AB}$ and $\overleftrightarrow{CD}$ at G and H, respectively. If $\overleftrightarrow{AB} \parallel \overleftrightarrow{CD}$, $m\angle BGH = 3x - 20$, and $m\angle GHC = 2x + 10$:

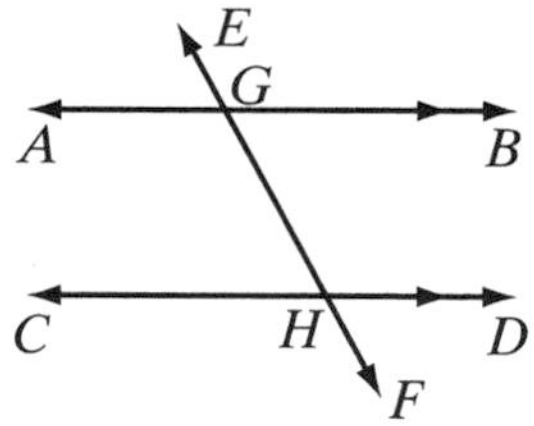

    a. Find the value of x.   **b.** Find $m\angle GHC$.

    c. Find $m\angle GHD$.

Solution **a.** Since $\overleftrightarrow{AB} \parallel \overleftrightarrow{CD}$ and these lines are cut by transversal $\overleftrightarrow{EF}$, the alternate interior angles are congruent: $m\angle BGH = m\angle GHC$

$$3x - 20 = 2x + 10$$
$$3x - 2x = 10 + 20$$
$$x = 30$$

b. $m\angle GHC = 2x + 10$
$$= 2(30) + 10$$
$$= 70$$

c. Since $\angle GHC$ and $\angle GHD$ form a linear pair and are supplementary,

$$m\angle GHD = 180 - m\angle GHC$$
$$= 180 - 70$$
$$= 110$$

Answers **a.** $x = 70$ **b.** $m\angle GHC = 70$ **c.** $m\angle GHD = 110$

Using Theorem 9.1, we may also prove the following theorems:

Theorem 9.6

If a transversal is perpendicular to one of two parallel lines, it is perpendicular to the other.

Given $\overleftrightarrow{AB} \parallel \overleftrightarrow{CD}, \overleftrightarrow{EF} \perp \overleftrightarrow{AB}$

Prove $\overleftrightarrow{EF} \perp \overleftrightarrow{CD}$

Strategy Show that alternate interior angles are right angles.

The proof of this theorem is left to the student. (See exercise 19.)

Theorem 9.7

If two of three lines in the same plane are each parallel to the third line, then they are parallel to each other.

Given $\overleftrightarrow{AB} \parallel \overleftrightarrow{LM}$ and $\overleftrightarrow{CD} \parallel \overleftrightarrow{LM}$

Prove $\overleftrightarrow{AB} \parallel \overleftrightarrow{CD}$

Proof Draw transversal $\overleftrightarrow{EJ}$ intersecting $\overleftrightarrow{LM}$ at H. Since $\overleftrightarrow{AB} \parallel \overleftrightarrow{LM}$, this transversal also intersects $\overleftrightarrow{AB}$. Call this point F. Similarly, since $\overleftrightarrow{CD} \parallel \overleftrightarrow{LM}$, this transversal also intersects $\overleftrightarrow{CD}$ at a point G.

Since $\overleftrightarrow{AB} \parallel \overleftrightarrow{LM}$, alternate interior angles formed are congruent. Therefore, $\angle AFG \cong \angle GHM$. Similarly, since $\overleftrightarrow{CD} \parallel \overleftrightarrow{LM}$, $\angle CGH \cong \angle GHM$. By the transitive property of congruence, $\angle AFG \cong \angle CGH$. Angles AFG and CGH are congruent corresponding angles when $\overleftrightarrow{AB}$ and $\overleftrightarrow{CD}$ are intersected by transversal

$\overleftrightarrow{EJ}$. Therefore, $\overleftrightarrow{AB} \parallel \overleftrightarrow{CD}$ because if two coplanar lines are cut by a transversal so that the corresponding angles formed are congruent, then the two lines are parallel.

SUMMARY OF PROPERTIES OF PARALLEL LINES
If two lines are parallel:

1. A transversal forms congruent alternate interior angles.

2. A transversal forms congruent corresponding angles.

3. A transversal forms supplementary interior angles on the same side of the transversal.

4. A transversal perpendicular to one line is also perpendicular to the other.

5. A third line in the same plane that is parallel to one of the lines is parallel to the other.

EXAMPLE 2

Given: Quadrilateral $ABCD$, $\overline{BC} \cong \overline{DA}$, and $\overline{BC} \parallel \overline{DA}$

Prove: $\overline{AB} \parallel \overline{CD}$

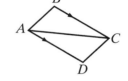

Proof Use congruent triangles to prove congruent alternate interior angles.

Statements	Reasons
1. $\overline{BC} \cong \overline{DA}$	1. Given.
2. $\overline{BC} \parallel \overline{DA}$	2. Given.
3. $\angle BCA \cong \angle DAC$	3. If two parallel lines are cut by a transversal, the alternate interior angles are congruent.
4. $\overline{AC} \cong \overline{AC}$	4. Reflexive property of congruence.
5. $\triangle BAC \cong \triangle DCA$	5. SAS.
6. $\angle BAC \cong \angle DCA$	6. Corresponding parts of congruent triangles are congruent.
7. $\overline{AB} \parallel \overline{CD}$	7. If two lines cut by a transversal form congruent alternate interior angles, the lines are parallel.

Note: In the diagram for Example 2, you may have noticed that two parallel lines, $\overleftrightarrow{BC}$ and $\overleftrightarrow{DA}$, each contained a single arrowhead in the same direction. Such pairs of arrowheads are used on diagrams to indicate that two lines are parallel.

Exercises

Writing About Mathematics

1. a. Is the inverse of Theorem 9.1a always true? Explain why or why not.

   b. Is the inverse of Theorem 9.6 always true? Explain why or why not.

2. Two parallel lines are cut by a transversal forming alternate interior angles that are supplementary. What conclusion can you draw about the measures of the angles formed by the parallel lines and the transversal. Justify your answer.

Developing Skills

In 3–12, $\overleftrightarrow{AB} \parallel \overleftrightarrow{CD}$ are cut by transversal $\overleftrightarrow{EF}$ as shown in the diagram. Find:

3. m∠5 when m∠3 = 80.　　　**4.** m∠2 when m∠6 = 150.

5. m∠4 when m∠5 = 60.　　　**6.** m∠7 when m∠1 = 75.

7. m∠8 when m∠3 = 65.　　　**8.** m∠5 when m∠2 = 130.

9. m∠3 when m∠3 = 3x and m∠5 = x + 28.

10. m∠5 when m∠3 = x and m∠4 = x + 20.

11. m∠7 when m∠1 = x + 40 and m∠2 = 5x − 10.

12. m∠5 when m∠2 = 7x − 20 and m∠8 = x + 100.

13. Two parallel lines are cut by a transversal. For each pair of interior angles on the same side of the transversal, the measure of one angle exceeds the measure of twice the other by 48 degrees. Find the measures of one pair of interior angles.

14. Two parallel lines are cut by a transversal. The measure of one of the angles of a pair of corresponding angles can be represented by 42 less than three times the other. Find the measures of the angles of this pair of corresponding angles.

15. In the diagram, $\overleftrightarrow{AFB} \parallel \overleftrightarrow{CD}$ and $\overleftrightarrow{EF}$ and $\overleftrightarrow{GF}$ intersect $\overleftrightarrow{AB}$ at F.

   a. If m∠FGD = 110 and m∠FEC = 130, find the measures of each of the angles numbered 1 through 9.

   b. What is the measure of an exterior angle of △EFG at F?

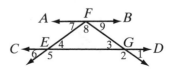

c. Is the measure of an exterior angle at F greater than the measure of either of the nonadjacent interior angles?

d. What is the sum of the measures of the nonadjacent interior angles of an exterior angle at F?

e. What is the sum of the measures of the nonadjacent interior angles of the exterior angle, $\angle FGD$?

f. What is the sum of the measures of the nonadjacent interior angles of the exterior angle, $\angle FEC$?

g. What is the sum of the measures of the angles of $\triangle EFG$?

16. Two pairs of parallel lines are drawn; $\overleftrightarrow{ABE} \parallel \overleftrightarrow{DC}$ and $\overleftrightarrow{AD} \parallel \overleftrightarrow{BC}$. If $m\angle CBE = 75$, find the measure of each angle of quadrilateral $ABCD$.

Applying Skills

17. Prove Theorem 9.3b, "If two parallel lines are cut by a transversal, then two interior angles on the same side of the transversal are supplementary."

18. Prove Theorem 9.6, "If a transversal is perpendicular to one of two parallel lines, it is perpendicular to the other."

19. Prove that if two parallel lines are cut by a transversal, the alternate exterior angles are congruent.

20. *Given:* $\triangle ABC, \overrightarrow{CE}$ bisects exterior $\angle BCD$, and $\overrightarrow{CE} \parallel \overline{AB}$.

Prove: $\angle A \cong \angle B$.

21. *Given:* $\angle CAB \cong \angle DCA$ and $\angle DCA \cong \angle ECB$

Prove: **a.** $\overleftrightarrow{AB} \parallel \overleftrightarrow{DCE}$.

   b. $\angle CAB$ is the supplement of $\angle CBG$.

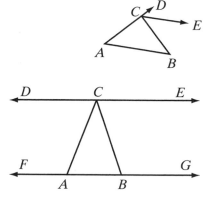

22. The opposite sides of quadrilateral $PQRS$ are parallel, that is, $\overline{PQ} \parallel \overline{RS}$ and $\overline{QR} \parallel \overline{SP}$. If $\angle P$ is a right angle, prove that $\angle Q$, $\angle R$, and $\angle S$ are right angles.

23. The opposite sides of quadrilateral $KLMN$ are parallel, that is, $\overline{KL} \parallel \overline{MN}$ and $\overline{LM} \parallel \overline{NK}$. If $\angle K$ is an acute angle, prove that $\angle M$ is an acute angle and that $\angle L$ and $\angle N$ are obtuse angles.

9-3 PARALLEL LINES IN THE COORDINATE PLANE

In Chapter 6 we stated postulates about horizontal and vertical lines in the coordinate plane. One of these postulates states that each vertical line is perpendicular to each horizontal line. We can use this postulate to prove the following theorem:

Theorem 9.8 | If two lines are vertical lines, then they are parallel.

Proof: Since each vertical line is perpendicular to each horizontal line, each vertical line is perpendicular to the *x*-axis, a horizontal line. Theorem 9.6 states that if two coplanar lines are each perpendicular to the same line, then they are parallel. Therefore, all vertical lines are parallel.

A similar theorem can be proved about horizontal lines:

Theorem 9.9 | If two lines are horizontal lines, then they are parallel.

Proof: Since each horizontal line is perpendicular to each vertical line, each horizontal line is perpendicular to the *y*-axis, a vertical line. Theorem 9.6 states that if two coplanar lines are each perpendicular to the same line, then they are parallel. Therefore, all horizontal lines are parallel.

We know that all horizontal lines have the same slope, 0. We also know that all vertical lines have no slope.

Do parallel lines that are neither horizontal nor vertical have the same slope? When we draw parallel lines in the coordinate plane, it appears that this is true.

Theorem 9.10a | If two non-vertical lines in the same plane are parallel, then they have the same slope.

Given $l_1 \parallel l_2$

Prove The slope of l_1 is equal to slope of l_2.

Proof In the coordinate plane, let the slope of l_1 be $m \neq 0$. Choose any point on l_1. Through a given point, one and only one line can be drawn perpendicular to a given line. Through that point, draw *k*, a line perpendicular to l_1.

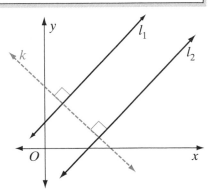

If two lines are perpendicular, the slope of one is the negative reciprocal of the slope of the other. Therefore, the slope of k is $-\frac{1}{m}$. It is given that $l_1 \parallel l_2$. Then, k is perpendicular to l_2 because if a line is perpendicular to one of two parallel lines, then it is perpendicular to the other. The slope of l_2 is the negative reciprocal of the slope of k. The negative reciprocal of $-\frac{1}{m}$ is m. Therefore, the slope of l_1 is equal to the slope of l_2.

Is the converse of this statement true? We will again use the fact that two lines are perpendicular if and only if the slope of one is the negative reciprocal of the slope of the other to prove that it is.

Theorem 9.10b

> If the slopes of two non-vertical lines in the coordinate plane are equal, then the lines are parallel.

Given Lines l_1 and l_2 with slope m

Prove $l_1 \parallel l_2$

Proof Choose any point on l_1. Through a given point, one and only one line can be drawn perpendicular to a given line. Through that point, draw k, a line perpendicular to l_1. The slope of k is $-\frac{1}{m}$ since two non-vertical lines are perpendicular if and only if the slope of one is the negative reciprocal of the slope of the other. But this means that $l_2 \perp k$ because the slope of l_2 is also the negative reciprocal of the slope of k. Therefore, $l_1 \parallel l_2$ because two lines perpendicular to the same line are parallel.

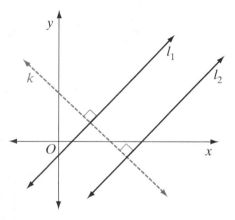

We can write the statements that we have proved as a biconditional:

Theorem 9.10

> Two non-vertical lines in the coordinate plane are parallel if and only if they have the same slope.

EXAMPLE 1

The vertices of quadrilateral $ABCD$ are $A(2, -4)$, $B(6, -2)$, $C(2, 6)$, and $D(-1, 2)$.

a. Show that two sides of the quadrilateral are parallel.

b. Show that the quadrilateral has two right angles.

Solution The slope of $\overline{AB} = \frac{-2 - (-4)}{6 - 2} = \frac{2}{4} = \frac{1}{2}.$

The slope of $\overline{BC} = \frac{6 - (-2)}{2 - 6} = \frac{8}{-4} = -2.$

The slope of $\overline{CD} = \frac{2 - 6}{-1 - 2} = \frac{-4}{-3} = \frac{4}{3}.$

The slope of $\overline{DA} = \frac{-4 - 2}{2 - (-1)} = \frac{-6}{3} = -2.$

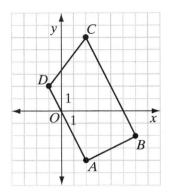

a. $\overline{BC}$ and $\overline{DA}$ are parallel because they have equal slopes.

b. The slope of $\overline{AB}$ is the negative reciprocal of the slope of $\overline{BC}$, so they are perpendicular. Therefore, $\angle B$ is a right angle.

The slope of $\overline{AB}$ is the negative reciprocal of the slope of $\overline{DA}$, so they are perpendicular. Therefore, $\angle A$ is a right angle.

Answers **a.** $\overline{BC} \parallel \overline{DA}$ **b.** $\angle A$ and $\angle B$ are right angles.

EXAMPLE 2

Write an equation for l_1, the line through $(-2, 5)$ that is parallel to the line l_2 whose equation is $2x + y = 7$.

Solution (1) Solve the equation of l_2 for y:

$2x + y = 7$

$y = -2x + 7$

(2) Find the slope of l_2. The slope of a line in slope-intercept form is the coefficient of x:

$y = -2x + 7$

slope

(3) Find the slope of l_1, which is equal to the slope of l_2:

slope of $l_1 = -2$

(4) Use the definition of slope to write an equation of l_1. Let (x, y) and $(-2, 5)$ be two points on l_1:

$\frac{y_1 - y_2}{x_1 - x_2} = $ slope

$\frac{y - 5}{x - (-2)} = -2$

$y - 5 = -2(x + 2)$

$y - 5 = -2x - 4$

$y = -2x + 1$

Answer $y = -2x + 1$ or $2x + y = 1$

Exercises

Writing About Mathematics

1. If l_1 and l_2 have the same slope and have a common point, what must be true about l_1 and l_2?

2. Theorem 9.10 is true for all lines that are not vertical. Do vertical lines have the same slope? Explain your answer.

Developing Skills

In 3–8, for each pair of lines whose equations are given, tell whether the lines are parallel, perpendicular, or neither parallel nor perpendicular.

3. $x + y = 7$

   $x - y = 3$

4. $2x - y = 5$

   $y = 2x - 3$

5. $x = \frac{1}{3}y + 2$

   $y = 3x - 2$

6. $2x + y = 6$

   $2x - y = 3$

7. $x = 2$

   $x = 5$

8. $x = 2$

   $y = 3$

In 9–12, write an equation of the line that satisfies the given conditions.

9. Parallel to $y = -3x + 1$ with y-intercept 4.

10. Perpendicular to $y = -3x + 1$ with y-intercept 4.

11. Parallel to $x - 2y = 4$ and through the point $(4, 5)$.

12. Parallel to and 3 units below the x-axis.

Applying Skills

13. Quadrilateral $ABCD$ has two pairs of parallel sides, $\overline{AB} \parallel \overline{CD}$ and $\overline{BC} \parallel \overline{DA}$. The coordinates of A are $(1, 2)$, the coordinates of B are $(7, -1)$ and the coordinates of C are $(8, 2)$.

   a. What is the slope of $\overline{AB}$?

   b. What is the slope of $\overline{CD}$?

   c. Write an equation for $\overleftrightarrow{CD}$.

   d. What is the slope of $\overline{BC}$?

   e. What is the slope of $\overline{AD}$?

   f. Write an equation for $\overleftrightarrow{AD}$.

   g. Use the equation of $\overleftrightarrow{CD}$ and the equation of $\overleftrightarrow{AD}$ to find the coordinates of D.

14. In quadrilateral $ABCD$, $\overline{BC} \perp \overline{AB}$, $\overline{DA} \perp \overline{AB}$, and $\overline{DA} \perp \overline{DC}$. The coordinates of A are $(1, -1)$, the coordinates of B are $(4, 2)$, and the coordinates of C are $(2, 4)$.

  a. What is the slope of $\overline{AB}$?

  b. What is the slope of $\overline{BC}$?

  c. What is the slope of $\overline{AD}$? Justify your answer.

  d. What is the slope of $\overline{DC}$? Justify your answer.

  e. Write an equation for $\overleftrightarrow{DC}$.

  f. Write an equation for $\overleftrightarrow{AD}$.

  g. Use the equation of $\overleftrightarrow{DC}$ and the equation of $\overleftrightarrow{AD}$ to find the coordinates of D.

15. The coordinates of the vertices of quadrilateral $PQRS$ are $P(0, -1)$, $Q(4, 0)$, $R(2, 3)$, and $S(-2, 2)$.

  a. Show that $PQRS$ has two pairs of parallel sides.

  b. Show that $PQRS$ does not have a right angle.

16. The coordinates of the vertices of quadrilateral $KLMN$ are $K(-2, -1)$, $L(4, -3)$, $M(2, 1)$, and $N(-1, 2)$.

  a. Show that $KLMN$ has only one pair of parallel sides.

  b. Show that $KLMN$ has two right angles.

Hands-On Activity 1

In this activity, we will use a compass and a straightedge, or geometry software to construct a line parallel to a given line through a point not on the line.

STEP 1. Given a point, P, not on line, l. Through P, construct a line perpendicular to line l. Label this line n.

STEP 2. Through P, construct a line, p, perpendicular to line n.

Result: $l \parallel p$

a. Justify the construction given in the procedure.

b. In (1)–(3), construct a line parallel to the line through the given point.

  (1) $y = \frac{1}{4}x + 5$; $\left(3, 5\frac{1}{2}\right)$ (2) $-12x + y = 19$; $(12, -4)$ (3) $y = -\frac{1}{9}x - 3$; $(0, 4)$

Hands-On Activity 2

A **midsegment** is a segment formed by joining two midpoints of the sides of a triangle. In this activity, we will prove that a midsegment is parallel to the third side of the triangle using coordinate geometry.

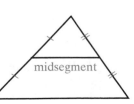

midsegment

  1. With a partner or in a small group, complete the following:

    a. Write the coordinates of a triangle using variables. These coordinates can be any convenient variables.

b. Find the midpoints of two sides of the triangle.

c. Prove that the midsegment formed is parallel to the third side of the triangle.

2. Compare your proof with the proofs of the other groups. Were different coordinates used? Which coordinates seem easiest to work with?

9-4 THE SUM OF THE MEASURES OF THE ANGLES OF A TRIANGLE

In previous courses, you have demonstrated that the sum of the measures of the angles of a triangle is 180 degrees. The congruent angles formed when parallel lines are cut by a transversal make it possible for us to prove this fact.

Theorem 9.11 | The sum of the measures of the angles of a triangle is 180°.

Given $\triangle ABC$

Prove $m\angle A + m\angle B + m\angle C = 180$

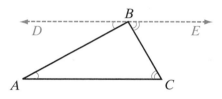

Proof

Statements	Reasons
1. Let $\overleftrightarrow{DE}$ be the line through B that is parallel to $\overline{AC}$.	**1.** Through a given point not on a given line, there exists one and only one line parallel to the given line.
2. $m\angle DBE = 180$	**2.** A straight angle is an angle whose degree measure is 180.
3. $m\angle DBA + m\angle ABC + m\angle CBE = 180$	**3.** The whole is equal to the sum of all its parts.
4. $\angle A \cong \angle DBA$ and $\angle C \cong \angle CBE$	**4.** If two parallel lines are cut by a transversal, the alternate interior angles are congruent.
5. $m\angle A = m\angle DBA$ and $m\angle C = \angle CBE$	**5.** Congruent angles are equal in measure.
6. $m\angle A + m\angle ABC + m\angle C = 180$	**6.** Substitution postulate.

Many corollaries to this important theorem exist.

Corollary 9.11a | If two angles of one triangle are congruent to two angles of another triangle, then the third angles are congruent.

Proof: Let $\triangle ABC$ and $\triangle DEF$ be two triangles in which $\angle A \cong D$ and $\angle B \cong \angle E$. Since the sum of the degree measures of the angles of a triangle is 180, then

$$m\angle A + m\angle B + m\angle C = m\angle D + m\angle E + m\angle F.$$

We use the subtraction postulate to prove that $m\angle C = m\angle F$ and therefore, that $\angle C \cong \angle F$. ∎

Corollary 9.11b | The acute angles of a right triangle are complementary.

Proof: In any triangle ABC, $m\angle A + m\angle B + m\angle C = 180$. If $\angle C$ is a right angle, $m\angle C = 90$,

$$m\angle A + m\angle B + 90 = 180$$
$$m\angle A + m\angle B = 90$$

Therefore, $\angle A$ and $\angle B$ are complementary. ∎

Corollary 9.11c | Each acute angle of an isosceles right triangle measures 45°.

Proof: In isosceles right triangle ABC, $m\angle C = 90$ and $\overline{AC} \cong \overline{BC}$. Therefore, $m\angle A = m\angle B$. Using Corollary 9.12b, we know that $\angle A$ and $\angle B$ are complementary. Therefore, the measure of each must be 45. ∎

Corollary 9.11d | Each angle of an equilateral triangle measures 60°.

Proof: In equilateral triangle ABC, $m\angle A = m\angle B = m\angle C$. We substitute $m\angle A$ for $m\angle B$ and $m\angle C$ in the equation $m\angle A + m\angle B + m\angle C = 180$, and then solve the resulting equation: $3m\angle A = 180$ so $m\angle A = 60$. ∎

Corollary 9.11e | The sum of the measures of the angles of a quadrilateral is 360°.

Proof: In quadrilateral $ABCD$, we draw $\overline{AC}$, forming two triangles. The sum of the measures of the angles of quadrilateral $ABCD$ is the sum of the measures of the angles of the two triangles:

$$180 + 180 = 360$$

Corollary 9.11f | The measure of an exterior angle of a triangle is equal to the sum of the measures of the nonadjacent interior angles.

The proof is left to the student. (See exercise 30.)

Note: Recall that the Exterior Angle Theorem of Section 7-5 gives an *inequality* that relates the exterior angle of a triangle to the nonadjacent interior angles: "The measure of an exterior angle of a triangle is *greater than* the measure of either nonadjacent interior angle." Corollary 9.11f is a version of the Exterior Angle Theorem involving *equality*.

EXAMPLE 1

The measure of the vertex angle of an isosceles triangle exceeds the measure of each base angle by 30 degrees. Find the degree measure of each angle of the triangle.

Solution   Let x = measure of each base angle.

Let $x + 30$ = measure of vertex angle.

The sum of the measures of the angles of a triangle is 180.

$$x + x + x + 30 = 180$$
$$3x + 30 = 180$$
$$3x = 150$$
$$x = 50$$
$$x + 30 = 80$$

Answer   The angle measures are 50°, 50°, and 80°.

EXAMPLE 2

In $\triangle ABC$, the measures of the three angles are represented by $9x$, $3x - 6$, and $11x + 2$. Show that $\triangle ABC$ is a right triangle.

Solution   Triangle ABC will be a right triangle if one of the angles is a right angle. Write and equation for the sum of the measures of the angles of $\triangle ABC$.

$$9x + 3x - 6 + 11x + 2 = 180$$
$$23x - 4 = 180$$
$$23x = 184$$
$$x = 8$$

Substitute $x = 8$ in the representations of the angle measures.

$$9x = 9(8) \qquad 3x - 6 = 3(8) - 6 \qquad 11x + 2 = 11(8) + 2$$
$$= 72 \qquad\qquad = 24 - 6 \qquad\qquad\quad = 88 + 2$$
$$= 18 \qquad\qquad\qquad = 90$$

Answer Triangle ABC is a right triangle because the degree measure of one of its angles is 90.

EXAMPLE 3

B is a not a point on $\overleftrightarrow{ACD}$. Ray $\overrightarrow{CE}$ bisects $\angle DCB$ and $\overline{AC} \cong \overline{BC}$. Prove that $\overleftrightarrow{AB} \parallel \overleftrightarrow{CE}$.

Solution Given: $\overrightarrow{CE}$ bisects $\angle DCB$ and $\overline{AC} \cong \overline{BC}$.

Prove: $\overleftrightarrow{AB} \parallel \overleftrightarrow{CE}$

Proof

Statements	Reasons
1. $\overrightarrow{CE}$ bisects $\angle DCB$.	1. Given.
2. $\angle DCE \cong \angle ECB$	2. Definition of an angle bisector.
3. $m\angle DCE = m\angle ECB$	3. Measures of congruent angles are equal.
4. $\overline{AC} \cong \overline{BC}$	4. Given.
5. $m\angle CAB = m\angle CBA$	5. Isosceles triangle theorem.
6. $m\angle DCB = m\angle CAB + m\angle CBA$	6. An exterior angle of a triangle is equal to the sum of the measures of the nonadjacent interior angles.
7. $m\angle DCB = m\angle DCE + m\angle ECB$	7. Partition postulate.
8. $m\angle DCE + m\angle ECB =$ $m\angle CAB + m\angle CBA$	8. Substitution postulate (steps 6 and 7).
9. $m\angle DCE + m\angle DCE =$ $m\angle CAB + m\angle CAB$ or $2m\angle DCE = 2\,m\angle CAB$	9. Substitution postulate (steps 3, 5, and 8).
10. $m\angle DCE = m\angle CAB$	10. Division postulate.
11. $\overleftrightarrow{AB} \parallel \overleftrightarrow{CE}$	11. If two lines are cut by a transversal forming equal corresponding angles, then the lines are parallel.

Exercises

Writing About Mathematics

1. McKenzie said that if a triangle is obtuse, two of the angles of the triangle are acute. Do you agree with McKenzie? Explain why or why not.

2. Giovanni said that since the sum of the measures of the angles of a triangle is 180, the angles of a triangle are supplementary. Do you agree with Giovanni? Explain why or why not.

Developing Skills

In 3–6, determine whether the given numbers can be the degree measures of the angles of a triangle.

3. 25, 100, 55   **4.** 95, 40, 45   **5.** 75, 75, 40   **6.** 12, 94, 74

In 7–10, the given numbers are the degree measures of two angles of a triangle. Find the measure of the third angle.

7. 80, 60   **8.** 45, 85   **9.** 90, 36   **10.** 65, 65

In 11–14, the measure of the vertex angle of an isosceles triangle is given. Find the measure of a base angle.

11. 20   **12.** 90   **13.** 76   **14.** 110

In 15–18, the measure of a base angle of an isosceles triangle is given. Find the measure of the vertex angle.

15. 80   **16.** 20   **17.** 45   **18.** 63

19. What is the measure of each exterior angle of an equilateral triangle?

In 20–23, the diagram shows $\triangle ABC$ and exterior $\angle ACD$.

20. If $m\angle A = 40$ and $m\angle B = 20$, find $m\angle ACD$ and $m\angle ACB$.

21. If $m\angle A = 40$ and $m\angle B = 50$, find $m\angle ACD$ and $m\angle ACB$.

22. If $m\angle A = 40$ and $m\angle ACB = 120$, find $m\angle ACD$ and $m\angle B$.

23. If $m\angle A = 40$, $m\angle B = 3x + 20$, and $m\angle ACD = 5x + 10$, find $m\angle B$, $m\angle ACD$, and $m\angle ACB$.

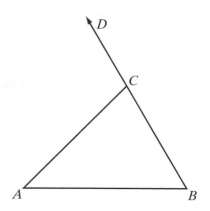

Applying Skills

24. The measure of each base angle of an isosceles triangle is $21°$ more than the measure of the vertex angle. Find the measure of each angle of the triangle.

25. The measure of an exterior angle at C of isosceles $\triangle ABC$ is 110°. If $AC = BC$, find the measure of each angle of the triangle.

26. The measure of an exterior angle at D of isosceles $\triangle DEF$ is 100°. If $DE = EF$, find the measure of each angle of the triangle.

27. Triangle LMN is a right triangle with $\angle M$ the right angle. If m$\angle L = 32$, find the measure of $\angle N$ and the measure of the exterior angle at N.

28. In $\triangle ABC$, m$\angle A = 2x + 18$, m$\angle B = x + 40$, and m$\angle C = 3x - 40$.

   a. Find the measure of each angle of the triangle.

   b. Which is the longest side of the triangle?

29. The measure of an exterior angle at B, the vertex of isosceles $\triangle ABC$, can be represented by $3x + 12$. If the measure of a base angle is $2x - 2$, find the measure of the exterior angle and of the interior angles of $\triangle ABC$.

30. Prove Corollary 9.11f, "The measure of an exterior angle of a triangle is equal to the sum of the measures of the nonadjacent interior angles."

31. a. In the coordinate plane, graph points $A(5, 2)$, $B(2, 2)$, $C(2, -1)$, $D(-1, -1)$.

   b. Draw $\overleftrightarrow{AB}$ and $\triangle BDC$.

   c. Explain how you know that $\triangle BDC$ is an isosceles right triangle.

   d. What is the measure of $\angle BDC$? Justify your answer.

   e. What is the measure of $\angle DBA$? Justify your answer.

32. Prove that the sum of the measures of the angles of hexagon $ABCDEF$ is 720°. (*Hint:* draw $\overline{AD}$.)

33. $ABCD$ is a quadrilateral with $\overrightarrow{BD}$ the bisector of $\angle ABC$ and $\overrightarrow{DB}$ the bisector of $\angle ADC$. Prove that $\angle A \cong \angle C$.

9-5 PROVING TRIANGLES CONGRUENT BY ANGLE, ANGLE, SIDE

When two angles of one triangle are congruent to two angles of another triangle, the third angles are congruent. This is not enough to prove that the two triangles are congruent. We must know that at least one pair of corresponding sides are congruent. We already know that if two angles and the side between them in one triangle are congruent to the corresponding angles and side in another triangle, then the triangles are congruent by ASA. Now we want to prove angle-angle-side or **AAS triangle congruence**. This would allow us to conclude that if any two angles and any side in one triangle are congruent to the corresponding angles and side in another triangle, then the triangles are congruent.

Theorem 9.12 | If two angles and the side opposite one of them in one triangle are congruent to the corresponding angles and side in another triangle, then the triangles are congruent. (AAS)

Given $\triangle ABC$ and $\triangle DEF$, $\angle A \cong \angle D$, $\angle C \cong \angle F$, and $\overline{AB} \cong \overline{DE}$

Prove $\triangle ABC \cong \triangle DEF$

Proof

Statements	Reasons
1. $\angle A \cong \angle D$	**1.** Given.
2. $\angle C \cong \angle F$	**2.** Given.
3. $\angle B \cong \angle E$	**3.** If two angles of one triangle are congruent to two angles of another triangle, then the third angles are congruent.
4. $\overline{AB} \cong \overline{DE}$	**4.** Given.
5. $\triangle ABC \cong \triangle DEF$	**5.** ASA.

Therefore, when two angles and any side in one triangle are congruent to the corresponding two angles and side of a second triangle, we may say that the triangles are congruent either by ASA or by AAS.

The following corollaries can proved using AAS. Note that in every right triangle, the hypotenuse is the side opposite the right angle.

Corollary 9.12a | Two right triangles are congruent if the hypotenuse and an acute angle of one right triangle are congruent to the hypotenuse and an acute angle of the other right triangle.

The proof uses AAS and is left to the student. (See exercise 15.)

Corollary 9.12b | If a point lies on the bisector of an angle, then it is equidistant from the sides of the angle.

Recall that the distance from a point to a line is the length of the perpendicular from the point to the line. The proof uses AAS and is left to the student. (See exercise 16.)

You now have four ways to prove two triangles congruent: SAS, ASA, SSS, and AAS.

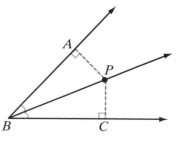

EXAMPLE I

Prove that the altitudes drawn to the legs of an isosceles triangle from the endpoints of the base are congruent.

Given: Isosceles triangle ABC with $\overline{BA} \cong \overline{BC}$, $\overline{AE} \perp \overline{BC}$, and $\overline{CD} \perp \overline{BA}$.

Prove: $\overline{CD} \cong \overline{AE}$

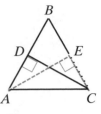

Proof	Statements	Reasons
	1. In $\triangle ABC$, $\overline{BA} \cong \overline{BC}$.	**1.** Given.
A **2.**	$\angle BAC \cong \angle BCA$	**2.** If two sides of a triangle are congruent, the angles opposite these sides are congruent.
	3. $\overline{AE} \perp \overline{BC}$, $\overline{CD} \perp \overline{BA}$	**3.** Given.
	4. $\angle CDA$ and $\angle AEC$ are right angles.	**4.** Perpendicular lines are two lines that intersect to form right angles.
A **5.**	$\angle CDA \cong \angle AEC$	**5.** All right angles are congruent.
S **6.**	$\overline{AC} \cong \overline{AC}$	**6.** Reflexive property of congruence.
	7. $\triangle DAC \cong \triangle ECA$	**7.** AAS (steps 2, 5, and 6).
	8. $\overline{CD} \cong \overline{AE}$	**8.** Corresponding parts of congruent triangles are congruent.

EXAMPLE 2

The coordinates of the vertices of $\triangle ABC$ are $A(-6, 0)$, $B(-1, 0)$ and $C(-5, 2)$. The coordinates of $\triangle DEF$ are $D(3, 0)$, $E(8, 0)$, and $F(4, 2)$. Prove that the triangles are congruent.

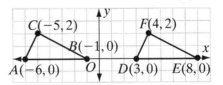

Solution (1) Prove that the triangles are right triangles.

In $\triangle ABC$:

The slope of $\overline{AC}$ is $\dfrac{2-0}{-5-(-6)} = \dfrac{2}{1} = 2$.

The slope of $\overline{CB}$ is $\dfrac{2-0}{-5-(-1)} = \dfrac{2}{-4} = -\dfrac{1}{2}$.

In $\triangle DEF$:

The slope of $\overline{DF}$ is $\dfrac{2-0}{4-3} = \dfrac{2}{1} = 2$.

The slope of $\overline{FE}$ is $\dfrac{2-0}{4-8} = \dfrac{2}{-4} = -\dfrac{1}{2}$.

Two lines are perpendicular if the slope of one is the negative reciprocal of the slope of the other. Therefore, $\overline{AC} \perp \overline{CB}$, $\angle ACB$ is a right angle, and $\triangle ACB$ is a right triangle. Also, $\overline{DF} \perp \overline{FE}$, $\angle DFE$ is a right angle, and $\triangle DFE$ is a right triangle.

(2) Prove that two acute angles are congruent.
Two lines are parallel if their slopes are equal. Therefore, $\overline{CB} \parallel \overline{FE}$. The x-axis is a transversal forming congruent corresponding angles, so $\angle CBA$ and $\angle FED$ are congruent.

(3) Prove that the hypotenuses are congruent.
The hypotenuse of $\triangle ABC$ is $\overline{AB}$, and $AB = |-6 - (-1)| = 5$. The hypotenuse of $\triangle DEF$ is $\overline{DE}$, and $DE = |3 - 8| = 5$. Line segments that have the same measure are congruent, and so $\overline{AB} \cong \overline{DE}$.

(4) Therefore, $\triangle ABC \cong \triangle DEF$ because the hypotenuse and an acute angle of one triangle are congruent to the hypotenuse and an acute angle of the other.

EXAMPLE 3

Show that if a triangle has two sides and an angle opposite one of the sides congruent to the corresponding sides and angle of another triangle, the triangles may not be congruent.

Solution (1) Draw an angle, $\angle ABC$.

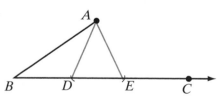

(2) Open a compass to a length that is smaller than AB but larger than the distance from A to $\overline{BC}$. Use the compass to mark two points, D and E, on $\overline{BC}$.

(3) Draw $\overline{AD}$ and $\overline{AE}$.

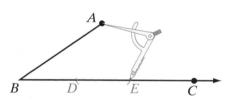

(4) In $\triangle ABD$ and $\triangle ABE$, $\overline{AB} \cong \overline{AB}$, $\angle B \cong \angle B$, and $\overline{AD} \cong \overline{AE}$. In these two triangles, two sides and the angle opposite one of the sides are congruent to the corresponding parts of the other triangle. But $\triangle ABD$ and $\triangle ABE$ are not congruent. This counterexample proves that SSA is not sufficient to prove triangles congruent.

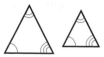

Note: Triangles in which two sides and an angle opposite one of them are congruent *may not* be congruent to each other. Therefore, SSA is *not* a valid proof of triangle congruence. Similarly, triangles in which all three angles are congruent *may not* be congruent to each other, so AAA is also *not* a valid proof of triangle congruence.

Exercises

Writing About Mathematics

1. In Example 3, we showed that SSA cannot be used to prove two triangles congruent. Does this mean that whenever two sides and an angle opposite one of the sides are congruent to the corresponding parts of another triangle the two triangles are not congruent? Explain your answer.

2. In the coordinate plane, points A and C are on the same horizontal line and C and B are on the same vertical line. Are $\angle CAB$ and $\angle CBA$ complementary angles? Justify your answer.

Developing Skills

In 3–8, each figure shows two triangles. Congruent parts of the triangles have been marked. Tell whether or not the given congruent parts are sufficient to prove that the triangles are congruent. Give a reason for your answer.

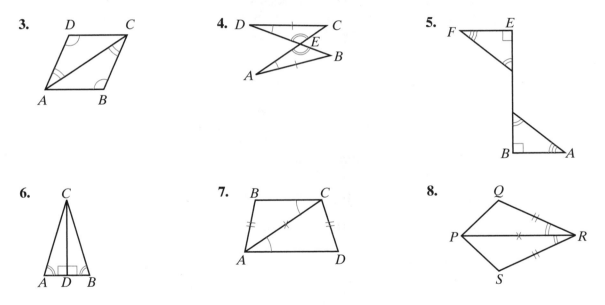

Applying Skills

9. Prove that if two triangles are congruent, then the altitudes drawn from corresponding vertices are congruent.

10. Prove that if two triangles are congruent, then the medians drawn from corresponding vertices are congruent.

11. Prove that if two triangles are congruent, then the angle bisectors drawn from corresponding vertices are congruent.

12. *Given:* Quadrilateral $ABCD$ with $\angle A \cong \angle C$ and $\overrightarrow{BD}$ the bisector of
     $\angle ABC$.

   Prove: $\overrightarrow{DB}$ bisects $\angle ADC$.

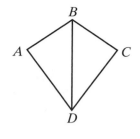

13. *Given:* $\overline{AB} \parallel \overline{CD}$, $\overline{AB} \cong \overline{CD}$, and $\overline{AB} \perp \overline{BEC}$.

   Prove: $\overline{AED}$ and $\overline{BEC}$ bisect each other.

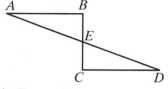

14. a. Use a translation to prove that $\triangle ABC$ and $\triangle DEF$ in Example 2 are congruent.

   b. Use two line reflections to prove that $\triangle ABC$ and $\triangle DEF$ in Example 2 are congruent.

15. Prove Corollary 9.12a, "Two right triangles are congruent if the hypotenuse and an acute
   angle of one right triangle are congruent to the hypotenuse and an acute angle of the other
   right triangle."

16. Prove Corollary 9.12b, "If a point lies on the bisector of an angle, it is equidistant from the
   sides of the angle."

17. Prove that if three angles of one triangle are congruent to the corresponding angles of
   another (AAA), the triangles may not be congruent. (Through any point on side $\overline{BC}$ of
   $\triangle ABC$, draw a line segment parallel to $\overline{AC}$.)

9-6 THE CONVERSE OF THE ISOSCELES TRIANGLE THEOREM

The Isosceles Triangle Theorem, proved in Section 5-3 of this book, is restated
here in its conditional form.

> ▶ **If two sides of a triangle are congruent, then the angles opposite these sides
> are congruent.**

When we proved the Isosceles Triangle Theorem, its converse would have
been very difficult to prove with the postulates and theorems that we had avail-
able at that time. Now that we can prove two triangles congruent by AAS, its
converse is relatively easy to prove.

Theorem 9.13

> If two angles of a triangle are congruent, then the sides opposite these angles
> are congruent.

Given △*ABC* with ∠*A* ≅ ∠*B*.

Prove $\overline{CA} \cong \overline{CB}$

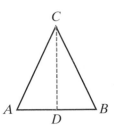

Proof We can use either the angle bisector or the altitude from *C* to separate the triangle into two congruent triangles. We will use the angle bisector.

Statements	Reasons
1. Draw $\overline{CD}$, the bisector of ∠*ACB*.	**1.** Every angle has one and only one bisector.
2. ∠*ACD* ≅ ∠*BCD*	**2.** An angle bisector of a triangle is a line segment that bisects an angle of the triangle.
3. ∠*A* ≅ ∠*B*	**3.** Given.
4. $\overline{CD} \cong \overline{CD}$	**4.** Reflexive property of congruence.
5. △*ACD* ≅ △*BCD*	**5.** AAS.
6. $\overline{CA} \cong \overline{CB}$	**6.** Corresponding parts of congruent triangles are congruent.

The statement of the Isosceles Triangle Theorem (Theorem 5.1) and its converse (Theorem 9.14) can now be written in biconditional form:

▶ **Two angles of a triangle are congruent if and only if the sides opposite these angles are congruent.**

To prove that a triangle is isosceles, we may now prove that either of the following two statements is true:

1. Two sides of the triangle are congruent.

2. Two angles of the triangle are congruent.

Corollary 9.13a | If a triangle is equiangular, then it is equilateral.

Given △*ABC* with ∠*A* ≅ ∠*B* ≅ ∠*C*.

Prove △*ABC* is equilateral.

Proof We are given equiangular $\triangle ABC$. Then since $\angle A \cong \angle B$, the sides opposite these angles are congruent, that is, $\overline{BC} \cong \overline{AC}$. Also, since $\angle B \cong \angle C$, $\overline{AC} \cong \overline{AB}$ for the same reason. Therefore, $\overline{AC} \cong \overline{AB}$ by the transitive property of congruence, $\overline{AB} \cong \overline{BC} \cong \overline{CA}$, and $\triangle ABC$ is equilateral.

EXAMPLE I

In $\triangle PQR$, $\angle Q \cong \angle R$. If $PQ = 6x - 7$ and $PR = 3x + 11$, find:

a. the value of x **b.** PQ **c.** PR

Solution

a. Since two angles of $\triangle PQR$ are congruent, the sides opposite these angles are congruent. Thus, $PQ = PR$.

$6x - 7 = 3x + 11$

$6x - 3x = 11 + 7$

$3x = 18$

$x = 6$ *Answer*

b. $PQ = 6x - 7$

$= 6(6) - 7$

$= 36 - 7$

$= 29$ *Answer*

c. $PR = 3x + 11$

$= 3(6) + 11$

$= 18 + 11$

$= 29$ *Answer*

EXAMPLE 2

The degree measures of the three angles of $\triangle ABC$ are represented by $m\angle A = x + 30$, $m\angle B = 3x$, and $m\angle C = 4x + 30$. Describe the triangle as acute, right, or obtuse, and as scalene, isosceles, or equilateral.

Solution The sum of the degree measures of the angles of a triangle is 180.

$$x + 30 + 3x + 4x + 30 = 180$$
$$8x + 60 = 180$$
$$8x = 120$$
$$x = 15$$

Substitute $x = 15$ in the representations given for the three angle measures.

$m\angle A = x + 30$

$= 15 + 30$

$= 45$

$m\angle B = 3x$

$= 3(15)$

$= 45$

$m\angle C = 4x + 30$

$= 4(15) + 30$

$= 60 + 30$

$= 90$

Since $\angle A$ and $\angle B$ each measure $45°$, the triangle has two congruent angles and therefore two congruent sides. The triangle is isosceles. Also, since one angle measures $90°$, the triangle is a right triangle.

Answer $\triangle ABC$ is an isosceles right triangle.

EXAMPLE 3

Given: Quadrilateral $ABCD$ with $\overline{AB} \parallel \overline{CD}$
and $\overrightarrow{AC}$ bisects $\angle DAB$.

Prove: $\overline{AD} \cong \overline{CD}$

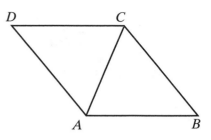

Proof	**Statements**	**Reasons**
	1. $\overline{AB} \parallel \overline{CD}$	**1.** Given.
	2. $\angle DCA \cong \angle CAB$	**2.** If two parallel lines are cut by a transversal, the alternate interior angles are congruent.
	3. $\overrightarrow{AC}$ bisects $\angle DAB$.	**3.** Given.
	4. $\angle CAB \cong \angle DAC$	**4.** A bisector of an angle divides the angle into two congruent parts.
	5. $\angle DCA \cong \angle DAC$ congruence.	**5.** Transitive property of
	6. $\overline{AD} \cong \overline{CD}$	**6.** If two angles of a triangle are congruent, the sides opposite these angles are congruent.

Exercises

Writing About Mathematics

1. Julian said that the converse of the Isosceles Triangle Theorem could have been proved as a corollary to Theorem 7.3, "If the lengths of two sides of a triangle are unequal, then the measures of the angles opposite these sides are unequal." Do you agree with Julian? Explain why or why not.

2. Rosa said that if the measure of one angle of a right triangle is 45 degrees, then the triangle is an isosceles right triangle. Do you agree with Rosa? Explain why or why not.

In 3–6, in each case the degree measures of two angles of a triangle are given.

a. Find the degree measure of the third angle of the triangle.

b. Tell whether the triangle is isosceles or is not isosceles.

3. 70, 40   **4.** 30, 120

5. 50, 65   **6.** 80, 40

7. In $\triangle ABC$, $m\angle A = m\angle C$, $AB = 5x + 6$, and $BC = 3x + 14$. Find the value of x.

8. In $\triangle PQR$, $m\angle Q = m\angle P$, $PR = 3x$, and $RQ = 2x + 7$. Find PR and RQ.

9. In $\triangle MNR$, $MN = NR$, $m\angle M = 72$, and $m\angle R = 2x$. Find the measures of $\angle R$ and of $\angle N$.

10. In $\triangle ABC$, $m\angle A = 80$ and $m\angle B = 50$. If $AB = 4x - 4$, $AC = 2x + 16$, and $BC = 4x + 6$, find the measure of each side of the triangle.

11. The degree measures of the angles of $\triangle ABC$ are represented by $x + 10$, $2x$, and $2x - 30$. Show that $\triangle ABC$ is an isosceles triangle.

12. The degree measures of the angles of $\triangle ABC$ are represented by $x + 35$, $2x + 10$, and $3x - 15$. Show that $\triangle ABC$ is an equilateral triangle.

13. The degree measures of the angles of $\triangle ABC$ are represented by $3x + 18$, $4x + 9$, and $10x$. Show that $\triangle ABC$ is an isosceles right triangle.

14. What is the measure of each exterior angle of an equilateral triangle?

15. What is the sum of the measures of the exterior angles of any triangle?

Applying Skills

16. *Given:* P is not on $\overleftrightarrow{ABCD}$ and
$\quad\quad \angle ABP \cong \angle PCD$.

Prove: $\triangle BPC$ is isosceles.

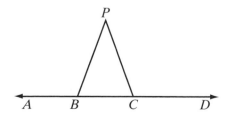

17. *Given:* P is not on $\overline{AB}$ and
$\quad\quad \angle PAB \cong \angle PBA$.

Prove: P is on the perpendicular bisector of $\overline{AB}$.

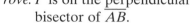
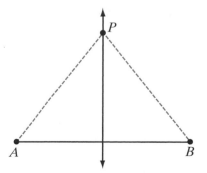

18. *Given:* $\overrightarrow{BE}$ bisects $\angle DBC$, an exterior
angle of $\triangle ABC$, and $\overrightarrow{BE} \parallel \overline{AC}$.

Prove: $\overline{AB} \cong \overline{CB}$

19. *Given:* P is not on $\overline{ABCD}$,
$\angle PBC \cong \angle PCB$, and
$\angle APB \cong \angle DPC$

Prove: $\overline{AP} \cong \overline{DP}$

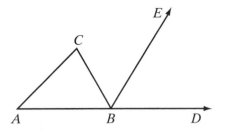

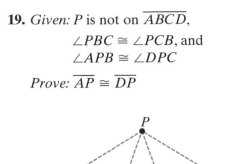

20. Prove Theorem 9.13 by drawing the altitude from C.

9-7 PROVING RIGHT TRIANGLES CONGRUENT BY HYPOTENUSE, LEG

We showed in Section 5 of this chapter that, when two sides and an angle opposite one of these sides in one triangle are congruent to the corresponding two sides and angle in another triangle, the two triangles may or may not be congruent. When the triangles are right triangles, however, it is possible to prove that they are congruent. The congruent angles are the right angles, and each right angle is opposite the hypotenuse of the triangle.

Theorem 9.14

If the hypotenuse and a leg of one triangle are congruent to the corresponding parts of the other, then the two right triangles are congruent. (HL)

Given Right $\triangle ABC$ with right angle B and right $\triangle DEF$ with right angle E, $\overline{AC} \cong \overline{DF}$, $\overline{BC} \cong \overline{EF}$

Prove $\triangle ABC \cong \triangle DEF$

Proof To prove this theorem, we will construct a third triangle, $\triangle GEF$, that shares a common side with $\triangle DEF$ and prove that each of the two given triangles is congruent to $\triangle GEF$ and, thus, to each other.

We first show that $\triangle ABC$ is congruent to $\triangle GEF$:

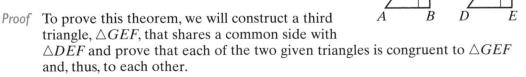

(1) Since any line segment may be extended any required length, extend $\overline{DE}$ to G so that $\overline{EG} \cong \overline{AB}$. Draw $\overline{FG}$.

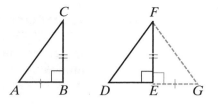

(2) $\angle GEF$ and $\angle DEF$ form a linear pair, and $\angle DEF$ is a right angle. Therefore, $\angle GEF$ is a right angle. We are given that $\angle B$ is a right angle. All right angles are congruent, so $\angle B \cong \angle GEF$.

(3) We are also given $\overline{BC} \cong \overline{EF}$.

(4) Therefore, $\triangle ABC \cong \triangle GEF$ by SAS.

We now show that $\triangle DEF$ is also congruent to the constructed triangle, $\triangle GEF$:

(5) Since corresponding sides of congruent triangles are congruent, $\overline{AC} \cong \overline{GF}$. Since we are given $\overline{AC} \cong \overline{DF}$, $\overline{GF} \cong \overline{DF}$ by the transitive property of congruence.

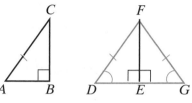

(6) If two sides of a triangle are congruent, the angles opposite these sides are congruent. In $\triangle DFG$, $\overline{GF} \cong \overline{DF}$, so $\angle D \cong \angle G$. Also, $\angle DEF \cong \angle GEF$ since all right angles are congruent.

(7) Therefore, $\triangle DEF \cong \triangle GEF$ by AAS.

(8) Therefore, $\triangle ABC \cong \triangle DEF$ by the transitive property of congruence (steps 4 and 7).

    This theorem is called the **hypotenuse-leg triangle congruence theorem**, abbreviated **HL**. Therefore, from this point on, when the hypotenuse and a leg of one right triangle are congruent to the corresponding parts of a second right triangle, we may say that the triangles are congruent.

    A corollary of this theorem is the converse of Corollary 9.12b.

Corollary 9.14a

If a point is equidistant from the sides of an angle, then it lies on the bisector of the angle.

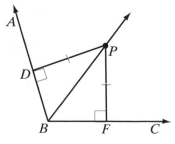

Given $\angle ABC$, $\overline{PD} \perp \overrightarrow{BA}$ at D, $\overline{PF} \perp \overrightarrow{BC}$ at F, and $PD = PF$

Prove $\angle ABP \cong \angle CBP$

Strategy   Use HL to prove $\triangle PDB \cong \triangle PFB$.

    The proof of this theorem is left to the student. (See exercise 8.)

Concurrence of Angle Bisectors of a Triangle

In earlier chapters, we saw that the perpendicular bisectors of the sides of a triangle intersect in a point and that the altitudes of a triangle intersect in a point. Now we can prove that the angle bisectors of a triangle intersect in a point.

Theorem 9.15 | The angle bisectors of a triangle are concurrent.

Given $\triangle ABC$ with $\overline{AL}$ the bisector of $\angle A$, $\overline{BM}$ the bisector of $\angle B$, and $\overline{CN}$ the bisector of $\angle C$.

Prove $\overline{AL}$, $\overline{BM}$, and $\overline{CN}$ intersect in a point, P.

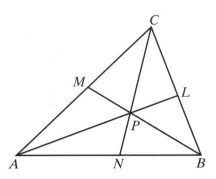

Proof Let P be the point at which $\overline{AL}$ and $\overline{BM}$ intersect. If a point lies on the bisector of an angle, then it is equidistant from the sides of the angle. Therefore, P is equidistant from $\overline{AC}$ and $\overline{AB}$ because it lies on the bisector of $\angle A$, and P is equidistant from $\overline{AB}$ and $\overline{BC}$ because it lies on the bisector of $\angle B$. Therefore, P is equidistant from $\overline{AC}$, $\overline{AB}$, and $\overline{BC}$. If a point is equidistant from the sides of an angle, then it lies on the bisector of the angle. Since P is equidistant from $\overline{AC}$ and $\overline{BC}$, then it lies of the bisector of $\angle C$. Therefore, the three angle bisectors of $\triangle ABC$ intersect at a point, P.

The point where the angle bisectors of a triangle are concurrent is called the **incenter**.

EXAMPLE I

Given: $\triangle ABC$, $\overline{AB} \perp \overline{BD}$, $\overline{AB} \parallel \overline{DC}$, and $\overline{AD} \cong \overline{BC}$.

Prove: $\angle DAB \cong \angle BCD$

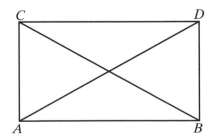

Proof We can show that $\triangle ADB$ and $\triangle CBD$ are right triangles and use HL to prove them congruent.

Statements	Reasons
1. $\overline{AB} \perp \overline{BD}$	1. Given.
2. $\overline{AB} \parallel \overline{DC}$	2. Given.
3. $\overline{BD} \perp \overline{DC}$	3. If a line is perpendicular to one of two parallel lines it is perpendicular to the other.
4. $\angle ABD$ and $\angle CDB$ are right angles.	4. Perpendicular lines intersect to form right angles.
5. $\overline{AD} \cong \overline{BC}$	5. Given.
6. $\overline{BD} \cong \overline{BD}$	6. Reflexive property of congruence.
7. $\triangle ADB \cong \triangle CBD$	7. HL (steps 5 and 6).
8. $\angle DAB \cong \angle BCD$	8. Corresponding parts of congruent triangles are congruent.

Exercises

Writing About Mathematics

1. In two right triangles, the right angles are congruent. What other pairs of corresponding parts must be known to be congruent in order to prove these two right triangles congruent?

2. The incenter of $\triangle ABC$ is P. If PD is the distance from P to $\overline{AB}$ and Q is any other point on $\overline{AB}$, is PD greater than PQ, equal to PQ, or less than PQ? Justify your answer.

Developing Skills

3. In $\triangle ABC$, m$\angle CAB = 40$ and m$\angle ABC = 60$. The angle bisectors of $\triangle ABC$ intersect at P.

a. Find m$\angle BCA$.

b. Find the measure of each angle of $\triangle APB$.

c. Find the measure of each angle of $\triangle BPC$.

d. Find the measure of each angle of $\triangle CPA$.

e. Does the bisector of $\angle CAB$ also bisect $\angle CPB$? Explain your answer.

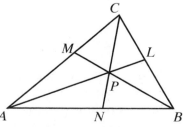

4. Triangle ABC is an isosceles right triangle with the right angle at C. Let P be the incenter of $\triangle ABC$.

a. Find the measure of each acute angle of $\triangle ABC$.

b. Find the measure of each angle of $\triangle APB$.

c. Find the measure of each angle of $\triangle BPC$.

d. Find the measure of each angle of $\triangle CPA$.

e. Does the bisector of $\angle ACB$ also bisect $\angle APB$? Explain your answer.

5. Triangle ABC is an isosceles triangle with $m\angle C = 140$. Let P be the incenter of $\triangle ABC$.

a. Find the measure of each acute angle of $\triangle ABC$.

b. Find the measure of each angle of $\triangle APB$.

c. Find the measure of each angle of $\triangle BPC$.

d. Find the measure of each angle of $\triangle CPA$.

e. Does the bisector of $\angle ACB$ also bisect $\angle APB$? Explain your answer.

6. In $\triangle RST$, the angle bisectors intersect at P. If $m\angle RTS = 50$, $m\angle TPR = 120$, and $m\angle RPS = 115$, find the measures of $\angle TRS$, $\angle RST$, and $\angle SPT$.

7. a. Draw a scalene triangle on a piece of paper or using geometry software. Label the triangle $\triangle ABC$.

b. Using compass and straightedge or geometry software, construct the angle bisectors of the angles of the triangle. Let $\overline{AL}$ be the bisector of $\angle A$, $\overline{BM}$ be the bisector of $\angle B$, and $\overline{CN}$ be the bisector of $\angle C$, such that L, M, and N are points on the triangle.

c. Label the incenter P.

d. In $\triangle ABC$, does $AP = BP = CP$? Explain why or why not.

e. If the incenter is equidistant from the vertices of $\triangle DEF$, what kind of a triangle is $\triangle DEF$?

Applying Skills

8. Prove Corollary 9.14a, "If a point is equidistant from the sides of an angle, then it lies on the bisector of the angle."

9. Given $\overline{DB} \perp \overline{ABC}$ and $\overline{AD} \perp \overline{DC}$, when is $\triangle ABD$ congruent to $\triangle DBC$? Explain.

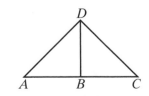

10. When we proved that the bisectors of the angles of a triangle intersect in a point, we began by stating that two of the angle bisectors, $\overline{AL}$ and $\overline{BM}$, intersect at P. To prove that they intersect, show that they are not parallel. (*Hint*: $\overline{AL}$ and $\overline{BM}$ are cut by transversal $\overline{AB}$. Show that a pair of interior angles on the same side of the transversal cannot be supplementary.)

11. *Given:* Quadrilateral $ABCD$, $\overline{AB} \perp \overline{BD}$, $\overline{BD} \perp \overline{DC}$, and $\overline{AD} \cong \overline{CB}$.

　Prove: $\angle A \cong \angle C$ and $\overline{AD} \parallel \overline{CB}$

12. In $\triangle QRS$, the bisector of $\angle QRS$ is perpendicular to $\overline{QS}$ at P.

　a. Prove that $\triangle QRS$ is isosceles.

　b. Prove that P is the midpoint of $\overline{QS}$.

13. Each of two lines from the midpoint of the base of an isosceles triangle is perpendicular to one of the legs of the triangle. Prove that these lines are congruent.

14. In quadrilateral $ABCD$, $\angle A$ and $\angle C$ are right angles and $AB = CD$. Prove that:

　a. $AD = BC$　　**b.** $\angle ABD \cong \angle CDB$　　**c.** $\angle ADC$ is a right angle.

15. In quadrilateral $ABCD$, $\angle ABC$ and $\angle BCD$ are right angles, and $AC = BD$. Prove that $AB = CD$.

16. *Given:* $\overrightarrow{ABC}$, $\overrightarrow{APS}$, and $\overrightarrow{ADE}$ with $\overrightarrow{PB} \perp \overrightarrow{ABC}$, $\overrightarrow{PD} \perp \overrightarrow{ADE}$, and $PB = PD$.

　Prove: $\overrightarrow{APS}$ bisects $\angle CAE$.

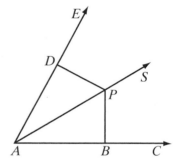

9-8 INTERIOR AND EXTERIOR ANGLES OF POLYGONS

Polygons

Recall that a polygon is a closed figure that is the union of line segments in a plane. Each vertex of a polygon is the endpoint of two line segments. We have proved many theorems about triangles and have used what we know about triangles to prove statements about the sides and angles of quadrilaterals, polygons with four sides. Other common polygons are:

- A **pentagon** is a polygon that is the union of five line segments.
- A **hexagon** is a polygon that is the union of six line segments.
- An **octagon** is a polygon that is the union of eight line segments.
- A **decagon** is a polygon that is the union of ten line segments.
- In general, an *n*-**gon** is a polygon with *n* sides.

A **convex polygon** is a polygon in which each of the interior angles measures less than 180 degrees. Polygon *PQRST* is a convex polygon and a pentagon. A **concave polygon** is a polygon in which at least one interior angle measures more than 180 degrees. Polygon *ABCD* is a concave polygon and a quadrilateral. In the rest of this textbook, unless otherwise stated, all polygons are convex.

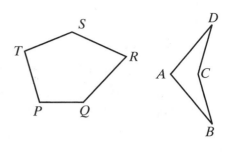

Interior Angles of a Polygon

A pair of angles whose vertices are the endpoints of a common side are called **consecutive angles**. And the vertices of consecutive angles are called **consecutive vertices** or **adjacent vertices**. For example, in *PQRST*, ∠*P* and ∠*Q* are consecutive angles and *P* and *Q* are consecutive or adjacent vertices. Another pair of consecutive angles are ∠*T* and ∠*P*. Vertices *R* and *T* are nonadjacent vertices.

A **diagonal** of a polygon is a line segment whose endpoints are two nonadjacent vertices. In hexagon *ABCDEF*, the vertices adjacent to *B* are *A* and *C* and the vertices nonadjacent to *B* are *D*, *E*, and *F*. Therefore, there are three diagonals with endpoint *B*: $\overline{BD}, \overline{BE}$, and $\overline{BF}$.

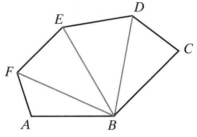

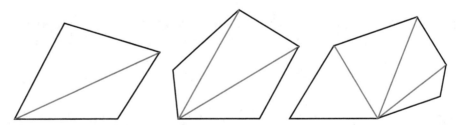

The polygons shown above have four, five, and six sides. In each polygon, all possible diagonals from a vertex are drawn. In the quadrilateral, two triangles are formed. In the pentagon, three triangles are formed, and in the hexagon, four triangles are formed. Note that in each polygon, the number of triangles formed is two less than the number of sides.

- In a quadrilateral: the sum of the measures of the angles is 2(180) = 360.
- In a pentagon: the sum of the measures of the angles is 3(180) = 540.
- In a hexagon: the sum of the measures of the angles is 4(180) = 720.

In general, the number of triangles into which the diagonals from a vertex separate a polygon of n sides is two less than the number of sides, or $n - 2$. The sum of the interior angles of the polygon is the sum of the interior angles of the triangles formed, or $180(n - 2)$. We have just proved the following theorem:

Theorem 9.16

> The sum of the measures of the interior angles of a polygon of n sides is $180(n - 2)°$.

Exterior Angles of a Polygon

At any vertex of a polygon, an exterior angle forms a linear pair with the interior angle. The interior angle and the exterior angle are supplementary. Therefore, the sum of their measures is $180°$. If a polygon has n sides, the sum of the interior and exterior angles of the polygon is $180n$. Therefore, in a polygon with n sides:

$$\text{The measures of the exterior angles} = 180n - \text{the measures of the interior angles}$$
$$= 180n - 180(n - 2)$$
$$= 180n - 180n + 360$$
$$= 360$$

We have just proved the following theorem:

Theorem 9.17

> The sum of the measures of the exterior angles of a polygon is $360°$.

DEFINITION

A **regular polygon** is a polygon that is both equilateral and equiangular.

If a triangle is equilateral, then it is equiangular. For polygons that have more than three sides, the polygon can be equiangular and not be equilateral, or can be equilateral and not be equiangular.

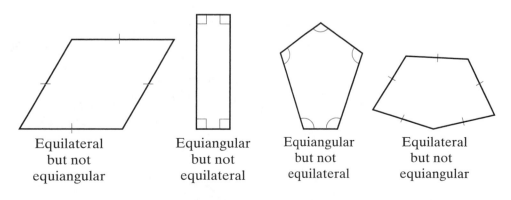

Equilateral but not equiangular   Equiangular but not equilateral   Equiangular but not equilateral   Equilateral but not equiangular

EXAMPLE 1

The measure of an exterior angle of a regular polygon is 45 degrees.

   a. Find the number of sides of the polygon.

   b. Find the measure of each interior angle.

   c. Find the sum of the measures of the interior angles.

Solution   **a.** Let n be the number of sides of the polygon. Then the sum of the measures of the exterior angles is n times the measure of one exterior angle.

$$45n = 360$$
$$n = \tfrac{360}{45}$$
$$n = 8 \ \textit{Answer}$$

   b. Each interior angle is the supplement of each exterior angle.

$$\text{Measure of each interior angle} = 180 - 45$$
$$= 135 \ \textit{Answer}$$

   c. Use the sum of the measures of the interior angles, $180(n - 2)$.

$$180(n - 2) = 180(8 - 2)$$
$$= 180(6)$$
$$= 1{,}080 \ \textit{Answer}$$

or

Multiply the measure of each interior angle by the number of sides.

$$8(135) = 1{,}080 \ \textit{Answer}$$

Answers **a.** 8 sides **b.** $135°$ **c.** $1{,}080°$

EXAMPLE 2

In quadrilateral $ABCD$, $m\angle A = x$, $m\angle B = 2x - 12$, $m\angle C = x + 22$, and $m\angle D = 3x$.

   a. Find the measure of each interior angle of the quadrilateral.

   b. Find the measure of each exterior angle of the quadrilateral.

Solution   **a.** $m\angle A + m\angle B + m\angle C + m\angle D = 18(n - 2)$

$$x + 2x - 12 + x + 22 + 3x = 180(4 - 2)$$
$$7x + 10 = 360$$
$$7x = 350$$
$$x = 50$$

$$m\angle A = x$$
$$= 50$$

$$m\angle B = 2x - 12$$
$$= 2(50) - 12$$
$$= 88$$

$$m\angle C = x + 22$$
$$= 50 + 22$$
$$= 72$$

$$m\angle D = 3x$$
$$= 3(50)$$
$$= 150$$

b. Each exterior angle is the supplement of the interior angle with the same vertex.

The measure of the exterior angle at A is $180 - 50 = 130$.

The measure of the exterior angle at B is $180 - 88 = 92$.

The measure of the exterior angle at C is $180 - 72 = 108$.

The measure of the exterior angle at D is $180 - 150 = 30$.

Answers **a.** $50°, 88°, 72°, 150°$ **b.** $130°, 92°, 108°, 30°$

Exercises

Writing About Mathematics

1. Taylor said that each vertex of a polygon with n sides is the endpoint of $(n - 3)$ diagonals. Do you agree with Taylor? Justify your answer.

2. Ryan said that every polygon with n sides has $\frac{n}{2}(n - 3)$ diagonals. Do you agree with Ryan? Justify your answer.

Developing Skills

3. Find the sum of the degree measures of the interior angles of a polygon that has:

a. 3 sides   **b.** 7 sides   **c.** 9 sides   **d.** 12 sides

4. Find the sum of the degree measures of the interior angles of:

a. a hexagon   **b.** an octagon   **c.** a pentagon   **d.** a quadrilateral

5. Find the sum of the measures of the exterior angles of a polygon that has:

a. 4 sides   **b.** 8 sides   **c.** 10 sides   **d.** 36 sides

In 6–14, for each *regular* polygon with the given number of sides, find the degree measures of: **a.** one exterior angle   **b.** one interior angle

6. 4 sides   **7.** 5 sides   **8.** 6 sides

9. 8 sides   **10.** 9 sides   **11.** 12 sides

12. 20 sides   **13.** 36 sides   **14.** 42 sides

15. Find the number of sides of a *regular* polygon each of whose exterior angles contains:

  a. 30°   **b.** 45°   **c.** 60°   **d.** 120°

16. Find the number of sides of a *regular* polygon each of whose interior angles contains:

  a. 90°   **b.** 120°   **c.** 140°   **d.** 160°

17. Find the number of sides a polygon if the sum of the degree measures of its interior angles is:

  a. 180   **b.** 360   **c.** 540   **d.** 900

  e. 1,440   **f.** 2,700   **g.** 1,800   **h.** 3,600

Applying Skills

18. The measure of each interior angle of a regular polygon is three times the measure of each exterior angle. How many sides does the polygon have?

19. The measure of each interior angle of a regular polygon is 20 degrees more than three times the measure of each exterior angle. How many sides does the polygon have?

20. The sum of the measures of the interior angles of a *concave* polygon is also $180(n - 2)$, where n is the number of sides. Is it possible for a concave quadrilateral to have two interior angles that are both more than 180°? Explain why or why not.

21. From vertex A of regular pentagon $ABCDE$, two diagonals are drawn, forming three triangles.

  a. Prove that two of the triangles formed by the diagonals are congruent.

  b. Prove that the congruent triangles are isosceles.

  c. Prove that the third triangle is isosceles.

22. From vertex L of regular hexagon $LMNRST$, three diagonals are drawn, forming four triangles.

  a. Prove that two of the triangles formed by the diagonals are congruent.

  b. Prove that the other two triangles formed by the diagonals are congruent.

  c. Find the measures of each of the angles in each of the four triangles.

23. The coordinates of the vertices of quadrilateral $ABCD$ are $A(-2, 0)$, $B(0, -2)$, $C(2, 0)$, and $D(0, 2)$.

  a. Prove that each angle of the quadrilateral is a right angle.

  b. Segments of the x-axis and the y-axis are diagonals of the quadrilateral. Prove that the four triangles into which the diagonals separate the quadrilateral are congruent.

  c. Prove that $ABCD$ is a regular quadrilateral.

Hands-On Activity

 In Section 9-7, we saw that the angle bisectors of a triangle are concurrent in a point called the *incenter*. In this activity, we will study the intersection of the angle bisectors of polygons.

a. Draw various polygons that are *not* regular of different sizes and numbers of sides. Construct the angle bisector of each interior angle. Do the angle bisectors appear to intersect in a single point?

b. Draw various *regular* polygons of different sizes and numbers of sides. Construct the angle bisector of each interior angle. Do the angle bisectors appear to intersect in a single point?

c. Based on the results of part **a** and **b**, state a conjecture regarding the intersection of the angle bisector of polygons.

CHAPTER SUMMARY

Definitions to Know

- **Parallel lines** are coplanar lines that have no points in common, or have all points in common and, therefore, coincide.

- A **transversal** is a line that intersects two other coplanar lines in two different points.

- The **incenter** is the point of intersection of the bisectors of the angles of a triangle.

- A **convex polygon** is a polygon in which each of the interior angles measures less than 180 degrees.

- A **concave polygon** is a polygon in which at least one of the interior angles measures more than 180 degrees.

- A **regular polygon** is a polygon that is both equilateral and equiangular.

Postulates

9.1 Two distinct coplanar lines are either parallel or intersecting.

9.2 Through a given point not on a given line, there exists one and only one line parallel to the given line.

Theorems and Corollaries

9.1 Two coplanar lines cut by a transversal are parallel if and only if the alternate interior angles formed are congruent.

9.2 Two coplanar lines cut by a transversal are parallel if and only if corresponding angles are congruent.

9.3 Two coplanar lines cut by a transversal are parallel if and only if interior angles on the same side of the transversal are supplementary.

9.4 If two coplanar lines are each perpendicular to the same line, then they are parallel.

9.5 If, in a plane, a line intersects one of two parallel lines, it intersects the other.

9.6 If a transversal is perpendicular to one of two parallel lines, it is perpendicular to the other.

9.7 If two of three lines in the same plane are each parallel to the third line, then they are parallel to each other.

9.8 If two lines are vertical lines, then they are parallel.

9.9 If two lines are horizontal lines, then they are parallel.

9.10 Two non-vertical lines in the coordinate plane are parallel if and only if they have the same slope.

9.11 The sum of the measures of the angles of a triangle is $180°$.

9.11a If two angles of one triangle are congruent to two angles of another triangle, then the third angles are congruent.

9.11b The acute angles of a right triangle are complementary.

9.11c Each acute angle of an isosceles right triangle measures $45°$.

9.11d Each angle of an equilateral triangle measures $60°$.

9.11e The sum of the measures of the angles of a quadrilateral is $360°$.

9.11f The measure of an exterior angle of a triangle is equal to the sum of the measures of the nonadjacent interior angles.

9.12 If two angles and the side opposite one of them in one triangle are congruent to the corresponding angles and side in another triangle, then the triangles are congruent. (AAS)

9.12a Two right triangles are congruent if the hypotenuse and an acute angle of one right triangle are congruent to the hypotenuse and an acute angle of the other right triangle.

9.12b If a point lies on the bisector of an angle, then it is equidistant from the sides of the angle.

9.13 If two angles of a triangle are congruent, then the sides opposite these angles are congruent.

9.13a If a triangle is equiangular, then it is equilateral.

9.14 If the hypotenuse and a leg of one right triangle are congruent to the corresponding parts of the other, then the two right triangles are congruent. (HL)

9.14a If a point is equidistant from the sides of an angle, then it lies on the bisector of the angle.

9.15 The angle bisectors of a triangle are concurrent.

9.16 The sum of the measures of the interior angles of a polygon of n sides is $180(n - 2)°$.

9.17 The sum of the measures of the exterior angles of a polygon is $360°$.

VOCABULARY

9-1 Euclid's parallel postulate • Playfair's postulate • Coplanar • Parallel lines • Transversal • Interior angles • Exterior angles • Alternate interior angles • Alternate exterior angles • Interior angles on the same side of the transversal • Corresponding angles

9-3 Midsegment

9-5 AAS triangle congruence

9-7 Hypotenuse-leg triangle congruence theorem (HL) • Incenter

9-8 Pentagon • Hexagon • Octagon • Decagon • n-gon • Convex polygon • Concave polygon • Consecutive angles • Consecutive vertices • Adjacent vertices • Diagonal of a polygon • Regular polygon

REVIEW EXERCISES

In 1–5, $\overleftrightarrow{AB} \parallel \overleftrightarrow{CD}$ and these lines are cut by transversal $\overleftrightarrow{GH}$ at points E and F, respectively.

1. If m$\angle AEF = 5x$ and m$\angle DFE = 75$, find x.

2. If m$\angle CFE = 3y + 20$ and m$\angle AEG = 4y - 10$, find y.

3. If m$\angle BEF = 5x$ and m$\angle CFE = 7x - 48$, find x.

4. If m$\angle DFE = y$ and m$\angle BEF = 3y - 40$, find m$\angle DFE$.

5. If m$\angle AEF = 4x$ and m$\angle EFD = 3x + 18$, find:

   a. the value of x **b.** m$\angle AEF$

   c. m$\angle EFD$ **d.** m$\angle BEF$ **e.** m$\angle CFH$

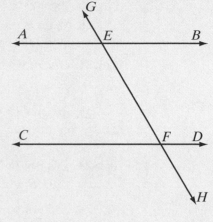

6. The degree measure of the vertex angle of an isosceles triangle is 120. Find the measure of a base angle of the triangle.

7. In $\triangle ABC$, $\angle A \cong \angle C$. If $AB = 8x + 4$ and $CB = 3x + 34$, find x.

8. In an isosceles triangle, if the measure of the vertex angle is 3 times the measure of a base angle, find the degree measure of a base angle.

9. In a triangle, the degree measures of the three angles are represented by x, $x + 42$, and $x - 6$. Find the angle measures.

10. In $\triangle PQR$, if m$\angle P = 35$ and m$\angle Q = 85$, what is the degree measure of an exterior angle of the triangle at vertex R?

11. An exterior angle at the base of an isosceles triangle measures $130°$. Find the measure of the vertex angle.

12. In $\triangle ABC$, if $\overline{AB} \cong \overline{AC}$ and m$\angle A = 70$, find m$\angle B$.

13. In $\triangle DEF$, if $\overline{DE} \cong \overline{DF}$ and m$\angle E = 13$, find m$\angle D$.

14. In $\triangle PQR$, $\overline{PQ}$ is extended through Q to point T, forming exterior $\angle RQT$. If m$\angle RQT = 70$ and m$\angle R = 10$, find m$\angle P$.

15. In $\triangle ABC$, $\overline{AC} \cong \overline{BC}$. The degree measure of an exterior angle at vertex C is represented by $5x + 10$. If m$\angle A = 30$, find x.

16. The degree measures of the angles of a triangle are represented by $x - 10$, $2x + 20$, and $3x - 10$. Find the measure of each angle of the triangle.

17. If the degree measures of the angles of a triangle are represented by x, y, and $x + y$, what is the measure of the largest angle of the triangle?

18. If parallel lines are cut by a transversal so that the degree measures of two corresponding angles are represented by $2x + 50$ and $3x + 20$, what is the value of x?

19. The measure of one exterior angle of a regular polygon is 30°. How many sides does the regular polygon have?

20. What is the sum of the degree measures of the interior angles of a polygon with nine sides?

21. *Given:* Right triangle ABC with $\angle C$ the right angle.

    Prove: $AB > AC$

22. *Given:* $\overline{AEB}$ and $\overline{CED}$ bisect each other at E.

    Prove: $\overline{AC} \parallel \overline{BD}$

23. P is not on $\overline{ABCD}$ and $\overline{PA}, \overline{PB}, \overline{PC},$ and $\overline{PD}$ are drawn. If $\overline{PB} \cong \overline{PC}$ and $\angle APB \cong \angle DPC$, prove that $\overline{PA} \cong \overline{PD}$.

24. P is not on $\overline{ABCD}$ and $\overline{PA}, \overline{PB}, \overline{PC},$ and $\overline{PD}$ are drawn. If $\angle PBC \cong \angle PCB$ and $\overline{AB} \cong \overline{DC}$, prove that $\overline{PA} \cong \overline{PD}$.

25. Herbie wanted to draw pentagon $ABCDE$ with m$\angle A$ = m$\angle B$ = 120 and m$\angle C$ = m$\angle D$ = 150. Is such a pentagon possible? Explain your answer.

Exploration

The geometry that you have been studying is called plane Euclidean geometry. Investigate a non-Euclidean geometry. How do the postulates of a non-Euclidean geometry differ from the postulates of Euclid? How can the postulates from this chapter be rewritten to fit the non-Euclidean geometry you investigated? What theorems from this chapter are not valid in the non-Euclidean geometry that you investigated? One possible non-Euclidean geometry is the geometry of the sphere suggested in the Chapter 1 Exploration.

CUMULATIVE REVIEW   Chapters 1–9

Part I

Answer all questions in this part. Each correct answer will receive 2 credits. No partial credit will be allowed.

1. If M is the midpoint of $\overline{AB}$, which of the following may be false?

   (1) M is between A and B.

   (2) $AM = MB$

   (3) A, B, and M are collinear.

   (4) $\overleftrightarrow{MN}$, a line that intersects $\overline{AB}$ at M, is the perpendicular bisector of $\overline{AB}$.

2. The statement "If two angles form a linear pair, then they are supplementary" is true. Which of the following statements must also be true?

(1) If two angles do not form a linear pair, then they are not supplementary.

(2) If two angles are not supplementary, then they do not form a linear pair.

(3) If two angles are supplementary, then they form a linear pair.

(4) Two angles form a linear pair if and only if they are supplementary.

3. Which of the following is a statement of the reflexive property of equality for all real numbers a, b, and c?

(1) $a = a$ (3) If $a = b$ and $b = c$, then $a = c$.

(2) If $a = b$, then $b = a$.   (4) If $a = b$, then $ac = bc$.

4. Two angles are complementary. If the measure of the larger angle is 10 degrees less than three times the measure of the smaller, what is the measure of the larger angle?

(1) $20°$ (2) $25°$ (3) $65°$ (4) $70°$

5. Under the transformation $r_{x\text{-axis}} \circ R_{90°}$, the image of $(-2, 5)$ is

(1) $(-5, -2)$ (2) $(-5, 2)$ (3) $(5, -2)$ (4) $(2, -5)$

6. An equation of the line through $(0, -1)$ and perpendicular to the line $x + 3y = 4$ is

(1) $3x + y = 1$ (3) $3x - y = 1$

(2) $x - 3y = 1$ (4) $x + 3y = -1$

7. The coordinates of the midpoint of the line segment whose endpoints are $(-3, 4)$ and $(5, -6)$ are

(1) $(1, -1)$ (2) $(-4, 5)$ (3) $(4, 5)$ (4) $(-4, -5)$

8. If a, b, c, and d are real numbers and $a > b$ and $c > d$, which of the following must be true?

(1) $a + c > b + d$ (3) $ac > bc$

(2) $a - c > b - d$ (4) $\frac{a}{c} > \frac{b}{d}$

9. The measure of each base angle of an isosceles triangle is 5 more than twice the measure of the vertex angle. The measure of the vertex angle is

(1) $34°$ (2) $73°$ (3) $43.75°$ (4) $136.25°$

10. Which of the following properties is not preserved under a line reflection?

(1) distance   (3) angle measure

(2) orientation   (4) midpoint

Part II

Answer all questions in this part. Each correct answer will receive 2 credits. Clearly indicate the necessary steps, including appropriate formula substitutions, diagrams, graphs, charts, etc. For all questions in this part, a correct numerical answer with no work shown will receive only 1 credit.

11. C is a point on $\overline{AD}$ and B is a point that is not on $\overleftrightarrow{AD}$. If m$\angle CAB = 65$, m$\angle CBD = 20$, and m$\angle BCD = 135$, which is the longest side of $\triangle ABC$?

12. If P is a point on the perpendicular bisector of $\overline{AB}$, prove that $\triangle ABP$ is isosceles.

Part III

Answer all questions in this part. Each correct answer will receive 4 credits. Clearly indicate the necessary steps, including appropriate formula substitutions, diagrams, graphs, charts, etc. For all questions in this part, a correct numerical answer with no work shown will receive only 1 credit.

13. $ABCD$ is an equilateral quadrilateral. Prove that the diagonal, $\overline{AC}$, bisects $\angle DAB$ and $\angle DCB$.

14. $\overleftrightarrow{AEB}$ and $\overleftrightarrow{CED}$ intersect at E and $\overleftrightarrow{AD} \parallel \overleftrightarrow{CB}$.

Prove that m$\angle DEB = $ m$\angle EBC + $ m$\angle EDA$.

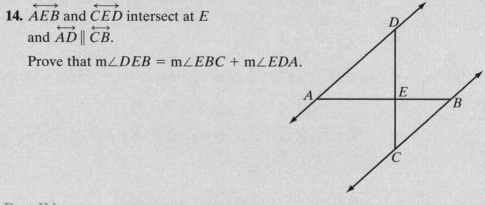

Part IV

Answer all questions in this part. Each correct answer will receive 6 credits. Clearly indicate the necessary steps, including appropriate formula substitutions, diagrams, graphs, charts, etc. For all questions in this part, a correct numerical answer with no work shown will receive only 1 credit.

15. The measures of the angles of a triangle are in the ratio $3 : 4 : 8$. Find the measure of the smallest exterior angle.

16. Write an equation of the perpendicular bisector of $\overline{AB}$ if the coordinates of the endpoints of $\overline{AB}$ are $A(-1, -2)$ and $B(7, 6)$.

QUADRILATERALS

Euclid's fifth postulate was often considered to be a "flaw" in his development of geometry. Girolamo Saccheri (1667–1733) was convinced that by the application of rigorous logical reasoning, this postulate could be proved. He proceeded to develop a geometry based on an isosceles quadrilateral with two base angles that are right angles. This isosceles quadrilateral had been proposed by the Persian mathematician Nasir al-Din al-Tusi (1201–1274). Using this quadrilateral, Saccheri attempted to prove Euclid's fifth postulate by reasoning to a contradiction. After his death, his work was published under the title *Euclid Freed of Every Flaw*. Saccheri did not, as he set out to do, prove the parallel postulate, but his work laid the foundations for new geometries. János Bolyai (1802–1860) and Nicolai Lobachevsky (1793–1856) developed a geometry that allowed two lines parallel to the given line, through a point not on a given line. Georg Riemann (1826–1866) developed a geometry in which there is no line parallel to a given line through a point not on the given line.

10-1 THE GENERAL QUADRILATERAL

Patchwork is an authentic American craft, developed by our frugal ancestors in a time when nothing was wasted and useful parts of discarded clothing were stitched into warm and decorative quilts.

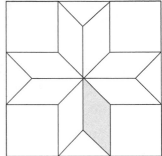

Quilt patterns, many of which acquired names as they were handed down from one generation to the next, were the product of creative and industrious people. In the Lone Star pattern, *quadrilaterals* are arranged to form larger *quadrilaterals* that form a star. The creators of this pattern were perhaps more aware of the pleasing effect of the design than of the mathematical relationships that were the basis of the pattern.

A **quadrilateral** is a polygon with four sides. In this chapter we will study the various special quadrilaterals and the properties of each. Let us first name the general parts and state properties of any quadrilateral, using *PQRS* as an example.

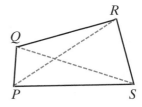

- **Consecutive vertices** or **adjacent vertices** are vertices that are endpoints of the same side such as *P* and *Q*, *Q* and *R*, *R* and *S*, *S* and *P*.

- **Consecutive sides** or **adjacent sides** are sides that have a common endpoint, such as $\overline{PQ}$ and $\overline{QR}$, $\overline{QR}$ and $\overline{RS}$, $\overline{RS}$ and $\overline{SP}$, $\overline{SP}$ and $\overline{PQ}$.

- **Opposite sides of a quadrilateral** are sides that do not have a common endpoint, such as $\overline{PQ}$ and $\overline{RS}$, $\overline{SP}$ and $\overline{QR}$.

- **Consecutive angles of a quadrilateral** are angles whose vertices are consecutive, such as ∠*P* and ∠*Q*, ∠*Q* and ∠*R*, ∠*R* and ∠*S*, ∠*S* and ∠*P*.

- **Opposite angles of a quadrilateral** are angles whose vertices are not consecutive, such as ∠*P* and ∠*R*, ∠*Q* and ∠*S*.

- A **diagonal of a quadrilateral** is a line segment whose endpoints are two nonadjacent vertices of the quadrilateral, such as $\overline{PR}$ and $\overline{QS}$.

- The sum of the measures of the angles of a quadrilateral is 360 degrees. Therefore, m∠*P* + m∠*Q* + m∠*R* + m∠*S* = 360.

10-2 THE PARALLELOGRAM

DEFINITION

A **parallelogram** is a quadrilateral in which two pairs of opposite sides are parallel.

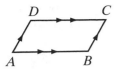

Quadrilateral $ABCD$ is a parallelogram because $\overline{AB} \parallel \overline{CD}$ and $\overline{BC} \parallel \overline{DA}$. The symbol for parallelogram $ABCD$ is $\square ABCD$.

Note the use of arrowheads, pointing in the same direction, to show sides that are parallel in the figure.

| **Theorem 10.1** | A diagonal divides a parallelogram into two congruent triangles. |

Given Parallelogram $ABCD$ with diagonal $\overline{AC}$

Prove $\triangle ABC \cong \triangle CDA$

Proof Since opposite sides of a parallelogram are parallel, alternate interior angles can be proved congruent using the diagonal as the transversal.

Statements	Reasons
1. $ABCD$ is a parallelogram.	**1.** Given.
2. $\overline{AB} \parallel \overline{CD}$ and $\overline{BC} \parallel \overline{DA}$	**2.** A parallelogram is a quadrilateral in which two pairs of opposite sides are parallel.
3. $\angle BAC \cong \angle DCA$ and $\angle BCA \cong \angle DAC$	**3.** If two parallel lines are cut by a transversal, alternate interior angles are congruent.
4. $\overline{AC} \cong \overline{AC}$	**4.** Reflexive property of congruence.
5. $\triangle ABC \cong \triangle CDA$	**5.** ASA.

We have proved that the diagonal $\overline{AC}$ divides parallelogram $ABCD$ into two congruent triangles. An identical proof could be used to show that $\overline{BD}$ divides the parallelogram into two congruent triangles, $\triangle ABD \cong \triangle CDB$.

The following corollaries result from this theorem.

| **Corollary 10.1a** | Opposite sides of a parallelogram are congruent. |

| **Corollary 10.1b** | Opposite angles of a parallelogram are congruent. |

The proofs of these corollaries are left to the student. (See exercises 14 and 15.)

We can think of each side of a parallelogram as a segment of a transversal that intersects a pair of parallel lines. This enables us to prove the following theorem.

Theorem 10.2 | Two consecutive angles of a parallelogram are supplementary.

Proof In $\square ABCD$, opposite sides are parallel. If two parallel lines are cut by a transversal, then two interior angles on the same side of the transversal are supplementary. Therefore, $\angle A$ is supplementary to $\angle B$, $\angle B$ is supplementary to $\angle C$, $\angle C$ is supplementary to $\angle D$, and $\angle D$ is supplementary to $\angle A$.

Theorem 10.3 | The diagonals of a parallelogram bisect each other.

Given $\square ABCD$ with diagonals $\overline{AC}$ and $\overline{BD}$ intersecting at E.

Prove $\overline{AC}$ and $\overline{BD}$ bisect each other.

Proof

Statements	**Reasons**
1. $\overline{AB} \parallel \overline{CD}$	1. Opposite sides of a parallelogram are parallel.
2. $\angle BAE \cong \angle DCE$ and $\angle ABE \cong \angle CDE$	2. If two parallel lines are cut by a transversal, the alternate interior angles are congruent.
3. $\overline{AB} \cong \overline{CD}$	3. Opposite sides of a parallelogram are congruent.
4. $\triangle ABE \cong \triangle CDE$	4. ASA.
5. $\overline{AE} \cong \overline{CE}$ and $\overline{BE} \cong \overline{DE}$	5. Corresponding part of congruent triangles are congruent.
6. E is the midpoint of $\overline{AC}$ and of $\overline{BD}$.	6. The midpoint of a line segment divides the segment into two congruent segments.
7. $\overline{AC}$ and $\overline{BD}$ bisect each other.	7. The bisector of a line segment intersects the segment at its midpoint.

DEFINITION

The **distance between two parallel lines** is the length of the perpendicular from any point on one line to the other line.

Properties of a Parallelogram

1. Opposite sides are parallel.

2. A diagonal divides a parallelogram into two congruent triangles.

3. Opposite sides are congruent.

4. Opposite angles are congruent.

5. Consecutive angles are supplementary.

6. The diagonals bisect each other.

EXAMPLE 1

In $\square ABCD$, $\text{m}\angle B$ exceeds $\text{m}\angle A$ by 46 degrees. Find $\text{m}\angle B$.

Solution Let $x = \text{m}\angle A$.

Then $x + 46 = \text{m}\angle B$.

Two consecutive angles of a parallelogram are supplementary. Therefore,

$$\text{m}\angle A + \text{m}\angle B = 180 \qquad\qquad \text{m}\angle B = x + 46$$
$$x + x + 46 = 180 \qquad\qquad\qquad = 67 + 46$$
$$2x + 46 = 180 \qquad\qquad\qquad\quad = 113$$
$$2x = 134$$
$$x = 67$$

Answer $\text{m}\angle B = 113$

Exercises

Writing About Mathematics

1. Theorem 10.2 states that two consecutive angles of a parallelogram are supplementary. If two opposite angles of a quadrilateral are supplementary, is the quadrilateral a parallelogram? Justify your answer.

2. A diagonal divides a parallelogram into two congruent triangles. Do two diagonals divide a parallelogram into four congruent triangles? Justify your answer.

Developing Skills

3. Find the degree measures of the other three angles of a parallelogram if one angle measures:

    a. 70   **b.** 65   **c.** 90   **d.** 130   **e.** 155   **f.** 168

In 4–11, *ABCD* is a parallelogram.

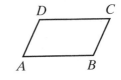

4. The degree measure of $\angle A$ is represented by $2x - 20$ and the degree measure of $\angle B$ by $2x$. Find the value of x, of m$\angle A$, and of m$\angle B$.

5. The degree measure of $\angle A$ is represented by $2x + 10$ and the degree measure of $\angle B$ by $3x$. Find the value of x, of m$\angle A$, and of m$\angle B$.

6. The measure of $\angle A$ is 30 degrees less than twice the measure of $\angle B$. Find the measure of each angle of the parallelogram.

7. The measure of $\angle A$ is represented by $x + 44$ and the measure of $\angle C$ by $3x$. Find the measure of each angle of the parallelogram.

8. The measure of $\angle B$ is represented by $7x$ and m$\angle D$ by $5x + 30$. Find the measure of each angle of the parallelogram.

9. The measure of $\angle C$ is one-half the measure of $\angle B$. Find the measure of each angle of the parallelogram.

10. If $AB = 4x + 7$ and $CD = 3x + 12$, find AB and CD.

11. If $AB = 4x + y$, $BC = y + 4$, $CD = 3x + 6$, and $DA = 2x + y$, find the lengths of the sides of the parallelogram.

12. The diagonals of $\square ABCD$ intersect at E. If $AE = 5x - 3$ and $EC = 15 - x$, find AC.

13. The diagonals of $\square ABCD$ intersect at E. If $DE = 4y + 1$ and $EB = 5y - 1$, find DB.

Applying Skills

14. Prove Corollary 10.1a, "Opposite sides of a parallelogram are congruent."

15. Prove Corollary 10.1b, "Opposite angles of a parallelogram are congruent."

16. *Given:* Parallelogram *EBFD* and parallelogram *ABCD* with
$\overline{EAB}$ and $\overline{DCF}$

   Prove: $\triangle EAD \cong \triangle FCB$

17. Petrina said that the floor of her bedroom is in the shape of a parallelogram and that at least one of the angles is a right angle. Show that the floor of Petrina's bedroom has four right angles.

18. The deck that Jeremiah is building is in the shape of a quadrilateral, *ABCD*. The measure of the angle at *A* is not equal to the measure of the angle at *C*. Prove that the deck is not in the shape of a parallelogram.

19. Quadrilaterals $ABCD$ and $PQRS$ are parallelograms with $\overline{AB} \cong \overline{PQ}$ and $\overline{BC} \cong \overline{QR}$. Prove that $ABCD \cong PQRS$ or draw a counterexample to show that they may not be congruent.

20. Quadrilaterals $ABCD$ and $PQRS$ are parallelograms with $\overline{AB} \cong \overline{PQ}, \overline{BC} \cong \overline{QR}$, and $\angle B \cong \angle Q$. Prove that $ABCD \cong PQRS$ or draw a counterexample to show that they may not be congruent.

10-3 PROVING THAT A QUADRILATERAL IS A PARALLELOGRAM

If we wish to prove that a certain quadrilateral is a parallelogram, we can do so by proving its opposite sides are parallel, thus satisfying the definition of a parallelogram. Now we want to determine other ways of proving that a quadrilateral is a parallelogram.

Theorem 10.4
> If both pairs of opposite sides of a quadrilateral are congruent, the quadrilateral is a parallelogram.

Given   Quadrilateral $ABCD$ with $\overline{AB} \cong \overline{CD}, \overline{AD} \cong \overline{BC}$

Prove $ABCD$ is a parallelogram.

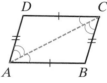

Proof   In $ABCD$, $\overline{AC}$ is a diagonal. Triangles ABC and CDA are congruent by SSS. Corresponding parts of congruent triangles are congruent, so $\angle BAC \cong \angle DCA$ and $\angle DAC \cong \angle ACB$. $\overline{AC}$ is a transversal that cuts $\overline{AB}$ and $\overline{DC}$. Alternate interior angles $\angle BAC$ and $\angle DCA$ are congruent, so $\overline{AB} \parallel \overline{DC}$. $\overline{AC}$ is also a transversal that cuts $\overline{AD}$ and $\overline{BC}$. Alternate interior angles $\angle DAC$ and $\angle ACB$ are congruent so $\overline{AD} \parallel \overline{BC}$. Therefore, $ABCD$ is a parallelogram.   ◻

Theorem 10.5
> If one pair of opposite sides of a quadrilateral is both congruent and parallel, the quadrilateral is a parallelogram.

Given   Quadrilateral $ABCD$ with $\overline{AB} \parallel \overline{CD}$ and $\overline{AB} \cong \overline{CD}$

Prove $ABCD$ is a parallelogram.

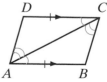

Proof   Since $\overline{AB}$ is parallel to $\overline{CD}$, $\angle BAC$ and $\angle DCA$ are a pair of congruent alternate interior angles. Therefore, by SAS, $\triangle DCA \cong \triangle BAC$. Corresponding parts of congruent triangles are congruent, so $\angle DAC \cong \angle ACB$. $\overline{AC}$ is a transversal that cuts $\overline{AD}$ and $\overline{BC}$, forming congruent alternate interior angles $\angle DAC$ and $\angle ACB$. Therefore, $\overline{AD} \parallel \overline{BC}$, and $ABCD$ is a parallelogram.   ◻

Theorem 10.6 | If both pairs of opposite angles of a quadrilateral are congruent, the quadrilateral is a parallelogram.

Given Quadrilateral $ABCD$ with $\angle A \cong \angle C$ and $\angle B \cong \angle D$

Prove $ABCD$ is a parallelogram.

Proof The sum of the measures of the angles of a quadrilateral is 360 degrees. Therefore, $m\angle A + m\angle B + m\angle C + m\angle D = 360$. It is given that $\angle A \cong \angle C$ and $\angle B \cong \angle D$. Congruent angles have equal measures so $m\angle A = m\angle C$ and $m\angle B = m\angle D$.

By substitution, $m\angle A + m\angle D + m\angle A + m\angle D = 360$. Then, $2m\angle A + 2m\angle D = 360$ or $m\angle A + m\angle D = 180$. Similarly, $m\angle A + m\angle B = 180$. If the sum of the measures of two angles is 180, the angles are supplementary. Therefore, $\angle A$ and $\angle D$ are supplementary and $\angle A$ and $\angle B$ are supplementary.

Two coplanar lines are parallel if a pair of interior angles on the same side of the transversal are supplementary. Therefore, $\overline{AB} \parallel \overline{DC}$ and $\overline{AD} \parallel \overline{BC}$. Quadrilateral $ABCD$ is a parallelogram because it has two pairs of parallel sides.

Theorem 10.7 | If the diagonals of a quadrilateral bisect each other, the quadrilateral is a parallelogram.

Given Quadrilateral $ABCD$ with $\overline{AC}$ and $\overline{BD}$ intersecting at E, $\overline{AE} \cong \overline{EC}, \overline{BE} \cong \overline{ED}$

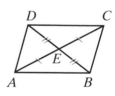

Prove $ABCD$ is a parallelogram.

Strategy Prove that $\triangle ABE \cong \triangle CDE$ to show that one pair of opposite sides is congruent and parallel.

The proof of Theorem 10.7 is left to the student. (See exercise 15.)

SUMMARY To prove that a quadrilateral is a parallelogram, prove that any one of the following statements is true:

1. Both pairs of opposite sides are parallel.

2. Both pairs of opposite sides are congruent.

3. One pair of opposite sides is both congruent and parallel.

4. Both pairs of opposite angles are congruent.

5. The diagonals bisect each other.

EXAMPLE I

Given: ABCD is a parallelogram.

   E is on $\overline{AB}$, *F* is on $\overline{DC}$, and $\overline{EB} \cong \overline{DF}$.

Prove: $\overline{DE} \parallel \overline{FB}$

Proof We will prove that *EBFD* is a parallelogram.

Statements	Reasons
1. *ABCD* is a parallelogram.	**1.** Given.
2. $\overline{AB} \parallel \overline{DC}$	**2.** Opposite sides of a parallelogram are parallel.
3. $\overline{EB} \parallel \overline{DF}$	**3.** Segments of parallel lines are parallel.
4. $\overline{EB} \cong \overline{DF}$	**4.** Given.
5. *EBFD* is a parallelogram.	**5.** If one pair of opposite sides of a quadrilateral is both congruent and parallel, the quadrilateral is a parallelogram.
6. $\overline{DE} \parallel \overline{FB}$	**6.** Opposite sides of a parallelogram are parallel.

Exercises

Writing About Mathematics

1. What statement and reason can be added to the proof in Example 1 to prove that $\overline{DE} \cong \overline{FB}$?

2. What statement and reason can be added to the proof in Example 1 to prove that $\angle DEB \cong \angle BFD$?

Developing Skills

In 3–7, in each case, the *given* is marked on the figure. Tell why each quadrilateral *ABCD* is a parallelogram.

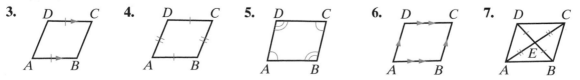

3. **4.** **5.** **6.** **7.**

8. *ABCD* is a quadrilateral with $\overline{AB} \parallel \overline{CD}$ and $\angle A \cong \angle C$. Prove that *ABCD* is a parallelogram.

9. *PQRS* is a quadrilateral with $\angle P \cong \angle R$ and $\angle P$ the supplement of $\angle Q$. Prove that *PQRS* is a parallelogram.

10. *DEFG* is a quadrilateral with $\overline{DF}$ drawn so that $\angle FDE \cong \angle DFG$ and $\angle GDF \cong \angle EFD$. Prove that *DEFG* is a parallelogram.

11. *ABCD* is a parallelogram. *E* is the midpoint of $\overline{AB}$ and *F* is the midpoint of $\overline{CD}$. Prove that *AEFD* is a parallelogram.

12. *EFGH* is a parallelogram and *J* is a point on $\overrightarrow{EF}$ such that *F* is the midpoint of $\overline{EJ}$. Prove that *FJGH* is a parallelogram.

13. *ABCD* is a parallelogram. The midpoint of $\overline{AB}$ is *P*, the midpoint of $\overline{BC}$ is *Q*, the midpoint of $\overline{CD}$ is *R*, and the midpoint of $\overline{DA}$ is *S*.

   a. Prove that $\triangle APS \cong \triangle CRQ$ and that $\triangle BQP \cong \triangle DSR$.

   b. Prove that *PQRS* is a parallelogram.

14. A quadrilateral has three right angles. Is the quadrilateral a parallelogram? Justify your answer.

Applying Skills

15. Prove Theorem 10.7, "If the diagonals of a quadrilateral bisect each other, the quadrilateral is a parallelogram."

16. Prove that a parallelogram can be drawn by joining the endpoints of two line segments that bisect each other.

17. The vertices of quadrilateral *ABCD* are $A(-2, 1)$, $B(4, -2)$, $C(8, 2)$, and $D(2, 5)$. Is *ABCD* a parallelogram? Justify your answer.

18. Farmer Brown's pasture is in the shape of a quadrilateral, *PQRS*. The pasture is crossed by two diagonal paths, $\overline{PR}$ and $\overline{QS}$. The quadrilateral is not a parallelogram. Show that the paths do not bisect each other.

19. Toni cut two congruent scalene triangles out of cardboard. She labeled one triangle $\triangle ABC$ and the other $\triangle A'B'C'$ so that $\triangle ABC \cong \triangle A'B'C'$. She placed the two triangles next to each other so that *A* coincided with *B'* and *B* coincided with *A'*. Was the resulting quadrilateral *ACBC'* a parallelogram? Prove your answer.

20. Quadrilateral *ABCD* is a parallelogram with *M* the midpoint of $\overline{AB}$ and *N* the midpoint of $\overline{CD}$.

   a. Prove that *AMND* and *MBCN* are parallelograms.

   b. Prove that *AMND* and *MBCN* are congruent.

10-4 THE RECTANGLE

DEFINITION _____

A **rectangle** is a parallelogram containing one right angle.

_____

If one angle, $\angle A$, of a parallelogram $ABCD$ is a right angle, then $\square ABCD$ is a rectangle.

Any side of a rectangle may be called the *base* of the rectangle. Thus, if side $\overline{AB}$ is taken as the base, then either consecutive side, $\overline{AD}$ or $\overline{BC}$, is called the *altitude* of the rectangle.

Since a rectangle is a special kind of parallelogram, a rectangle has all the properties of a parallelogram. In addition, we can prove two special properties for the rectangle.

Theorem 10.8 | All angles of a rectangle are right angles.

Given Rectangle $ABCD$ with $\angle A$ a right angle.

Prove $\angle B$, $\angle C$, and $\angle D$ are right angles.

Proof By definition, rectangle $ABCD$ is a parallelogram. Opposite angles of a parallelogram are congruent, so $\angle A \cong \angle C$. Angle A is a right angle so $\angle C$ is a right angle. Consecutive angles of a parallelogram are supplementary. Therefore, since $\angle A$ and $\angle B$ are supplementary and $\angle A$ is right angle, $\angle B$ is also a right angle. Similarly, since $\angle C$ and $\angle D$ are supplementary and $\angle C$ is a right angle, $\angle D$ is also a right angle.

Theorem 10.9 | The diagonals of a rectangle are congruent.

Given $ABCD$ is a rectangle.

Prove $\overline{AC} \cong \overline{BD}$

Strategy Prove $\triangle DAB \cong \triangle CBA$.

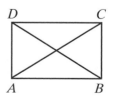

The proof of Theorem 10.9 is left to the student. (See exercise 12.)

Properties of a Rectangle

1. A rectangle has all the properties of a parallelogram.

2. A rectangle has four right angles and is therefore equiangular.

3. The diagonals of a rectangle are congruent.

Proving That a Quadrilateral Is a Rectangle

We prove that a quadrilateral is a rectangle by showing that it has the special properties of a rectangle. For example:

Theorem 10.10 | If a quadrilateral is equiangular, then it is a rectangle.

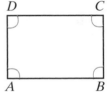

Given Quadrilateral $ABCD$ with $\angle A \cong \angle B \cong \angle C \cong \angle D$.

Prove $ABCD$ is a rectangle.

Proof Quadrilateral $ABCD$ is a parallelogram because the opposite angles are congruent. The sum of the measures of the angles of a quadrilateral is 360 degrees. Thus, the measure of each of the four congruent angles is 90 degrees. Therefore, $ABCD$ is a rectangle because it is a parallelogram with a right angle.

Theorem 10.11 | If the diagonals of a parallelogram are congruent, the parallelogram is a rectangle.

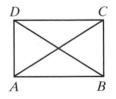

Given Parallelogram $ABCD$ with $\overline{AC} \cong \overline{BD}$

Prove $ABCD$ is a rectangle.

Strategy Prove that $\triangle DAB \cong \triangle CBA$. Therefore, $\angle DAB$ and $\angle CBA$ are both congruent and supplementary.

The proof of Theorem 10.11 is left to the student. (See exercise 13.)

SUMMARY To prove that a quadrilateral is a rectangle, prove that any one of the following statements is true:

1. The quadrilateral is a parallelogram with one right angle.

2. The quadrilateral is equiangular.

3. The quadrilateral is a parallelogram whose diagonals are congruent.

EXAMPLE 1

Given: ABCD is a parallelogram with m∠*A* = m∠*D*.

Prove: ABCD is a rectangle.

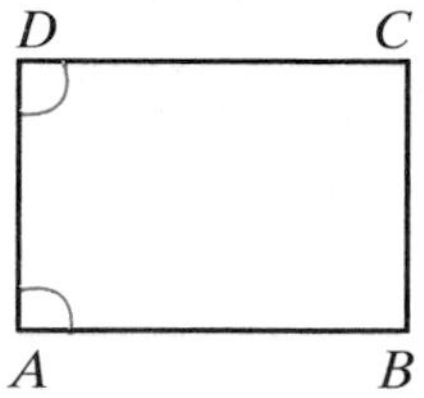

Proof We will prove that *ABCD* has a right angle.

Statements	Reasons
1. *ABCD* is a parallelogram.	**1.** Given.
2. $\overline{AB} \parallel \overline{CD}$	**2.** A parallelogram is a quadrilateral in which two pairs of opposite sides are parallel.
3. ∠*A* and ∠*D* are supplementary.	**3.** Two consecutive angles of a parallelogram are supplementary.
4. m∠*A* + m∠*D* = 180	**4.** Supplementary angles are two angles the sum of whose measures is 180.
5. m∠*A* = m∠*D*	**5.** Given.
6. m∠*A* + m∠*A* = 180 or 2m∠*A* = 180	**6.** A quantity may be substituted for its equal.
7. m∠*A* = 90	**7.** Division postulate.
8. *ABCD* is a rectangle.	**8.** A rectangle is a parallelogram with a right angle.

EXAMPLE 2

The lengths of diagonals of a rectangle are represented by $7x$ centimeters and $5x + 12$ centimeters. Find the length of each diagonal.

Solution The diagonals of a rectangle are congruent and therefore equal in length.

$$7x = 5x + 12$$
$$2x = 12$$
$$x = 6$$

$$7x = 7(6) \qquad 5x + 12 = 5(6) + 12$$
$$= 42 \qquad\qquad\qquad = 30 + 12$$
$$= 42$$

Answer The length of each diagonal is 42 centimeters.

Exercises

Writing About Mathematics

1. Pauli said that if one angle of a parallelogram is not a right angle, then the parallelogram is not a rectangle. Do you agree with Pauli? Explain why or why not.

2. Cindy said that if two congruent line segments intersect at their midpoints, then the quadrilateral formed by joining the endpoints of the line segments in order is a rectangle. Do you agree with Cindy? Explain why or why not.

Developing Skills

In 3–10, the diagonals of rectangle $ABCD$ intersect at E.

3. Prove that $\triangle ABE$ is isosceles.

4. $AC = 4x + 6$ and $BD = 5x - 2$. Find AC, BD, AE, and BE.

5. $AE = y + 12$ and $DE = 3y - 8$. Find AE, DE, AC, and BD.

6. $BE = 3a + 1$ and $ED = 6a - 11$. Find BE, ED, BD, and AC.

7. $AE = x + 5$ and $BD = 3x - 2$. Find AE, BD, AC, and BE.

8. $m\angle CAB = 35$. Find $m\angle CAD$, $m\angle ACB$, $m\angle AEB$, and $\angle AED$.

9. $m\angle AEB = 3x$ and $m\angle DEC = x + 80$. Find $m\angle AEB$, $m\angle DEC$, $m\angle CAB$, and $m\angle CAD$.

10. $m\angle AED = y + 10$ and $m\angle AEB = 4y - 30$. Find $m\angle AED$, $m\angle AEB$, $m\angle BAC$, and $m\angle CAD$.

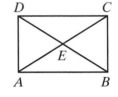

Applying Skills

11. Write a coordinate proof of Theorem 10.8, "All angles of a rectangle are right angles." Let the vertices of the rectangle be $A(0, 0)$, $B(b, 0)$, $C(b, c)$, and $D(0, c)$.

12. Prove Theorem 10.9, "The diagonals of a rectangle are congruent."

13. Prove Theorem 10.11, "If the diagonals of a parallelogram are congruent, the parallelogram is a rectangle."

14. If $PQRS$ is a rectangle and M is the midpoint of $\overline{RS}$, prove that $\overline{PM} \cong \overline{QM}$.

15. The coordinates of the vertices of $ABCD$ are $A(-2, 0)$, $B(2, -2)$, $C(5, 4)$, and $D(1, 6)$.

   a. Prove that $ABCD$ is a rectangle.

   b. What are the coordinates of the point of intersection of the diagonals?

   c. The vertices of $A'B'C'D'$ are $A'(0, -2)$, $B'(2, 2)$, $C'(-4, 5)$, and $D'(-6, 1)$. Under what specific transformation is $A'B'C'D'$ the image of $ABCD$?

   d. Prove that $A'B'C'D'$ is a rectangle.

16. The coordinates of the vertices of *PQRS* are $P(-2, 1)$, $Q(1, -3)$, $R(5, 1)$, and $S(2, 5)$.

    a. Prove that *PQRS* is a parallelogram.

    b. Prove that *PQRS* is not a rectangle.

    c. $P'Q'R'S'$ is the image of *PQRS* under $T_{-3,-3} \circ r_{y = x}$. What are the coordinates of $P'Q'R'S'$?

    d. Prove that $P'Q'R'S'$ is congruent to *PQRS*.

17. Angle *A* in quadrilateral *ABCD* is a right angle and quadrilateral *ABCD* is not a rectangle. Prove that *ABCD* is not a parallelogram.

18. Tracy wants to build a rectangular pen for her dog. She has a tape measure, which enables her to make accurate measurements of distance, but has no way of measuring angles. She places two stakes in the ground to represent opposite corners of the pen. How can she find two points at which to place stakes for the other two corners?

19. Archie has a piece of cardboard from which he wants to cut a rectangle with a diagonal that measures 12 inches. On one edge of the cardboard Archie marks two points that are less than 12 inches apart to be the endpoints of one side of the rectangle. Explain how Archie can use two pieces of string that are each 12 inches long to find the other two vertices of the rectangle.

10-5 THE RHOMBUS

DEFINITION

A **rhombus** is a parallelogram that has two congruent consecutive sides.

    If the consecutive sides $\overline{AB}$ and $\overline{AD}$ of parallelogram *ABCD* are congruent (that is, if $\overline{AB} \cong \overline{AD}$), then $\square ABCD$ is a rhombus.

    Since a rhombus is a special kind of parallelogram, a rhombus has all the properties of a parallelogram. In addition, we can prove three special properties for the rhombus.

Theorem 10.12 | All sides of a rhombus are congruent.

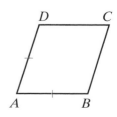

Given   *ABCD* is a rhombus with $\overline{AB} \cong \overline{DA}$.

Prove $\overline{AB} \cong \overline{BC} \cong \overline{CD} \cong \overline{DA}$

Proof   By definition, rhombus *ABCD* is a parallelogram. It is given that $\overline{AB} \cong \overline{DA}$. Opposite sides of a parallelogram are congruent, so $\overline{AB} \cong \overline{CD}$ and $\overline{BC} \cong \overline{DA}$. Using the transitive property of congruence, $\overline{AB} \cong \overline{BC} \cong \overline{CD} \cong \overline{DA}$.

Theorem 10.13 | The diagonals of a rhombus are perpendicular to each other.

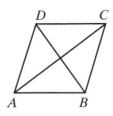

Given Rhombus $ABCD$

Prove $\overline{AC} \perp \overline{BD}$

Proof By Theorem 10.12, all sides of a rhombus are congruent. Segments that are congruent are equal. Consider the diagonal $\overline{AC}$. Point B is equidistant from the endpoints A and C since $BA = BC$. Point D is also equidistant from the endpoints A and C since $DA = DC$. If two points are each equidistant from the endpoints of a line segment, the points determine the perpendicular bisector of the line segment. Therefore, $\overline{AC} \perp \overline{BD}$.

Theorem 10.14 | The diagonals of a rhombus bisect its angles.

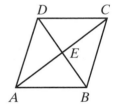

Given Rhombus $ABCD$

Prove $\overline{AC}$ bisects $\angle DAB$ and $\angle DCB$ and $\overline{DB}$ bisects $\angle CDA$ and $\angle CBA$.

Strategy Show that the diagonals separate the rhombus into four congruent triangles.

The proof of this theorem is left to the student. (See exercise 16.)

Properties of a Rhombus

1. A rhombus has all the properties of a parallelogram.

2. A rhombus is equilateral.

3. The diagonals of a rhombus are perpendicular to each other.

4. The diagonals of a rhombus bisect its angles.

Methods of Proving That a Quadrilateral Is a Rhombus

We prove that a quadrilateral is a rhombus by showing that it has the special properties of a rhombus.

Theorem 10.15 | If a quadrilateral is equilateral, then it is a rhombus.

Given Quadrilateral $ABCD$ with $\overline{AB} \cong \overline{BC} \cong \overline{CD} \cong \overline{DA}$

Prove $ABCD$ is a rhombus.

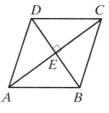

Proof It is given that in $ABCD$, $\overline{AB} \cong \overline{BC} \cong \overline{CD} \cong \overline{DA}$. Since both pairs of opposite sides are congruent, $ABCD$ is a parallelogram. Two consecutive sides of parallelogram $ABCD$ are congruent, so by definition, $ABCD$ is a rhombus.

Theorem 10.16 | If the diagonals of a parallelogram are perpendicular to each other, the parallelogram is a rhombus.

Given Parallelogram $ABCD$ with $\overline{AC} \perp \overline{BD}$

Prove $ABCD$ is a rhombus.

Strategy The diagonals divide parallelogram $ABCD$ into four triangles. Prove that two of the triangles that share a common side are congruent. Then use the fact that corresponding parts of congruent triangles are congruent to show that parallelogram $ABCD$ has two congruent consecutive sides.

The proof of this theorem is left to the student. (See exercise 17.)

SUMMARY To prove that a quadrilateral is a rhombus, prove that any one of the following statements is true:

1. The quadrilateral is a parallelogram with two congruent consecutive sides.

2. The quadrilateral is equilateral.

3. The quadrilateral is a parallelogram whose diagonals are perpendicular to each other.

EXAMPLE I

$PQRS$ is a rhombus and m$\angle PQR = 60$. Prove that $\overline{PR}$ divides the rhombus into two equilateral triangles.

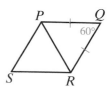

Proof   Since all sides of a rhombus are congruent, we know that $\overline{PQ} \cong \overline{RQ}$. Thus, $\triangle PQR$ is isosceles and its base angles are equal in measure. Let $m\angle PRQ = m\angle RPQ = x$.

$$x + x + 60 = 180$$
$$2x + 60 = 180$$
$$2x = 120$$
$$x = 60$$

Therefore, $m\angle PRQ = 60$, $m\angle RPQ = 60$, and $m\angle PQR = 60$. Triangle PQR is equilateral since an equiangular triangle is equilateral.

Since opposite angles of a rhombus have equal measures, $m\angle RSP = 60$. By similar reasoning, $\triangle RSP$ is equilateral.

EXAMPLE 2

Given: $ABCD$ is a parallelogram.

$AB = 2x + 1$, $DC = 3x - 11$, $AD = x + 13$.

Prove: $ABCD$ is a rhombus.

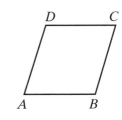

Proof   (1) Since $ABCD$ is a parallelogram, opposite sides are equal in length:

$$DC = AB$$
$$3x - 11 = 2x + 1$$
$$x = 12$$

(2) Substitute $x = 12$ to find the lengths of sides $\overline{AB}$ and $\overline{AD}$:

$AB = 2x + 1$	$AD = x + 13$
$= 2(12) + 1$	$= 12 + 13$
$= 25$	$= 25$

(3) Since $ABCD$ is a parallelogram with two congruent consecutive sides, $ABCD$ is a rhombus.

Exercises

Writing About Mathematics

1. Rochelle said that the diagonals of a rhombus separate the rhombus into four congruent right triangles. Do you agree with Rochelle? Explain why or why not.

2. Concepta said that if the lengths of the diagonals of a rhombus are represented by d_1 and d_2, then a formula for the area of a rhombus is $A = \frac{1}{2}d_1d_2$. Do you agree with Concepta? Explain why or why not.

Developing Skills

In 3–10, the diagonals of rhombus $ABCD$ intersect at E.

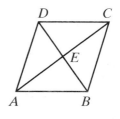

3. Name four congruent line segments.

4. Name two pairs of congruent line segments.

5. Name a pair of perpendicular line segments.

6. Name four right angles.

7. Under a rotation of 90° about E, does A map to B? does B map to C? Justify your answer.

8. Does rhombus $ABCD$ have rotational symmetry under a rotation of 90° about E?

9. Under a reflection in E, name the image of A, of B, of C, of D, and of E.

10. Does rhombus $ABCD$ have point symmetry under a reflection in E?

11. In rhombus $PQRS$, m$\angle P$ is 120.

    a. Prove that the diagonal $\overline{PR}$ separates the rhombus into two equilateral triangles.

    b. If $PR = 24$ cm, what is the length of each side of the rhombus?

In 12–15, tell whether each conclusion follows from the given premises. If not, draw a counter-example.

12. In a parallelogram, opposite sides are congruent.
A rhombus is a parallelogram.
Conclusion: In a rhombus, opposite sides are congruent.

13. In a rhombus, diagonals are perpendicular to each other.
A rhombus is a parallelogram.
Conclusion: In a parallelogram, diagonals are perpendicular to each other.

14. The diagonals of a rhombus bisect the angles of the rhombus.
A rhombus is a parallelogram.
Conclusion: The diagonals of a parallelogram bisect its angles.

15. Consecutive angles of a parallelogram are supplementary.
A rhombus is a parallelogram.
Conclusion: Consecutive angles of a rhombus are supplementary.

Applying Skills

16. Prove Theorem 10.14, "The diagonals of a rhombus bisect its angles."

17. Prove Theorem 10.16, "If the diagonals of a parallelogram are perpendicular to each other, the parallelogram is a rhombus."

18. The vertices of quadrilateral $ABCD$ are $A(-1, -1)$, $B(4, 0)$, $C(5, 5)$, and $D(0, 4)$.

    a. Prove that $ABCD$ is a parallelogram.

    b. Prove that the diagonals of $ABCD$ are perpendicular.

    c. Is $ABCD$ a rhombus? Justify your answer.

19. If a diagonal separates a quadrilateral $KLMN$ into two equilateral triangles, prove that $KLMN$ is a rhombus with the measure of one angle equal to 60 degrees.

20. Prove that the diagonals of a rhombus separate the rhombus into four congruent right triangles.

21. Prove that if the diagonals of a quadrilateral are the perpendicular bisectors of each other, the quadrilateral is a rhombus.

22. Prove that if the diagonals of a parallelogram bisect the angles of the parallelogram, then the parallelogram is a rhombus.

23. Let P be any point on diagonal $\overline{BD}$ of rhombus $ABCD$.

    Prove that $\overline{AP} \cong \overline{CP}$.

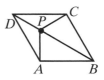

24. The vertices of quadrilateral $ABCD$ are $A(-2, -1)$, $B(2, -4)$, $C(2, 1)$, $D(-2, 4)$.

    a. Prove that $ABCD$ is a parallelogram.

    b. Find the coordinates of E, the point of intersection of the diagonals.

    c. Prove that the diagonals are perpendicular.

    d. Is $ABCD$ a rhombus? Justify your answer.

25. $ABCD$ is a parallelogram. The midpoint of $\overline{AB}$ is M, the midpoint of $\overline{CD}$ is N, and $AM = AD$.

    a. Prove that $AMND$ is a rhombus.

    b. Prove that $MBCN$ is a rhombus.

    c. Prove that $AMND$ is congruent to $MBCN$.

Hands-On Activity

In this activity, you will construct a rhombus given the diagonal. You may use geometry software or compass and straightedge.

    1. First draw a segment measuring 12 centimeters. This will be the diagonal of the rhombus. The endpoints of this segment are opposite vertices of the rhombus.

    2. Now construct the perpendicular bisector of this segment.

    3. Show that you can choose any point on that perpendicular bisector as a third vertex of the rhombus.

    4. How can you determine the fourth vertex of the rhombus?

    5. Compare the rhombus you constructed with rhombuses constructed by your classmates. How are they alike? How are they different?

10-6 THE SQUARE

DEFINITION _____

A **square** is a rectangle that has two congruent consecutive sides.

If consecutive sides $\overline{AB}$ and $\overline{AD}$ of rectangle $ABCD$ are congruent (that is, if $\overline{AB} \cong \overline{AD}$), then rectangle $ABCD$ is a square.

Theorem 10.17 | A square is an equilateral quadrilateral.

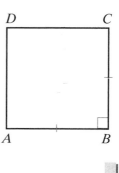

Given $ABCD$ is a square with $\overline{AB} \cong \overline{BC}$.

Prove $\overline{AB} \cong \overline{BC} \cong \overline{CD} \cong \overline{DA}$

Proof A square is a rectangle and a rectangle is a parallelogram, so $ABCD$ is a parallelogram. It is given that $\overline{AB} \cong \overline{BC}$. Opposite sides of a parallelogram are congruent, so $\overline{AB} \cong \overline{CD}$ and $\overline{BC} \cong \overline{DA}$. Using the transitive property of congruence, $\overline{AB} \cong \overline{BC} \cong \overline{CD} \cong \overline{DA}$.

Theorem 10.18 | A square is a rhombus.

Given Square $ABCD$

Prove $ABCD$ is a rhombus.

Proof A square is an equilateral quadrilateral. If a quadrilateral is equilateral, then it is a rhombus. Therefore, $ABCD$ is a rhombus.

Properties of a Square

1. A square has all the properties of a rectangle.

2. A square has all the properties of a rhombus.

Methods of Proving That a Quadrilateral Is a Square

We prove that a quadrilateral is a square by showing that it has the special properties of a square.

Theorem 10.19

> If one of the angles of a rhombus is a right angle, then the rhombus is a square.

Given $ABCD$ is a rhombus with $\angle A$ a right angle.

Prove $ABCD$ is a square.

Strategy   Show that $ABCD$ is a rectangle and that $\overline{AB} \cong \overline{BC}$.

The proof of this theorem is left to the student. (See exercise 12.)

SUMMARY

To prove that a quadrilateral is a square, prove either of the following statements:

1. The quadrilateral is a rectangle in which two consecutive sides are congruent.

2. The quadrilateral is a rhombus one of whose angles is a right angle.

EXAMPLE I

Given: Quadrilateral $ABCD$ is equilateral and $\angle ABC$ is a right angle.

Prove: $ABCD$ is a square.

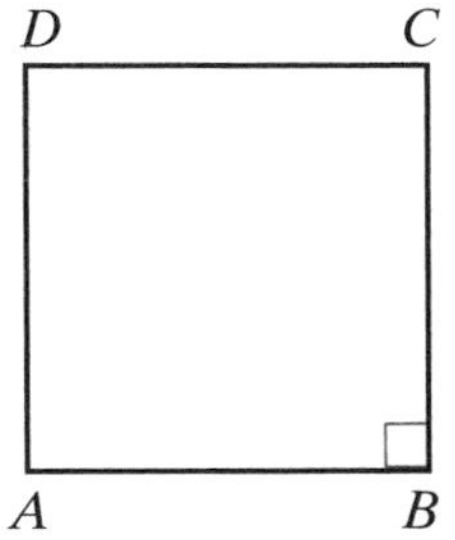

Proof

Statements	Reasons
1. $ABCD$ is equilateral.	1. Given
2. $ABCD$ is a rhombus.	2. If a quadrilateral is equilateral, then it is a rhombus.
3. $\angle ABC$ is a right angle.	3. Given.
4. $ABCD$ is a square.	4. If one of the angles of a rhombus is a right angle, then the rhombus is a square.

EXAMPLE 2

In square $PQRS$, $\overline{SQ}$ is a diagonal.
Find m$\angle PSQ$.

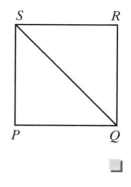

Solution A square is a rhombus, and the diagonal of a rhombus bisects its angles. Therefore, the diagonal $\overline{SQ}$ bisects $\angle PSR$. Since a square is a rectangle, $\angle PSR$ is a right angle and m$\angle PSR = 90$. Therefore, m$\angle PSQ = \frac{1}{2}(90) = 45$.

Exercises

Writing About Mathematics

1. Ava said that the diagonals of a square separate the square into four congruent isosceles right triangles. Do you agree with Ava? Justify your answer.

2. Raphael said that a square could be defined as a quadrilateral that is both equiangular and equilateral. Do you agree with Raphael? Justify your answer.

Developing Skills

In 3–6, the diagonals of square $ABCD$ intersect at M.

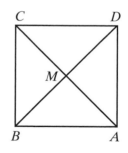

3. Prove that $\overline{AC}$ is the perpendicular bisector of $\overline{BD}$.

4. If $AC = 3x + 2$ and $BD = 7x - 10$, find AC, BD, AM, BM.

5. If $AB = a + b$, $BC = 2a$, and $CD = 3b - 5$, find AB, BC, CD, and DA.

6. If m$\angle AMD = x + 2y$ and m$\angle ABC = 2x - y$, find the values of x and y.

In 7–11, tell whether each conclusion follows from the given premises. If not, draw a counter-example.

7. If a quadrilateral is a square, then all sides are congruent.
If all sides of a quadrilateral are congruent then it is a rhombus.
Conclusion: If a quadrilateral is a square, then it is a rhombus.

8. A diagonal of a parallelogram separates the parallelogram into two congruent triangles.
A square is a parallelogram.
Conclusion: A diagonal of a square separates the square into two congruent triangles.

9. If a quadrilateral is a square, then all angles are right angles.
If a quadrilateral is a square, then it is a rhombus.
Conclusion: In a rhombus, all angles are right angles.

10. If a quadrilateral is a square, then it is a rectangle.
If a quadrilateral is a rectangle, then it is a parallelogram.
Conclusion: If a quadrilateral is a square, then it is a parallelogram.

11. If a quadrilateral is a square, then it is equilateral.
If a quadrilateral is a square, then it is a rectangle.
Conclusion: If a quadrilateral is equilateral, then it is a rectangle.

Applying Skills

12. Prove Theorem 10.19, "If one of the angles of a rhombus is a right angle, then the rhombus is a square."

13. Prove that the diagonals of a square are perpendicular to each other.

14. Prove that the diagonals of a square divide the square into four congruent isosceles right triangles.

15. Two line segments, $\overline{AEC}$ and $\overline{BED}$, are congruent. Each is the perpendicular bisector of the other. Prove that $ABCD$ is a square.

16. Prove that if the midpoints of the sides of a square are joined in order, another square is formed.

17. The vertices of quadrilateral $PQRS$ are $P(1, 1)$, $Q(4, -2)$, $R(7, 1)$, and $S(4, 4)$.

   a. Prove that the diagonals of the quadrilateral bisect each other.

   b. Prove that the diagonals of the quadrilateral are perpendicular to each other.

   c. Is the quadrilateral a square? Justify your answer.

   d. The vertices of $P'Q'R'S'$ are $P'(-1, 1)$, $Q'(2, 4)$, $R'(-1, 7)$, and $S'(-4, 4)$. Under what specific transformation is $P'Q'R'S'$ the image of $PQRS$?

18. The vertices of quadrilateral $ABCD$ are $A(-3, 2)$, $B(1, -2)$, $C(5, 2)$, and $D(1, 6)$.

   a. Prove that the diagonals of the quadrilateral bisect each other.

   b. Prove that the diagonals of the quadrilateral are perpendicular to each other.

   c. Is the quadrilateral a square? Justify your answer.

   d. The vertices of $A'B'C'D'$ are $A'(-6, 3)$, $B'(-2, -1)$, $C'(2, 3)$, and $D'(-2, 7)$. Under what specific transformation is $A'B'C'D'$ the image of $ABCD$?

10-7 THE TRAPEZOID

DEFINITION

A **trapezoid** is a quadrilateral in which two and only two sides are parallel.

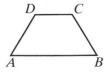

If $\overline{AB} \parallel \overline{DC}$ and $\overline{AD}$ is not parallel to $\overline{BC}$, then quadrilateral $ABCD$ is a trapezoid. The parallel sides, $\overline{AB}$ and $\overline{DC}$, are called the **bases** of the trapezoid; the nonparallel sides, $\overline{AD}$ and $\overline{BC}$, are called the **legs** of the trapezoid.

The Isosceles Trapezoid and Its Properties

DEFINITION

An **isosceles trapezoid** is a trapezoid in which the nonparallel sides are congruent.

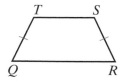

If $\overline{TS} \parallel \overline{QR}$ and $\overline{QT} \cong \overline{RS}$, then $QRST$ is an isosceles trapezoid. The angles whose vertices are the endpoints of a base are called **base angles**. Here, $\angle Q$ and $\angle R$ are one pair of base angles because Q and R are endpoints of base $\overline{QR}$. Also, $\angle T$ and $\angle S$ are a second pair of base angles because T and S are endpoints of base $\overline{TS}$.

Proving That a Quadrilateral Is an Isosceles Trapezoid

We prove that a quadrilateral is an isosceles trapezoid by showing that it satisfies the conditions of the definition of an isosceles trapezoid: only two sides are parallel and the nonparallel sides are congruent.

We may also prove special theorems for an isosceles trapezoid.

Theorem 10.20a | If a trapezoid is isosceles, then the base angles are congruent.

Given   Isosceles trapezoid $QRST$ with $\overline{QR} \parallel \overline{ST}$ and $\overline{TQ} \cong \overline{SR}$

Prove $\angle Q \cong \angle R$

Proof

Statements	**Reasons**
1. Through S, draw a line parallel to $\overline{QT}$ that intersects $\overline{QR}$ at P: $\overline{SP} \parallel \overline{QT}$	1. Through a given point, only one line can be drawn parallel to a given line.
2. $\overline{QR} \parallel \overline{ST}$	2. Given.
3. $QPST$ is a parallelogram.	3. A parallelogram is a quadrilateral with two pairs of parallel sides.
4. $\overline{QT} \cong \overline{SP}$	4. Opposite sides of a parallelogram are congruent.
5. $\overline{QT} \cong \overline{SR}$	5. Given.
6. $\overline{SP} \cong \overline{SR}$	6. Transitive property of congruence.

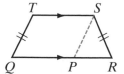

(Continued)	**Statements**	**Reasons**
	7. $\angle SPR \cong \angle R$	**7.** If two sides of a triangle are congruent, the angles opposite these sides are congruent.
	8. $\angle Q \cong \angle SPR$	**8.** If two parallel lines are cut by a transversal, the corresponding angles are congruent.
	9. $\angle Q \cong \angle R$	**9.** Transitive property of congruence.

We have proved Theorem 10.20a for $\angle Q \cong \angle R$ but $\angle S$ and $\angle T$ are also congruent base angles. We often refer to $\angle Q$ and $\angle R$ as the **lower base angles** and $\angle S$ and $\angle T$ as the **upper base angles**. The proof of this theorem for $\angle S$ and $\angle T$ is left to the student. (See exercise 15.)

Theorem 10.20b | If the base angles of a trapezoid are congruent, then the trapezoid is isosceles.

Given   Trapezoid $QRST$ with $\overline{QR} \parallel \overline{ST}$ and $\angle Q \cong \angle R$

Prove $\overline{QT} \cong \overline{RS}$

Strategy   Draw $\overline{SP} \parallel \overline{TQ}$. Prove $\angle SPR \cong \angle R$. Then use the converse of the isosceles triangle theorem.

The proof of this theorem is left to the student. (See exercise 16.) Theorems 10.20a and 10.20b can be written as a biconditional.

Theorem 10.20 | A trapezoid is isosceles if and only if the base angles are congruent.

We can also prove theorems about the diagonals of an isosceles trapezoid.

Theorem 10.21a | If a trapezoid is isosceles, then the diagonals are congruent.

Given   Isosceles trapezoid $ABCD$ with $\overline{AB} \parallel \overline{CD}$ and $\overline{AD} \cong \overline{BC}$

Prove $\overline{AC} \cong \overline{BD}$

Proof We will show $\triangle DAB \cong CBA$. It is given that in trapezoid $ABCD$, $\overline{AD} \cong \overline{BC}$. It is given that $ABCD$ is an isosceles trapezoid. In an isosceles trapezoid, base angles are congruent, so $\angle DAB \cong \angle CBA$. By the reflexive property, $\overline{AB} \cong \overline{AB}$. Therefore, $\triangle DAB \cong \triangle CBA$ by SAS. Corresponding parts of congruent triangles are congruent, so $\overline{AC} \cong \overline{BD}$.

Theorem 10.21b If the diagonals of a trapezoid are congruent, then the trapezoid is isosceles.

Given Trapezoid $ABCD$ with $\overline{AB} \parallel \overline{CD}$ and $\overline{AC} \cong \overline{BD}$

Prove $\overline{AD} \cong \overline{BC}$

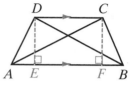

Strategy Draw $\overline{DE} \perp \overline{AB}$ and $\overline{CF} \perp \overline{AB}$. First prove that $\triangle DEB$ and $\triangle CFA$ are congruent by HL. Therefore, $\angle CAB \cong \angle DBA$. Now, prove that $\triangle ACB \cong \triangle BDA$ by SAS. Then $\overline{AD}$ and $\overline{BC}$ are congruent corresponding parts of congruent triangles.

The proof of this theorem is left to the student. (See exercise 17.) Theorems 10.21a and 10.21b can also be written as a biconditional.

Theorem 10.21 A trapezoid is isosceles if and only if the diagonals are congruent.

Recall that the median of a triangle is a line segment from a vertex to the midpoint of the opposite sides. A triangle has three medians. A trapezoid has only one median, and it joins two midpoints.

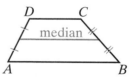

DEFINITION

The **median of a trapezoid** is a line segment whose endpoints are the midpoints of the nonparallel sides of the trapezoid.

We can prove two theorems about the median of a trapezoid.

Theorem 10.22 The median of a trapezoid is parallel to the bases.

Given Trapezoid $ABCD$ with $\overline{AB} \parallel \overline{CD}$, M the midpoint of $\overline{AD}$, and N the midpoint of $\overline{BC}$

Prove $\overline{MN} \parallel \overline{AB}$ and $\overline{MN} \parallel \overline{CD}$

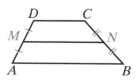

Proof We will use a coordinate proof.

Place the trapezoid so that the parallel sides are on horizontal lines. Place $\overline{AB}$ on the x-axis with $(0, 0)$ the coordinates of A and $(b, 0)$ the coordinates of B, and $b \neq 0$.

Place $\overline{CD}$ on a line parallel to the x-axis. Every point on a line parallel to the x-axis has the same y-coordinate. Let (c, e) be the coordinates of C and (d, e) be the coordinates of D, and $c, d, e \neq 0$.

Since M is the midpoint of $\overline{AD}$, the coordinates of M are

$$\left(\frac{0 + d}{2}, \frac{0 + e}{2} \right) = \left(\frac{d}{2}, \frac{e}{2} \right).$$

Since N is the midpoint of $\overline{BC}$, the coordinates of N are

$$\left(\frac{b + c}{2}, \frac{0 + e}{2} \right) = \left(\frac{b + c}{2}, \frac{e}{2} \right).$$

Points that have the same y-coordinate are on the same horizontal line. Therefore, $\overline{MN}$ is a horizontal line. All horizontal lines are parallel. Therefore, $\overline{MN} \parallel \overline{AB}$ and $\overline{MN} \parallel \overline{CD}$.   ◾

Theorem 10.23 | The length of the median of a trapezoid is equal to one-half the sum of the lengths of the bases.

Given   Trapezoid $ABCD$ with $\overline{AB} \parallel \overline{CD}$, M the midpoint of $\overline{AD}$, and N the midpoint of $\overline{CD}$

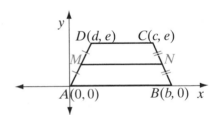

Prove $MN = \frac{1}{2}(AB + CD)$

Strategy   Use a coordinate proof. Let the coordinates of A, B, C, D, M, and N be those used in the proof of Theorem 10.22. The length of a horizontal line segment is the absolute value of the difference of the x-coordinates of the endpoints.

The proof of this theorem is left to the student. (See exercise 19.)

EXAMPLE 1

The coordinates of the vertices of $ABCD$ are $A(0, 0)$, $B(4, -1)$, $C(5, 2)$, and $D(2, 6)$. Prove that $ABCD$ is a trapezoid.

Proof

$$\text{Slope of } \overline{AB} = \frac{-1 - 0}{4 - 0} = \frac{-1}{4} = -\frac{1}{4}$$

$$\text{Slope of } \overline{BC} = \frac{2 - (-1)}{5 - 4} = \frac{3}{1} = 3$$

$$\text{Slope of } \overline{CD} = \frac{6 - 2}{2 - 5} = \frac{4}{-3} = -\frac{4}{3}$$

$$\text{Slope of } \overline{DA} = \frac{0 - 6}{0 - 2} = \frac{-6}{-2} = 3$$

The slopes of $\overline{BC}$ and $\overline{DA}$ are equal. Therefore, $\overline{BC}$ and $\overline{DA}$ are parallel. The slopes of $\overline{AB}$ and $\overline{CD}$ are not equal. Therefore, $\overline{AB}$ and $\overline{CD}$ are not parallel. Because quadrilateral $ABCD$ has only one pair of parallel sides, it is a trapezoid.

EXAMPLE 2

In quadrilateral $ABCD$, m$\angle A = 105$, m$\angle B = 75$, and m$\angle C = 75$.

a. Is $ABCD$ a parallelogram? Justify your answer.

b. Is $ABCD$ a trapezoid? Justify your answer.

c. Is $ABCD$ an isosceles trapezoid? Justify your answer.

d. If $AB = x$, $BC = 2x - 1$, $CD = 3x - 8$, and $DA = x + 1$, find the measure of each side of the quadrilateral.

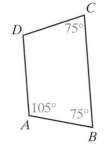

Solution   **a.** One pair of opposite angles of $ABCD$ are $\angle A$ and $\angle C$. Since these angles are not congruent, the quadrilateral is not a parallelogram. The quadrilateral does not have two pairs of parallel sides.

b. $\angle A$ and $\angle B$ are interior angles on the same side of transversal $\overline{AB}$ and they are supplementary. Therefore, $\overline{AD}$ and $\overline{BC}$ are parallel. The quadrilateral has only one pair of parallel sides and is therefore a trapezoid.

c. Because $\angle B$ and $\angle C$ are congruent base angles of the trapezoid, the trapezoid is isosceles.

d. The congruent legs of the trapezoid are $\overline{AB}$ and $\overline{CD}$.

$$AB = CD$$
$$x = 3x - 8$$
$$-2x = -8$$
$$x = 4$$

$AB = x$	$BC = 2x - 1$	$CD = 3x - 8$	$DA = x + 1$
$= 4$	$= 2(4) - 1$	$= 3(4) - 8$	$= 4 + 1$
	$= 8 - 1$	$= 12 - 8$	$= 5$
	$= 7$	$= 4$	

Exercises

Writing About Mathematics

1. Can a trapezoid have exactly one right angle? Justify your answer.

2. Can a trapezoid have three obtuse angles? Justify your answer.

Developing Skills

In 3–8, $ABCD$ is an isosceles trapezoid, $\overline{AB} \parallel \overline{DC}$, and $\overline{AD} \cong \overline{BC}$.

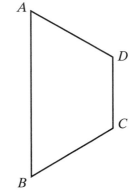

3. If $m\angle ADC = 110$, find: **a.** $m\angle BCD$ **b.** $m\angle ABC$ **c.** $m\angle DAB$.

4. If $AD = 3x + 7$ and $BC = 25$, find the value of x.

5. If $AD = 2y - 5$ and $BC = y + 3$, find AD.

6. If $m\angle DAB = 4x - 5$ and $m\angle ABC = 3x + 15$, find the measure of each angle of the trapezoid.

7. If $m\angle ADC = 4x + 20$ and $m\angle DAB = 8x - 20$, find the measure of each angle of the trapezoid.

8. The perimeter of $ABCD$ is 55 centimeters. If $AD = DC = BC$ and $AB = 2AD$, find the measure of each side of the trapezoid.

In 9–14, determine whether each statement is true or false. Justify your answer with an appropriate definition or theorem, or draw a counterexample.

9. In an isosceles trapezoid, nonparallel sides are congruent.

10. In a trapezoid, at most two sides can be congruent.

11. In a trapezoid, the base angles are always congruent.

12. The diagonals of a trapezoid are congruent if and only if the nonparallel sides of the trapezoid are congruent.

13. The sum of the measures of the angles of a trapezoid is 360°.

14. In a trapezoid, there are always two pairs of supplementary angles.

Applying Skills

15. In Theorem 10.20a, we proved that the lower base angles of $QRST$, $\angle Q$ and $\angle R$, are congruent. Use this fact to prove that the upper base angles of $QRST$, $\angle S$ and $\angle T$, are congruent.

16. Prove Theorem 10.20b, "If the base angles of a trapezoid are congruent, then the trapezoid is isosceles."

17. a. Prove Theorem 10.21b, "If the diagonals of a trapezoid are congruent, then the trapezoid is isosceles."

b. Why can't Theorem 10.21b be proved using the same method as in 10.21a?

18. Prove Theorem 10.23, "The length of the median of a trapezoid is equal to one-half the sum of the lengths of the bases."

19. Let the coordinates of the vertices of $ABCD$ be $A(2, -6)$, $B(6, 2)$, $C(0, 8)$, and $D(-2, 4)$.

    a. Find the slopes of $\overline{AB}$, $\overline{BC}$, $\overline{CD}$, and $\overline{DA}$.

    b. Prove that $ABCD$ is a trapezoid.

    c. Find the coordinates of E and F, the midpoints of the nonparallel sides.

    d. Find the slope of $\overline{EF}$.

    e. Show that the median is parallel to the bases.

20. Prove that the diagonals of a trapezoid do not bisect each other.

21. Prove that if the diagonals of a trapezoid are unequal, then the trapezoid is not isosceles.

22. Prove that if a quadrilateral does not have a pair of consecutive angles that are supplementary, the quadrilateral is not a trapezoid.

10-8 AREAS OF POLYGONS

DEFINITION

The **area of a polygon** is the unique real number assigned to any polygon that indicates the number of non-overlapping square units contained in the polygon's interior.

We know that the area of a rectangle is the product of the lengths of two adjacent sides. For example, the rectangle to the right contains mn unit squares or has an area of $m \times n$ square units.

In rectangle $ABCD$, $AB = b$, the length of the base, and $BC = h$, the length of the altitude, a line segment perpendicular to the base.

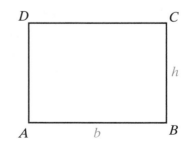

$$\text{Area of } ABCD = (AB)(BC) = bh$$

The formula for the area of every other polygon can be derived from this formula. In order to derive the formulas for the areas of other polygons from the formula for the area of a rectangle, we will use the following postulate.

Postulate 10.1 | The areas of congruent figures are equal.

EXAMPLE 1

$ABCD$ is a parallelogram and E is a point on $\overline{AB}$ such that $\overline{DE} \perp \overline{AB}$. Prove that if $DC = b$ and $DE = h$, the area of parallelogram $ABCD = bh$.

Proof Let F be a point on $\overleftrightarrow{AB}$ such that $\overline{CF} \perp \overleftrightarrow{AB}$. Since two lines perpendicular to the same line are parallel, $\overline{DE} \parallel \overline{CF}$. Therefore, $EFCD$ is a parallelogram with a right angle, that is, a rectangle.

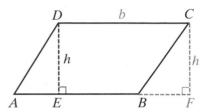

Perpendicular lines intersect to form right angles. Therefore, $\angle DEA$ and $\angle CFB$ are right angles and $\triangle DEA$ and $\triangle CFB$ are right triangles. In these right triangles, $\overline{AD} \cong \overline{BC}$ and $\overline{DE} \cong \overline{CF}$ because the opposite sides of a parallelogram are congruent. Therefore, $\triangle DEA \cong \triangle CFB$ by HL.

The base of rectangle $EFCD$ is DC. Since $ABCD$ is a parallelogram, $DC = AB = b$. The height is $DE = h$. Therefore:

$$\text{Area of rectangle } EFCD = (DC)(DE) = bh$$

$$\text{Area of parallelogram } ABCD = \text{area of } \triangle AED + \text{area of trapezoid } EBCD$$

$$\text{Area of rectangle } EFCD = \text{area of } \triangle BFC + \text{area of trapezoid } EBCD$$

$$\text{Area of } \triangle AED = \text{area of } \triangle BFC$$

Therefore, the area of parallelogram $ABCD$ is equal to the area of rectangle $EFCD$ or bh.

Exercises

Writing About Mathematics

1. If $ABCD$ and $PQRS$ are rectangles with $AB = PQ$ and $BC = QR$, do $ABCD$ and $PQRS$ have equal areas? Justify your answer.

2. If $ABCD$ and $PQRS$ are parallelograms with $AB = PQ$ and $BC = QR$, do $ABCD$ and $PQRS$ have equal areas? Justify your answer.

Developing Skills

3. Find the area of a rectangle whose vertices are $(0, 0)$, $(8, 0)$, $(0, 5)$, and $(8, 5)$.

4. a. Draw $\triangle ABC$. Through C, draw a line parallel to $\overleftrightarrow{AB}$, and through B, draw a line parallel to $\overleftrightarrow{AC}$. Let the point of intersection of these lines be D.

   b. Prove that $\triangle ABC \cong \triangle DBC$.

   c. Let E be a point on $\overleftrightarrow{AB}$ such that $\overline{CE} \perp \overleftrightarrow{AB}$. Let $AB = b$ and $CE = h$. Use the results of Example 1 to prove that the area of $\triangle ABC = \frac{1}{2}bh$.

5. Find the area of a triangle whose vertices are $(-1, -1), (7, -1)$, and $(3, 5)$.

6. a. Draw trapezoid $ABCD$. Let E and F be points on $\overleftrightarrow{AB}$ such that $\overline{CE} \perp \overleftrightarrow{AB}$ and $\overline{DF} \perp \overleftrightarrow{AB}$.

    b. Prove that $CE = DF$.

    c. Let $AB = b_1, CD = b_2$, and $CE = DF = h$. Prove that the area of trapezoid $ABCD$ is $\frac{h}{2}(b_1 + b_2)$.

7. Find the area of a trapezoid whose vertices are $(-2, 1), (2, 2), (2, 7)$, and $(-2, 4)$.

8. a. Draw rhombus $ABCD$. Let the diagonals of $ABCD$ intersect at E.

    b. Prove that $\triangle ABE \cong \triangle CBE \cong \triangle CDE \cong \triangle ADE$.

    c. Let $AC = d_1$ and $BD = d_2$. Prove that the area of $\triangle ABE = \frac{1}{8}d_1 d_2$.

    d. Prove that the area of rhombus $ABCD = \frac{1}{2}d_1 d_2$.

9. Find the area of a rhombus whose vertices are $(0, 2), (2, -1), (4, 2)$, and $(2, 5)$.

Applying Skills

10. a. The vertices of $ABCD$ are $A(-2, 1), B(2, -2), C(6, 1)$, and $D(2, 4)$. Prove that $ABCD$ is a rhombus.

    b. Find the area of $ABCD$.

11. a. The vertices of $ABCD$ are $A(-2, 2), B(1, -1), C(4, 2)$, and $D(1, 5)$. Prove that $ABCD$ is a square.

    b. Find the area of $ABCD$.

    c. Find the coordinates of the vertices of $A'B'C'D'$, the image of $ABCD$ under a reflection in the y-axis.

    d. What is the area of $A'B'C'D'$?

    e. Let E and F be the coordinates of the fixed points under the reflection in the y-axis. Prove that $AEA'F$ is a square.

    f. What is the area of $AEA'F$?

12. $\overline{KM}$ is a diagonal of parallelogram $KLMN$. The area of $\triangle KLM$ is 94.5 square inches.

    a. What is the area of parallelogram $KLMN$?

    b. If $MN = 21.0$ inches, what is the length of $\overline{NP}$, the perpendicular line segment from N to $\overline{KL}$?

13. $ABCD$ is a parallelogram and S and T are two points on $\overleftrightarrow{CD}$. Prove that the area of $\triangle ABS$ is equal to the area of $\triangle ABT$.

14. The vertices of $ABCD$ are $A(1, -2), B(4, 2), C(4, 6)$, and $D(-4, 2)$. Draw the polygon on graph paper and draw the diagonal, $\overline{DB}$.

    a. Find the area of $\triangle DBC$.

    b. Find the area of $\triangle DBA$.

    c. Find the area of polygon $ABCD$.

15. The altitude to a base $\overline{AB}$ of trapezoid $ABCD$ is $\overline{DH}$ and the median is $\overline{EF}$. Prove that the area of a trapezoid is equal to the product, $(DH)(EF)$.

16. The vertices of polygon $PQRS$ are $P(1, 2)$, $Q(9, 1)$, $R(8, 4)$, and $S(3, 4)$. Draw a vertical line through P that intersects a horizontal line through S at M and a horizontal line through Q at N. Draw a vertical line through Q that intersects a horizontal line through R at L.

   a. Find the area of rectangle $NQLM$.

   b. Find the areas of $\triangle PNQ$, $\triangle QLR$, and $\triangle SMP$.

   c. Find the area of polygon $PQRS$.

17. Find the area of polygon $ABCD$ if the coordinates of the vertices are $A(5, 0)$, $B(8, 2)$, $C(8, 8)$, and $D(0, 4)$.

Hands-On Activity

1. Draw square $ABCD$ with diagonals that intersect at E. Let $AC = d$. Represent BD, AE, EC, BE, and ED in terms of d.

2. Express the area of $\triangle ACD$ and of $\triangle ACB$ in terms of d.

3. Find the area of $ABCD$ in terms of d.

4. Let $AB = s$. Express the area of $ABCD$ in terms of s.

5. Write an equation that expresses the relationship between d and s.

6. Solve the equation that you wrote in step 5 for d in terms of s.

7. Use the result of step 6 to express the length of the hypotenuse of an isosceles right triangle in terms of the length of a leg.

CHAPTER SUMMARY

Definitions to Know

- A **parallelogram** is a quadrilateral in which two pairs of opposite sides are parallel.
- The **distance between two parallel lines** is the length of the perpendicular from any point on one line to the other line.
- A **rectangle** is a parallelogram containing one right angle.
- A **rhombus** is a parallelogram that has two congruent consecutive sides.
- A **square** is a rectangle that has two congruent consecutive sides.
- A **trapezoid** is a quadrilateral in which two and only two sides are parallel.
- An **isosceles trapezoid** is a trapezoid in which the nonparallel sides are congruent.
- The **area of a polygon** is the unique real number assigned to any polygon that indicates the number of non-overlapping square units contained in the polygon's interior.

Postulate	**10.1** The areas of congruent figures are equal.

Theorems and Corollaries

10.1   A diagonal divides a parallelogram into two congruent triangles.

10.1a Opposite sides of a parallelogram are congruent.

10.1b Opposite angles of a parallelogram are congruent.

10.2   Two consecutive angles of a parallelogram are supplementary.

10.3   The diagonals of a parallelogram bisect each other.

10.4   If both pairs of opposite sides of a quadrilateral are congruent, the quadrilateral is a parallelogram.

10.5   If one pair of opposite sides of a quadrilateral is both congruent and parallel, the quadrilateral is a parallelogram.

10.6   If both pairs of opposite angles of a quadrilateral are congruent, the quadrilateral is a parallelogram.

10.7   If the diagonals of a quadrilateral bisect each other, the quadrilateral is a parallelogram.

10.8   All angles of a rectangle are right angles.

10.9   The diagonals of a rectangle are congruent.

10.10 If a quadrilateral is equiangular, then it is a rectangle.

10.11 If the diagonals of a parallelogram are congruent, the parallelogram is a rectangle.

10.12 All sides of a rhombus are congruent.

10.13 The diagonals of a rhombus are perpendicular to each other.

10.14 The diagonals of a rhombus bisect its angles.

10.15 If a quadrilateral is equilateral, then it is a rhombus.

10.16 If the diagonals of a parallelogram are perpendicular to each other, the parallelogram is a rhombus.

10.17 A square is an equilateral quadrilateral.

10.18 A square is a rhombus.

10.19 If one of the angles of a rhombus is a right angle, then the rhombus is a square.

10.20 A trapezoid is isosceles if and only if the base angles are congruent.

10.21 A trapezoid is isosceles if and only if the diagonals are congruent.

10.22 The median of a trapezoid is parallel to the bases.

10.23 The length of the median of a trapezoid is equal to one-half the sum of the lengths of the bases.

VOCABULARY

10-1 Quadrilateral • Consecutive vertices of a quadrilateral • Adjacent vertices of a quadrilateral • Consecutive sides of a quadrilateral • Adjacent sides of a quadrilateral • Opposite sides of a quadrilateral • Consecutive angles of a quadrilateral • Opposite angles of a quadrilateral • Diagonal of a quadrilateral

10-2 Parallelogram • Distance between two parallel lines

REVIEW EXERCISES

1. Is it possible to draw a parallelogram that has only one right angle? Explain why or why not.

2. The measure of two consecutive angles of a parallelogram are represented by $3x$ and $5x - 12$. Find the measure of each angle of the parallelogram.

3. Point P is on side $\overline{BC}$ of rectangle $ABCD$. If $AB = BP$, find m$\angle APC$.

4. Quadrilateral $ABCD$ is a parallelogram, M is a point on $\overline{AB}$, and N is a point on $\overline{CD}$. If $\overline{DM} \perp \overline{AB}$ and $\overline{BN} \perp \overline{CD}$, prove that $\triangle AMD \cong \triangle CNB$.

5. Quadrilateral $ABCD$ is a parallelogram, M is a point on $\overline{AB}$, and N is a point on $\overline{CD}$. If $\overline{CM} \parallel \overline{AN}$, prove that $\triangle AND \cong \triangle CMB$.

6. The diagonals of rhombus $ABCD$ intersect at E.

   a. Name three angles that are congruent to $\angle EAB$.

   b. Name four angles that are the complements of $\angle EAB$.

7. The diagonals of parallelogram $PQRS$ intersect at T. If $\triangle PTQ$ is isosceles with $\angle PTQ$ the vertex angle, prove that $PQRS$ is a rectangle.

8. Point P is the midpoint of side $\overline{BC}$ of rectangle $ABCD$. Prove that $\overline{AP} \cong \overline{DP}$.

9. A regular polygon is a polygon that is both equilateral and equiangular.

   a. Is an equilateral triangle a regular polygon? Justify your answer.

   b. Is an equilateral quadrilateral a regular polygon? Justify your answer.

10. The diagonals of rhombus $ABEF$ intersect at G and the diagonals of rhombus $BCDE$ intersect at H. Prove that $BHEG$ is a rectangle.

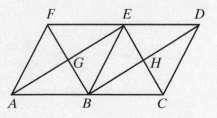

11. The diagonals of square *ABEF* intersect at *G* and the diagonals of square *BCDE* intersect at *H*. Prove that *BHEG* is a square.

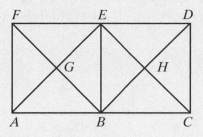

12. Points $A(-3, -2)$, $B(3, -2)$, $C(5, 3)$, and $D(-1, 3)$ are the vertices of quadrilateral *ABCD*.

a. Plot these points on graph paper and draw the quadrilateral.

b. What kind of quadrilateral is *ABCD*? Justify your answer.

c. Find the area of quadrilateral *ABCD*.

13. The vertices of quadrilateral *DEFG* are $D(1, -1)$, $E(4, 1)$, $F(1, 3)$, and $G(-2, 1)$.

a. Is the quadrilateral a parallelogram? Justify your answer.

b. Is the quadrilateral a rhombus? Justify your answer.

c. Is the quadrilateral a square? Justify your answer.

d. Explain how the diagonals can be used to find the area of the quadrilateral.

14. The coordinates of the vertices of quadrilateral *ABCD* are $A(0, 0)$, $B(2b, 0)$, $C(2b + 2d, 2a)$, and $D(2d, 2a)$.

a. Prove that *ABCD* is a parallelogram.

b. The midpoints of the sides of *ABCD* are *P*, *Q*, *R*, and *S*. Find the coordinates of these midpoints.

c. Prove that *PQRS* is a parallelogram.

15. The area of a rectangle is 12 square centimeters and the perimeter is 16 centimeters.

a. Write an equation for the area of the rectangle in terms of the length, x, and the width, y.

b. Write an equation for the perimeter of the rectangle in terms of the length, x, and the width, y.

c. Solve the equations that you wrote in **a** and **b** to find the length and the width of the rectangle.

16. Each of the four sides of quadrilateral *ABCD* is congruent to the corresponding side of quadrilateral *PQRS* and $\angle A \cong \angle P$. Prove that *ABCD* and *PQRS* are congruent quadrilaterals or draw a counterexample to prove that they may not be congruent.

Exploration

In Chapter 5 we found that the perpendicular bisectors of the sides of a triangle intersect in a point called the circumcenter. Do quadrilaterals also have circumcenters?

 In this activity, you will explore the perpendicular bisectors of the sides of quadrilaterals. You may use compass and straightedge or geometry software.

a. Construct the perpendicular bisectors of two adjacent sides of each of the following quadrilaterals.

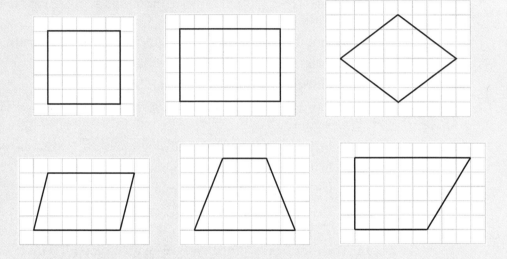

b. Construct the third and fourth perpendicular bisectors of the sides of each of the quadrilaterals. For which of the above quadrilaterals is the intersection of the first two perpendicular bisectors the same point as the intersection of the last two perpendicular bisectors?

c. When all four perpendicular bisectors of a quadrilateral intersect in the same point, that point is called the circumcenter. Of the specific quadrilaterals studied in this chapter, which types do you expect to have a circumcenter and which types do you expect not to have a circumcenter?

d. For each of the quadrilaterals above that has a circumcenter, place the point of your compass on the circumcenter and the pencil on any of the vertices. Draw a circle.

e. Each circle drawn in **d** is called the **circumcircle** of the polygon. Based on your observations, write a definition of circumcircle.

CUMULATIVE REVIEW

Part I

Answer all questions in this part. Each correct answer will receive 2 credits. No partial credit will be allowed.

1. The measure of an exterior angle at vertex B of isosceles $\triangle ABC$ is $70°$. The measure of a base angle of the triangle is

(1) $110°$ (2) $70°$ (3) $55°$ (4) $35°$

2. The measure of $\angle A$ is 12 degrees less than twice the measure of its complement. The measure of the $\angle A$ is

(1) $34°$ (2) $56°$ (3) $64°$ (4) $116°$

3. The slope of the line determined by $A(2, -3)$ and $B(-1, 3)$ is

(1) -2 (2) $-\frac{1}{2}$ (3) $\frac{1}{2}$ (4) 2

4. What is the slope of a line that is parallel to the line whose equation is $3x + y = 5$?

(1) -3 (2) $-\frac{5}{3}$ (3) $\frac{5}{3}$ (4) 3

5. The coordinates of the image of $A(3, -2)$ under a reflection in the x-axis are

(1) $(3, 2)$ (2) $(-3, 2)$ (3) $(-3, -2)$ (4) $(-2, 3)$

6. The measures of two sides of a triangle are 8 and 12. The measure of the third side *cannot* be

(1) 16 (2) 12 (3) 8 (4) 4

7. The line segment $\overline{BD}$ is the median and the altitude of $\triangle ABC$. Which of the following statements must be false?

(1) $\overline{BD}$ bisects $\overline{AC}$.   (3) $m\angle A = 90$

(2) $\triangle BDA$ is a right triangle.   (4) B is equidistant from A and C.

8. What is the equation of the line through $(0, -1)$ and perpendicular to the line whose equation is $y = 2x + 5$?

(1) $y = 2x - 1$ (3) $2y - 1 = x$

(2) $x + 2y + 2 = 0$ (4) $2x + y + 2 = 0$

9. Which of the following transformations is not a direct isometry?

(1) line reflection   (3) translation

(2) point reflection   (4) rotation

10. If $\overline{AC} \cong \overline{DF}$ and $\angle A \cong \angle D$, which of the following is not sufficient to prove that $\triangle ABC \cong \triangle DEF$?

(1) $\overline{AB} \cong \overline{DE}$ (2) $\overline{BC} \cong \overline{EF}$ (3) $\angle C \cong \angle F$ (4) $\angle B \cong \angle E$

Part II

Answer all questions in this part. Each correct answer will receive 2 credits. Clearly indicate the necessary steps, including appropriate formula substitutions, diagrams, graphs, charts, etc. For all questions in this part, a correct numerical answer with no work shown will receive only 1 credit.

11. The measures of the angles of a triangle can be represented by $3x, 4x + 5$, and $5x - 17$. Find the measure of each angle of the triangle.

12. In the diagram, $\overleftrightarrow{AB} \parallel \overleftrightarrow{CD}$. Prove that

$m\angle x + m\angle y = m\angle z.$

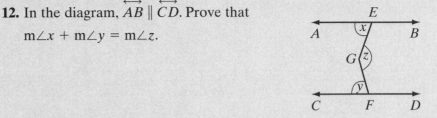

Part III

Answer all questions in this part. Each correct answer will receive 4 credits. Clearly indicate the necessary steps, including appropriate formula substitutions, diagrams, graphs, charts, etc. For all questions in this part, a correct numerical answer with no work shown will receive only 1 credit.

13. The bases, $\overline{AB}$ and $\overline{DE}$, of two isosceles triangles, $\triangle ABC$ and $\triangle DEF$, are congruent. If $\angle CAB \cong \angle FDE$, prove that $\triangle ABC \cong \triangle DEF$.

14. Points A, B, C, and D are on a circle with center at O and diameter $\overline{AOB}$. $ABCD$ is a trapezoid with $\overline{AB} \parallel \overline{CD}$ and $\overline{AO} \cong \overline{CD}$. Prove that $\overline{OC}$ and $\overline{OD}$ separate the trapezoid into three equilateral triangles. (*Hint:* All radii of a circle are congruent.)

Part IV

Answer all questions in this part. Each correct answer will receive 6 credits. Clearly indicate the necessary steps, including appropriate formula substitutions, diagrams, graphs, charts, etc. For all questions in this part, a correct numerical answer with no work shown will receive only 1 credit.

15. The vertices of quadrilateral $KLMN$ are $K(-3, 1)$, $L(2, 0)$, $M(6, 4)$, and $N(2, 6)$. Show that $KLMN$ is a trapezoid.

16. The coordinates of the vertices of $\triangle ABC$ are $A(2, 2)$, $B(4, 0)$, and $C(4, 2)$. Find the coordinates of the vertices of $\triangle A'B'C'$, the image of $\triangle ABC$ under the composition $r_{y\text{-axis}} \circ r_{y=x}$.

THE GEOMETRY OF THREE DIMENSIONS

Bonaventura Cavalieri (1598–1647) was a follower of Galileo and a mathematician best known for his work on areas and volumes. In this aspect, he proved to be a forerunner of the development of integral calculus. His name is associated with *Cavalieri's Principle* which is a fundamental principle for the determination of the volume of a solid.

Cavalieri's Principle can be stated as follows:

■ Given two geometric solids and a plane, if every plane parallel to the given plane that intersects both solids intersects them in surfaces of equal areas, then the volumes of the two solids are equal.

This means that two solids have equal volume when their corresponding cross-sections are in all cases equal.

11-1 POINTS, LINES, AND PLANES

In this text, we have been studying points and lines in a plane, that is, the geometry of two dimensions. But the world around us is three-dimensional. The geometry of three dimensions is called **solid geometry**. We begin this study with some postulates that we can accept as true based on our observations.

We know that two points determine a line. How many points are needed to determine a plane? A table or chair that has four legs will sometimes be unsteady on a flat surface. But a tripod or a stool with three legs always sits firmly. This observation suggests the following postulate.

Postulate 11.1

There is one and only one plane containing three non-collinear points.

For a set of the three non-collinear points that determine a plane, each pair of points determines a line and all of the points on that line are points of the plane.

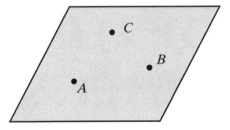

Postulate 11.2

A plane containing any two points contains all of the points on the line determined by those two points.

These two postulates make it possible for us to prove the following theorems.

Theorem 11.1

There is exactly one plane containing a line and a point not on the line.

Given Line *l* and point *P* not on *l*.

Prove There is exactly one plane containing *l* and *P*.

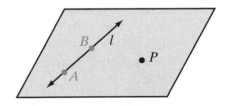

Proof Choose two points *A* and *B* on line *l*. The three points, *A*, *B*, and *P*, determine one and only one plane. If the two points *A* and *B* on line *l* are on the plane, then all of the points of *l* are on the plane, that is, the plane contains line *l*. Therefore, there is exactly one plane that contains the given line and point.

Theorem 11.2 | If two lines intersect, then there is exactly one plane containing them.

This theorem can be stated in another way:

Theorem 11.2 | Two intersecting lines determine a plane.

Given Lines *l* and *m* intersecting at point *P*.

Prove There is exactly one plane containing *l* and *m*.

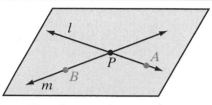

Proof Choose two points, *A* on line *l* and *B* on line *m*. The three points, *A*, *B*, and *P*, determine one and only one plane. A plane containing any two points contains all of the points on the line determined by those two points. Since the two points *A* and *P* on line *l* are on the plane, then all of the points of *l* are on the plane. Since the two points *B* and *P* on line *m* are on the plane, then all of the points of *m* are on the plane. Therefore, there is exactly one plane that contains the given intersecting lines. ∎

The definition of parallel lines gives us another set of points that must lie in a plane.

DEFINITION _____

Parallel lines in space are lines in the same plane that have no points in common.

This definition can be written as a biconditional: Two lines are parallel if and only if they are coplanar and have no points in common. If $\overleftrightarrow{AB}$ and $\overleftrightarrow{CD}$ are parallel lines in space, then they determine a plane if and only if they are two distinct lines.

We have seen that intersecting lines lie in a plane and parallel lines lie in a plane. For example, in the diagram, $\overleftrightarrow{AB} \parallel \overleftrightarrow{CD}$ so $\overleftrightarrow{AB}$ and $\overleftrightarrow{CD}$ lie in a plane. Also, $\overleftrightarrow{AB}$ and $\overleftrightarrow{BF}$ are intersecting lines so they lie in a plane. But there are some pairs of lines that do not intersect and are not parallel. In the diagram, $\overleftrightarrow{AB}$ and $\overleftrightarrow{CG}$ are neither intersecting nor parallel lines. $\overleftrightarrow{AB}$ and $\overleftrightarrow{CG}$ are called *skew lines*. $\overleftrightarrow{BF}$ and $\overleftrightarrow{EH}$ are another pair of skew lines.

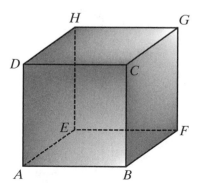

> **DEFINITION** _____
> **Skew lines** are lines in space that are neither parallel nor intersecting.

EXAMPLE I

Does a triangle determine a plane?

Solution A triangle consists of three line segments and the three non-collinear points that are the endpoints of each pair of segments. Three non-collinear points determine a plane. That plane contains all of the points of the lines determined by the points. Therefore, a triangle determines a plane.

Exercises

Writing About Mathematics

1. Joel said that another definition for skew lines could be two lines that do not lie in the same plane. Do you agree with Joel? Explain why or why not.

2. Angelina said that if $\overleftrightarrow{AC}$ and $\overleftrightarrow{BD}$ intersect at a point, then $A, B, C,$ and D lie in a plane and form a quadrilateral. Do you agree with Angelina? Explain why or why not.

Developing Skills

3. $\overleftrightarrow{AB}$ is parallel to $\overleftrightarrow{CD}$ and $AB \neq CD$. Prove that $A, B, C,$ and D must lie in a plane and form a trapezoid.

4. $\overleftrightarrow{AB} \parallel \overleftrightarrow{CD}$ and $\overleftrightarrow{AD} \parallel \overleftrightarrow{BC}$. Prove that $A, B, C,$ and D must lie in a plane and form a parallelogram.

5. $\overline{AEC} \cong \overline{BED}$ and each segment is the perpendicular bisector of the other. Prove that $A, B,$ $C,$ and D must lie in a plane and form a square.

In 6–9, use the diagram at the right.

6. Name two pairs of intersecting lines.

7. Name two pairs of skew lines.

8. Name two pairs of parallel lines.

9. Which pairs of lines that you named in exercises 6, 7, and 8 are not lines in the same plane?

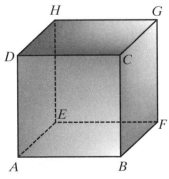

10. Let p represent "Two lines are parallel."

Let q represent "Two lines are coplanar."

Let r represent "Two lines have no point in common."

a. Write the biconditional "Two lines are parallel if and only if they are coplanar and have no points in common" in terms of p, q, r, and logical symbols.

b. The biconditional is true. Show that q is true when p is true.

Applying Skills

11. A photographer wants to have a steady base for his camera. Should he choose a base with four legs or with three? Explain your answer.

12. Ken is building a tool shed in his backyard. He begins by driving four stakes into the ground to be the corners of a rectangular floor. He stretches strings from two of the stakes to the opposite stakes and adjusts the height of the stakes until the strings intersect. Explain how the strings assure him that the four stakes will all be in the same plane.

13. Each of four equilateral triangles has a common side with each of the three other triangles and form a solid called a *tetrahedron*. Prove that the triangles are congruent.

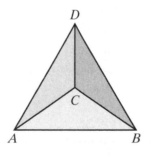

11-2 PERPENDICULAR LINES AND PLANES

Look at the floor, walls, and ceiling of the classroom. Each of these surfaces can be represented by a plane. Many of these planes intersect. For example, each wall intersects the ceiling and each wall intersects the floor. Each intersection can be represented by a line segment. This observation allows us to state the following postulate.

Postulate 11.3 | If two planes intersect, then they intersect in exactly one line.

The Angle Formed by Two Intersecting Planes

Fold a piece of paper. The part of the paper on one side of the crease represents a half-plane and the crease represents the edge of the half-plane. The folded paper forms a dihedral angle.

DEFINITION _____

A **dihedral angle** is the union of two half-planes with a common edge.

Each half-plane of a dihedral angle can be compared to a ray of an angle in a plane (or a **plane angle**) and the edge to the vertex of a plane angle. If we choose some point on the edge of a dihedral angle and draw, from this point, a ray in each half-plane perpendicular to the edge, we have drawn a plane angle.

DEFINITION _____

The **measure of a dihedral angle** is the measure of the plane angle formed by two rays each in a different half-plane of the angle and each perpendicular to the common edge at the same point of the edge.

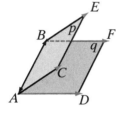

A plane angle whose measure is the same as that of the dihedral angle can be drawn at any points on the edge of the dihedral angle. Each plane angle of a dihedral angle has the same measure. In the figure, planes p and q intersect at $\overleftrightarrow{AB}$. In plane p, $\overrightarrow{AC} \perp \overleftrightarrow{AB}$ and in plane q, $\overrightarrow{AD} \perp \overleftrightarrow{AB}$. The measure of the dihedral angle is equal to the measure of $\angle CAD$. Also, in plane p, $\overrightarrow{BE} \perp \overleftrightarrow{AB}$ and in plane q, $\overrightarrow{BF} \perp \overleftrightarrow{AB}$. The measure of the dihedral angle is equal to the measure of $\angle EBF$.

DEFINITION _____

Perpendicular planes are two planes that intersect to form a right dihedral angle.

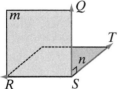

In the diagram, $\angle QST$ is a right angle. In plane m, $\overrightarrow{SQ} \perp \overrightarrow{SR}$ and in plane n, $\overrightarrow{ST} \perp \overrightarrow{SR}$. The dihedral angle formed by half-planes of planes m and n with edge $\overleftrightarrow{RS}$ has the same measure as $\angle QST$. Therefore, the dihedral angle is a right angle, and $m \perp n$.

The floor and a wall of a room usually form a right dihedral angle. Look at the line that is the intersection of two adjacent walls of the classroom. This line intersects the ceiling in one point and intersects the floor in one point. This observation suggests the following theorem.

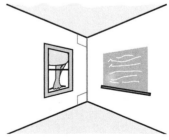

Theorem 11.3

If a line not in a plane intersects the plane, then it intersects in exactly one point.

Given Line *l* is not in plane *p* and *l* intersects *p*.

Prove Line *l* intersects *p* in exactly one point.

Proof Use an indirect proof.

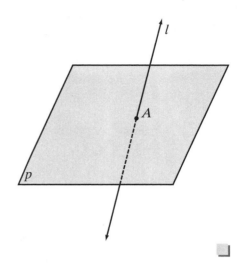

Assume that line *l* intersects the plane in two points. Then all of the points on line *l* lie in plane *p*, that is, the line lies in the plane. Because this contradicts the hypothesis that line *l* is not in plane *p*, the assumption must be false. A line not in a plane that intersects the plane, intersects it in exactly one point.

Again, look at the corner of the classroom in the figure on the previous page. The line that is the intersection of two adjacent walls intersects the ceiling so that the line is perpendicular to any line in the ceiling through the point of intersection. We say that this line is perpendicular to the plane of the ceiling.

DEFINITION _____

A line is perpendicular to a plane if and only if it is perpendicular to each line in the plane through the intersection of the line and the plane.

A **plane is perpendicular to a line** if the line is perpendicular to the plane.

It is easy to demonstrate that a line that is perpendicular to one line in a plane may not be perpendicular to the plane. For example, fold a rectangular sheet of paper. Draw a ray perpendicular to the crease with its endpoint on the crease. Keep the half of the folded sheet that does not contain the ray in contact with your desk. This is the plane. Move the other half to different positions. The ray that you drew is always perpendicular to the crease but is not always perpendicular to the plane. Based on this observation, we can state the following postulate.

Postulate 11.4

At a given point on a line, there are infinitely many lines perpendicular to the given line.

In order to prove that a line is perpendicular to a plane, the definition requires that we show that *every* line through the point of intersection is perpendicular to the given line. However, it is possible to prove that if line *l* is known to be perpendicular to each of two lines in plane *p* that intersect at point *A*, then *l* is perpendicular to plane *p* at *A*.

Theorem 11.4

If a line is perpendicular to each of two intersecting lines at their point of intersection, then the line is perpendicular to the plane determined by these lines.

Given A plane p determined by $\overleftrightarrow{AP}$ and $\overleftrightarrow{BP}$, two lines that intersect at P. Line l such that $l \perp \overleftrightarrow{AP}$ and $l \perp \overleftrightarrow{BP}$.

Prove $l \perp p$

Proof To begin, let R and S be points on l such that P is the midpoint of $\overline{RS}$. Since it is given that $l \perp \overleftrightarrow{AP}$ and $l \perp \overleftrightarrow{BP}$, $\angle RPA$, $\angle SPA$, $\angle RPB$, and $\angle SPB$ are right angles and therefore congruent. Then let $\overleftrightarrow{PT}$ be any other line through P in plane p. Draw $\overleftrightarrow{AB}$ intersecting $\overleftrightarrow{PT}$ at Q. To prove this theorem, we need to show three different pairs of congruent triangles: $\triangle RPA \cong \triangle SPA$, $\triangle RPB \cong \triangle SPB$, and $\triangle RPQ \cong \triangle SPQ$. However, to establish the last congruence we must prove that $\triangle RAB \cong \triangle SAB$ and $\triangle RAQ \cong \triangle SAQ$.

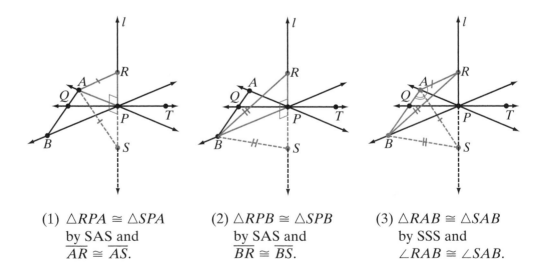

(1) $\triangle RPA \cong \triangle SPA$
by SAS and
$\overline{AR} \cong \overline{AS}$.

(2) $\triangle RPB \cong \triangle SPB$
by SAS and
$\overline{BR} \cong \overline{BS}$.

(3) $\triangle RAB \cong \triangle SAB$
by SSS and
$\angle RAB \cong \angle SAB$.

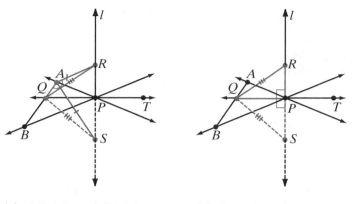

(4) $\triangle RAQ \cong \triangle SAQ$ by
by SAS since
$\angle RAB \cong \angle RAQ$ and
$\angle SAB \cong \angle SAQ$,
and $\overline{RQ} \cong \overline{SQ}$.

(5) $\triangle RPQ \cong \triangle SPQ$
by SSS and
$\angle RPQ \cong \angle SPQ$.

Now since $\angle RPQ$ and $\angle SPQ$ are a congruent linear pair of angles, they are right angles, and $l \perp \overleftrightarrow{PQ}$. Since $\overleftrightarrow{PQ}$, that is, $\overleftrightarrow{PT}$ can be *any* line in p through P, l is perpendicular to *every* line in plane p through point P. ∎

Theorem 11.5a | If two planes are perpendicular to each other, one plane contains a line perpendicular to the other plane.

Given Plane $p \perp$ plane q

Prove A line in p is perpendicular to q and a line in q is perpendicular to p.

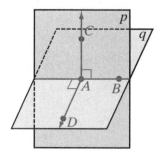

Proof If planes p and q are perpendicular to each other then they form a right dihedral angle. Let $\overleftrightarrow{AB}$ be the edge of the dihedral angle. In plane p, construct $\overleftrightarrow{AC} \perp \overleftrightarrow{AB}$, and in plane q, construct $\overleftrightarrow{AD} \perp \overleftrightarrow{AB}$. Since p and q from a right dihedral angle, $\angle CAD$ is a right angle and $\overleftrightarrow{AC} \perp \overleftrightarrow{AD}$. Two lines in plane p, $\overleftrightarrow{AB}$ and $\overleftrightarrow{AC}$, are each perpendicular to $\overleftrightarrow{AD}$. Therefore, $\overleftrightarrow{AD} \perp p$. Similarly, two lines in plane q, $\overleftrightarrow{AB}$ and $\overleftrightarrow{AD}$, are each perpendicular to $\overleftrightarrow{AC}$. Therefore, $\overleftrightarrow{AC} \perp q$. ∎

The converse of Theorem 11.5a is also true.

Theorem 11.5b

If a plane contains a line perpendicular to another plane, then the planes are perpendicular.

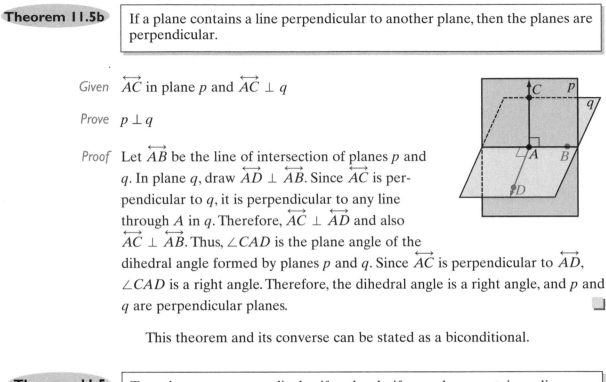

Given $\overleftrightarrow{AC}$ in plane p and $\overleftrightarrow{AC} \perp q$

Prove $p \perp q$

Proof Let $\overleftrightarrow{AB}$ be the line of intersection of planes p and q. In plane q, draw $\overleftrightarrow{AD} \perp \overleftrightarrow{AB}$. Since $\overleftrightarrow{AC}$ is perpendicular to q, it is perpendicular to any line through A in q. Therefore, $\overleftrightarrow{AC} \perp \overleftrightarrow{AD}$ and also $\overleftrightarrow{AC} \perp \overleftrightarrow{AB}$. Thus, $\angle CAD$ is the plane angle of the dihedral angle formed by planes p and q. Since $\overleftrightarrow{AC}$ is perpendicular to $\overleftrightarrow{AD}$, $\angle CAD$ is a right angle. Therefore, the dihedral angle is a right angle, and p and q are perpendicular planes.

This theorem and its converse can be stated as a biconditional.

Theorem 11.5

Two planes are perpendicular if and only if one plane contains a line perpendicular to the other.

We know that in a plane, there is only one line perpendicular to a given line at a given point on the line. In space, there are infinitely many lines perpendicular to a given line at a given point on the line. These perpendicular lines are all in the same plane. However, only one line is perpendicular to a plane at a given point.

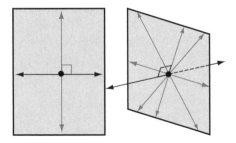

Theorem 11.6

Through a given point on a plane, there is only one line perpendicular to the given plane.

Given Plane p and $\overleftrightarrow{AB} \perp p$ at A.

Prove $\overleftrightarrow{AB}$ is the only line perpendicular to p at A.

Proof Use an indirect proof.

Assume that there exists another line, $\overleftrightarrow{AC} \perp p$ at A. Points A, B, and C determine a plane, q, that intersects plane p at $\overleftrightarrow{AD}$. Therefore, in plane q, $\overleftrightarrow{AB} \perp \overleftrightarrow{AD}$ and $\overleftrightarrow{AC} \perp \overleftrightarrow{AD}$. But in a given plane, there is only one line perpendicular to a given line at a given point. Our assumption is false, and there is only one line perpendicular to a given plane at a given point.

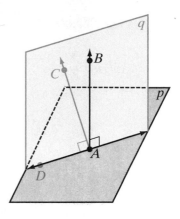

As we noted above, in space, there are infinitely many lines perpendicular to a given line at a given point. Any two of those intersecting lines determine a plane perpendicular to the given line. Each of these pairs of lines determine the same plane perpendicular to the given line.

> **Theorem 11.7** Through a given point on a line, there can be only one plane perpendicular to the given line.

Given Any point P on $\overleftrightarrow{AB}$.

Prove There is only one plane perpendicular to $\overleftrightarrow{AB}$.

Proof Use an indirect proof.

Assume that there are two planes, m and n, that are each perpendicular to $\overleftrightarrow{AB}$. Choose any point Q in m. Since $m \perp \overleftrightarrow{APB}$, $\overleftrightarrow{AP} \perp \overleftrightarrow{PQ}$. Points A, P, and Q determine a plane p that intersects plane n in a line $\overleftrightarrow{PR}$. Since $n \perp \overleftrightarrow{APB}$, $\overleftrightarrow{AP} \perp \overleftrightarrow{PR}$. Therefore, in plane p, $\overleftrightarrow{AP} \perp \overleftrightarrow{PQ}$ and $\overleftrightarrow{AP} \perp \overleftrightarrow{PR}$. But in a plane, at a given point there is one and only one line perpendicular to a given line. Our assumption must be false, and there is only one plane perpendicular to $\overleftrightarrow{AB}$ at P.

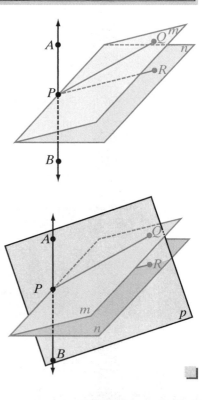

Theorem 11.8 | If a line is perpendicular to a plane, then any line perpendicular to the given line at its point of intersection with the given plane is in the plane.

Given $\overleftrightarrow{AB} \perp$ plane p at A and $\overleftrightarrow{AB} \perp \overleftrightarrow{AC}$.

Prove $\overleftrightarrow{AC}$ is in plane p.

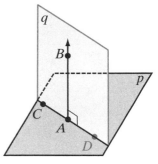

Proof Points A, B, and C determine a plane q. Plane q intersects plane p in a line, $\overleftrightarrow{AD}$. $\overleftrightarrow{AB} \perp \overleftrightarrow{AD}$ because $\overleftrightarrow{AB}$ is perpendicular to every line in p through A. It is given that $\overleftrightarrow{AB} \perp \overleftrightarrow{AC}$. Therefore, $\overleftrightarrow{AD}$ and $\overleftrightarrow{AC}$ in plane q are perpendicular to $\overleftrightarrow{AB}$ at A. But at a given point in a plane, only one line can be drawn perpendicular to a given line. Therefore, $\overleftrightarrow{AD}$ and $\overleftrightarrow{AC}$ are the same line, that is, C is on $\overleftrightarrow{AD}$. Since $\overleftrightarrow{AD}$ is the intersection of planes p and q, $\overleftrightarrow{AC}$ is in plane p.

Theorem 11.9 | If a line is perpendicular to a plane, then every plane containing the line is perpendicular to the given plane.

Given Plane p with $\overleftrightarrow{AB} \perp p$ at A, and C any point not on p.

Prove The plane q determined by A, B, and C is perpendicular to p.

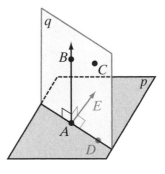

Proof Let the intersection of p and q be $\overleftrightarrow{AD}$, so $\overleftrightarrow{AD}$ is the edge of the dihedral angle formed by p and q. Let $\overleftrightarrow{AE}$ be a line in p that is perpendicular to $\overleftrightarrow{AD}$. Since $\overleftrightarrow{AB} \perp p$, $\overleftrightarrow{AB}$ is perpendicular to every line in p through A. Therefore, $\overleftrightarrow{AB} \perp \overleftrightarrow{AD}$ and $\overleftrightarrow{AB} \perp \overleftrightarrow{AE}$. $\angle BAE$ is a plane angle whose measure is the measure of the dihedral angle. Since $\overleftrightarrow{AB} \perp \overleftrightarrow{AE}$, m$\angle BAE = 90$. Therefore, the dihedral angle is a right angle, and $q \perp p$.

EXAMPLE 1

Show that the following statement is false.

Two planes perpendicular to the same plane have no points in common.

Solution Recall that a statement that is sometimes true and sometimes false is regarded to be false. Consider the adjacent walls of a room. Each wall is perpendicular to the floor but the walls intersect in a line. This counterexample shows that the given statement is false.

EXAMPLE 2

Planes p and q intersect in $\overleftrightarrow{AB}$. In p, $\overleftrightarrow{AC} \perp \overleftrightarrow{AB}$ and in q, $\overleftrightarrow{AD} \perp \overleftrightarrow{AB}$. If m$\angle CAD < 90$, is $p \perp q$?

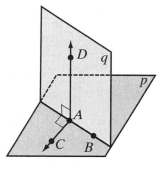

Solution Since in p, $\overleftrightarrow{AC} \perp \overleftrightarrow{AB}$, and in q, $\overleftrightarrow{AD} \perp \overleftrightarrow{AB}$, $\angle CAD$ is a plane angle whose measure is equal to the measure of the dihedral angle formed by the planes. Since $\angle CAD$ is not a right angle, then the dihedral angle is not a right angle, and the planes are not perpendicular.

EXAMPLE 3

Given: Line l intersects plane p at A, and l is not perpendicular to p.

Prove: There is at least one line through A in plane p that is not perpendicular to l.

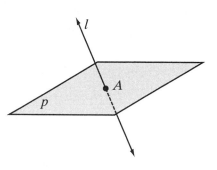

Proof Use an indirect proof.

Let $\overleftrightarrow{AB}$ and $\overleftrightarrow{AC}$ be two lines through A in p. Assume that $l \perp \overleftrightarrow{AB}$ and $l \perp \overleftrightarrow{AC}$. Therefore, $l \perp p$ because if a line is perpendicular to each of two lines at their point of intersection, then the line is perpendicular to the plane determined by these lines. But it is given that l is not perpendicular to p. Therefore, our assumption is false, and l is not perpendicular to at least one of the lines $\overleftrightarrow{AB}$ and $\overleftrightarrow{AC}$.

Exercises

Writing About Mathematics

1. Carmen said if two planes intersect to form four dihedral angles that have equal measures, then the planes are perpendicular to each other. Do you agree with Carmen? Explain why or why not.

2. Each of three lines is perpendicular to the plane determined by the other two.

   a. Is each line perpendicular to each of the other two lines? Justify your answer.

   b. Name a physical object that justifies your answer.

Developing Skills

In 3–11, state whether each of the statements is true or false. If it is true, state a postulate or theorem that supports your answer. If it is false, describe or draw a counterexample.

3. At a given point on a given line, only one line is perpendicular to the line.

4. If A is a point in plane p and B is a point not in p, then no other point on $\overleftrightarrow{AB}$ is in plane p.

5. A line perpendicular to a plane is perpendicular to every line in the plane.

6. A line and a plane perpendicular to the same line at two different points have no points in common.

7. Two intersecting planes that are each perpendicular to a third plane are perpendicular to each other.

8. If $\overleftrightarrow{AB}$ is perpendicular to plane p at A and $\overleftrightarrow{AB}$ is in plane q, then $p \perp q$.

9. At a given point on a given plane, only one plane is perpendicular to the given plane.

10. If a plane is perpendicular to one of two intersecting lines, it is perpendicular to the other.

11. If a line is perpendicular to one of two intersecting planes, it is perpendicular to the other.

Applying Skills

12. Prove step 1 of Theorem 11.4.

13. Prove step 3 of Theorem 11.4.

14. Prove step 5 of Theorem 11.4.

15. Prove that if a line segment is perpendicular to a plane at the midpoint of the line segment, then every point in the plane is equidistant from the endpoints of the line segment.

    Given: $\overline{AB} \perp$ plane p at M, the midpoint of $\overline{AB}$, and R is any point in plane p.

    Prove: $AR = BR$

16. Prove that if two points are each equidistant from the endpoints of a line segment, then the line segment is perpendicular to the plane determined by the two points and the midpoint of the line segment.

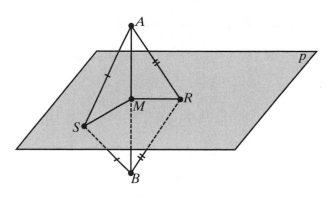

Given: M is the midpoint of $\overline{AB}$, $\overline{RA} \cong \overline{RB}$, and $\overline{SA} \cong \overline{SB}$.

Prove: $\overline{AB}$ is perpendicular to the plane determined by M, R, and S.

17. Equilateral triangle ABC is in plane p and $\overleftrightarrow{AD}$ is perpendicular to plane p. Prove that $\overline{BD} \cong \overline{CD}$.

18. $\overleftrightarrow{AB}$ and $\overleftrightarrow{AC}$ intersect at A and determine plane p. $\overleftrightarrow{AD}$ is perpendicular to plane p at A. If $AB = AC$, prove that $\triangle ABD \cong \triangle ACD$.

19. Triangle QRS is in plane p, $\overline{ST}$ is perpendicular to plane p, and $\angle QTS \cong \angle RTS$. Prove that $\overline{TQ} \cong \overline{TR}$.

20. Workers who are installing a new telephone pole position the pole so that it is perpendicular to the ground along two different lines. Prove that this is sufficient to prove that the telephone pole is perpendicular to the ground.

21. A telephone pole is perpendicular to the level ground. Prove that two wires of equal length attached to the pole at the same point and fastened to the ground are at equal distances from the pole.

11-3 PARALLEL LINES AND PLANES

Look at the floor, walls, and ceiling of the classroom. Each of these surfaces can be represented by a plane. Some of these surfaces, such as the floor and the ceiling, do not intersect. These can be represented as portions of *parallel planes*.

DEFINITION

Parallel planes are planes that have no points in common.

A line is parallel to a plane if it has no points in common with the plane.

EXAMPLE I

Plane p intersects plane q in $\overleftrightarrow{AB}$ and plane r in $\overleftrightarrow{CD}$. Prove that if $\overleftrightarrow{AB}$ and $\overleftrightarrow{CD}$ intersect, then planes q and r are not parallel.

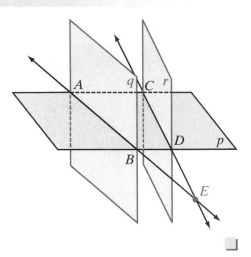

Proof Let E be the point at which $\overleftrightarrow{AB}$ and $\overleftrightarrow{CD}$ intersect. Then E is a point on q and E is a point on r. Therefore, planes q and r intersect in at least one point and are not parallel.

Theorem 11.10 If a plane intersects two parallel planes, then the intersection is two parallel lines.

Given Plane p intersects plane m at $\overleftrightarrow{AB}$ and plane n at $\overleftrightarrow{CD}$, $m \parallel n$.

Prove $\overleftrightarrow{AB} \parallel \overleftrightarrow{CD}$

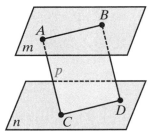

Proof Use an indirect proof.

Lines $\overleftrightarrow{AB}$ and $\overleftrightarrow{CD}$ are two lines of plane p. Two lines in the same plane either intersect or are parallel. If $\overleftrightarrow{AB}$ is not parallel to $\overleftrightarrow{CD}$ then they intersect in some point E. Since E is a point of $\overleftrightarrow{AB}$, then it is a point of plane m. Since E is a point of $\overleftrightarrow{CD}$, then it is a point of plane n. But $m \parallel n$ and have no points in common. Therefore, $\overleftrightarrow{AB}$ and $\overleftrightarrow{CD}$ are two lines in the same plane that do not intersect, and $\overleftrightarrow{AB} \parallel \overleftrightarrow{CD}$.

In a plane, two lines perpendicular to a given line are parallel. Can we prove that two lines perpendicular to a given plane are also parallel?

Theorem 11.11 | Two lines perpendicular to the same plane are parallel.

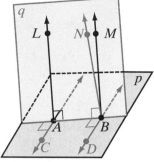

Given   Plane p, line $\overleftrightarrow{LA} \perp p$ at A, and line $\overleftrightarrow{MB} \perp p$ at B.

Prove $\overleftrightarrow{LA} \parallel \overleftrightarrow{MB}$

Proof   We will construct line $\overleftrightarrow{NB}$ at B that is parallel to $\overleftrightarrow{LA}$, and show that $\overleftrightarrow{MB}$ and $\overleftrightarrow{NB}$ are the same line.

(1) Since it is given that $\overleftrightarrow{LA} \perp p$ at A, $\overleftrightarrow{LA}$ is perpendicular to any line in p through A, so $\overleftrightarrow{LA} \perp \overleftrightarrow{AB}$. Let q be the plane determined by $\overleftrightarrow{LA}$ and $\overleftrightarrow{AB}$. In plane p, draw $\overleftrightarrow{AC} \perp \overleftrightarrow{AB}$. Then $\angle LAC$ is a right angle, and p and q form a right dihedral angle.

(2) At point B, there exists a line $\overleftrightarrow{NB}$ in q that is parallel to $\overleftrightarrow{LA}$. If one of two parallel lines is perpendicular to a third line, then the other is perpendicular to the third line, that is, since $\overleftrightarrow{LA} \perp \overleftrightarrow{AB}$, then $\overleftrightarrow{NB} \perp \overleftrightarrow{AB}$.

(3) Draw $\overleftrightarrow{BD} \perp \overleftrightarrow{AB}$ in p. Because p and q form a right dihedral angle, $\angle NBD$ is a right angle, and so $\overleftrightarrow{NB} \perp \overleftrightarrow{BD}$.

(4) Therefore, $\overleftrightarrow{NB}$ is perpendicular to two lines in p at B (steps 2 and 3), so $\overleftrightarrow{NB}$ is perpendicular to p at B.

(5) But it is given that $\overleftrightarrow{MB} \perp p$ at B and there is only one line perpendicular to a given plane at a given point. Therefore, $\overleftrightarrow{MB}$ and $\overleftrightarrow{NB}$ are the same line, and $\overleftrightarrow{LA} \parallel \overleftrightarrow{MB}$. ∎

We have shown that two lines perpendicular to the same plane are parallel. Since parallel lines lie in the same plane, we have just proved the following corollary to this theorem:

Corollary 11.11a | Two lines perpendicular to the same plane are coplanar.

Theorem 11.12a | If two planes are perpendicular to the same line, then they are parallel.

Given   Plane $p \perp \overleftrightarrow{AB}$ at A and $q \perp \overleftrightarrow{AB}$ at B.

Prove $p \parallel q$

Proof   Use an indirect proof.

Assume that p is not parallel to q. Then p and q
intersect in a line. Let R be any point on the line of
intersection. Then A, B, and R determine a plane, s.
In plane s, $\overleftrightarrow{AR} \perp \overleftrightarrow{AB}$ and $\overleftrightarrow{BR} \perp \overleftrightarrow{AB}$. But two lines
in a plane that are perpendicular to the same line are
parallel. Therefore, our assumption must be false,
and $p \parallel q$.

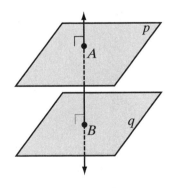

Theorem 11.12b | If two planes are parallel, then a line perpendicular to one of the planes is perpendicular to the other.

Given   Plane p parallel to plane q, and $\overleftrightarrow{AB} \perp$ plane p and
intersecting plane q at B

Prove $\overleftrightarrow{AB} \perp$ plane q

Proof   To prove this theorem, we will construct two lines
$\overleftrightarrow{BE}$ and $\overleftrightarrow{FB}$ in q that are both perpendicular to
$\overleftrightarrow{AB}$. From this, we will conclude that $\overleftrightarrow{AB}$ is perpendicular to q.

(1) Let C be a point in p. Let r be the
plane determined by A, B, and C
intersecting q at $\overleftrightarrow{BE}$. Since p and q
are parallel, $\overleftrightarrow{AC} \parallel \overleftrightarrow{BE}$. It is given that
$\overleftrightarrow{AB} \perp$ plane p. Therefore, $\overleftrightarrow{AB} \perp \overleftrightarrow{AC}$.
Then, in plane r, $\overleftrightarrow{AB} \perp \overleftrightarrow{BE}$.

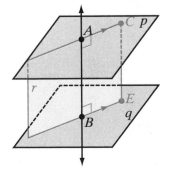

(2) Let D be a point in p. Let s be the
plane determined by A, B, and D
intersecting q at $\overleftrightarrow{BF}$. Since p and q
are parallel, $\overleftrightarrow{AD} \parallel \overleftrightarrow{BF}$. It is given
that $\overleftrightarrow{AB} \perp$ plane p. Therefore,
$\overleftrightarrow{AB} \perp \overleftrightarrow{AD}$. Then, in plane s,
$\overleftrightarrow{AB} \perp \overleftrightarrow{BF}$.

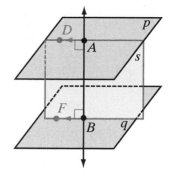

(3) If a line is perpendicular to each of two intersecting lines at their point of
intersection, then the line is perpendicular to the plane determined by
these lines. Therefore, $\overleftrightarrow{AB} \perp$ plane q.

Theorem 11.12a and 11.2b are converse statements. Therefore, we may write
these two theorems as a biconditional.

Theorem 11.12

> Two planes are perpendicular to the same line if and only if the planes are
> parallel.

Let p and q be two parallel planes. From A in p, draw $\overline{AB} \perp q$ at B.
Therefore, $\overline{AB} \perp p$ at A. The distance from p to q is AB.

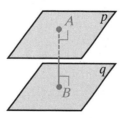

DEFINITION

The **distance between two planes** is the length of the line segment perpendicular
to both planes with an endpoint on each plane.

Theorem 11.13 | Parallel planes are everywhere equidistant.

Given Parallel planes p and q, with $\overline{AC}$ and $\overline{BD}$ each perpendicular to p and q with an endpoint on each plane.

Prove $AC = BD$

Proof Two lines perpendicular to the same plane are both parallel and coplanar. Therefore, $\overline{AC} \parallel \overline{BD}$ and lie on the same plane. That plane intersects parallel planes p and q in parallel lines $\overleftrightarrow{AB}$ and $\overleftrightarrow{CD}$. In the plane of $\overleftrightarrow{AC}$ and $\overleftrightarrow{BD}$, $ABDC$ is a parallelogram with a right angle, that is, a rectangle. Therefore, $\overline{AC}$ and $\overline{BD}$ are congruent and $AC = BD$.

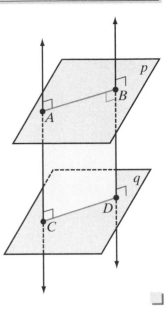

EXAMPLE 2

Line l is perpendicular to plane p and line l is not perpendicular to plane q. Is $p \parallel q$?

Solution Assume that $p \parallel q$. If two planes are parallel, then a line perpendicular to one is perpendicular to the other. Therefore, since l is perpendicular to plane p, l must be perpendicular to plane q. This contradicts the given statement that l is not perpendicular to q, and the assumption is false. Therefore, p is not parallel to q.

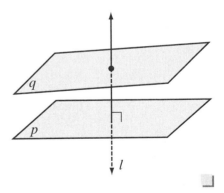

EXAMPLE 3

Given: $\overleftrightarrow{AB} \perp$ plane p at A, $\overleftrightarrow{CD} \perp$ plane p at C, and $AB = CD$.

Prove: A, B, C, and D are the vertices of a parallelogram.

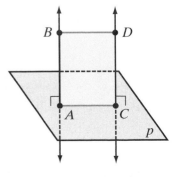

Proof Two lines perpendicular to the same plane are parallel and coplanar. Therefore, since it is given that $\overleftrightarrow{AB}$ and $\overleftrightarrow{CD}$ are each perpendicular to p, they are parallel and coplanar. Since $AB = CD$ and segments of equal length are congruent, $ABCD$ is a quadrilateral with one pair of sides congruent and parallel. Therefore, $ABCD$ is a parallelogram.

Exercises

Writing About Mathematics

1. Two planes are perpendicular to the same plane. Are the planes parallel? Justify your answer.

2. Two planes are parallel to the same plane. Are the planes parallel? Justify your answer.

Developing Skills

In 3–9, each of the given statements is sometimes true and sometimes false. **a.** Give an example from the diagram to show that the statement can be true. **b.** Give a counterexample from the diagram to show that the statement can be false. In the diagram, each quadrilateral is a rectangle.

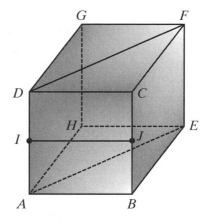

3. If two planes are perpendicular, a line parallel to one plane is perpendicular to the other.

4. Two planes parallel to the same line are parallel to each other.

5. Two lines perpendicular to the same line are parallel.

6. Two lines that do not intersect are parallel.

7. Two planes perpendicular to the same plane are parallel to each other.

8. Two lines parallel to the same plane are parallel.

9. If two lines are parallel, then a line that is skew to one line is skew to the other.

Applying Skills

10. ABC is an isosceles triangle with base $\overline{BC}$ in plane p. Plane $q \parallel p$ through point D on $\overline{AB}$ and point E on $\overline{AC}$. Prove that $\triangle ADE$ is isosceles.

11. Plane p is perpendicular to $\overleftrightarrow{PQ}$ at Q and two points in p, A and B, are equidistant from P. Prove that $\overline{AQ} \cong \overline{BQ}$.

12. Noah is building a tool shed. He has a rectangular floor in place and wants to be sure that the posts that he erects at each corner of the floor as the ends of the walls are parallel. He erects each post perpendicular to the floor. Are the posts parallel to each other? Justify your answer.

13. Noah wants the flat ceiling on his tool shed to be parallel to the floor. Two of the posts are 80 inches long and two are 78 inches long. Will the ceiling be parallel to the floor? Justify your answer. What must Noah do to make the ceiling parallel to the floor?

11-4 SURFACE AREA OF A PRISM

Polyhedron

In the plane, a polygon is a closed figure that is the union of line segments. In space, a *polyhedron* is a figure that is the union of polygons.

DEFINITION

A **polyhedron** is a three-dimensional figure formed by the union of the surfaces enclosed by plane figures.

The portions of the planes enclosed by a plane figure are called the **faces** of the polyhedron. The intersections of the faces are the **edges** of the polyhedron and the intersections of the edges are the **vertices** of the polyhedron.

DEFINITION

A **prism** is a polyhedron in which two of the faces, called the **bases** of the prism, are congruent polygons in parallel planes.

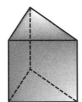 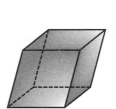 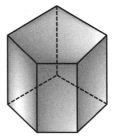

Examples of prisms

The surfaces between corresponding sides of the bases are called the **lateral sides** of the prism and the common edges of the lateral sides are called the **lateral edges**. An **altitude** of a prism is a line segment perpendicular to each of the bases with an endpoint on each base. The **height** of a prism is the length of an altitude.

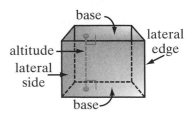

Since the bases are parallel, the corresponding sides of the bases are congruent, parallel line segments. Therefore, each lateral side has a pair of congruent, parallel sides, the corresponding edges of the bases, and are thus parallelograms. The other pair of sides of these parallelograms, the lateral edges, are also congruent and parallel. Therefore, we can make the following statement:

▶ **The lateral edges of a prism are congruent and parallel.**

DEFINITION _____

A **right prism** is a prism in which the lateral sides are all perpendicular to the bases. All of the lateral sides of a right prism are rectangles.

Using graph paper, cut two 7-by-5 rectangles and two 7-by-4 rectangles. Use tape to join the 7-by-4 rectangles to opposite sides of one of the 7-by-5 rectangles along the congruent edges. Then join the other 7-by-5 rectangle to the congruent edges forming four of the six sides of a prism. Place the prism on your desk on its

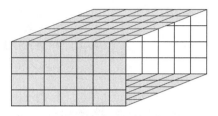

side so that one pair of congruent rectangles are the bases and the lateral edges are perpendicular to the bases. Are the opposite faces parallel? What would be the shape and size of the two missing sides? Then move the top base so that the lateral edges are not perpendicular to the bases. The figure is no longer a right prism. Are the opposite faces parallel? What would be the shape and size of the two missing sides?

Cut two more 7-by-5 rectangles and two parallelograms that are not rectangles. Let the lengths of two of the sides of the parallelograms be 7 and the length of the altitude to these sides be 4. Join the parallelograms to opposite sides of one of the rectangles along congruent sides. Then join the other rectangle to congru-

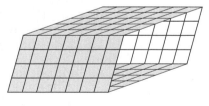

ent edges forming four of the six sides of a prism. Place the prism on your desk so that the rectangles are the bases and the lateral edges are perpendicular to the bases. Are the opposite faces parallel? Is the prism a right prism? What would be the shape and size of the two missing lateral sides? Move the top base of the prism so that the sides are not perpendicular to the bases. Are the opposite faces parallel? What would be the shape and size of the two missing sides? Now turn this prism so that the parallelograms are the bases and the edges of the rectangles are perpendicular to the bases. What is the shape of the two missing sides? Is this a right prism?

The solids that you have made are called *parallelepipeds*.

DEFINITION _____

A **parallelepiped** is a prism that has parallelograms as bases.

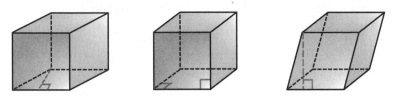

Examples of parallelepipeds

Rectangular Solids

DEFINITION _____

A **rectangular parallelepiped** is a parallelepiped that has rectangular bases and lateral edges perpendicular to the bases.

A rectangular parallelepiped is usually called a **rectangular solid**. It is the most common prism and is the union of six rectangles. Any two parallel rectangles of a rectangular solid can be the bases.

In the figure, $ABCDEFGH$ is a rectangular solid. The bases $ABCD$ and $EFGH$ are congruent rectangles with $AB = EF = CD = GH = 7$ and $BC = FG = DA = HE = 5$. Two of the lateral sides are rectangles $ABFE$ and $DCGH$ whose dimensions are 7 by 4. The other two lateral sides are $BCGF$ and $ADHE$ whose dimensions are 5 by 4.

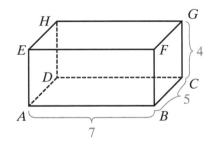

- The area of each base: $7 \times 5 = 35$
- The area of each of two lateral sides: $7 \times 4 = 28$
- The area of each of the other two lateral sides: $5 \times 4 = 20$

The **lateral area** of the prism is the sum of the areas of the lateral faces. The **total surface area** is the sum of the lateral area and the areas of the bases.

- The lateral area of the prism is $2(28) + 2(20) = 96$.
- The area of the bases are each $2(35) = 70$.
- The surface area of the prism is $96 + 70 = 166$.

EXAMPLE 1

The bases of a right prism are regular hexagons. The length of each side of a base is 5 centimeters and the height of the prism is 8 centimeters. Describe the number, shape, and size of the lateral sides.

Solution A hexagon has six sides. Because the base is a regular hexagon, it has six congruent sides, and therefore, the prism has six lateral sides. Since it is a right prism, the lateral sides are rectangles. The length of each of two edges of a lateral side is the length of an edge of a base, 5 centimeters. The length of each of the other two edges of a lateral side is the height of the prism, 8 centimeters.

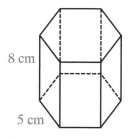

8 cm

5 cm

Answer There are six lateral sides, each is a rectangle that is 8 centimeters by 5 centimeters.

EXAMPLE 2

The bases of a right prism are equilateral triangles. The length of one edge of a base is 4 inches and the height of the prism is 5 inches.

a. How many lateral sides does this prism have and what is their shape?

b. What is the lateral area of the prism?

Solution **a.** Because this is a prism with a triangular base, the prism has three lateral sides. Because it is a right prism, the lateral sides are rectangles.

b. For each rectangular side, the length of one pair of edges is the length of an edge of the base, 4 inches. The height of the prism, 5 inches, is the length of a lateral edge. Therefore, the area of each lateral side is 4(5) or 20 square inches. The lateral area of the prism is 3(20) = 60 square inches.

Answers **a.** 3   **b.** 60 square inches

EXAMPLE 3

The lateral sides of a prism are five congruent rectangles. Prove that the bases are equilateral pentagons.

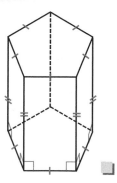

Proof The five lateral sides are congruent rectangles. Two parallel sides of each rectangle are lateral edges. The other two parallel sides of each rectangle are edges of the bases. Each edge of a base is a side of a rectangle. Therefore, the base has five sides. The corresponding sides of the congruent rectangles are congruent. Therefore, the edges of each base are congruent. The bases are equilateral pentagons.

Exercises

Writing About Mathematics

1. Cut a 12-by-5 rectangle from graph paper, fold it into three 4-by-5 rectangles and fasten the two sides of length 5 with tape. Then cut a 16-by-5 rectangle from graph paper, fold it into four 4-by-5 rectangles and fasten the two sides of length 5 with tape. Let the open ends of each figure be the bases of a prism.

   a. What is the shape of a base of the prism formed from the 12-by-5 paper? Can the shape of this base be changed? Explain your answer.

   b. What is the shape of a base of the prism formed from the 16-by-5 paper? Can the shape of this base be changed? Explain your answer.

   c. Are both figures always right prisms? Explain your answer.

2. A prism has bases that are rectangles, two lateral faces that are rectangles and two lateral faces that are parallelograms that are not rectangles.

   a. Is an altitude of the solid congruent to an altitude of one of the rectangular faces? Explain your answer.

   b. Is an altitude of the solid congruent to an altitude of one of the faces that are parallelograms? Explain your answer.

Developing Skills

In 3–6, find the surface area of each of the rectangular solid with the given dimensions.

3. 5.0 cm by 8.0 cm by 3.0 cm

4. 15 in. by 10.0 in. by 2.0 ft

5. 2.5 ft by 8.0 ft by 12 ft

6. 56.3 cm by 18.7 cm by 0.500 m

7. The bases of a prism are right triangles whose edges measure 9.00 centimeters, 40.0 centimeters, and 41.0 centimeters. The lateral sides of the prism are perpendicular to the bases. The height of the prism is 14.5 centimeters.

   a. What is the shape of the lateral sides of the prism?

   b. What are the dimensions of each lateral side of the prism?

   c. What is the total surface area of the prism?

8. The bases of a right prism are isosceles triangles. The lengths of the sides of the bases are 5 centimeters, 5 centimeters, and 6 centimeters. The length of the altitude to the longest side of a base is 4 centimeters. The height of the prism is 12 centimeters.

   a. What is the shape of the lateral sides of the prism?

   b. What are the dimensions of each lateral side of the prism?

   c. What is the total surface area of the prism?

9. How many faces does a parallelepiped have? Justify your answer.

10. A prism has bases that are trapezoids. Is the prism a parallelepiped? Justify your answer.

11. The length of an edge of a cube is 5.20 inches. What is the total surface area of the cube to the nearest square inch?

Applying Skills

12. The bases of a parallelepiped are $ABCD$ and $EFGH$, and $ABCD \cong EFGH$. Prove that $\overline{AE} \parallel \overline{BF} \parallel \overline{CG} \parallel \overline{DH}$ and that $AE = BF = CG = DH$.

13. The bases of a prism are $\triangle ABC$ and $\triangle DEF$, and $\triangle ABC \cong \triangle DEF$. The line through A perpendicular to the plane of $\triangle ABC$ intersects the plane of $\triangle DEF$ at D, and the line through B perpendicular to the plane of $\triangle ABC$ intersects the plane of $\triangle DEF$ at E.

 a. Prove that the lateral faces of the prism are rectangles.

 b. When are the lateral faces of the prism congruent polygons? Justify your answer.

14. A right prism has bases that are squares. The area of one base is 81 square feet. The lateral area of the prism is 144 square feet. What is the length of the altitude of the prism?

15. Show that the edges of a parallelepiped form three sets of parallel line segments.

16. The lateral faces of a parallelepiped are squares. What must be the shape of the bases? Justify your answer.

17. The lateral faces of a parallelepiped are squares. One angle of one of the bases is a right angle. Prove that the parallelepiped is a *cube*, that is, a rectangular parallelepiped with congruent faces.

18. The walls, floor, and ceiling of a room form a rectangular solid. The total surface area of the room is 992 square feet. The dimensions of the floor are 12 feet by 20 feet.

 a. What is the lateral area of the room?

 b. What is the height of the room?

Hands-On Activity

Let the bases of a prism be $ABCD$ and $A'B'C'D'$. Use the prisms that you made out of graph paper for page 441 to demonstrate each of the following.

 1. When $\overline{AA'}$ is perpendicular to the planes of the bases, the lateral faces are rectangles and the height of the each lateral face is the height of the prism.

 2. When a line through A perpendicular to the bases intersects the plane of $A'B'C'D'$ at a point on $\overleftrightarrow{A'B'}$, two of the lateral faces are rectangles and two are parallelograms. The height of the prism is the height of the lateral faces that are parallelograms but the height of the rectangles is not equal to the height of the prism.

 3. When a line through A perpendicular to the bases intersects the plane of $A'B'C'D'$ at a point that is not on a side of $A'B'C'D'$, then the lateral faces are parallelograms and the height of the prism is not equal to the heights of the parallelogram.

11-5 VOLUME OF A PRISM

A cube whose edges each measure 1 centimeter is a unit of volume called a **cubic centimeter**. If the bases of a rectangular solid measure 8 centimeters by 5 centimeters, we know that the area of a base is 8×5 or 40 square centimeters and that 40 cubes each with a volume of 1 cubic centimeter can fill one base. If the height of the solid is 3 centimeters, we know that we can place 3 layers with 40 cubic centimeters in each layer to fill the rectangular solid. The volume of the solid is 40×3 or 120 cubic centimeters. The volume of the rectangular solid is the area of the base times the height. This can be applied to any prism and suggests the following postulate.

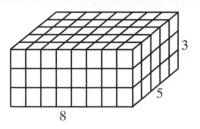

Postulate 11.5

The volume of a prism is equal to the area of the base times the height.

If V represents the volume of a prism, B represents the area of the base, and h the height of the prism, then:

$$V = Bh$$

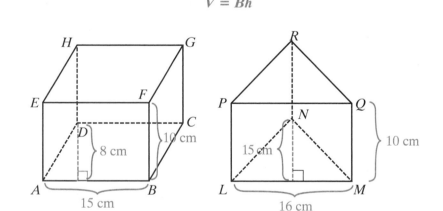

The figure shows two prisms. One is a parallelepiped with parallelograms *ABCD* and *EFGH* as bases and rectangular faces *ABFE*, *BCGF*, *CDHG*, and *DAEH*. If *AB* is 15 centimeters and the length of the altitude from *D* to $\overline{AB}$ is 8 centimeters, then the area of the base *ABCD* is 15×8 or 120 square centimeters. If *BF*, the height of the parallelepiped, is 10 centimeters, then:

Volume of the parallelepiped $= Bh$

$= 120 \times 10$

$= 1{,}200$ cubic centimeters

The other prism has bases that are triangles, $\triangle LMN$ and $\triangle PQR$. If LM is 16 centimeters and the length of the altitude to $\overline{LM}$ is 15 centimeters, then the area of a base is $\frac{1}{2}(16)(15)$ or 120 square centimeters. If the height of this prism is 10 centimeters, then:

Volume of the triangular prism $= Bh$

$$= 120 \times 10$$

$$= 1{,}200 \text{ cubic centimeters}$$

Note that for these two prisms, the areas of the bases are equal and the heights of the prisms are equal. Therefore, the volumes of the prisms are equal. This is true in general since volume is defined as the area of the base times the height of the prism.

The terms "base" and "height" are used in more than one way when describing a prism. For example, each of the congruent polygons in parallel planes is a base of the prism. The distance between the parallel planes is the height of the prism. In order to find the area of a base that is a triangle or a parallelogram, we use the length of a base and the height of the triangle or parallelogram. When finding the area of a lateral face that is a parallelogram, we use the length of the base and the height of that parallelogram. Care must be taken in distinguishing to what line segments the words "base" and "height" refer.

EXAMPLE I

The bases of a right prism are $\triangle ABC$ and $\triangle A'B'C'$ with D a point on $\overline{CB}$, $\overline{AD} \perp \overline{BC}$, $AB = 10$ cm, $AC = 10$ cm, $BC = 12$ cm, $AD = 8$ cm, and $BB' = 15$ cm. Find the volume of the prism.

Solution Since this is a right prism, all of the lateral faces are rectangles and the height of the prism, AA', is the height of each face.

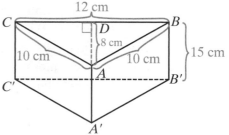

Each base is an isosceles triangle. The length of the base of the isosceles triangle is $BC = 12$ cm, and the length of the altitude to the base of the triangle is $AD = 8$ cm.

Area of $\triangle ABC = \frac{1}{2}(BC)(AD)$

$$= \frac{1}{2}(12)(8)$$

$$= 48$$

Since the prism is a right prism, the height of the prism is $BB' = 15$.

Volume of the prism $=$ (area of a base)(height of the prism)

$$= (48)(15)$$

$$= 720 \text{ cubic centimeters } \textit{Answer}$$

Exercises

Writing About Mathematics

1. Zoe said that if two solids have equal volumes and equal heights, then they must have congruent bases. Do you agree with Zoe? Justify your answer.

2. Piper said that the height of a prism is equal to the height of each of its lateral sides only if all of the lateral sides of the prism are rectangles. Do you agree with Piper? Explain why or why not.

Developing Skills

In 3–7, find the volume of each prism.

3. The area of the base is 48 square feet and the height is 18 inches.

4. The prism is a rectangular solid whose dimensions are 2.0 feet by 8.5 feet by 1.6 feet.

5. One base is a right triangle whose legs measure 5 inches and 7 inches. The height of the prism is 9 inches.

6. One base is a square whose sides measure 12 centimeters and the height of an altitude is 75 millimeters.

7. One base is parallelogram $ABCD$ and the other is parallelogram $A'B'C'D'$, $AB = 47$ cm, the length of the perpendicular from D to $\overline{AB} = 56$ cm, $\overline{AA'} \perp \overline{AB}$, $\overline{AA'} \perp \overline{AD}$, and $AA' = 19$ cm.

Applying Skills

8. A fish tank in the form of a rectangular solid is to accommodate 6 fish, and each fish needs at least 7,500 cubic centimeters of space. The dimensions of the base are to be 30 centimeters by 60 centimeters. What is the minimum height that the tank needs to be?

9. A parallelepiped and a rectangular solid have equal volume and equal height. The bases of the rectangular solid measure 15 centimeters by 24 centimeters. If the length of one side of a base of the parallelepiped measures 20 centimeters, what must be the length of the altitude to that base?

10. A prism whose bases are triangles and one whose bases are squares have equal volume and equal height. Triangle ABC is one base of the triangular prism and $PQRS$ is one base of the square prism. If $\overline{CD}$ is the altitude from C to $\overline{AB}$ and $AB = PQ$, what is the ratio of AB to CD? Justify your answer.

11. Prove that a plane that lies between the bases of a triangular prism and is parallel to the bases intersects the lateral sides of the prism in a triangle congruent to the bases.

12. Prove that the lateral area of a right prism is equal to the perimeter of a base times the height of the prism.

11-6 PYRAMIDS

A **pyramid** is a solid figure with a base that is a polygon and lateral faces that are triangles. Each lateral face shares a common edge with the base and a common edge with two other lateral faces. All lateral edges meet in a point called the **vertex**. The **altitude** of a pyramid is the perpendicular line segment from the vertex to the base ($\overline{PC}$ in the diagram on the right.) The **height** of a pyramid is the length of the altitude.

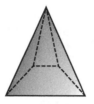

Examples of pyramids

Regular Pyramids

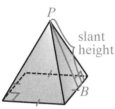

regular pyramid

A **regular pyramid** is a pyramid whose base is a regular polygon and whose altitude is perpendicular to the base at its center. The lateral edges of a regular polygon are congruent. Therefore, the lateral faces of a regular pyramid are isosceles triangles. The length of the altitude of a triangular lateral face of a regular pyramid, PB, is the **slant height** of the pyramid.

Surface Area and Volume of a Pyramid

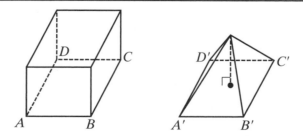

The figure shows a prism and a pyramid that have congruent bases and equal heights. If we were to fill the pyramid with water and empty the water into the prism, we would need to do this three times to fill the prism. Thus, since the volume of a prism is given by Bh:

$$\text{Volume of a pyramid} = \tfrac{1}{3}Bh$$

The lateral area of a pyramid is the sum of the areas of the faces. The total surface area is the lateral area plus the area of the bases.

EXAMPLE 1

A regular pyramid has a square base and four lateral sides that are isosceles triangles. The length of an edge of the base is 10 centimeters and the height of the pyramid is 12 centimeters. The length of the altitude to the base of each lateral side is 13 centimeters.

a. What is the total surface area of the pyramid?

b. What is the volume of the pyramid?

Solution   Let e be the length of a side of the square base: $e = 10$ cm

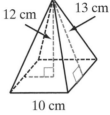

Let h_p be the height of the pyramid: $h_p = 12$ cm

Let h_s be the slant height of the pyramid: $h_s = 13$ cm

a. The base is a square with $e = 10$ cm.

The area of the base is $e^2 = (10)^2 = 100$ cm^2.

Each lateral side is an isosceles triangle. The length of each base, e, is 10 centimeters and the height, h_s, is 13 centimeters.

The area of each lateral side is $\frac{1}{2}eh_s = \frac{1}{2}(10)(13) = 65$ cm^2.

The total surface area of the pyramid is $100 + 4(65) = 360$ cm^2.

b. The volume of the prism is one-third of the area of the base times the height of the pyramid.

$$V = \tfrac{1}{3}Bh_p$$
$$= \tfrac{1}{3}(100)(12)$$
$$= 400 \text{ cm}^3$$

Answers   **a.** 360 cm^2 **b.** 400 cm^3

Properties of Regular Pyramids

The base of a regular pyramid is a regular polygon and the altitude is perpendicular to the base at its center. The *center of a regular polygon* is defined as the point that is equidistant to its vertices. In a regular polygon with three sides, an equilateral triangle, we proved that the perpendicular bisector of the sides of the triangle meet in a point that is equidistant from the vertices of the triangle. In a regular polygon with four sides, a square, we know that the diagonals are congruent. Therefore, the point at which the diagonals bisect each other is equidistant from the vertices. In Chapter 13, we will show that for any regular polygon, a point equidistant from the vertices exists. For now, we can use this fact to show that the lateral sides of a regular pyramid are isosceles triangles.

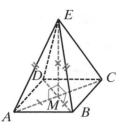

For example, consider a regular pyramid with square $ABCD$ for a base and vertex E. The diagonals of $ABCD$ intersect at M and $AM = BM = CM = DM$. Since $\overline{EM}$ is perpendicular to the base, it is perpendicular to any line in the base through M. Therefore, $\overline{EM} \perp \overline{MA}$ and $\angle EMA$ is a right angle. Also, $\overline{EM} \perp \overline{MB}$ and $\angle EMB$ is a right angle. Since the diagonals of a square are congruent and bisect each other, $\overline{MA} \cong \overline{MB}$. Then since $\overline{EM} \cong \overline{EM}$, $\triangle EMA \cong \triangle EMB$ by SAS, and $\overline{EA} \cong \overline{EB}$ because they are corresponding parts of congruent triangles. Similar reasoning will lead us to conclude that $\overline{EB} \cong \overline{EC}$, $\overline{EC} \cong \overline{ED}$, and $\overline{ED} \cong \overline{EA}$. A similar proof can be given for any base that is a regular polygon. Therefore, we can make the following statement:

▶ **The lateral faces of a regular pyramid are isosceles triangles.**

In the regular pyramid with base $ABCD$ and vertex E,

$$AB = BC = CD = DA \quad \text{and} \quad AE = BE = CE = DE$$

Therefore, $\triangle ABE \cong \triangle BCE \cong \triangle CDE \cong \triangle DAE$, that is, the lateral faces of the pyramid are congruent.

▶ **The lateral faces of a regular pyramid are congruent.**

EXAMPLE 2

A regular pyramid has a base that is the hexagon $ABCDEF$ and vertex at V. If the length $\overline{AB}$ is 2.5 centimeters, and the slant height of the pyramid is 6 centimeters, find the lateral area of the pyramid.

Solution The slant height of the pyramid is the height of a lateral face. Therefore:

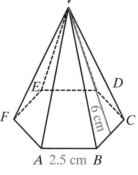

$$\text{Area of } \triangle ABV = \tfrac{1}{2}bh$$
$$= \tfrac{1}{2}(2.5)(6)$$
$$= \tfrac{15}{2} \text{ cm}^2$$

The lateral faces of the regular pyramid are congruent. Therefore, they have equal areas. There are six lateral faces.

$$\text{Lateral area of the pyramid} = 6\left(\tfrac{15}{2}\right)$$
$$= 45 \text{ cm}^2 \text{ } \textit{Answer}$$

Exercises

Writing About Mathematics

1. Martin said that if the base of a regular pyramid is an equilateral triangle, then the foot of the altitude of the pyramid is the point at which the altitudes of the base intersect. Sarah said that it is the point at which the medians intersect. Who is correct? Justify your answer.

2. Are the lateral faces of a pyramid always congruent triangles? Explain your answer.

Developing Skills

In 3–5, the information refers to a regular pyramid. Let e be the length of an edge of the base and h_s be the slant height. Find lateral area of each pyramid.

3. The pyramid has a square base; $e = 12$ cm, $h_s = 10$ cm

4. The pyramid has a triangular base; $e = 8.0$ ft, $h_s = 10$ ft

5. The pyramid has a base that is a hexagon; $e = 48$ cm, $h_s = 32$ cm

In 6–9, the information refers to a regular pyramid. Let e be the length of an edge of the base and h_p be the height of the pyramid. Find the volume of each pyramid.

6. The area of the base is 144 square centimeters; $h_p = 12$ cm

7. The area of the base is 27.6 square inches; $h_p = 5.0$ in.

8. The pyramid has a square base; $e = 2$ ft, $h_p = 1.5$ ft

9. The pyramid has a square base; $e = 22$ cm, $h_p = 14$ cm

10. The volume of a pyramid is 576 cubic inches and the height of the pyramid is 18 inches. Find the area of the base.

Applying Skills

11. A tetrahedron is a solid figure made up of four congruent equilateral triangles. Any one of the triangles can be considered to be the base and the other three to be the lateral sides of a regular pyramid. The length of a side of a triangle is 10.7 centimeters, the slant height is 9.27 centimeters, and the height of the prism is 8.74 centimeters.

  a. Find the area of the base of the tetrahedron.

  b. Find the lateral area of the tetrahedron.

  c. Find the total surface area of the tetrahedron.

  d. Find the volume of the tetrahedron.

12. When Connie camps, she uses a tent that is in the form of a regular pyramid with a square base. The length of an edge of the base is 9 feet and the height of the tent at its center is 8 feet. Find the volume of the space enclosed by the tent.

13. Prove that the lateral edges of a regular pyramid with a base that is an equilateral triangle are congruent.

14. Let F be the vertex of a pyramid with square base $ABCD$. If $\overline{AF} \cong \overline{CF}$, prove that the pyramid is regular.

15. Prove that the altitudes of the lateral faces of a regular pyramid with a base that is an equilateral triangle are congruent.

16. Let p be the perimeter of the base of a regular pyramid and h_s be the slant height. Prove that the lateral area of a regular pyramid is equal to $\frac{1}{2}ph_s$.

17. a. How does the lateral area of a regular pyramid change when both the slant height and the perimeter are doubled? tripled? Use the formula derived in exercise 16.

b. How does the volume of a regular pyramid with a triangle for a base change when both the sides of the base and the height of the pyramid are doubled? tripled?

11-7 CYLINDERS

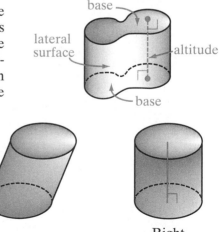

A prism has bases that are congruent polygons in parallel planes. What if the bases were congruent closed curves instead of polygons? Let $\overline{PP'}$ be a line segment joining corresponding points of two congruent curves. Imagine the surface generated as $\overline{PP'}$ moves along the curves, always joining corresponding points of the bases. The solid figure formed by the congruent parallel curves and the surface that joins them is called a **cylinder**.

The closed curves form the **bases** of the cylinder and the surface that joins the bases is the **lateral surface** of the cylinder. The **altitude** of a cylinder is a line segment perpendicular to the bases with endpoints on the bases. The **height** of a cylinder is the length of an altitude.

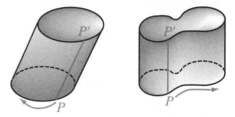

The most common cylinder is one that has bases that are congruent circles. This cylinder is a **circular cylinder**. If the line segment joining the centers of the circular bases is perpendicular to the bases, the cylinder is a **right circular cylinder**.

Circular cylinder

Right circular cylinder

Surface Area and Volume of a Circular Cylinder

The label on a cylindrical can of soup is a rectangle whose length is the circumference of the base of the can and whose width is the height of the can. This label is equal in area to the lateral surface of the cylindrical can. In Exercise 12 of Section 11-5, you proved

$2\pi r$

that the lateral area, A, of a right prism is the product of the perimeter, p, of the prism and the height, h_p, of the prism ($A = ph_p$). The circumference of the base of a cylinder corresponds to the perimeter of the base of a prism. Therefore, we can say that the area of the lateral surface of a right circular cylinder is equal to the circumference of the circular base times the height of the cylinder.

If a right circular cylinder has bases that are circles of radius r and height h, then:

$$\text{The lateral area of the cylinder} = 2\pi rh$$

$$\text{The total surface area of the cylinder} = 2\pi rh + 2\pi r^2$$

$$\text{The volume of the cylinder} = Bh = \pi r^2 h$$

Note: The volume of *any* circular cylindar is $\pi r^2 h$.

EXAMPLE I

Jenny wants to build a right circular cylinder out of cardboard with bases that have a radius of 6.0 centimeters and a height of 14 centimeters.

a. How many square centimeters of cardboard are needed for the cylinder *to the nearest square centimeter*?

b. What will be the volume of the cylinder *to the nearest cubic centimeter*?

Solution   **a.** $A = 2\pi rh + 2\pi r^2$

$= 2\pi(6.0)(14) + 2\pi(6.0)^2$

$= 168\pi + 72\pi$

$= 240\pi$ cm^2

Express this result as a rational approximation rounded to the nearest square centimeter.

$240\pi \approx 753.9822369 \approx 754$ cm^2

b. $V = \pi r^2 h$

$= \pi(6.0)^2(14)$

$= 504\pi$ cm^3

Express this result as a rational approximation rounded to the nearest cubic centimeter.

$$504\pi \approx 1{,}583.362697 \approx 1{,}584 \text{ cm}^3$$

Answers   **a.** 754 cm²   **b.** 1,584 cm³

Exercises

Writing About Mathematics

1. Amy said that if the radius of a circular cylinder were doubled and the height decreased by one-half, the volume of the cylinder would remain unchanged. Do you agree with Amy? Explain why or why not.

2. Cindy said that if the radius of a right circular cylinder were doubled and the height decreased by one-half, the lateral area of the cylinder would remain unchanged. Do you agree with Cindy? Explain why or why not.

Developing Skills

In 3–6, the radius of a base, r, and the height, h, of a right circular cylinder are given. Find for each cylinder: **a.** the lateral area, **b.** the total surface area, **c.** the volume.
   Express each measure as an exact value in terms of π.

3. $r = 34.0$ cm, $h = 60.0$ cm

4. $r = 4.0$ in., $h = 12$ in.

5. $r = 18.0$ in., $h = 2.00$ ft

6. $r = 1.00$ m, $h = 75.0$ cm

7. The volume of a right circular cylinder is 252 cubic centimeters and the radius of the base is 3.6 centimeters. What is the height of the cylinder to the nearest tenth?

8. The volume of a right circular cylinder is 586 cubic centimeters and the height of the cylinder is 4.6 centimeters. What is the radius of the base to the nearest tenth?

9. The areas of the bases of a cylinder are each 124 square inches and the volume of the cylinder is 1,116 cubic inches. What is the height, h, of the cylinder?

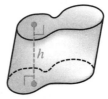

10. A circular cylinder has a base with a radius of 7.5 centimeters and a height of 12 centimeters. A rectangular prism has a square base and a height of 8.0 centimeters. If the cylinder and the prism have equal volumes, what is the length of the base of the prism to the nearest tenth?

Applying Skills

11. A can of beets has a top with a diameter of 2.9 inches and a height of 4.2 inches. What is the volume of the can to the nearest tenth?

12. A truck that delivers gasoline has a circular cylindrical storage space. The diameter of the bases of the cylinder is 11 feet, and the length (the height of the cylinder) is 17 feet. How many whole gallons of gasoline does the truck hold? (Use 1 cubic foot = 7.5 gallons.)

13. Karen makes pottery on a potter's wheel. Today she is making vases that are in the shape of a circular cylinder that is open at the top, that is, it has only one base. The base has a radius of 4.5 centimeters and is 0.75 centimeters thick. The lateral surface of the cylinder will be 0.4 centimeters thick. She uses 206 cubic centimeters of clay for each vase.

    a. How much clay is used for the base of the vase to the nearest tenth?

    b. How much clay will be used for the lateral surface of the vase to the nearest tenth?

    c. How tall will a vase be to the nearest tenth?

    d. What will be the area of the lateral surface to the nearest tenth? (Use the value of the height of the vase found in part **c.**)

14. Mrs. Taggart sells basic cookie dough mix. She has been using circular cylindrical containers to package the mix but wants to change to rectangular prisms that will pack in cartons for shipping more efficiently. Her present packaging has a circular base with a diameter of 4.0 inches and a height of 5.8 inches. She wants the height of the new package to be 6.0 inches and the dimensions of the base to be in the ratio 2 : 5. Find, to the nearest tenth, the dimensions of the new package if the volume is to be the same as the volume of the cylindrical containers.

11-8 CONES

Think of $\overleftrightarrow{OQ}$ perpendicular to plane p at O. Think of a point P on plane p. Keeping point Q fixed, move P through a circle on p with center at O. The surface generated by $\overleftrightarrow{PQ}$ is a **right circular conical surface**. Note that a conical surface extends infinitely.

In our discussion, we will consider the part of the conical surface generated by $\overline{PQ}$ from plane p to Q, called a **right circular cone**. The point Q is the **vertex** of the cone. The circle in plane p with radius OP is the **base** of the cone, $\overline{OQ}$ is the **altitude** of the cone, OQ is the **height** of the cone, and PQ is the **slant height** of the cone.

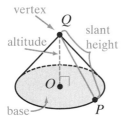

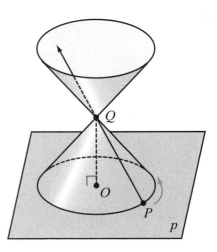

We can make a model of a right circular cone. Draw a large circle on a piece of paper and draw two radii. Cut out the circle and remove the part of the circle between the two radii. Join the two cut edges of the remaining part of the circle with tape.

Surface Area and Volume of a Cone

For a pyramid, we proved that the lateral area is equal to one-half the product of the perimeter of the base and the slant height. A similar relationship is true for a cone. The lateral area of a cone is equal to one-half the product of the circumference of the base and the slant height. Let L be the lateral area of the cone, C be the circumference of the base, S be the total surface area of the cone, h_s be the slant height, and r be the radius of the base. Then:

$$L = \tfrac{1}{2}Ch_s = \tfrac{1}{2}(2\pi r)h_s = \pi r h_s$$

$$S = L + \pi r^2 = \pi r h_s + \pi r^2$$

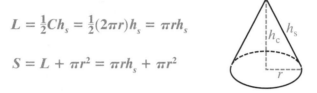

We can also use the relationship between the volume of a prism and the volume of a pyramid to write a formula for the volume of a cone. The volume of a cone is equal to one-third the product of the area of the base and the height of the cone. Let V be the volume of the cone, B be the area of the base with radius r, and h_c be the height of the cone. Then:

$$V = \tfrac{1}{3}Bh_c = \tfrac{1}{3}\pi r^2 h_c$$

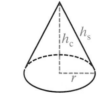

EXAMPLE I

A right circular cone has a base with a radius of 10 inches, a height of 24 inches, and a slant height of 26 inches. Find the exact values of:

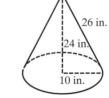

a. the lateral area   **b.** the area of the base
c. the total surface area of the cone

Solution The radius of the base, r, is 10 inches. Therefore, the diameter of the base is 20 inches, and the circumference of the base, C, is 20π, the slant height, h_s, is 26 inches, and the height of the cone, h_c, is 24 inches.

   a. Lateral area $= \tfrac{1}{2}Ch_s$

   $= \tfrac{1}{2}(20\pi)(26)$

   $= 260\pi$ in.^{2} *Answer*

b. Area of the base $= \pi(10)^2$

$\qquad = 100\pi$ cm^2 *Answer*

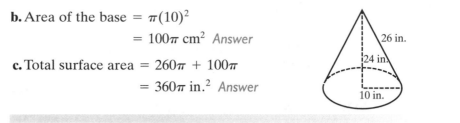

c. Total surface area $= 260\pi + 100\pi$

$\qquad = 360\pi$ in.^{2} *Answer*

EXAMPLE 2

A cone and a cylinder have equal volumes and equal heights. If the radius of the base of the cone is 3 centimeters, what is the radius of the base of the cylinder?

Solution Let r be the radius of the base of the cylinder and h be the height of both the cylinder and the cone.

$$\text{Volume of the cylinder} = \pi r^2 h$$
$$\text{Volume of the cone} = \tfrac{1}{3}\pi(3)^2 h$$

$$\text{Volume of the cylinder} = \text{Volume of the cone}$$
$$\pi r^2 h = \tfrac{1}{3}\pi(3)^2 h$$
$$r^2 = 3$$
$$r = \sqrt{3} \ \ \textit{Answer}$$

Exercises

Writing About Mathematics

1. Elaine said that if a pyramid and a cone have equal heights and bases that have equal areas, then they have equal lateral areas. Do you agree with Elaine? Justify your answer.

2. Josephus said that if two cones have equal heights and the radius of one cone is equal to the diameter of the other, then the volume of the larger cone is twice the volume of the smaller. Do you agree with Josephus? Explain why or why not.

Developing Skills

In 3–6, the radius of the base, r, the slant height of the cone, h_s, and the height of the cone, h_c, are given. Find: **a.** the lateral area of the cone, **b.** the total surface area of the cone, **c.** the volume of the cone.

Express each measure as an exact value in terms of π and rounded to the nearest tenth.

3. $r = 3.00$ cm, $h_s = 5.00$ cm, $h_c = 4.00$ cm

4. $r = 5.00$ cm, $h_s = 13.0$ cm, $h_c = 12.0$ cm

5. $r = 24$ cm, $h_s = 25$ cm, $h_c = 7.0$ cm

6. $r = 8.00$ cm, $h_s = 10.0$ cm, $h_c = 6.00$ cm

7. The volume of a cone is 127 cubic inches and the height of the cone is 6.0 inches. What is the radius of the base to the nearest tenth?

8. The volume of a cone is 56 cubic centimeters and the area of the base is 48 square centimeters. What is the height of the cone to the nearest tenth?

9. The area of the base of a cone is equal to the area of the base of a cylinder, and their volumes are equal. If the height of the cylinder is 2 feet, what is the height of the cone?

Applying Skills

10. The highway department has a supply of road salt for use in the coming winter. The salt forms a cone that has a height of 10 feet and a circular base with a diameter of 12 feet. How many cubic feet of salt does the department have stored? Round to the nearest foot.

11. The spire of the city hall is in the shape of a cone that has a circular base that is 20 feet in diameter. The slant height of the cone is 40 feet. How many whole gallons of paint will be needed to paint the spire if a gallon of paint will cover 350 square feet?

12. A cone with a height of 10 inches and a base with a radius of 6 inches is cut into two parts by a plane parallel to the base. The upper part is a cone with a height of 4 inches and a base with a radius of 2.4 inches. Find the volume of the lower part, the **frustum** of the cone, in terms of π.

13. When a cone is cut by a plane perpendicular to the base through the center of the base, the cut surface is a triangle whose base is the diameter of the base of the cone and whose altitude is the altitude of the cone. If the radius of the base is equal to the height of the cone, prove that the cut surface is an isosceles right triangle.

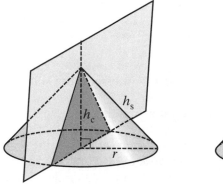
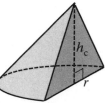

11-9 SPHERES

In a plane, the set of all points at a given distance from a fixed point is a circle. In space, this set of points is a *sphere*.

DEFINITION _____

A **sphere** is the set of all points equidistant from a fixed point called the **center**.

The **radius** of a sphere is the length of the line segment from the center of the sphere to any point on the sphere.

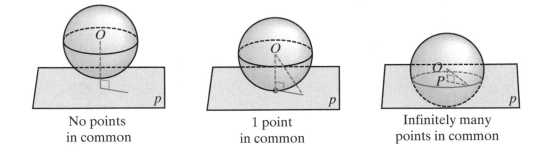

No points
in common

1 point
in common

Infinitely many
points in common

If the distance of a plane from the center of a sphere is greater than the radius of the sphere, the plane will have no points in common with the sphere. If the distance of a plane from the center of a sphere is equal to the radius of the sphere, the plane will have one point in common with the sphere. If the distance of a plane from the center of a sphere is less than the radius of the sphere, the plane will have infinitely many points in common with the sphere. We can prove that these points form a circle. Recall the definition of a circle.

▶ **A** *circle* **is the set of all points in a plane equidistant from a fixed point in the plane called the** *center.*

Theorem 11.14a   The intersection of a sphere and a plane through the center of the sphere is a circle whose radius is equal to the radius of the sphere.

Given   A sphere with center at O and radius r. Plane p through O intersects the sphere.

Prove   The intersection is a circle with radius r.

Proof   Let A be any point on the intersection. Since A is on the sphere, $OA = r$. Therefore, every point on the intersection is at the same distance from O and the intersection is a circle with radius r.

DEFINITION _____
A **great circle of a sphere** is the intersection of a sphere and a plane through the center of the sphere.

Theorem 11.14b   If the intersection of a sphere and a plane does not contain the center of the sphere, then the intersection is a circle.

Given   A sphere with center at O plane p intersecting the sphere at A and B.

Prove   The intersection is a circle.

Proof	Statements	Reasons
	1. Draw a line through O, perpendicular to plane p at C.	**1.** Through a given point there is one line perpendicular to a given plane.
	2. $\angle OCA$ and $\angle OCB$ are right angles.	**2.** A line perpendicular to a plane is perpendicular to every line in the plane through the intersection of the line and the plane.
	3. $\overline{OA} \cong \overline{OB}$	**3.** A sphere is the set of points in space equidistant from a fixed point.
	4. $\overline{OC} \cong \overline{OC}$	**4.** Reflexive property.
	5. $\triangle OAC \cong \triangle OBC$	**5.** HL.
	6. $\overline{CA} \cong \overline{CB}$	**6.** Corresponding sides of congruent triangles are congruent.
	7. The intersection is a circle.	**7.** A circle is the set of all points in a plane equidistant from a fixed point.

We can write Theorems 11.14a and 11.14b as a single theorem.

Theorem 11.14

> The intersection of a plane and a sphere is a circle.

In the proof of Theorem 11.14b, we drew right triangle OAC with OA the radius of the sphere and AC the radius of the circle at which the plane and the sphere intersect. Since $\angle OCA$ is the right angle, it is the largest angle of $\triangle OAC$ and $OA > AC$. Therefore, a great circle, whose radius is equal to the radius of the sphere, is larger than any other circle that can be drawn on the sphere. We have just proved the following corollary:

Corollary 11.14a

> A great circle is the largest circle that can be drawn on a sphere.

Let p and q be any two planes that intersect the sphere with center at O. In the proof of Theorem 11.14b, the radius of the circle is the length of a leg of a right triangle whose hypotenuse is the radius of the sphere and whose other leg is the distance from the center of the circle to the plane. This suggests that if two planes are equidistant from the center of a sphere, they intersect the sphere in congruent circles.

Theorem 11.15

> If two planes are equidistant from the center of a sphere and intersect the sphere, then the intersections are congruent circles.

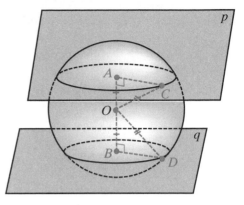

Given A sphere with center at O intersected by planes p and q, $OA = OB$, $\overline{OA} \perp p$ and $\overline{OB} \perp q$.

Prove The intersections are congruent circles.

Proof Let C be any point on the intersection with p and D be any point on the intersection with q. Then $OA = OB$ and $OC = OD$ (they are both radii of the sphere). Therefore, $\triangle OAC$ and $\triangle OBD$ are congruent right triangles by HL. Since the corresponding sides of congruent triangles are congruent, the radii of the circles, AC and BD, are equal and the circles are congruent.

Surface Area and Volume of a Sphere

The formulas for the surface area and volume of a sphere are derived in advanced courses in mathematics. We can state and make use of these formulas.

The surface area of a sphere is equal to the area of four great circles. Let S be the surface area of a sphere of radius r. Then the surface area of the sphere is:

$$S = 4\pi r^2$$

The volume of a sphere is equal to four-thirds the product of π and the cube of the radius. Let V be the volume of a sphere of radius r. Then the volume of the sphere is:

$$V = \tfrac{4}{3}\pi r^3$$

EXAMPLE 1

Find the surface area and the volume of a sphere whose radius is 5.25 centimeters to the nearest centimeter.

Solution

$$
\begin{aligned}
S &= 4\pi r^2 \\
&= 4\pi(5.25)^2 \\
&= 110.25\pi \\
&\approx 346.3605901 \text{ cm}^2
\end{aligned}
$$

When we round to the nearest centimeter to express the answer,
$S = 346$ cm². *Answer*

$$V = \tfrac{4}{3}\pi r^3$$
$$= \tfrac{4}{3}\pi (5.25)^3$$
$$= 192.9375\pi$$
$$\approx 606.1310326 \text{ cm}^3$$

When we round to the nearest centimeter to express the answer,
$V = 606$ cm³. *Answer*

Exercises

Writing About Mathematics

1. Meg said that if d is the diameter of a sphere, then the surface area of a sphere is equal to πd^2. Do you agree with Meg? Justify your answer.

2. Tim said that if the base a cone is congruent to a great circle of a sphere and the height of the cone is the radius of the sphere, then the volume of the cone is one-half the volume of the sphere. Do you agree with Tim? Justify your answer.

Developing Skills

In 3–6, find the surface area and the volume of each sphere whose radius, r, is given. Express each answer in terms of π and as a rational approximation to the nearest unit.

3. $r = 7.50$ in.

4. $r = 13.2$ cm

5. $r = 2.00$ ft

6. $r = 22.3$ cm

7. Find the radius of a sphere whose surface area is 100π square feet.

8. Find, to the nearest tenth, the radius of a sphere whose surface area is 84 square centimeters.

9. Find, to the nearest tenth, the radius of a sphere whose volume is 897 cubic inches.

10. Express, in terms of π, the volume of a sphere whose surface area is 196π square inches.

Applying Skills

11. A vase is in the shape of a sphere with a radius of 3 inches. How many whole cups of water will come closest to filling the vase? (1 cup = 14.4 cubic inches)

12. The radius of a ball is 5.0 inches. The ball is made of a soft foam that weighs 1 ounce per 40 cubic inches. How much does the ball weigh to the nearest tenth?

13. The diameter of the earth is about 7,960 miles. What is the surface area of the earth in terms of π?

14. The diameter of the moon is about 2,160 miles. What is the surface area of the moon in terms of π?

15. A cylinder has a base congruent to a great circle of a sphere and a height equal to the diameter of the sphere. If the radius of the sphere is 16 centimeters, compare the lateral area of the cylinder and the surface area of the sphere.

16. A cylinder has a base congruent to a great circle of a sphere and a height equal to the diameter of the sphere. If the diameter of the sphere is r, compare the lateral area of the cylinder and the surface area of the sphere.

Hands-On Activity

A **symmetry plane** is a plane that divides a solid into two congruent parts. For each solid shown below, determine the number of symmetry planes and describe their position relative to the solid.

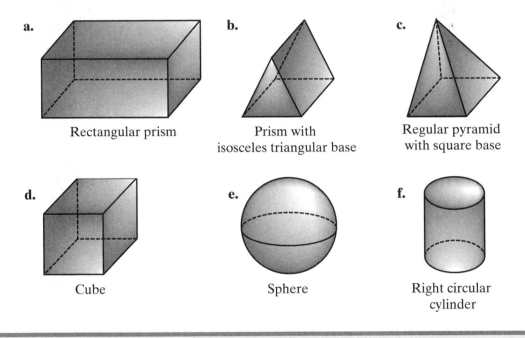

a. Rectangular prism

b. Prism with isosceles triangular base

c. Regular pyramid with square base

d. Cube

e. Sphere

f. Right circular cylinder

CHAPTER SUMMARY

Definitions to Know

- **Parallel lines in space** are lines in the same plane that have no points in common.
- **Skew lines** are lines in space that are neither parallel nor intersecting.
- A **dihedral angle** is the union of two half-planes with a common edge.
- The **measure of a dihedral angle** is the measure of the plane angle formed by two rays each in a different half-plane of the angle and each perpendicular to the common edge at the same point of the edge.

- **Perpendicular planes** are two planes that intersect to form a right dihedral angle.

- A **line is perpendicular to a plane** if and only if it is perpendicular to each line in the plane through the intersection of the line and the plane.

- A **plane is perpendicular to a line** if the line is perpendicular to the plane.

- **Parallel planes** are planes that have no points in common.

- A **line is parallel to a plane** if it has no points in common with the plane.

- The **distance between two planes** is the length of the line segment perpendicular to both planes with an endpoint on each plane.

- A **polyhedron** is a three-dimensional figure formed by the union of the surfaces enclosed by plane figures.

- The portions of the planes enclosed by a plane figure are called the **faces** of the polyhedron.

- The intersections of the faces are the **edges** of the polyhedron, and the intersections of the edges are the **vertices** of the polyhedron.

- A **prism** is a polyhedron in which two of the faces, called the **bases** of the prism, are congruent polygons in parallel planes.

- The sides of a prism that are not bases are called the **lateral sides**.

- The union of two lateral sides is a **lateral edge**.

- If the line segments joining the corresponding vertices of the bases of a prism are perpendicular to the planes of the bases, then the prism is a **right prism**.

- The **altitude** of a prism is a line segment perpendicular to each of the bases with an endpoint on each base.

- The **height** of a prism is the length of an altitude.

- A **parallelepiped** is a prism that has parallelograms as bases.

- A **rectangular parallelepiped** is a parallelepiped that has rectangular bases and lateral edges perpendicular to the bases.

- The **lateral area** of the prism is the sum of the areas of the lateral faces.

- The **total surface area** of a solid figure is the sum of the lateral area and the areas of the bases.

- A **pyramid** is a solid figure with a base that is a polygon and lateral faces that are triangles.

- A **regular pyramid** is a pyramid whose base is a regular polygon and whose altitude is perpendicular to the base at its center.

- The length of the altitude of a triangular lateral face of a regular pyramid is the **slant height** of the pyramid.

- A **cylinder** is a solid figure formed by congruent parallel curves and the surface that joins them.

- If the line segment joining the centers of circular bases of a cylinder is perpendicular to the bases, the cylinder is a **right circular cylinder**.

- Let $\overline{OP}$ be a line segment perpendicular to a plane at O and A be a point on a circle in the plane with center at O. A **right circular cone** is the solid figure that is the union of a circular base and the surface generated by line segment $\overline{AP}$ as P moves around the circle.

- A **sphere** is the set of all points equidistant from a fixed point called the **center**.

- The **radius** of a sphere is the length of the line segment from the center of the sphere to any point on the sphere.

- A **great circle of a sphere** is the intersection of a sphere and a plane through the center of the sphere.

Postulates   **11.1**   There is one and only one plane containing three non-collinear points.

11.2   A plane containing any two points contains all of the points on the line determined by those two points.

11.3   If two planes intersect, then they intersect in exactly one line.

11.4   At a given point on a line, there are infinitely many lines perpendicular to the given line.

11.5   The volume of a prism is equal to the area of the base times the height.

Theorems and   **11.1**   There is exactly one plane containing a line and a point not on the line.
Corollaries   **11.2**   Two intersecting lines determine a plane.

11.3   If a line not in a plane intersects the plane, then it intersects in exactly one point.

11.4   If a line is perpendicular to each of two intersecting lines at their point of intersection, then the line is perpendicular to the plane determined by these lines.

11.5   Two planes are perpendicular if and only if one plane contains a line perpendicular to the other.

11.6   Through a given point on a plane, there is only one line perpendicular to the given plane.

11.7   Through a given point on a line, there can be only one plane perpendicular to the given line.

11.8   If a line is perpendicular to a plane, then any line perpendicular to the given line at its point of intersection with the given plane is in the plane.

11.9   If a line is perpendicular to a plane, then every plane containing the line is perpendicular to the given plane.

11.10   If a plane intersects two parallel planes, then the intersection is two parallel lines.

11.11   Two lines perpendicular to the same plane are parallel.

11.11a   Two lines perpendicular to the same plane are coplanar.

11.12   Two planes are perpendicular to the same line if and only if the planes are parallel.

11.13   Parallel planes are everywhere equidistant.

11.14 The intersection of a plane and a sphere is a circle.

11.14a A great circle is the largest circle that can be drawn on a sphere.

11.15 If two planes are equidistant from the center of a sphere and intersect the sphere, then the intersections are congruent circles.

Formulas

L = lateral area
S = surface area
C = circumference of the base of a cone
V = volume

p = perimeter of a base
r = radius of a base of a cylinder or a cone; radius of a sphere
B = area of a base

h_s = height of a lateral surface
h_p = height of a prism or a pyramid
h_c = height of a cylinder or a cone

Right Prism	Regular Pyramid	Right Circular Cylinder
$L = ph_s$	$L = \frac{1}{2}ph_s$	$L = 2\pi rh_c$
$S = L + 2B$	$S = L + B$	$S = 2\pi rh_c + 2\pi r^2$
$V = Bh_p$	$V = \frac{1}{3}Bh_p$	$V = Bh_c = \pi r^2 h_c$

Cone	Sphere
$L = \frac{1}{2}Ch_s = \pi rh_s$	$S = 4\pi r^2$
$V = \frac{1}{3}Bh_c = \frac{1}{3}\pi r^2 h_c$	$V = \frac{4}{3}\pi r^3$

VOCABULARY

11-1 Cavalieri's Principle • Solid geometry • Parallel lines in space • Skew lines

11-2 Dihedral angle • Plane angle • Measure of a dihedral angle • Perpendicular planes • Line perpendicular to a plane • Plane perpendicular to a line

11-3 Parallel planes • Line parallel to a plane • Distance between two planes

11-4 Polyhedron • Faces of a polyhedron • Edges of a polyhedron • Vertices of a polyhedron • Prism • Bases of a prism • Lateral sides of a prism • Lateral edge of a prism • Altitude of a prism • Height of a prism • Right prism • Parallelepiped • Rectangular parallelepiped • Rectangular solid • Lateral area • Total surface area

11-5 Cubic centimeter

11-6 Pyramid • Vertex of a pyramid • Altitude of a pyramid • Height of a pyramid • Regular pyramid • Slant height of a pyramid

11-7 Cylinder • Base of a cylinder • Lateral surface of a cylinder • Altitude of a cylinder • Height of a cylinder • Right circular cylinder

11-8 Right circular conical surface • Right circular cone • Vertex of a cone • Base of a cone • Altitude of a cone • Height of a cone • Slant height of a cone • Frustum of a cone

11-9 Sphere • Center of a sphere • Radius of a sphere • Great circle of a sphere • Symmetry plane

REVIEW EXERCISES

In 1–16, answer each question and state the postulate, theorem, or definition that justifies your answer or draw a counterexample.

1. Lines $\overleftrightarrow{AB}$ and $\overleftrightarrow{CD}$ intersect at E. A line, $\overleftrightarrow{EF}$, is perpendicular to $\overleftrightarrow{AB}$ and to $\overleftrightarrow{CD}$. Is $\overleftrightarrow{EF}$ perpendicular to the plane determined by the intersecting lines?

2. Plane p is perpendicular to $\overleftrightarrow{AB}$ at B. Plane q intersects $\overleftrightarrow{AB}$ at B. Can q be perpendicular to $\overleftrightarrow{AB}$?

3. A line, $\overleftrightarrow{RS}$, is perpendicular to plane p at R. If T is a second point not on p, can $\overleftrightarrow{RT}$ be perpendicular to plane p?

4. Lines $\overleftrightarrow{AB}$ and $\overleftrightarrow{LM}$ are each perpendicular to plane p. Are $\overleftrightarrow{AB}$ and $\overleftrightarrow{LM}$ coplanar?

5. A line $\overleftrightarrow{AB}$ is in plane q and $\overleftrightarrow{AB}$ is perpendicular to plane p. Are planes p and q perpendicular?

6. Two planes, p and q, are perpendicular to each other. Does p contain a line perpendicular to q?

7. A line $\overleftrightarrow{AB}$ is perpendicular to plane p at B and $\overleftrightarrow{BC} \perp \overleftrightarrow{AB}$. Is $\overleftrightarrow{BC}$ in plane p?

8. A line $\overleftrightarrow{RS}$ is perpendicular to plane p and $\overleftrightarrow{RS}$ is in plane q. Is plane p perpendicular to plane q?

9. Plane r intersects plane p in $\overleftrightarrow{AB}$ and plane r intersects plane q in $\overleftrightarrow{CD}$. Can $\overleftrightarrow{AB}$ intersect $\overleftrightarrow{CD}$?

10. Planes p and q are each perpendicular to $\overleftrightarrow{AB}$. Are p and q parallel?

11. Two lateral edges of a prism are $\overline{AB}$ and $\overline{CD}$. Can $\overleftrightarrow{AB}$ and $\overleftrightarrow{CD}$ intersect?

12. Two lateral edges of a prism are $\overline{AB}$ and $\overline{CD}$. Is $AB = CD$?

13. Two prisms have equal heights and bases with equal areas. Do the prisms have equal volumes?

14. A prism and a pyramid have equal heights and bases with equal areas. Do the prism and the pyramid have equal volumes?

15. Two planes intersect a sphere at equal distances from the center of the sphere. Are the circles at which the planes intersect the sphere congruent?

16. Plane p intersects a sphere 2 centimeters from the center of the sphere and plane q contains the center of the sphere and intersects the sphere. Are the circles at which the planes intersect the sphere congruent circles?

In 17–22, find the lateral area, total surface area, and volume of each solid figure to the nearest unit.

17. The length of each side of the square base of a rectangular prism is 8 centimeters and the height is 12 centimeters.

18. The height of a prism with bases that are right triangles is 5 inches. The lengths of the sides of the bases are 9, 12, and 15 inches.

19. The base of a rectangular solid measures 9 feet by 7 feet and the height of the solid is 4 feet.

20. A pyramid has a square base with an edge that measures 6 inches. The slant height of a lateral side is 5 inches and the height of the pyramid is 4 inches.

21. The diameter of the base of a cone is 10 feet, its height is 12 feet, and its slant height is 13 feet.

22. The radius of the base of a right circular cylinder is 7 centimeters and the height of the cylinder is 9 centimeters.

23. A cone and a pyramid have equal volumes and equal heights. Each side of the square base of the pyramid measures 5 meters. What is the radius of the base of the cone? Round to the nearest tenth.

24. Two prisms with square bases have equal volumes. The height of one prism is twice the height of the other. If the measure of a side of the base of the prism with the shorter height is 14 centimeters, find the measure of a side of the base of the other prism in simplest radical form.

25. Ice cream is sold by street vendors in containers that are right circular cylinders. The base of the cylinder has a diameter of 5 inches and the cylinder has a height of 6 inches.

    a. Find, to the nearest tenth, the amount of ice cream that a container can hold.

    b. If a scoop of ice cream is a sphere with a diameter of 2.4 inches, find, to the nearest tenth, the amount of ice cream in a single scoop.

    c. If the ice cream is packed down into the container, how many whole scoops of ice cream will come closest to filling the container?

Exploration

A **regular polyhedron** is a solid, all of whose faces are congruent regular polygons with the sides of the same number of polygons meeting at each vertex. There are five regular polyhedra: a tetrahedron, a cube, an octahedron, a dodecahedron, and an icosahedron. These regular polyhedra are called the **Platonic solids**.

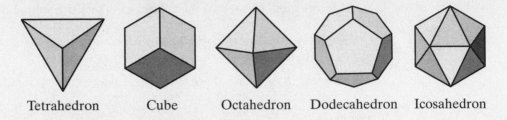

Tetrahedron   Cube   Octahedron   Dodecahedron   Icosahedron

a. Make a paper model of the Platonic solids as follows:
   (1) Draw the diagrams below on paper.
   (2) Cut out the diagrams along the solid lines and fold along the dotted lines.
   (3) Tape the folded sides together.

Note: You may wish to enlarge to the diagrams for easier folding.

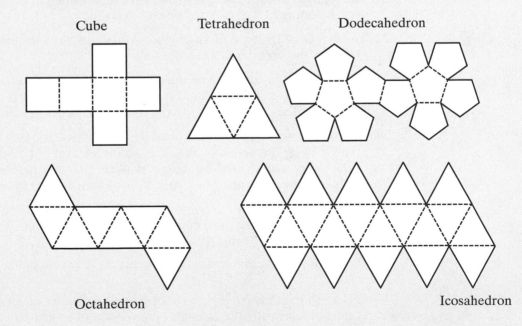

Cube   Tetrahedron   Dodecahedron

Octahedron   Icosahedron

b. Using the solids you constructed in part **a**, fill in the table below:

	Number of vertices	Number of faces	Number of edges
Tetrahedron			
Cube			
Octahedron			
Dodecahedron			
Icosahedron			

Do you observe a relationship among the number of vertices, faces, and edges for each Platonic solid? If so, state this relationship.

c. Research the five Platonic solids and investigate why there are only five regular polyhedra. Share your findings with your classmates.

CUMULATIVE REVIEW — Chapters 1–11

Part I

Answer all questions in this part. Each correct answer will receive 2 credits. No partial credit will be allowed.

1. If the measures of the angles of a triangle are represented by x, $2x - 20$, and $2x$, what is the measure of the smallest angle?

(1) 40 (2) 60 (3) 80 (4) 90

2. In $\triangle ABC$, $m\angle A = 40$ and the measure of an exterior angle at B is 130. The triangle is

(1) scalene and acute (3) isosceles and right
(2) scalene and right (4) isosceles and acute

3. The coordinates of the midpoint of a segment with endpoints at $(2, -3)$ and $(-6, 1)$ are

(1) $(4, 2)$ (2) $(4, -2)$ (3) $(-4, -2)$ (4) $(-2, -1)$

4. The lengths of the diagonals of a rhombus are 8 centimeters and 12 centimeters. The area of the rhombus is

(1) 24 cm^2 (2) 32 cm^2 (3) 48 cm^2 (4) 96 cm^2

5. Two parallel lines are cut by a transversal. The measure of one interior angle is $x + 7$ and the measure of another interior angle on the same side of the transversal is $3x - 3$. What is the value of x?

(1) 5 (2) 12 (3) 44 (4) 51

6. If "Today is Monday" is true and "It is May 5" is false, which of the following is true?
   (1) Today is Monday and it is May 5.
   (2) If today is Monday, then it is May 5.
   (3) Today is Monday only if it is May 5.
   (4) Today is Monday or it is May 5.

7. Which of the following do not always lie in the same plane?
   (1) three points   (3) two intersecting lines
   (2) two parallel lines   (4) three parallel lines

8. What is the slope of a line perpendicular to the line whose equation is $x - 2y = 3$?
   (1) $\frac{1}{2}$ (2) 2   (3) $-\frac{1}{2}$ (4) -2

9. A quadrilateral has diagonals that are not congruent and are perpendicular bisectors of each other. The quadrilateral is a
   (1) square   (2) rectangle   (3) trapezoid   (4) rhombus

10. The base of a right prism is a square whose area is 36 square centimeters. The height of the prism is 5 centimeters. The lateral area of a prism is
    (1) 30 cm^2 (2) 60 cm^2 (3) 120 cm^2 (4) 180 cm^2

Part II

Answer all questions in this part. Each correct answer will receive 2 credits. Clearly indicate the necessary steps, including appropriate formula substitutions, diagrams, graphs, charts, etc. For all questions in this part, a correct numerical answer with no work shown will receive only 1 credit.

11. A leg, $\overline{AB}$, of isosceles $\triangle ABC$ is congruent to a leg, $\overline{DE}$, of isosceles $\triangle DEF$. The vertex angle, $\angle B$, of isosceles $\triangle ABC$ is congruent to the vertex angle, $\angle E$, of isosceles $\triangle DEF$. Prove that $\triangle ABC \cong \triangle DEF$.

12. In triangle ABC, altitude $\overline{CD}$ bisects $\angle C$. Prove that the triangle is isosceles.

Part III

Answer all questions in this part. Each correct answer will receive 4 credits. Clearly indicate the necessary steps, including appropriate formula substitutions, diagrams, graphs, charts, etc. For all questions in this part, a correct numerical answer with no work shown will receive only 1 credit.

13. Quadrilateral $BCDE$ is a parallelogram and B is the midpoint of $\overline{ABC}$. Prove that $ABDE$ is a parallelogram.

14. Line segment $\overline{ABC}$ is perpendicular to plane p at B, the midpoint of $\overline{ABC}$. Prove that any point on p is equidistant from A and C.

Part IV

Answer all questions in this part. Each correct answer will receive 6 credits. Clearly indicate the necessary steps, including appropriate formula substitutions, diagrams, graphs, charts, etc. For all questions in this part, a correct numerical answer with no work shown will receive only 1 credit.

15. Write the equation of the median from B to $\overline{AC}$ of $\triangle ABC$ if the coordinates of the vertices are $A(-3, -2)$, $B(-1, 4)$, and $C(5, -2)$.

16. The coordinates of the endpoints of $\overline{AB}$ are $A(1, 3)$ and $B(5, -1)$. Find the coordinates of the endpoints of $\overline{A'B'}$, the image of $\overline{AB}$ under the composition $r_{x\text{-axis}} \circ R_{90°}$.

CHAPTER

12

RATIO, PROPORTION, AND SIMILARITY

The relationship that we know as the Pythagorean Theorem was known by philosophers and mathematicians before the time of Pythagoras (c. 582–507 B.C.). The Indian mathematician Baudhāyana discovered the theorem more than 300 years before Pythagoras. The Egyptians made use of the 3-4-5 right triangle to determine a right angle. It may have been in Egypt, where Pythagoras studied, that he become aware of this relationship. Ancient sources agree that Pythagoras gave a proof of this theorem but no original documents exist from that time period.

Early Greek statements of this theorem did not use the algebraic form $c^2 = a^2 + b^2$ with which we are familiar. Proposition 47 in Book 1 of Euclid's *Elements* states the theorem as follows:

"In right angled triangles, the square on the side subtending the right angle is equal to the sum of the squares on the sides containing the right angle."

Euclid's proof drew squares on the sides of the right triangle and proved that the area of the square drawn on the hypotenuse was equal to the sum of the areas of the squares drawn on the legs.

There exist today hundreds of proofs of this theorem.

12-1 RATIO AND PROPORTION

People often use a computer to share pictures with one another. At first the pictures may be shown on the computer screen as many small frames. These small pictures can be enlarged on the screen so that it is easier to see detail. Or a picture may be printed and then enlarged. Each picture, whether on the screen or printed, is *similar* to the original. When a picture is enlarged, the dimensions of each shape are in proportion to each other and with angle measure remaining the same. Shapes that are related in this way are said to be **similar**.

In this chapter we will review what we already know about ratio and proportion and apply those ideas to geometric figures.

The Meaning of Ratio

DEFINITION _____

The **ratio of two numbers**, a and b, where b is not zero, is the number $\frac{a}{b}$.

The ratio $\frac{a}{b}$ can also be written as $a : b$.

The two triangles $\triangle ABC$ and $\triangle DEF$ have the same shape but not the same size:

$AB = 20$ millimeters and $DE = 10$ millimeters. We can compare these lengths by means of a ratio, $\frac{20}{10}$ or $20 : 10$.

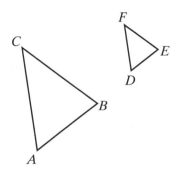

Since a ratio, like a fraction, is a comparison of two numbers by division, a ratio can be simplified by dividing each term of the ratio by a common factor. Therefore, the ratio of AB to DE can be written as $10 : 5$ or as $4 : 2$ or as $2 : 1$. A ratio is in simplest form when the terms of the ratio have no common factor greater than 1.

When the numbers represent lengths such as AB and DE, the lengths must be expressed in terms of the same unit of measure for the ratio to be meaningful.

For example, if AB had been given as 2 centimeters, it would have been necessary to change 2 centimeters to 20 millimeters before writing the ratio of AB to DE. Or we could have changed the length of $\overline{DE}$, 10 millimeters, to 1 centimeter before writing the ratio of AB to DE as $2 : 1$.

- When using millimeters, the ratio 20 mm : 10 mm = $2 : 1$.

- When using centimeters, the ratio 2 cm : 1 cm = $2 : 1$.

A ratio can also be used to express the relationship among three or more numbers. For example, if the measures of the angles of a triangle are 45, 60, and 75, the ratio of these measures can be written as $45 : 60 : 75$ or, in lowest terms, $3 : 4 : 5$.

When we do not know the actual values of two or more measures that are in a given ratio, we use a variable factor to express these measures. For example, if the lengths of the sides of a triangle are in the ratio $3:3:4$, we can let x be the greatest common factor of the measures of the sides. Then the measures of the sides may be expressed as $3x$, $3x$, and $4x$. If the perimeter of the triangle is 120 centimeters, this use of the variable x allows us to write and solve an equation.

$$3x + 3x + 4x = 120$$
$$10x = 120$$
$$x = 12$$

The measures of the sides of the triangle are $3(12)$, $3(12)$, and $4(12)$ or 36 centimeters, 36 centimeters, and 48 centimeters.

The Meaning of Proportion

Since the ratio $12:16$ is equal to the ratio $3:4$, we may write $\frac{12}{16} = \frac{3}{4}$. The equation $\frac{12}{16} = \frac{3}{4}$ is called a proportion. The proportion can also be written as $12:16 = 3:4$.

DEFINITION

A **proportion** is an equation that states that two ratios are equal.

The proportion $\frac{a}{b} = \frac{c}{d}$ can be written also as $a:b = c:d$. The four numbers a, b, c, and d are the terms of the proportion. The first and fourth terms, a and d, are the **extremes** of the proportion, and the second and third terms, b and c, are the **means**.

$$
\overset{\text{extremes}}{\overbrace{a:b = c:\!d}}
$$
$$
\underset{\text{means}}{\underbrace{}}
$$

Theorem 12.1

In a proportion, the product of the means is equal to the product of the extremes.

Given $\frac{a}{b} = \frac{c}{d}$ with $b \neq 0$ and $d \neq 0$

Prove $ad = bc$

Proof We can give an algebraic proof of this theorem.

	Statements	Reasons
1.	$\frac{a}{b} = \frac{c}{d}$	**1.** Given.
2.	$bd\left(\frac{a}{b}\right) = bd\left(\frac{c}{d}\right)$	**2.** Multiplication postulate.
3.	$\frac{b}{b}(ad) = \frac{d}{d}(bc)$	**3.** Associative property of multiplication.
4.	$1(ad) = 1(bc)$	**4.** A quantity may be substituted for its equal.
5.	$ad = bc$	**5.** Multiplicative identity.

Corollary 12.1a In a proportion, the means may be interchanged.

Given $\frac{a}{b} = \frac{c}{d}$ with $b \neq 0$, $c \neq 0$, and $d \neq 0$

Prove $\frac{a}{c} = \frac{b}{d}$

Proof It is given that $\frac{a}{b} = \frac{c}{d}$ and that $c \neq 0$. Then $\frac{b}{c} \times \frac{a}{b} = \frac{b}{c} \times \frac{c}{d}$ by the multiplication postulate of equality. Therefore, $\frac{a}{c} = \frac{b}{d}$.

Corollary 12.1b In a proportion, the extremes may be interchanged.

Given $\frac{a}{b} = \frac{c}{d}$ with $a \neq 0$, $b \neq 0$, and $d \neq 0$

Prove $\frac{d}{b} = \frac{c}{a}$

Proof It is given that $\frac{a}{b} = \frac{c}{d}$ and that $a \neq 0$. Then $\frac{d}{a} \times \frac{a}{b} = \frac{d}{a} \times \frac{c}{d}$ by the multiplication postulate of equality. Therefore, $\frac{d}{b} = \frac{c}{a}$.

These two corollaries tell us that:

If $3 : 5 = 12 : 20$, then $3 : 12 = 5 : 20$ and $20 : 5 = 12 : 3$.

Any two pairs of factors of the same number can be the means and the extremes of a proportion. For example, since $2(12) = 3(8)$, 2 and 12 can be the means of a proportion and 3 and 8 can be the extremes. We can write several proportions:

$$\frac{3}{2} = \frac{12}{8} \qquad \frac{8}{2} = \frac{12}{3} \qquad \frac{3}{12} = \frac{2}{8} \qquad \frac{8}{12} = \frac{2}{3}$$

The four proportions at the bottom of page 477 demonstrate the following corollary:

Corollary 12.1c

> If the products of two pairs of factors are equal, the factors of one pair can be the means and the factors of the other the extremes of a proportion.

The Mean Proportional

DEFINITION

If the two means of a proportion are equal, either mean is called the **mean proportional** between the extremes of the proportion.

In the proportion $\frac{2}{6} = \frac{6}{18}$, 6 is the mean proportional between 2 and 18. The mean proportional is also called the **geometric mean**.

EXAMPLE 1

Solve for x: $\frac{27}{x + 1} = \frac{9}{2}$

Solution Use that the product of the means is equal to the product of the extremes.

$$\frac{27}{x + 1} = \frac{9}{2}$$

$$9(x + 1) = 27(2)$$

$$9x + 9 = 54$$

$$9x = 45$$

$$x = 5$$

Check

$$\frac{27}{x + 1} = \frac{9}{2}$$

$$\frac{27}{5 + 1} \stackrel{?}{=} \frac{9}{2} \quad \text{Substitute 5 for } x.$$

$$\frac{27}{6} \stackrel{?}{=} \frac{9}{2} \quad \text{Simplify.}$$

$$\frac{9}{2} = \frac{9}{2} \checkmark$$

Answer $x = 5$

EXAMPLE 2

Find the mean proportional between 9 and 8.

Solution Let x represent the mean proportional.

$$\frac{9}{x} = \frac{x}{8}$$

$$x^2 = 72$$

$$x = \pm\sqrt{72} = \pm\sqrt{36}\sqrt{2} = \pm 6\sqrt{2}$$

Note that there are two solutions, one positive and one negative.

Answer $\pm 6\sqrt{2}$

EXAMPLE 3

The measures of an exterior angle of a triangle and the adjacent interior angle are in the ratio 7 : 3. Find the measure of the exterior angle.

Solution An exterior angle and the adjacent interior angle are supplementary.

Let $7x$ = the measure of the exterior angle,

and $3x$ = the measure of the interior angle.

$$7x + 3x = 180$$
$$10x = 180$$
$$x = 18$$
$$7x = 126$$

Answer The measure of the exterior angle is 126°.

Exercises

Writing About Mathematics

1. Carter said that a proportion can be rewritten by using the means as extremes and the extremes as means. Do you agree with Carter? Explain why or why not.

2. Ethan said that the mean proportional will be a rational number only if the extremes are both perfect squares. Do you agree with Ethan? Explain why or why not.

Developing Skills

In 3–8, determine whether each pair of ratios can form a proportion.

3. 6 : 15, 4 : 10

4. 8 : 7, 56 : 49

5. 49 : 7, 1 : 7

6. 10 : 15, 8 : 12

7. 9 : 3, 16 : 4

8. $3a : 5a, 12 : 20 \ (a \neq 0)$

In 9–11, use each set of numbers to form two proportions.

9. 30, 6, 5, 1

10. 18, 12, 6, 4

11. 3, 10, 15, 2

12. Find the exact value of the geometric mean between 10 and 40.

13. Find the exact value of the geometric mean between 6 and 18.

In 14–19, find the value of x in each proportion.

14. $4 : x = 10 : 15$

15. $\frac{9}{8} = \frac{x}{36}$

16. $\frac{12}{x + 1} = \frac{8}{x}$

17. $12 : x = x : 75$

18. $3x : 15 = 20 : x$

19. $x + 3 : 6 = 4 : x - 2$

Applying Skills

20. B is a point on $\overline{ABC}$ such that $AB : BC = 4 : 7$. If $AC = 33$, find AB and BC.

21. A line segment 48 centimeters long is divided into two segments in the ratio $1 : 5$. Find the measures of the segments.

22. A line segment is divided into two segments that are in the ratio $3 : 5$. The measure of one segment is 12 centimeters longer than the measure of the other. Find the measure of each segment.

23. The measures of the sides of a triangle are in the ratio $5 : 6 : 7$. Find the measure of each side if the perimeter of the triangle is 72 inches.

24. Can the measures of the sides of a triangle be in the ratio $2 : 3 : 7$? Explain why or why not.

25. The length and width of a rectangle are in the ratio $5 : 8$. If the perimeter of the rectangle is 156 feet, what are the length and width of the rectangle?

26. The measures of two consecutive angles of a parallelogram are in the ratio $2 : 7$. Find the measure of each angle.

12-2 PROPORTIONS INVOLVING LINE SEGMENTS

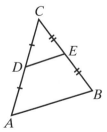

The midpoint of any line segment divides the segment into two congruent parts. In $\triangle ABC$, let D be the midpoint of $\overline{AC}$ and E be the midpoint of $\overline{BC}$. Draw the midsegment, $\overline{DE}$.

The line segment joining the midpoints of $\triangle ABC$ forms a new triangle, $\triangle DEC$. What are the ratios of the sides of these triangles?

- D is the midpoint of $\overline{AC}$. Therefore, $DC = \frac{1}{2}AC$ and $\frac{DC}{AC} = \frac{1}{2}$.
- E is the midpoint of $\overline{BC}$. Therefore, $EC = \frac{1}{2}BC$ and $\frac{EC}{BC} = \frac{1}{2}$.

If we measure $\overline{AB}$ and $\overline{DE}$, it appears that $DE = \frac{1}{2}AB$ and $\frac{DE}{AB} = \frac{1}{2}$. It also appears that $\overline{AB} \parallel \overline{DE}$. We can prove these last two observations as a theorem called the **midsegment theorem**.

Theorem 12.2 A line segment joining the midpoints of two sides of a triangle is parallel to the third side and its length is one-half the length of the third side.

Given $\triangle ABC$, D is the midpoint of $\overline{AC}$, and E is the midpoint of $\overline{BC}$.

Prove $\overline{DE} \parallel \overline{AB}$ and $DE = \frac{1}{2}AB$

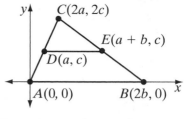

Proof   We will use a coordinate proof for this theorem. The triangle can be placed at any convenient position. We will place A at the origin and B on the x-axis. Let the coordinates of the vertices of $\triangle ABC$ be $A(0, 0)$, $B(2b, 0)$, and $C(2a, 2c)$. Then:

- The coordinates of D are $\left(\frac{2a + 0}{2}, \frac{2c + 0}{2}\right) = (a, c)$.
- The coordinates of E are $\left(\frac{2a + 2b}{2}, \frac{2c + 0}{2}\right) = (a + b, c)$.
- The slope of $\overline{AB}$ is $\frac{0 - 0}{2b - 0} = 0$. $\overline{AB}$ is a horizontal line segment.
- The slope of $\overline{DE}$ is $\frac{c - c}{a + b - a} = 0$. $\overline{DE}$ is a horizontal line segment.

Therefore, $\overline{AB}$ and $\overline{DE}$ are parallel line segments because horizontal line segments are parallel.

The length of a horizontal line segment is the absolute value of the difference of the x-coordinates of the endpoints.

$$AB = |2b - 0| \qquad \text{and} \qquad DE = |(a + b) - a|$$
$$= |2b| \qquad\qquad\qquad\qquad = |b|$$

Therefore, $DE = \frac{1}{2}AB$. ◻

Now that we know that our observations are correct, that $DE = \frac{1}{2}AB$ and that $\overline{AB} \parallel \overline{DE}$, we know that $\angle A \cong \angle EDC$ and $\angle B \cong \angle DEC$ because they are corresponding angles of parallel lines. We also know that $\angle C \cong \angle C$. Therefore, for $\triangle ABC$ and $\triangle DEC$, the corresponding angles are congruent and the ratios of the lengths of corresponding sides are equal.

Again, in $\triangle ABC$, let D be the midpoint of $\overline{AC}$ and E be the midpoint of $\overline{BC}$. Draw $\overline{DE}$. Now let F be the midpoint of $\overline{DC}$ and G be the midpoint of $\overline{EC}$. Draw $\overline{FG}$. We can derive the following information from the segments formed:

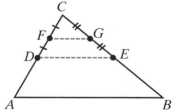

- $FC = \frac{1}{2}DC = \frac{1}{2}\left(\frac{1}{2}AC\right) = \frac{1}{4}AC$ or $\frac{FC}{AC} = \frac{1}{4}$
- $GC = \frac{1}{2}EC = \frac{1}{2}\left(\frac{1}{2}BC\right) = \frac{1}{4}BC$ or $\frac{GC}{BC} = \frac{1}{4}$
- $FG = \frac{1}{2}DE = \frac{1}{2}\left(\frac{1}{2}AB\right) = \frac{1}{4}AB$ or $\frac{FG}{AB} = \frac{1}{4}$ (by Theorem 12.2)
- Let $AC = 4x$. Then $AD = 2x$, $DC = 2x$, $DF = x$, and $FC = x$.

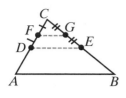

- Let $BC = 4y$. Then $BE = 2y$, $EC = 2y$, $EG = y$, and $GC = x$. Also, $BG = 2y + y = 3y$.
- Therefore, $\frac{FC}{AF} = \frac{x}{3x} = \frac{1}{3}$ and $\frac{GC}{BG} = \frac{y}{3y} = \frac{1}{3}$.

Since $\frac{FC}{AF}$ and $\frac{GC}{BG}$ are each equal to $\frac{1}{3}$, $\frac{FC}{AF} = \frac{GC}{BG}$. We say that the points F and G divide $\overline{AC}$ and $\overline{BC}$ proportionally because these points separate the segments into parts whose ratios form a proportion.

DEFINITION _____

Two line segments are **divided proportionally** when the ratio of the lengths of the parts of one segment is equal to the ratio of the lengths of the parts of the other.

The points D and E also divide $\overline{AC}$ and $\overline{BC}$ proportionally because these points also separate the segments into parts whose ratios form a proportion.

$$\frac{AD}{DC} = \frac{2x}{2x} = 1 \quad \text{and} \quad \frac{BE}{EC} = \frac{2y}{2y} = 1$$

Therefore, $\frac{AD}{DC} = \frac{BE}{EC}$.

Theorem 12.3a

If two line segments are divided proportionally, then the ratio of the length of a part of one segment to the length of the whole is equal to the ratio of the corresponding lengths of the other segment.

Given $\overline{ABC}$ and $\overline{DEF}$ with $\frac{AB}{BC} = \frac{DE}{EF}$.

Prove $\frac{AB}{AC} = \frac{DE}{DF}$

	Statements	Reasons
Proof		
1.	$\frac{AB}{BC} = \frac{DE}{EF}$	**1.** Given.
2.	$(AB)(EF) = (BC)(DE)$	**2.** The product of the means equals the product of the extremes.
3.	$(AB)(EF) = (BC)(DE)$ $+ (AB)(DE) \qquad + (AB)(DE)$	**3.** Addition postulate.
4.	$(AB)(EF + DE) = (DE)(BC + AB)$	**4.** Distributive property.
5.	$(AB)(DF) = (DE)(AC)$	**5.** Substitution postulate.
6.	$\frac{AB}{AC} = \frac{DE}{DF}$	**6.** If the products of two pairs of factors are equal, one pair of factors can be the means and the other the extremes of a proportion.

Theorem 12.3b

> If the ratio of the length of a part of one line segment to the length of the whole is equal to the ratio of the corresponding lengths of another line segment, then the two segments are divided proportionally.

The proof of this theorem is left to the student. (See exercise 21.) Theorems 12.3a and 12.3b can be written as a biconditional.

Theorem 12.3

> Two line segments are divided proportionally if and only if the ratio of the length of a part of one segment to the length of the whole is equal to the ratio of the corresponding lengths of the other segment.

EXAMPLE 1

In $\triangle PQR$, S is the midpoint of $\overline{RQ}$ and T is the midpoint of $\overline{PQ}$.

$RP = 7x + 5$ $ST = 4x - 2$ $SR = 2x + 1$ $PQ = 9x + 1$

Find ST, RP, SR, RQ, PQ, and TQ.

Solution The length of the line joining the midpoints of two sides of a triangle is equal to one-half the length of the third side.

$$4x - 2 = \tfrac{1}{2}(7x + 5)$$
$$2(4x - 2) = 2\left(\tfrac{1}{2}\right)(7x + 5)$$
$$8x - 4 = 7x + 5$$
$$x = 9$$

$ST = 4(9) - 2 = 36 - 2 = 34$
$RP = 7(9) + 5 = 63 + 5 = 68$
$SR = 2(9) + 1 = 18 + 1 = 19$
$RQ = 2SR = 2(19) = 38$
$PQ = 9(9) + 1 = 81 + 1 = 82$
$TQ = \tfrac{1}{2}PQ = \tfrac{1}{2}(82) = 41$

EXAMPLE 2

$\overline{ABC}$ and $\overline{DEF}$ are line segments. If $AB = 10$, $AC = 15$, $DE = 8$, and $DF = 12$, do B and E divide $\overline{ABC}$ and $\overline{DEF}$ proportionally?

Solution If $AB = 10$ and $AC = 15$, then:

$BC = 15 - 10$ $AB:BC = 10:5$
$= 5$ $= 2:1$

If $DE = 8$ and $DF = 12$, then:

$EF = 12 - 8$ $DE:EF = 8:4$
$= 4$ $= 2:1$

Since the ratios of $AB : BC$ and $DE : EF$ are equal, B and E divide $\overline{ABC}$ and $\overline{DEF}$ proportionally.

EXAMPLE 3

In the diagram, $\overline{ADEC}$ and $\overline{BFGC}$ are two sides of $\triangle ABC$. If $AD = DE = EC$ and $BF = FG = GC$, prove that $EG = \frac{1}{2}DF$.

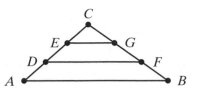

Solution   Since D and E are on $\overline{ADEC}$ and $DE = EC$, E is the midpoint of $\overline{DEC}$.

Since F and G are on $\overline{BFGC}$ and $FG = GC$, G is the midpoint of $\overline{FGC}$.

In $\triangle DFC$, $\overline{EG}$ is the line segment joining the midpoints of two sides of the triangle. By Theorem 12.2, $\overline{EG}$, a line segment joining the midpoints of two sides of a triangle, is parallel to the third side, $\overline{DF}$, and its length is one-half the length of the third side. Therefore, $EG = \frac{1}{2}DF$.

Exercises

Writing About Mathematics

1. Explain why the midpoints of two line segments always divide those segments proportionally.

2. Points B, C, D, and E divide $\overline{ABCDEF}$ into five equal parts. Emily said that $AB : BF = 1 : 5$. Do you agree with Emily? Explain why or why not.

Developing Skills

In 3–10, M is the midpoint of $\overline{DF}$ and N is the midpoint of $\overline{EF}$.

3. Find DE if $MN = 9$.

4. Find MN if $DE = 17$.

5. Find DM if $DF = 24$.

6. Find NF if $EF = 10$.

7. Find $DM : DF$.

8. Find $DP : PF$ if P is the midpoint of $\overline{MF}$.

9. Find $m\angle FMN$ if $m\angle D = 76$.

10. Find $m\angle ENM$ if $m\angle E = 42$.

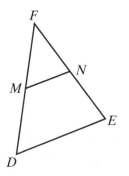

11. The length of the diagonal of a rectangle is 12 centimeters. What is the measure of a line segment that joins the midpoints of two consecutive sides of the rectangle?

In 12–15, the line segments $\overline{ABC}$ and $\overline{PQR}$ are divided proportionally by B and Q. $AB < BC$ and $PQ < QR$.

12. Find PQ when $AB = 15$, $BC = 25$, and $QR = 35$.

13. Find BC when $AB = 8$, $PQ = 20$, and $PR = 50$.

14. Find AC when $AB = 12$, $QR = 27$, and $BC = PQ$.

15. Find AB and BC when $AC = 21$, $PQ = 14$, and $QR = 35$.

16. Line segment $\overline{KLMN}$ is divided by L and M such that $KL : LM : MN = 2 : 4 : 3$. Find:
   a. $KL : KN$ **b.** $LN : MN$ **c.** $LM : LN$ **d.** $KM : LN$

17. Line segment $\overline{ABC}$ is divided by B such that $AB : BC = 2 : 3$ and line segment $\overline{DEF}$ is divided by E such that $DE : EF = 2 : 3$. Show that $AB : AC = DE : DF$.

Applying Skills

18. The midpoint the sides of $\triangle ABC$ are L, M, and N.

   a. Prove that quadrilateral $LMCN$ is a parallelogram.

   b. If $AB = 12$, $BC = 9$, and $CA = 15$, what is the perimeter of $LMCN$?

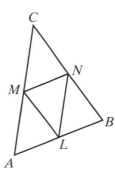

19. In right triangle ABC, the midpoint of the hypotenuse $\overline{AB}$ is M and the midpoints of the legs are P and Q. Prove that quadrilateral $PMQC$ is a rectangle.

20. In right triangle ABC, the midpoint of the hypotenuse $\overline{AB}$ is M, the midpoint of $\overline{BC}$ is P, and the midpoint of $\overline{CA}$ is Q. D is a point on $\overleftrightarrow{PM}$ such that $PM = MD$.

   a. Prove that $QADM$ is a rectangle.

   b. Prove that $\overline{CM} \cong \overline{AM}$.

   c. Prove that M is equidistant from the vertices of $\triangle ABC$.

21. Prove Theorem 12.3b, "If the ratio of the length of a part of one line segment to the length of the whole is equal to the ratio of the corresponding lengths of another line segment, then the two segments are divided proportionally."

22. The midpoints of the sides of quadrilateral $ABCD$ are M, N, P, and Q. Prove that quadrilateral $MNPQ$ is a parallelogram. (*Hint*: Draw $\overline{AC}$.)

12-3 SIMILAR POLYGONS

Two polygons that have the same shape but not the same size are called similar polygons. In the figure to the right, $ABCDE \sim PQRST$. The symbol $\sim$ is read "is similar to."

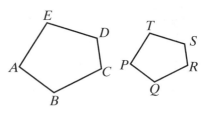

These polygons have the same shape because their corresponding angles are congruent and the ratios of the lengths of their corresponding sides are equal.

DEFINITION _____

Two polygons are similar if there is a one-to-one correspondence between their vertices such that:

1. All pairs of corresponding angles are congruent.
2. The ratios of the lengths of all pairs of corresponding sides are equal.

When the ratios of the lengths of the corresponding sides of two polygons are equal, as shown in the example above, we say that the corresponding sides of the two polygons are in proportion. The ratio of the lengths of corresponding sides of similar polygons is called the **ratio of similitude** of the polygons. The number represented by the ratio of similitude is called the **constant of proportionality**.

Both conditions mentioned in the definition must be true for polygons to be similar.

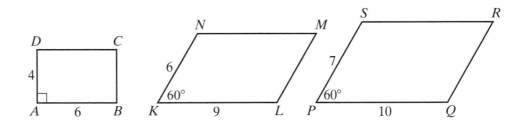

Rectangle $ABCD$ is *not* similar to parallelogram $KLMN$. The corresponding sides are in proportion, $\frac{4}{6} = \frac{6}{9}$, but the corresponding angles are not congruent.

Parallelogram $KLMN$ is *not* similar to parallelogram $PQRS$. The corresponding angles are congruent but the corresponding sides are not in proportion, $\frac{6}{9} \neq \frac{7}{10}$.

Recall that a mathematical definition is reversible:

▶ If two polygons are similar, then their corresponding angles are congruent and their corresponding sides are in proportion.

and

▶ If two polygons have corresponding angles that are congruent and corresponding sides that are in proportion, then the polygons are similar.

Since triangles are polygons, the definition given for two similar polygons will apply also to two similar triangles.

In the figures to the right, $\triangle ABC \sim \triangle A'B'C'$. We can draw the following conclusions about the two triangles:

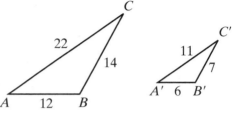

$\angle A \cong \angle A'$	$\angle B \cong \angle B'$	$\angle C \cong \angle C'$
$AB : A'B' = 12 : 6$	$BC : B'C' = 14 : 7$	$CA : C'A' = 22 : 11$
$= 2 : 1$	$= 2 : 1$	$= 2 : 1$

The ratio of similitude for the triangles is $2 : 1$.

Equivalence Relation of Similarity

The relation "is similar to" is true for polygons when their corresponding angles are congruent and their corresponding sides are in proportion. Thus, for a given set of triangles, we can test the following properties:

1. Reflexive property: $\triangle ABC \sim \triangle ABC$. (Here, the ratio of the lengths of corresponding sides is $1 : 1$.)

2. Symmetric property: If $\triangle ABC \sim \triangle DEF$, then $\triangle DEF \sim \triangle ABC$.

3. Transitive property: If $\triangle ABC \sim \triangle DEF$, and $\triangle DEF \sim \triangle RST$, then $\triangle ABC \sim \triangle RST$.

These properties for any similar geometric figures can be stated as postulates.

Postulate 12.1

Any geometric figure is similar to itself. (Reflexive property)

Postulate 12.2

A similarity between two geometric figures may be expressed in either order. (Symmetric property)

Postulate 12.3 | Two geometric figures similar to the same geometric figure are similar to each other. (Transitive property)

EXAMPLE I

In right triangle ABC, m$\angle A$ = 67.4, AB = 13.0, BC = 12.0, and CA = 5.00.

In right triangle DEF, m$\angle E$ = 22.6, DE = 19.5, EF = 18.0, and FD = 7.50.

Prove that $\triangle ABC \sim \triangle DEF$.

Proof Triangles ABC and DEF are right triangles. The angles opposite the longest sides are right angles. Therefore, m$\angle C$ = 90, m$\angle F$ = 90, and $\angle C \cong \angle F$.

The acute angles of a right triangle are complementary. Therefore, m$\angle B$ = 90 − m$\angle A$ = 90 − 67.4 = 22.6, and $\angle B \cong \angle E$.

Similarly, m$\angle D$ = 90 − m$\angle E$ = 90 − 22.6 = 67.4, and $\angle A \cong \angle D$.

$$\frac{AB}{DE} = \frac{13.0}{19.5} = \frac{2}{3} \qquad \frac{BC}{EF} = \frac{12.0}{18.0} = \frac{2}{3} \qquad \frac{CA}{FD} = \frac{5.00}{7.50} = \frac{2}{3}$$

Since the corresponding angles are congruent and the ratios of the lengths of corresponding sides are equal, the triangles are similar.

Exercises

Writing About Mathematics

1. Are all squares similar? Justify your answer.

2. Are *any* two regular polygons similar? Justify your answer.

Developing Skills

3. What is the ratio of the lengths of corresponding sides of two congruent polygons?

4. Are all congruent polygons similar? Explain your answer.

5. Are all similar polygons congruent? Explain your answer.

6. What must be the constant of proportionality of two similar polygons in order for the polygons to be congruent?

7. The sides of a triangle measure 4, 9, and 11. If the shortest side of a similar triangle measures 12, find the measures of the remaining sides of this triangle.

8. The sides of a quadrilateral measure 12, 18, 20, and 16. The longest side of a similar quadrilateral measures 5. Find the measures of the remaining sides of this quadrilateral.

9. Triangle $ABC \sim \triangle A'B'C'$, and their ratio of similitude is $1:3$. If the measures of the sides of $\triangle ABC$ are represented by a, b, and c, represent the measures of the sides of the larger triangle, $\triangle A'B'C'$.

Applying Skills

10. Prove that any two equilateral triangles are similar.

11. Prove that any two regular polygons that have the same number of sides are similar.

12. In $\triangle ABC$, the midpoint of $\overline{AC}$ is M and the midpoint of $\overline{BC}$ is N.

    a. Show that $\triangle ABC \sim \triangle MNC$.

    b. What is their ratio of similitude?

13. In $\triangle ABC$, the midpoint of $\overline{AC}$ is M, the midpoint of $\overline{MC}$ is P, the midpoint of $\overline{BC}$ is N, and the midpoint of $\overline{NC}$ is Q.

    a. Show that $\triangle ABC \sim \triangle PQC$.

    b. What is their ratio of similitude?

14. Show that rectangle $ABCD$ is similar to rectangle $EFGH$ if $\frac{AB}{EF} = \frac{BC}{FG}$.

15. Show that parallelogram $KLMN$ is similar to parallelogram $PQRS$ if $m\angle K \cong m\angle P$ and $\frac{KL}{PQ} = \frac{LM}{QR}$.

12-4 PROVING TRIANGLES SIMILAR

We have proved triangles similar by proving that the corresponding angles are congruent and that the ratios of the lengths of corresponding sides are equal. It is possible to prove that when some of these conditions exist, all of these conditions necessary for triangles to be similar exist.

Hands-On Activity

For this activity, you may use a compass and ruler, or geometry software.

STEP 1. Draw any triangle, $\triangle ABC$.

STEP 2. Draw any line $\overline{DE}$ with $\frac{DE}{AB} = \frac{3}{1}$, that is, $DE = 3AB$.

STEP 3. Construct $\angle GDE \cong \angle A$ and $\angle HED \cong \angle B$. Let F be the intersection of $\overrightarrow{DG}$ and $\overrightarrow{EH}$.

    a. Find the measures of $\overline{AC}$, $\overline{BC}$, $\overline{DF}$, and $\overline{EF}$.

    b. Is $\frac{DF}{AC} = \frac{3}{1}$? Is $\frac{EF}{CB} = \frac{3}{1}$?

    c. Is $\triangle DEF \sim \triangle ABC$?

    d. Repeat this construction using a different ratio of similitude. Are the triangles similar?

Our observations from the activity on page 489 seem to suggest the following **postulate of similarity**.

Postulate 12.4

> For any given triangle there exists a similar triangle with any given ratio of similitude.

We can also prove the angle-angle or **AA triangle similarity** theorem.

Theorem 12.4

> Two triangles are similar if two angles of one triangle are congruent to two corresponding angles of the other. (AA~)

Given $\triangle ABC$ and $\triangle A'B'C'$ with $\angle A \cong \angle A'$ and $\angle B \cong \angle B'$

Prove $\triangle A'B'C' \sim \triangle ABC$

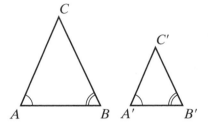

Proof

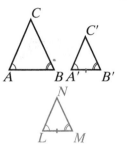

Statement	Reason
1. Draw $\triangle LMN \sim \triangle ABC$ with $\dfrac{LM}{AB} = \dfrac{A'B'}{AB}$.	**1.** Postulate of similarity.
2. $\angle L \cong \angle A$ and $\angle M \cong \angle B$	**2.** If two triangles are similar, then their corresponding angles are congruent.
3. $\angle A \cong \angle A'$ and $\angle B \cong \angle B'$	**3.** Given.
4. $\angle L \cong \angle A'$ and $\angle M \cong \angle B'$	**4.** Substitution postulate.
5. $\dfrac{LM}{AB} = \dfrac{A'B'}{AB}$	**5.** Step 1.
6. $(A'B')(AB) = (AB)(LM)$	**6.** In any proportion, the product of the means is equal to the product of the extremes.
7. $A'B' = LM$	**7.** Division postulate.
8. $\triangle A'B'C' \cong \triangle LMN$	**8.** ASA (steps 4, 7).
9. $\triangle A'B'C' \sim \triangle LMN$	**9.** If two triangles are congruent, then they are similar.
10. $\triangle A'B'C' \sim \triangle ABC$	**10.** Transitive property of similarity (steps 1, 9).

We can also prove other theorems about similar triangles by construction, such as the side-side-side or **SSS similarity theorem**.

Theorem 12.5
> Two triangles are similar if the three ratios of corresponding sides are equal. (SSS~)

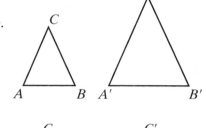

Given $\triangle ABC$ and $\triangle A'B'C'$ with $\frac{AB}{A'B'} = \frac{BC}{B'C'} = \frac{CA}{C'A'}$.

Prove $\triangle A'B'C' \sim \triangle ABC$

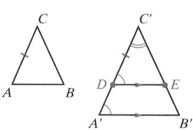

Proof We will construct a third triangle DEC' that is similar to both $\triangle ABC$ and $\triangle A'B'C'$. By the transitive property of similarity, we can conclude that $\triangle A'B'C' \sim \triangle ABC$.

Let $AC < A'C'$. Choose point D on $\overline{A'C'}$ so that $DC' = AC$. Choose point E on $\overline{B'C'}$ so that $\overline{DE} \parallel \overline{A'B'}$. Corresponding angles of parallel lines are congruent, so $\angle C'DE \cong \angle A'$ and $\angle C' \cong \angle C'$. Therefore, $\triangle A'B'C' \sim \triangle DEC'$ by AA~. If two polygons are similar, then their corresponding sides are in proportion, so $\frac{C'D}{C'A'} = \frac{C'E}{C'B'}$.

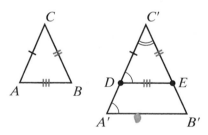

Substituting $C'D = CA$ into the proportion gives $\frac{CA}{C'A'} = \frac{C'E}{C'B'}$.

We are given that $\frac{CA}{C'A'} = \frac{BC}{B'C'}$ so by the transitive property, $\frac{C'E}{C'B'} = \frac{BC}{B'C'}$. Therefore, $(C'E)(B'C') = (C'B')(BC)$ or $C'E = BC$.

By similar reasoning, we find that $DE = AB$. Therefore, $\triangle DEC' \cong \triangle ABC$ by SSS and $\triangle DEC' \sim \triangle ABC$. Then by the transitive property of similarity, $\triangle A'B'C' \sim \triangle ABC$. ∎

Theorem 12.6
> Two triangles are similar if the ratios of two pairs of corresponding sides are equal and the corresponding angles included between these sides are congruent. (SAS~)

Given $\triangle ABC$ and $\triangle A'B'C'$ with $\frac{AB}{A'B'} = \frac{BC}{B'C'}$ and
$\angle B \cong \angle B'$

Prove $\triangle ABC \sim \triangle A'B'C'$

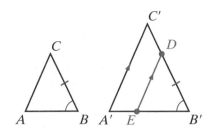

Strategy   The proof follows the same pattern as the
previous theorem. Let $BC < B'C'$. Choose
point D on $\overline{B'C'}$ so that $B'D = BC$. Choose
point E on $\overline{A'B'}$ so that $\overline{DE} \parallel \overline{A'C'}$. First prove that $\triangle A'B'C' \sim \triangle EB'D$. Then
use the given ratios to prove $EB' = AB$ and $\triangle EB'D \cong \triangle ABC$ by SAS.

We refer to Theorem 12.6 as the side-angle-side or **SAS similarity theorem**.
The proof of this theorem will be left to the student. (See exercise 17.) As a con-
sequence of these proofs, we have shown the following theorem to be true.

Theorem 12.7a | If a line is parallel to one side of a triangle and intersects the other two sides,
then the points of intersection divide the sides proportionally.

The converse of this theorem is also true.

Theorem 12.7b | If the points at which a line intersects two sides of a triangle divide those
sides proportionally, then the line is parallel to the third side.

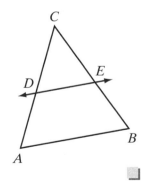

Given $\triangle ABC$ with $\frac{CD}{CA} = \frac{CE}{CB}$

Prove $\overline{DE} \parallel \overline{AB}$

Proof   Since $\frac{CD}{CA} = \frac{CE}{CB}$ and $\angle C \cong \angle C$, $\triangle ABC \sim \triangle DEC$ by
SAS$\sim$. Corresponding angles of similar triangles are
congruent, so $\angle CDE \cong \angle A$. As these are congruent
corresponding angles, $\overline{DE} \parallel \overline{AB}$.

Theorems 12.7a and 12.7b can be written as a biconditional.

Theorem 12.7 | A line is parallel to one side of a triangle and intersects the other two sides
if and only if the points of intersection divide the sides proportionally.

EXAMPLE 1

The lengths of the sides of $\triangle PQR$ are $PQ = 15$ cm, $QR = 8$ cm, and $RP = 12$ cm. If $\triangle PQR \sim \triangle DEF$ and the length of the smallest side of $\triangle DEF$ is 6 centimeters, find the measures of the other two sides of $\triangle DEF$.

Solution Since the smallest side of $\triangle PQR$ is $\overline{QR}$ and $\overline{QR}$ corresponds to $\overline{EF}$, $EF = 6$ cm.

$$\frac{QR}{EF} = \frac{PQ}{DE}$$

$$\frac{8}{6} = \frac{15}{DE}$$

$$8DE = 90$$

$$DE = \frac{90}{8} = \frac{45}{4} = 11\frac{1}{4}$$

$$\frac{QR}{EF} = \frac{RP}{FD}$$

$$\frac{8}{6} = \frac{12}{FD}$$

$$8FD = 72$$

$$FD = \frac{72}{8} = 9$$

Answer $DE = 11\frac{1}{4}$ cm and $FD = 9$ cm

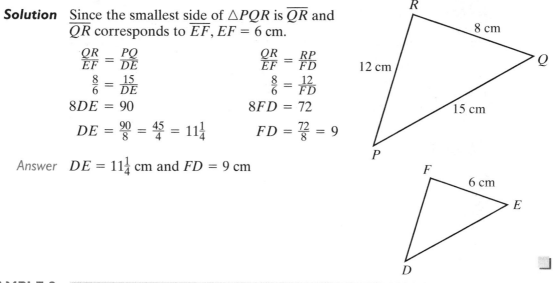

EXAMPLE 2

In $\triangle DEF$, a line is drawn parallel to $\overline{DE}$ that intersects $\overline{FD}$ at H and $\overline{FE}$ at G. If $FG = 8$, $GE = 12$, and $FD = 30$, find FH and HD.

Solution Since $\overline{GH}$ is parallel to $\overline{DE}$, H and G divide $\overline{FD}$ and $\overline{FE}$ proportionally, that is,

$$FH : HD = FG : GE.$$

Let $x = FH$.

Then $HD = FD - FH = 30 - x$.

$$\frac{FH}{HD} = \frac{FG}{GE}$$

$$\frac{x}{30 - x} = \frac{8}{12}$$

$$12x = 8(30 - x)$$

$$12x = 240 - 8x$$

$$20x = 240$$

$$x = 12$$

$$30 - x = 30 - 12$$

$$= 18$$

Check

$$\frac{FH}{HD} = \frac{FG}{GE}$$

$$\frac{12}{18} \overset{?}{=} \frac{8}{12}$$

$$\frac{2}{3} = \frac{2}{3} \checkmark$$

Answer $FH = 12$ and $HD = 18$

EXAMPLE 3

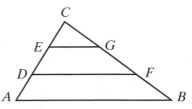

Given: $\overline{ADEC}$ and $\overline{BFGC}$ are two sides of $\triangle ABC$ with $AD = DE = EC$ and $BF = FG = GC$.

Prove: $\frac{AC}{DC} = \frac{BC}{FC}$ and $\triangle ABC \sim \triangle DFC$

Proof We are given $\overline{ADEC}$ and $AD = DE = EC$. Then $AC = AD + DE + EC$. By the substitution postulate, $AC = AD + AD + AD = 3AD$ and $DC = DE + EC = AD + AD = 2AD$.

We are also given $\overline{BFGC}$ and $BF = FG = GC$. Then $BC = BF + FG + GC$. By the substitution postulate, $BC = BF + BF + BF = 3BF$ and $FC = FG + FC = BF + BF = 2BF$.

Then, $\frac{AC}{DC} = \frac{3AD}{2AD} = \frac{3}{2}$ and $\frac{BC}{FC} = \frac{3BF}{2BF} = \frac{3}{2}$. Therefore, $\frac{AC}{DC} = \frac{BC}{FC}$.

In $\triangle ABC$ and $\triangle DFC$, $\frac{AC}{DC} = \frac{BC}{FC}$ and $\angle C \cong \angle C$. Therefore, $\triangle ABC \sim \triangle DFC$ by SAS$\sim$.

Exercises

Writing About Mathematics

1. Javier said that if an acute angle of one right triangle is congruent to an acute angle of another right trangle, the triangles are similar. Do you agree with Javier? Explain why or why not.

2. Fatima said that since two triangles can be proven similar by AA$\sim$, it follows that two triangles can be proven similar by SS$\sim$. Explain why Fatima is incorrect.

Developing Skills

In 3–15, D is a point on $\overline{AC}$ and E is a point on $\overline{BC}$ of $\triangle ABC$ such that $\overline{DE} \parallel \overline{AB}$. (The figure is not drawn to scale.)

3. Prove that $\triangle ABC \sim \triangle DEC$.

4. If $CA = 8$, $AB = 10$, and $CD = 4$, find DE.

5. If $CA = 24$, $AB = 16$, and $CD = 9$, find DE.

6. If $CA = 16$, $AB = 12$, and $CD = 12$, find DE.

7. If $CE = 3$, $DE = 4$, and $CB = 9$, find AB.

8. If $CD = 8$, $DA = 2$, and $CB = 7.5$, find CE.

9. If $CD = 6$, $DA = 4$, and $DE = 9$, find AB.

10. If $CA = 35$, $DA = 10$, and $CE = 15$, find EB.

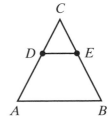

Dilations   **495**

11. If $CA = 48$, $DA = 12$, and $CE = 30$, find EB.

12. If $CD = 15$, $DA = 9$, and $DE = 10$, find AB.

13. If $CE = 20$, $EB = 10$, and $AB = 45$, find DE.

14. If $CD = x$, $DE = x$, $DA = 5$, and $AB = 14$, find DE.

15. If $CD = 6$, $DE = x$, $DA = x - 1$, and $AB = 6$, DE and DA.

Applying Skills

16. Complete the proof of Theorem 12.5 (SSS~) by showing that $DE = AB$.

17. Prove Theorem 12.6, "Two triangles are similar if the ratios of two pairs of corresponding sides are equal and the corresponding angles included between these sides are congruent. (SAS~)"

18. Triangle ABC is an isosceles right triangle with m$\angle C = 90$ and $\overline{CD}$ bisects $\angle C$ and intersects $\overline{AB}$ at D. Prove that $\triangle ABC \sim \triangle ACD$.

19. Quadrilateral $ABCD$ is a trapezoid with $\overline{AB} \parallel \overline{CD}$. The diagonals $\overline{AC}$ and $\overline{BD}$ intersect at E. Prove that $\triangle ABE \sim \triangle CDE$.

20. Lines $\overleftrightarrow{AEB}$ and $\overleftrightarrow{CED}$ intersect at E and $\angle DAE \cong \angle BCE$. Prove that $\triangle ADE \sim \triangle CBE$.

21. In parallelogram $ABCD$ on the right, $\overline{AE} \perp \overleftrightarrow{BC}$ and $\overline{AF} \perp \overleftrightarrow{CD}$. Prove that $\triangle ABE \sim \triangle ADF$.

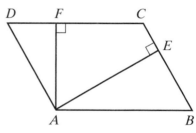

22. In the coordinate plane, the points $A(1, 2)$, $B(3, 2)$, $C(3, 6)$, $D(2, 6)$, and $E(2, 8)$ are the vertices of $\triangle ABC$ and $\triangle CDE$. Prove that $\triangle ABC \sim \triangle CDE$.

23. In the coordinate plane, the points $P(1, 1)$, $Q(3, 3)$, $R(3, 5)$, and $S(1, 5)$ are the vertices of $\triangle PQS$ and $\triangle QRS$ and $PQ = QS = 2\sqrt{2}$. Prove that $\triangle PQS \sim \triangle QRS$.

24. In the coordinate plane, the points $O(0, 0)$, $A(4, 0)$, and $B(0, 6)$ are the coordinates of $\triangle OAB$. The coordinates of C are $(4, 3)$, and D is the midpoint of $\overline{AB}$. Prove that $\triangle OAB \sim \triangle CDA$.

25. A pyramid with a triangular base is cut by a plane p parallel to the base. Prove that the triangle formed by the intersection of plane p with the lateral faces of the pyramid is similar to the base of the pyramid.

12-5 DILATIONS

In Chapter 6, we learned about dilations in the coordinate plane. In this section, we will continue to study dilations as they relate to similarity. Recall that a dilation is a transformation in the plane that preserves angle measure but not distance.

▶ A **dilation** of k is a transformation of the plane such that:

1. The image of point O, the center of dilation, is O.

2. When k is positive and the image of P is P', then $\overrightarrow{OP}$ and $\overrightarrow{OP'}$ are the same ray and $OP' = kOP$.

3. When k is negative and the image of P is P', then $\overrightarrow{OP}$ and $\overrightarrow{OP'}$ are opposite rays and $OP' = -kOP$.

When $|k| > 1$, the dilation is called an **enlargement**. When $0 < |k| < 1$, the dilation is called a **contraction**.

Recall also that in the coordinate plane, under a dilation of k with the center at the origin:

$$P(x, y) \rightarrow P'(kx, ky) \quad \text{or} \quad D_k(x, y) = (kx, ky)$$

For example, the image of $\triangle ABC$ is $\triangle A'B'C'$ under a dilation of $\frac{1}{2}$. The vertices of $\triangle ABC$ are $A(2, 6)$, $B(6, 4)$, and $C(4, 0)$. Under a dilation of $\frac{1}{2}$, the rule is

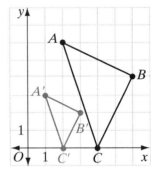

$$D_{\frac{1}{2}}(x,y) = \left(\tfrac{1}{2}x, \tfrac{1}{2}y\right)$$
$$A(2, 6) \rightarrow A'(1, 3)$$
$$B(6, 4) \rightarrow B'(3, 2)$$
$$C(4, 0) \rightarrow C'(2, 0)$$

Notice that $\triangle ABC$ and $\triangle A'B'C'$ appear to be similar. We can use a general triangle to prove that for any dilation, the image of a triangle is a similar triangle.

Let $\triangle ABC$ be any triangle in the coordinate plane with $A(a, 0)$, $B(b, d)$, and $C(c, e)$. Under a dilation of k through the origin, the image of $\triangle ABC$ is $\triangle A'B'C'$, and the coordinates of $\triangle A'B'C'$, are $A'(ka, 0)$, $B'(kb, kd)$, and $C'(kc, ke)$.

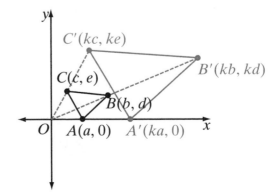

Slope of $\overline{AB} = \frac{d-0}{b-a}$	Slope of $\overline{AC} = \frac{e-0}{c-a}$	Slope of $\overline{BC} = \frac{d-e}{b-c}$
$= \frac{d}{b-a}$	$= \frac{e}{c-a}$	
Slope of $\overline{A'B'} = \frac{kd-0}{kb-ka}$	Slope of $\overline{A'C'} = \frac{ke-0}{kc-ka}$	Slope of $\overline{B'C'} = \frac{kd-ke}{kb-kc}$
$= \frac{k}{k}\left(\frac{d}{b-a}\right)$	$= \frac{k}{k}\left(\frac{e}{c-a}\right)$	$= \frac{k}{k}\left(\frac{d-e}{b-c}\right)$
$= \frac{d}{b-a}$	$= \frac{e}{c-a}$	$= \frac{d-e}{b-c}$
Therefore, $\overline{AB} \parallel \overline{A'B'}$.	Therefore, $\overline{AC} \parallel \overline{A'C'}$.	Therefore, $\overline{BC} \parallel \overline{B'C'}$.

We have shown that $\overline{AB} \parallel \overline{A'B'}$ and $\overline{AC} \parallel \overline{A'C'}$. Therefore, because they are corresponding angles of parallel lines:

$$m\angle OAB = m\angle OA'B'$$
$$m\angle OAC = m\angle OA'C'$$
$$m\angle OAB - m\angle OAC = m\angle OA'B' - m\angle OA'C'$$
$$m\angle BAC = m\angle B'A'C'$$

In a similar way we can prove that $\angle ACB \cong \angle A'C'B'$, and so $\triangle ABC \sim \triangle A'B'C'$ by AA~. Therefore, under a dilation, angle measure is preserved but distance is not preserved. Under a dilation of k, distance is changed by the factor k.

We have proved the following theorem:

Theorem 12.8 | Under a dilation, angle measure is preserved.

We will now prove that under a dilation, midpoint and collinearity are preserved.

Theorem 12.9 | Under a dilation, midpoint is preserved.

Proof: Under a dilation D_k:

$$A(a, c) \rightarrow A'(ka, kc)$$
$$B(b, d) \rightarrow (kb, kd)$$
$$M\left(\frac{a+b}{2}, \frac{c+d}{2}\right) \rightarrow M'\left(k\frac{a+b}{2}, k\frac{c+d}{2}\right)$$

The coordinates of the midpoint of $\overline{A'B'}$ are:

$$\left(\frac{ka+kb}{2}, \frac{kc+kd}{2}\right) \text{ or } \left(k\frac{a+b}{2}, k\frac{c+d}{2}\right)$$

Therefore, the image of M is the midpoint of the image of $\overline{AB}$, and midpoint is preserved.

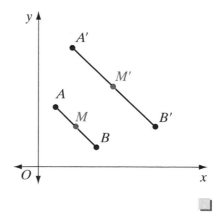

Theorem 12.10   Under a dilation, collinearity is preserved.

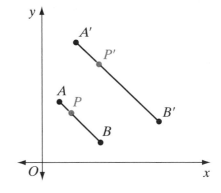

Proof: Under a dilation D_k:

$$A(a, c) \rightarrow A'(ka, kc)$$
$$B(b, d) \rightarrow B'(kb, kd)$$
$$P(p, q) \rightarrow P'(kp, kq)$$

Since P is on $\overline{AB}$, the slope of $\overline{AP}$ is equal to the slope of $\overline{PB}$. Therefore:

$$\text{slope of } \overline{AP} = \text{slope of } \overline{PB}$$

$$\frac{c - q}{a - p} = \frac{q - d}{p - b}$$

P' will be on $\overline{A'B'}$ if and only if the slope of $\overline{A'P'}$ is equal to the slope of $\overline{P'B'}$.

$$\text{slope of } \overline{A'P'} \overset{?}{=} \text{slope of } \overline{P'B'}$$

$$\frac{kc - kq}{ka - kp} \overset{?}{=} \frac{kq - kd}{kp - kb}$$

Since $\frac{c - q}{a - p} = \frac{d - q}{b - p}$ is true,

$$\frac{k}{k}\left(\frac{c - q}{a - p}\right) = \frac{k}{k}\left(\frac{q - d}{p - b}\right) \quad \text{or} \quad \frac{kc - kq}{ka - kp} = \frac{kd - kq}{kb - kp}$$

Thus, since we have shown that the slope of $\overline{A'P'}$ is equal to the slope of $\overline{P'B'}$, P' is on $\overline{A'B'}$ and collinearity is preserved.   ◻

EXAMPLE 1

The coordinates of parallelogram $EFGH$ are $E(0, 0)$, $F(3, 0)$, $G(4, 2)$, and $H(1, 2)$. Under D_3, the image of $EFGH$ is $E'F'G'H'$. Show that $E'F'G'H'$ is a parallelogram. Is parallelism preserved?

Solution $D_3(x, y) = (3x, 3y)$. Therefore, $E'(0, 0)$, $F'(9, 0)$, $G'(12, 6)$, and $H'(3, 6)$.

$$\text{slope of } \overline{E'F'} = \frac{0 - 0}{9 - 0} \quad \text{and} \quad \text{slope of } \overline{H'G'} = \frac{6 - 6}{12 - 3}$$
$$= 0 \qquad\qquad\qquad = 0$$

$$\text{slope of } \overline{F'G'} = \frac{6 - 0}{12 - 9} \quad \text{and} \quad \text{slope of } \overline{E'H'} = \frac{6 - 0}{3 - 0}$$
$$= \frac{6}{3} \qquad\qquad\qquad = \frac{6}{3}$$
$$= 2 \qquad\qquad\qquad = 2$$

Since the slopes of the opposite sides of $E'F'G'H'$ are equal, the opposite sides are parallel and $E'F'G'H'$ is a parallelogram. Parallelism is preserved because the images of parallel lines are parallel.   ◻

EXAMPLE 2

Find the coordinates of Q', the image of $Q(-3, 7)$ under the composition of transformations, $r_{y\text{-axis}} \circ D_{\frac{1}{2}}$.

Solution Perform the transformations from right to left.

The transformation at the right is to be performed first: $D_{\frac{1}{2}}(-3, 7) = \left(\frac{-3}{2}, \frac{7}{2}\right)$

Then perform the transformation on the left, using the result of the first transformation: $r_{y\text{-axis}}\left(\frac{-3}{2}, \frac{7}{2}\right) = \left(\frac{3}{2}, \frac{7}{2}\right)$

Answer $Q' = \left(\frac{3}{2}, \frac{7}{2}\right)$

Exercises

Writing About Mathematics

1. Under D_k, $k > 0$, the image of $\triangle ABC$ is $\triangle A'B'C'$. Is $\frac{AB}{A'B'} = \frac{BC}{B'C'} = \frac{AC}{A'C'}$? Justify your answer.

2. Under a dilation, the image of $A(3, 3)$ is $A'(4, 5)$ and the image of $B(4, 1)$ is $B'(6, 1)$. What are the coordinates of the center of dilation?

Developing Skills

In 3–6, use the rule $(x, y) \rightarrow \left(\frac{1}{3}x, \frac{1}{3}y\right)$ to find the coordinates of the image of each given point.

3. $(9, 6)$ **4.** $(-5, 0)$ **5.** $(18, 3)$ **6.** $(-1, -7)$

In 7–10, find the coordinates of the image of each given point under D_3.

7. $(8, 8)$ **8.** $(2, 13)$ **9.** $(-4, 7)$ **10.** $\left(\frac{1}{3}, \frac{5}{8}\right)$

In 11–14, each given point is the image under D_{-2}. Find the coordinates of each preimage.

11. $(4, -2)$ **12.** $(6, 8)$ **13.** $(-3, -2)$ **14.** $(20, 11)$

In 15–20, find the coordinates of the image of each given point under the given composition of transformations.

15. $D_3 \circ r_{x\text{-axis}}(2, 3)$ **16.** $R_{180°} \circ D_{-2\frac{1}{2}}(4, 3)$ **17.** $D_{\frac{5}{3}} \circ T_{5,3}(1, 1)$

18. $r_{y\text{-axis}} \circ D_3(1, 2)$ **19.** $T_{2,3} \circ D_{10\frac{1}{3}}(0, 0)$ **20.** $D_{-2} \circ r_{y=x}(-3, -5)$

In 21–24, each transformation is the composition of a dilation and a reflection in either the x-axis or the y-axis. In each case, write a rule for composition of transformations for which the image of A is A'.

21. $A(3, 3) \rightarrow A'\left(\frac{9}{2}, -\frac{9}{2}\right)$

22. $A(5, -1) \rightarrow A'(20, 4)$

23. $A(20, 12) \rightarrow A'(-5, 3)$

24. $A(-50, 35) \rightarrow A'(10, 7)$

25. In the diagram, $\triangle A'B'C'$ is the image of $\triangle ABC$. Identify three specific transformations, or compositions of transformations, that can map $\triangle ABC$ to $\triangle A'B'C'$. Justify your answer.

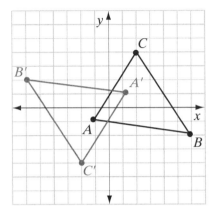

Applying Skills

26. If the coordinates of points A and B are $(0, 5)$ and $(5, 0)$, respectively, and A' and B' are the images of these points under D_{-3}, what type of quadrilateral is $ABA'B'$? Justify your answer.

27. Prove that if the sides of one angle are parallel to the sides of another angle, the angles are congruent.

Given: $\overrightarrow{BA} \parallel \overrightarrow{ED}, \overrightarrow{BC} \parallel \overrightarrow{EF}$, and $\overleftrightarrow{BEG}$

Prove: $\angle ABC \cong \angle DEF$

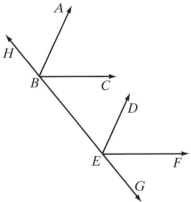

28. The vertices of rectangle $ABCD$ are $A(2, -3)$, $B(4, -3)$, $C(4, 1)$, and $D(2, 1)$.

a. Find the coordinates of the vertices of $A'B'C'D'$, the image of $ABCD$ under D_5.

b. Show that $A'B'C'D'$ is a parallelogram.

c. Show that $ABCD \sim A'B'C'D'$.

d. Show that $\triangle ABC \sim \triangle A'B'C'$.

29. The vertices of octagon $ABCDEFGH$ are $A(2, 1)$, $B(1, 2)$, $C(-1, 2)$, $D(-2, 1)$, $E(-2, -1)$, $F(-1, -2)$, $G(1, -2)$, $H(2, -1)$.

a. Draw $ABCDEFGH$ on graph paper.

b. Draw $A'B'C'D'E'F'G'H'$, the image of $ABCDEFGH$ under D_3, on graph paper and write the coordinates of its vertices.

c. Find HA, BC, DE, FG.

d. Find $H'A'$, $B'C'$, $D'E'$, $F'G'$.

e. If $AB = CD = EF = GH = \sqrt{2}$, find $A'B'$, $C'D'$, $E'F'$, $G'H'$.

f. Are $ABCDEFGH$ and $A'B'C'D'E'F'G'H'$ similar polygons? Justify your answer.

30. Let the vertices of $\triangle ABC$ be $A(-2, 3)$, $B(-2, -1)$, and $C(3, -1)$.

a. Find the area of $\triangle ABC$.

b. Find the area of the image of $\triangle ABC$ under D_3.

c. Find the area of the image of $\triangle ABC$ under D_4.

d. Find the area of the image of $\triangle ABC$ under D_5.

e. Make a conjecture regarding how the area of a figure under a dilation D_k is related to the **constant of dilation** k.

31. Complete the following to prove that dilations preserve parallelism, that is, if $\overleftrightarrow{AB} \parallel \overleftrightarrow{CD}$, then the images of each line under a dilation D_k are also parallel.

a. Let $\overline{AB}$ and $\overline{CD}$ be two vertical segments with endpoints $A(a, b)$, $B(a, b + d)$, $C(c, b)$, and $D(c, b + d)$. Under the dilation D_k, show that the images $\overline{A'B'}$ and $\overline{C'D'}$ are also parallel.

b. Let $\overline{AB}$ and $\overline{CD}$ be two nonvertical parallel segments with endpoints $A(a, b)$, $B(c, d)$, $C(a + e, b)$, and $D(c + e, d)$. Under the dilation D_k, show that the images $\overline{A'B'}$ and $\overline{C'D'}$ are also parallel.

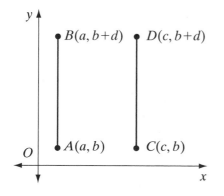

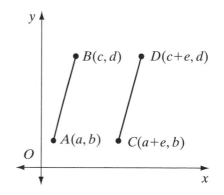

12-6 PROPORTIONAL RELATIONS AMONG SEGMENTS RELATED TO TRIANGLES

We have seen that, if two triangles are similar, their corresponding sides are in proportion. Other corresponding segments such as the altitudes, medians, and angle bisectors in similar triangles are also in proportion.

Theorem 12.11

> If two triangles are similar, the lengths of corresponding altitudes have the same ratio as the lengths of any two corresponding sides.

Given $\triangle ABC \sim \triangle A'B'C'$ with the ratio of similitude $k:1$, $\overline{BD} \perp \overline{AC}$, $\overline{B'D'} \perp \overline{A'C'}$, $BC = a$, $B'C' = a'$, $BD = h$, and $B'D' = h'$.

Prove $\dfrac{h}{h'} = \dfrac{a}{a'} = \dfrac{k}{1}$

Proof

Statements	Reasons
1. $\triangle ABC \sim \triangle A'B'C'$	**1.** Given.
2. $\angle C \cong \angle C'$	**2.** If two triangles are similar, then their corresponding angles are congruent.
3. $\overline{BD} \perp \overline{AC}$ and $\overline{B'D'} \perp \overline{A'C'}$	**3.** Given.
4. $\angle BDC \cong \angle B'D'C'$	**4.** Perpendicular lines form right angles and all right angles are congruent.
5. $\triangle DBC \sim \triangle D'B'C'$	**5.** AA~.
6. $\dfrac{a}{a'} = \dfrac{k}{1}$	**6.** Given.
7. $\dfrac{h}{h'} = \dfrac{a}{a'}$	**7.** If two triangles are similar, then their corresponding sides are in proportion.
8. $\dfrac{h}{h'} = \dfrac{a}{a'} = \dfrac{k}{1}$	**8.** Transitive property.

We can prove related theorems for medians and angle bisectors of similar triangles.

Theorem 12.12   If two triangles are similar, the lengths of corresponding medians have the same ratio as the lengths of any two corresponding sides.

Given $\triangle ABC \sim \triangle A'B'C'$ with the ratio of similitude $k : 1$, M is the midpoint of $\overline{AC}$, M' is the midpoint of $\overline{A'C'}$, $BC = a$, $B'C' = a'$, $BM = m$, and $B'M' = m'$.

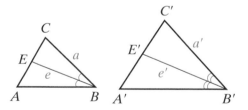

Prove $\dfrac{m}{m'} = \dfrac{a}{a'} = \dfrac{k}{1}$

Strategy   Here we can use SAS~ to prove $\triangle BCM \sim \triangle B'C'M'$.

Theorem 12.13   If two triangles are similar, the lengths of corresponding angle bisectors have the same ratio as the lengths of any two corresponding sides.

Given $\triangle ABC \sim \triangle A'B'C'$ with the ratio of similitude $k : 1$, E is the point at which the bisector of $\angle B$ intersects $\overline{AC}$, E' is the point at which the bisector of $\angle B'$ intersects $\overline{A'C'}$, $BC = a$, $B'C' = a'$, $BE = e$, and $B'E' = e'$.

Prove $\dfrac{e}{e'} = \dfrac{a}{a'} = \dfrac{k}{1}$

Strategy   Here we can use that halves of congruent angles are congruent and AA~ to prove $\triangle BCE \sim \triangle B'C'E'$.

The proofs of Theorems 12.12 and 12.13 are left to the student. (See exercises 10 and 11.)

EXAMPLE I

Two triangles are similar. The sides of the smaller triangle have lengths of 4 meters, 6 meters, and 8 meters. The perimeter of the larger triangle is 63 meters. Find the length of the shortest side of the larger triangle.

Solution   (1) In the smaller triangle, find the perimeter, p: $p = 4 + 6 + 8 = 18$

(2) Let k be the constant of proportionality of the larger triangle to the smaller triangle. Let the measures of the sides of the larger triangle be a, b, and c. Set up proportions and solve for a, b, and c:

$$\frac{a}{4} = \frac{k}{1} \qquad\qquad \frac{b}{6} = \frac{k}{1} \qquad\qquad \frac{c}{8} = \frac{k}{1}$$
$$a = 4k \qquad\qquad b = 6k \qquad\qquad c = 8k$$

(3) Solve for k:

$$4k + 6k + 8k = 63$$
$$18k = 63$$
$$k = 3.5$$

(4) Solve for a, b, and c:

$$a = 4k \qquad\qquad b = 6k \qquad\qquad c = 8k$$
$$= 4(3.5) \qquad\qquad = 6(3.5) \qquad\qquad = 8(3.5)$$
$$= 14 \qquad\qquad = 21 \qquad\qquad = 28$$

Answer   The length of the shortest side is 14 meters.

EXAMPLE 2

Given: $\overleftrightarrow{AFB} \parallel \overleftrightarrow{CGD}$, $\overline{AED}$ and $\overline{BEC}$ intersect at E, and $\overline{EF} \perp \overleftrightarrow{AFB}$.

Prove: $\triangle ABE \sim \triangle DCE$ and $\dfrac{AB}{DC} = \dfrac{EF}{EG}$.

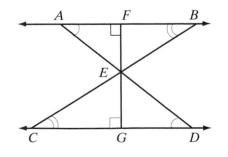

Proof	Statements	Reasons
	1. $\overleftrightarrow{AFB} \parallel \overleftrightarrow{CGD}$	1. Given.
	2. $\angle EAB \cong \angle EDC$ and $\angle EBA \cong \angle ECD$	2. If two parallel lines are cut by a transversal, the alternate interior angles are congruent.
	3. $\triangle ABE \sim \triangle DCE$	3. AA~.
	4. $\overline{EF} \perp \overleftrightarrow{AFB}$	4. Given.
	5. $\overline{EG} \perp \overleftrightarrow{CGD}$	5. If a line is perpendicular to one of two parallel lines, it is perpendicular to the other.
	6. $\overline{EF}$ is an altitude from E in $\triangle ABE$. $\overline{EG}$ is an altitude from E in $\triangle DCE$.	6. Definition of an altitude of a triangle.
	7. $\frac{AB}{DC} = \frac{EF}{EG}$	7. If two triangles are similar, the lengths of corresponding altitudes have the same ratio as the lengths of any two corresponding sides.

Exercises

Writing About Mathematics

1. The lengths of the corresponding sides of two similar triangles are 10 and 25. Irena said that the ratio of similitude is 2 : 5. Jeff said that it is $\frac{2}{5}$: 1. Who is correct? Justify your answer.

2. Maya said that if the constant of proportionality of two similar triangles is k, then the ratio of the perimeters will be $3k$: 1 because $\frac{k}{1} + \frac{k}{1} + \frac{k}{1} = \frac{3k}{1}$. Do you agree with Maya? Explain why or why not.

Developing Skills

3. The ratio of similitude in two similar triangles is 5 : 1. If a side in the larger triangle measures 30 centimeters, find the measure of the corresponding side in the smaller triangle.

4. If the lengths of the sides of two similar triangles are in the ratio 4 : 3, what is the ratio of the lengths of a pair of corresponding altitudes, in the order given?

5. The lengths of two corresponding sides of two similar triangles are 18 inches and 12 inches. If an altitude of the smaller triangle has a length of 6 inches, find the length of the corresponding altitude of the larger triangle.

6. The constant of proportionality of two similar triangles is $\frac{4}{5}$. If the length of a median in the larger triangle is 15 inches, find the length of the corresponding median in the smaller triangle.

7. The ratio of the lengths of the corresponding sides of two similar triangles is 6 : 7. What is the ratio of the altitudes of the triangles?

8. Corresponding altitudes of two similar triangles have lengths of 9 millimeters and 6 millimeters. If the length of a median of the larger triangle is 24 millimeters, what is the length of a median of the smaller triangle?

9. In meters, the sides of a triangle measure 14, 18, and 12. The length of the longest side of a similar triangle is 21 meters.

   a. Find the ratio of similitude of the two triangles.

   b. Find the lengths of the other two sides of the larger triangle.

   c. Find the perimeter of each triangle.

   d. Is the ratio of the perimeters equal to the ratio of the lengths of the sides of the triangle?

Applying Skills

10. Prove Theorem 12.12, "If two triangles are similar, the lengths of corresponding medians have the same ratio as the lengths of any two corresponding sides."

11. Prove Theorem 12.13, "If two triangles are similar, the lengths of corresponding angle bisectors have the same ratio as the lengths of any two corresponding sides."

12. Prove that if two parallelograms are similar, then the ratio of the lengths of the corresponding diagonals is equal to the ratio of the lengths of the corresponding sides.

13. Prove that if two triangles are similar, then the ratio of their areas is equal to the square of their ratio of similitude.

14. The diagonals of a trapezoid intersect to form four triangles that have no interior points in common.

   a. Prove that two of these four triangles are similar.

   b. Prove that the ratio of similitude is the ratio of the length of the parallel sides.

12-7 CONCURRENCE OF THE MEDIANS OF A TRIANGLE

We proved in earlier chapters that the altitudes of a triangle are concurrent and that the angle bisectors of a triangle are concurrent. If we draw the three medians of a triangle, we see that they also seem to intersect in a point. This point is called the **centroid** of the triangle.

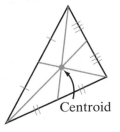

Centroid

Theorem 12.14

> Any two medians of a triangle intersect in a point that divides each median in the ratio 2 : 1.

Given $\overline{AM}$ and $\overline{BN}$ are medians of $\triangle ABC$ that intersect at P.

Prove $AP : MP = BP : NP = 2 : 1$

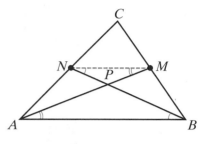

Proof

Statements	Reasons
1. $\overline{AM}$ and $\overline{BN}$ are the medians of $\triangle ABC$.	**1.** Given.
2. M is the midpoint of $\overline{BC}$ and N is the midpoint of $\overline{AC}$.	**2.** The median of a triangle is a line segment from a vertex to the midpoint of the opposite side.
3. Draw $\overline{MN}$.	**3.** Two points determine a line.
4. $\overline{MN} \parallel \overline{AB}$	**4.** The line joining the midpoints of two sides of a triangle is parallel to the third side.
5. $\angle MNB \cong \angle ABN$ and $\angle NMA \cong \angle BAM$	**5.** Alternate interior angles of parallel lines are congruent.
6. $\triangle MNP \sim \triangle ABP$	**6.** AA$\sim$.
7. $MN = \frac{1}{2}AB$	**7.** The length of the line joining the midpoints of two sides of a triangle is equal to one-half of the length of the third side.
8. $2MN = AB$	**8.** Multiplication postulate.
9. $AB : MN = 2 : 1$	**9.** If the products of two pairs of factors are equal, the factors of one pair can be the means and the factors of the other the extremes of a proportion.
10. $AP : MP = BP : NP = 2 : 1$	**10.** If two triangles are similar, the ratios of the lengths of the corresponding sides are equal.

Theorem 12.15	The medians of a triangle are concurrent.

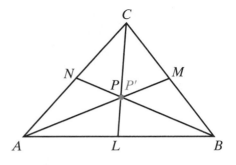

Given $\overline{AM}, \overline{BN}$, and $\overline{CL}$ are medians of $\triangle ABC$.

Prove $\overline{AM}, \overline{BN}$, and $\overline{CL}$ are concurrent.

Proof Let the intersection of $\overline{AM}$ and $\overline{BN}$ be P. Then P divides $\overline{AM}$ in the ratio $2:1$, that is, $AP:PM = 2:1$. Let $\overline{CL}$ intersect $\overline{AM}$ at P'. Then P' divides $\overline{AM}$ in the ratio $2:1$, that is, $AP':P'M = 2:1$. Both P and P' are on the same line segment, $\overline{AM}$, and divide that line segment in the ratio $2:1$. Therefore, P and P' are the same point and the three medians of $\triangle ABC$ are concurrent. ∎

EXAMPLE 1

Find the coordinates of the centroid of the triangle whose vertices are $A(-3, 6)$, $B(-9, 0)$, and $C(9, 0)$.

Solution

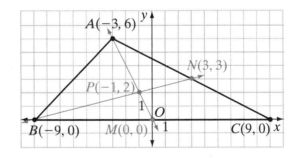

(1) Find the coordinates of the midpoint, M, of $\overline{BC}$ and of the midpoint, N, of $\overline{AC}$:

$$\text{coordinates of } M = \left(\frac{-9 + 9}{2}, \frac{0 + 0}{2}\right)$$
$$= (0, 0)$$
$$\text{coordinates of } N = \left(\frac{-3 + 9}{2}, \frac{6 + 0}{2}\right)$$
$$= (3, 3)$$

(2) Find the equation of $\overleftrightarrow{AM}$ and the equation of $\overleftrightarrow{BN}$.

Equation of $\overleftrightarrow{AM}$:	Equation of $\overleftrightarrow{BN}$:
$\dfrac{y-0}{x-0} = \dfrac{6-0}{-3-0}$	$\dfrac{y-0}{x-(-9)} = \dfrac{0-3}{-9-3}$
$\dfrac{y}{x} = -2$	$\dfrac{y}{x+9} = \dfrac{1}{4}$
$y = -2x$	$4y = x + 9$

(3) Find the coordinates of P, the point of intersection of $\overleftrightarrow{AM}$ and $\overleftrightarrow{BN}$:

Substitute $y = -2x$ into the equation $4y = x + 9$, and solve for x. Then find the corresponding value of y.

$4(-2x) = x + 9$	$y = -2x$
$-8x = x + 9$	$y = -2(-1)$
$-9x = 9$	$y = 2$
$x = -1$	

Answer The coordinates of the centroid are $P(-1, 2)$. ▢

We can verify the results of this example by showing that P is a point on the median from C:

(1) The coordinates of L, the midpoint of $\overline{AB}$, are:

$$\left(\frac{-3 + (-9)}{2}, \frac{6 + 0}{2}\right) = (-6, 3)$$

(2) The equation of $\overleftrightarrow{CL}$ is:

$$\frac{y-0}{x-9} = \frac{0-3}{9-(-6)}$$

$$\frac{y}{x-9} = \frac{-1}{5}$$

$$5y = -x + 9$$

(3) $P(-1, 2)$ is a point on $\overleftrightarrow{CL}$:

$$5y = -x + 9$$

$$5(2) \overset{?}{=} -(-1) + 9$$

$$10 = 10 \checkmark$$

Exercises

Writing About Mathematics

1. If $\overline{AM}$ and $\overline{BN}$ are two medians of $\triangle ABC$ that intersect at P, is P one of the points on $\overline{AM}$ that separate the segment into three congruent parts? Explain your answer.

2. Can the perpendicular bisector of a side of a triangle ever be the median to a side of a triangle? Explain your answer.

Developing Skills

In 3–10, find the coordinates of the centroid of each triangle with the given vertices.

3. $A(-3, 0)$, $B(1, 0)$, $C(-1, 6)$

4. $A(-5, -1)$, $B(1, -1)$, $C(1, 5)$

5. $A(-3, 3)$, $B(3, -3)$, $C(3, 9)$

6. $A(1, 2)$, $B(7, 0)$, $C(1, -2)$

7. $A(-1, 1)$, $B(3, 1)$, $C(1, 7)$

8. $A(-6, 2)$, $B(0, 0)$, $C(0, 10)$

9. $A(-2, -5)$, $B(0, 1)$, $C(-10, 1)$

10. $A(-1, -1)$, $B(17, -1)$, $C(5, 5)$

Applying Skills

11. The coordinates of a vertex of $\triangle ABC$ are $A(0, 6)$, and $AB = AC$.

    a. If B and C are on the x-axis and $BC = 4$, find the coordinates of B and C.

    b. Find the coordinates of the midpoint M of $\overline{AB}$ and of the midpoint N of $\overline{AC}$.

    c. Find the equation of $\overleftrightarrow{CM}$.

    d. Find the equation of $\overleftrightarrow{BN}$.

    e. Find the coordinates of the centroid of $\triangle ABC$.

12. The coordinates of the midpoint of $\overline{AB}$ of $\triangle ABC$ are $M(3, 0)$ and the coordinates of the centroid are $P(0, 0)$. If $\triangle ABC$ is isosceles and $AB = 6$, find the coordinates of A, B, and C.

12-8 PROPORTIONS IN A RIGHT TRIANGLE

Projection of a Point or of a Line Segment on a Line

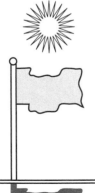

Whenever the sun is shining, any object casts a shadow. If the sun were directly overhead, the projection of an object would be suggested by the shadow of that object.

DEFINITION

The **projection of a point on a line** is the foot of the perpendicular drawn from that point to the line.

The **projection of a segment on a line**, when the segment is not perpendicular to the line, is the segment whose endpoints are the projections of the endpoints of the given line segment on the line.

In the figure, $\overline{MN}$ is the projection of $\overline{AB}$ on $\overleftrightarrow{PQ}$. The projection of R on $\overleftrightarrow{PQ}$ is P. If $\overline{PR} \perp \overleftrightarrow{PQ}$, the projection of $\overline{PR}$ on $\overleftrightarrow{PQ}$ is P.

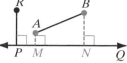

Proportions in the Right Triangle

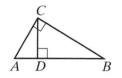

In the figure, $\triangle ABC$ is a right triangle, with the right angle at C. Altitude $\overline{CD}$ is drawn to hypotenuse $\overline{AB}$ so that two smaller triangles are formed, $\triangle ACD$ and $\triangle CBD$. Since $\overline{CD} \perp \overline{AB}$, $\angle CDA$ and $\angle CDB$ are right angles. The projection of $\overline{AC}$ on $\overline{AB}$ is $\overline{AD}$ and the projection of $\overline{BC}$ on $\overline{AB}$ is $\overline{BD}$. We want to prove that the three right triangles, $\triangle ABC$, $\triangle ACD$, and $\triangle CBD$, are similar triangles and, because they are similar triangles, the lengths of corresponding sides are in proportion.

Theorem 12.16

The altitude to the hypotenuse of a right triangle divides the triangle into two triangles that are similar to each other and to the original triangle.

Given $\triangle ABC$ with $\angle ACB$ a right angle and altitude $\overline{CD} \perp \overline{AB}$ at D.

Prove $\triangle ABC \sim \triangle ACD \sim \triangle CBD$

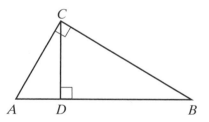

Proof

Statements	**Reasons**
1. $\angle ACB$ is a right angle.	**1.** Given.
2. $\overline{CD} \perp \overline{AB}$	**2.** Given.
3. $\angle ADC$ and $\angle BDC$ are right angles.	**3.** Perpendicular lines intersect to form right angles.
4. $\angle ACB \cong \angle ADC \cong \angle BDC$	**4.** All right angles are congruent.
5. $\angle A \cong \angle A$ and $\angle B \cong \angle B$	**5.** Reflexive property of congruence.
6. $\triangle ABC \sim \triangle ACD$ and $\triangle ABC \sim \triangle CBD$	**6.** AA~.
7. $\triangle ABC \sim \triangle ACD \sim \triangle CBD$	**7.** Transitive property of similarity.

Now that we have proved that these triangles are similar, we can prove that the lengths of corresponding sides are in proportion. Recall that if the means of a proportion are equal, either mean is called the *mean proportional* between the extremes.

Corollary 12.16a

The length of each leg of a right triangle is the mean proportional between the length of the projection of that leg on the hypotenuse and the length of the hypotenuse.

Given $\triangle ABC$ with $\angle ACB$ a right angle and altitude $\overline{CD} \perp \overline{AB}$ at D

Prove $\dfrac{AB}{AC} = \dfrac{AC}{AD}$ and $\dfrac{AB}{BC} = \dfrac{BC}{BD}$

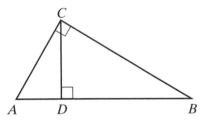

Proof   The lengths of the corresponding sides of similar triangles are in proportion.
Therefore, since $\triangle ABC \sim \triangle ACD$,
$\dfrac{AB}{AC} = \dfrac{AC}{AD}$ and since $\triangle ABC \sim \triangle CBD$,
$\dfrac{AB}{BC} = \dfrac{BC}{BD}$. ∎

Corollary 12.16b

The length of the altitude to the hypotenuse of a right triangle is the mean proportional between the lengths of the projections of the legs on the hypotenuse.

Proof: The lengths of the corresponding sides of similar triangles are in proportion. Therefore, since $\triangle ACD \sim \triangle CBD$, $\dfrac{AD}{CD} = \dfrac{CD}{BD}$. ∎

EXAMPLE 1

In right triangle ABC, altitude $\overline{CD}$ is drawn to hypotenuse $\overline{AB}$. If $AD = 8$ cm and $DB = 18$ cm, find: **a.** AC **b.** BC **c.** CD

Solution

$$AB = AD + DB$$
$$= 8 + 18$$
$$= 26$$

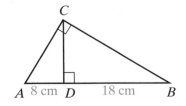

Since $\overline{CD}$ is the altitude to the hypotenuse of right $\triangle ABC$, then:

$$\frac{AB}{AC} = \frac{AC}{AD} \qquad\qquad \frac{AB}{BC} = \frac{BC}{BD} \qquad\qquad \frac{AD}{CD} = \frac{CD}{DB}$$

$$\frac{26}{AC} = \frac{AC}{8} \qquad\qquad \frac{26}{BC} = \frac{BC}{18} \qquad\qquad \frac{8}{CD} = \frac{CD}{18}$$

$$(AC)^2 = 208 \qquad\qquad (BC)^2 = 468 \qquad\qquad (CD)^2 = 144$$

$$AC = \sqrt{208} \qquad\qquad BC = \sqrt{468} \qquad\qquad CD = \sqrt{144}$$

$$= \sqrt{16}\sqrt{13} \qquad\qquad = \sqrt{36}\sqrt{13} \qquad\qquad = 12$$

$$= 4\sqrt{13} \qquad\qquad = 6\sqrt{13}$$

Answers   **a.** $4\sqrt{13}$ cm   **b.** $6\sqrt{13}$ cm   **c.** 12 cm

EXAMPLE 2

The altitude to the hypotenuse of right triangle ABC separates the hypotenuse into two segments. The length of one segment is 5 inches more than the measure of the other. If the length of the altitude is 6 inches, find the length of the hypotenuse.

Solution   Let x = the measure of the shorter segment.

Then $x + 5$ = the measure of the longer segment.

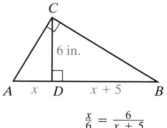

(1) The length of the altitude is the mean proportional between the lengths of the segments of the hypotenuse:

$$\frac{x}{6} = \frac{6}{x + 5}$$

(2) Set the product of the means equal to the product of the extremes:

$$x(x + 5) = 36$$
$$x^2 + 5x = 36$$

(3) Write the equation in standard form:

$$x^2 + 5x - 36 = 0$$

(4) Factor the left side:

$$(x - 4)(x + 9) = 0$$

(5) Set each factor equal to 0 and solve for x. Reject the negative root:

$$x - 4 = 0 \quad | \quad x + 9 = 0$$
$$x = 4 \quad | \qquad x = -9 \text{ reject}$$

(6) The length of the hypotenuse is the sum of the lengths of the segments:

$$x + x + 5 = 4 + 4 + 5$$
$$= 13 \text{ in.}$$

Answer   The length of the hypotenuse is 13 inches.

Exercises

Writing About Mathematics

1. When altitude $\overline{CD}$ is drawn to the hypotenuse of right triangle ABC, it is possible that $\triangle ACD$ and $\triangle BCD$ are congruent as well as similar. Explain when $\triangle ACD \cong \triangle BCD$.

2. The altitude to the hypotenuse of right $\triangle RST$ separates the hypotenuse, $\overline{RS}$, into two congruent segments. What must be true about $\triangle RST$?

Developing Skills

In 3–12, $\triangle ABC$ is a right triangle with $\angle ACB$ the right angle. Altitude $\overline{CD}$ intersects $\overline{AB}$ at D. In each case find the required length.

3. If $AD = 3$ and $CD = 6$, find DB.

4. If $AB = 8$ and $AC = 4$, find AD.

5. If $AC = 10$ and $AD = 5$, find AB.

6. If $AC = 6$ and $AB = 9$, find AD.

7. If $AD = 4$ and $DB = 9$, find CD.

8. If $DB = 4$ and $BC = 10$, find AB.

9. If $AD = 3$ and $DB = 27$, find CD.

10. If $AD = 2$ and $AB = 18$, find AC.

11. If $DB = 8$ and $AB = 18$, find BC.

12. If $AD = 3$ and $DB = 9$, find AC.

Applying Skills

In 13–21, the altitude to the hypotenuse of a right triangle divides the hypotenuse into two segments.

13. If the lengths of the segments are 5 inches and 20 inches, find the length of the altitude.

14. If the length of the altitude is 8 feet and the length of the shorter segment is 2 feet, find the length of the longer segment.

15. If the ratio of the lengths of the segments is 1:9 and the length of the altitude is 6 meters, find the lengths of the two segments.

16. The altitude drawn to the hypotenuse of a right triangle divides the hypotenuse into two segments of lengths 4 and 5. What is the length of the altitude?

17. If the length of the altitude to the hypotenuse of a right triangle is 8, and the length of the hypotenuse is 20, what are the lengths of the segments of the hypotenuse? (Let x and $20 - x$ be the lengths of the segments of the hypotenuse.)

18. The altitude drawn to the hypotenuse of a right triangle divides the hypotenuse into segments of lengths 2 and 16. What are the lengths of the legs of the triangle?

19. In a right triangle whose hypotenuse measures 50 centimeters, the shorter leg measures 30 centimeters. Find the measure of the projection of the shorter leg on the hypotenuse.

20. The segments formed by the altitude to the hypotenuse of right triangle ABC measure 8 inches and 10 inches. Find the length of the shorter leg of $\triangle ABC$.

21. The measures of the segments formed by the altitude to the hypotenuse of a right triangle are in the ratio $1 : 4$. The length of the altitude is 14.

   a. Find the measure of each segment.

   b. Express, in simplest radical form, the length of each leg.

12-9 PYTHAGOREAN THEOREM

The theorems that we proved in the last section give us a relationship between the length of a legs of a right triangle and the length of the hypotenuse. These proportions are the basis for a proof of the **Pythagorean Theorem**, which was studied in earlier courses.

Theorem 12.17a | If a triangle is a right triangle, then the square of the length of the longest side is equal to the sum of the squares of the lengths of the other two sides (the legs).

Given $\triangle ABC$ is a right triangle with $\angle ACB$ the right angle, c is the length of the hypotenuse, a and b are the lengths of the legs.

Prove $c^2 = a^2 + b^2$

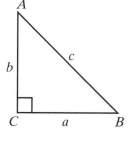

Proof	**Statements**	**Reasons**
	1. $\triangle ABC$ is a right triangle with $\angle ACB$ the right angle.	1. Given.
	2. Draw $\overline{CD} \perp \overline{AB}$. Let $BD = x$ and $AD = c - x$.	2. From a point not on a given line, one and only one perpendicular can be drawn to the given line.
	3. $\dfrac{c}{a} = \dfrac{a}{x}$ and $\dfrac{c}{b} = \dfrac{b}{c-x}$	3. The length of each leg of a right triangle is the mean proportional between the length of the projection of that leg on the hypotenuse and the length of the hypotenuse.
	4. $cx = a^2$ and $c(c - x) = b^2$ $\qquad c^2 - cx = b^2$	4. In a proportion, the product of the means is equal to the product of the extremes.
	5. $\qquad cx + c^2 - cx = a^2 + b^2$ $\qquad\qquad\quad c^2 = a^2 + b^2$	5. Addition postulate.

The Converse of the Pythagorean Theorem

If we know the lengths of the three sides of a triangle, we can determine whether the triangle is a right triangle by using the converse of the Pythagorean Theorem.

Theorem 12.17b

> If the square of the length of one side of a triangle is equal to the sum of the squares of the lengths of the other two sides, then the triangle is a right triangle.

Given $\triangle ABC$ with $AB = c$, $BC = a$, $CA = b$, and $c^2 = a^2 + b^2$

Prove $\triangle ABC$ is a right triangle with $\angle C$ the right angle.

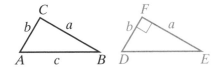

Proof Draw $\triangle DEF$ with $EF = a$, $FD = b$, and $\angle F$ a right angle. Then $DE^2 = a^2 + b^2$, $DE^2 = c^2$ and $DE = c$. Therefore, $\triangle ABC \cong \triangle DEF$ by SSS. Corresponding angles of congruent triangles are congruent, so $\angle C \cong \angle F$ and $\angle C$ is a right angle. Triangle ABC is a right triangle. ◾

We can state Theorems 12.17a and 12.17b as a single theorem.

Theorem 12.17

> A triangle is a right triangle if and only if the square of the length of the longest side is equal to the sum of the squares of the lengths of the other two sides.

EXAMPLE 1

What is the length of the altitude to the base of an isosceles triangle if the length of the base is 18 centimeters and the length of a leg is 21 centimeters? Round your answer to the nearest centimeter.

Solution The altitude to the base of an isosceles triangle is perpendicular to the base and bisects the base. In $\triangle ABC$, $\overline{CD}$ is the altitude to the base $\overline{AB}$, $\overline{AC}$ is the hypotenuse of right triangle ACD, $AD = 9.0$ cm, and $AC = 21$ cm.

$$AD^2 + CD^2 = AC^2$$
$$9^2 + CD^2 = 21^2$$
$$81 + CD^2 = 441$$
$$CD^2 = 360$$
$$CD = \sqrt{360} = \sqrt{36}\sqrt{10} = 6\sqrt{10} \approx 19$$

Answer The length of the altitude is approximately 19 centimeters. ◾

EXAMPLE 2

When a right circular cone is cut by a plane through the vertex and perpendicular to the base of the cone, the cut surface is an isosceles triangle. The length of the hypotenuse of the triangle is the slant height of the cone, the length one of the legs is the height of the cone, and the length of the other leg is the radius of the base of the cone. If a cone has a height of 24 centimeters and the radius of the base is 10 centimeters, what is the slant height of the cone?

Solution   Use the Pythagorean Theorem:

$$(h_s)^2 = (h_c)^2 + r^2$$
$$(h_s)^2 = 24^2 + 10^2$$
$$(h_s)^2 = 676$$
$$h_s = 26 \text{ cm} \ \textit{Answer}$$

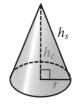

Pythagorean Triples

When three integers can be the lengths of the sides of a right triangle, this set of numbers is called a **Pythagorean triple**. The most common Pythagorean triple is 3, 4, 5:

$$3^2 + 4^2 = 5^2$$

If we multiply each number of a Pythagorean triple by some positive integer x, then the new triple created is also a Pythagorean triple because it will satisfy the relation $a^2 + b^2 = c^2$. For example:

If $\{3, 4, 5\}$ is a Pythagorean triple, then $\{3x, 4x, 5x\}$ is also a Pythagorean triple for a similar triangle where the ratio of similitude of the second triangle to the first triangle is $x : 1$.

Let $x = 2$. Then $\{3x, 4x, 5x\} = \{6, 8, 10\}$ and $6^2 + 8^2 = 10^2$.
Let $x = 3$. Then $\{3x, 4x, 5x\} = \{9, 12, 15\}$, and $9^2 + 12^2 = 15^2$.
Let $x = 10$. Then $\{3x, 4x, 5x\} = \{30, 40, 50\}$, and $30^2 + 40^2 = 50^2$.

Here are other examples of Pythagorean triples that occur frequently:

$\{5, 12, 13\}$ or, in general, $\{5x, 12x, 13x\}$ where x is a positive integer.
$\{8, 15, 17\}$ or, in general, $\{8x, 15x, 17x\}$ where x is a positive integer.

The 45-45-Degree Right Triangle

The legs of an isosceles right triangle are congruent. The measure of each acute angles of an isosceles right triangle is 45°. If two triangles are isosceles right triangles then they are similar by AA~. An isosceles right triangle is called a **45-45-degree right triangle**.

When a diagonal of a square is drawn, the square is separated into two isosceles right triangles. We can express the length of a leg of the isosceles right triangle in terms of the length of the hypotenuse or the length of the hypotenuse in terms of the length of a leg.

Let s be the length of the hypotenuse of an isosceles right triangle and x be the length of each leg. Use the Pythagorean Theorem to set up two equalities. Solve one for x and the other for s:

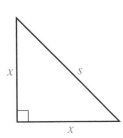

Solve for x:	*Solve for s:*
$a^2 + b^2 = c^2$	$c^2 = a^2 + b^2$
$x^2 + x^2 = s^2$	$s^2 = x^2 + x^2$
$2x^2 = s^2$	$s^2 = 2x^2$
$x^2 = \frac{s^2}{2}$	$s = x\sqrt{2}$
$x = \sqrt{\frac{s^2}{2}}$	
$x = \sqrt{\frac{s^2}{2} \cdot \frac{2}{2}}$	
$x = \frac{\sqrt{2}}{2}s$	

The 30-60-Degree Right Triangle

An altitude drawn to any side of an equilateral triangle bisects the base and separates the triangle into two congruent right triangles. Since the measure of each angle of an equilateral triangle is 60°, the measure of one of the acute angles of a right triangle formed by the altitude is 60° and the measure of the other acute angle is 30°. Each of the congruent right triangles formed by drawing an altitude to a side of an equilateral triangle is called a **30-60-degree right triangle**. If two triangles are 30-60-degree right triangles, then they are similar by AA~.

In the diagram, $\triangle ABC$ is an equilateral triangle with s the length of each side and h the length of an altitude. Then s is the length of the hypotenuse of the 30-60-degree triangle and h is the length of a leg. In the diagram, $\overline{CD} \perp \overline{AB}$, $AC = s$, $AD = \frac{s}{2}$, and $CD = h$.

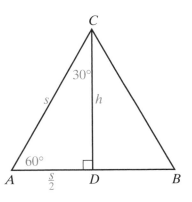

$$a^2 + b^2 = c^2$$
$$\left(\tfrac{s}{2}\right)^2 + h^2 = s^2$$
$$\tfrac{s^2}{4} + h^2 = \tfrac{4}{4}s^2$$
$$h^2 = \tfrac{3}{4}s^2$$
$$h = \tfrac{\sqrt{3}}{2}s$$

EXAMPLE 3

In right triangle ABC, the length of the hypotenuse, $\overline{AB}$, is 6 centimeters and the length of one leg is 3 centimeters. Find the length of the other leg.

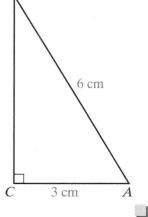

Solution

$$a^2 + b^2 = c^2$$
$$a^2 + (3)^2 = (6)^2$$
$$a^2 + 9 = 36$$
$$a^2 = 27$$
$$a = \sqrt{27} = \sqrt{9 \cdot 3} = 3\sqrt{3} \quad \textit{Answer}$$

Note: The measure of one leg is one-half the measure of the hypotenuse and the length of the other leg is $\frac{\sqrt{3}}{2}$ times the length of the hypotenuse. Therefore, this triangle is a 30-60-degree right triangle. We can use a calculator to verify this.

 Recall that $\tan A = \frac{\text{opp}}{\text{adj}} = \frac{BC}{AC} = \frac{3\sqrt{3}}{3} = \sqrt{3}$. Use a graphing calculator to find the measure of $\angle A$.

ENTER: [**2nd**] [**TAN⁻¹**] [**2nd**] [√] [**3**] [**ENTER**]

The calculator will return 60 as $m\angle A$. Therefore, $m\angle B = 30$.

Exercises

Writing About Mathematics

1. Ira said that if the lengths of the sides of an obtuse triangle $\triangle ABC$ are a, b, and c with c opposite the obtuse angle, then $c^2 > a^2 + b^2$. Do you agree with Ira? Explain why or why not. (*Hint:* Make use of the altitude from one of the acute angles.)

2. Sean said that if the measures of the diagonals of a parallelogram are 6 and 8 and the measure of one side of the parallelogram is 5 then the parallelogram is a rhombus. Do you agree with Sean? Explain why or why not.

Developing Skills

In 3–8, in each case the lengths of three sides of a triangle are given. Tell whether each triangle is a right triangle.

3. 6, 8, 10   **4.** 7, 8, 12   **5.** 5, 7, 8   **6.** 15, 36, 39   **7.** 14, 48, 50   **8.** 2, $2\sqrt{3}$, 4

9. Find, to the nearest tenth of a centimeter, the length of a diagonal of a square if the measure of one side is 8.0 centimeters.

10. Find the length of the side of a rhombus whose diagonals measure 40 centimeters and 96 centimeters.

11. The length of a side of a rhombus is 10 centimeters and the length of one diagonal is 120 millimeters. Find the length of the other diagonal.

12. The length of each side of a rhombus is 13 feet. If the length of the shorter diagonal is 10 feet, find the length of the longer diagonal.

13. Find the length of the diagonal of a rectangle whose sides measure 24 feet by 20 feet.

14. The diagonal of a square measures 12 feet.

   a. What is the exact measure of a side of the square?

   b. What is the area of the square?

15. What is the slant height of a cone whose height is 36 centimeters and whose radius is 15 centimeters?

16. One side of a rectangle is 9 feet longer than an adjacent side. The length of the diagonal is 45 feet. Find the dimensions of the rectangle.

17. One leg of a right triangle is 1 foot longer than the other leg. The hypotenuse is 9 feet longer than the shorter leg. Find the length of the sides of the triangle.

Applying Skills

18. Marvin wants to determine the edges of a rectangular garden that is to be 10 feet by 24 feet. He has no way of determining the measure of an angle but he can determine lengths very accurately. He takes a piece of cord that is 60 feet long and makes a mark at 10 feet and at 34 feet from one end. Explain how the cord can help him to make sure that his garden is a rectangle.

19. A plot of land is in the shape of an isosceles trapezoid. The lengths of the parallel sides are 109 feet and 95 feet. The length of each of the other two sides is 25 feet. What is the area of the plot of land?

20. From a piece of cardboard, Shanti cut a semicircle with a radius of 10 inches. Then she used tape to join one half of the diameter along which the cardboard had been cut to the other half, forming a cone. What is the height of the cone that Shanti made?

21. The lengths of two adjacent sides of a parallelogram are 21 feet and 28 feet. If the length of a diagonal of the parallelogram is 35 feet, show that the parallelogram is a rectangle.

22. The lengths of the diagonals of a parallelogram are 140 centimeters and 48 centimeters. The length of one side of the parallelogram is 74 centimeters. Show that the parallelogram is a rhombus.

23. A young tree is braced by wires that are 9 feet long and fastened at a point on the trunk of the tree 5 feet from the ground. Find to the nearest tenth of a foot how far from the foot of the tree the wires should be fastened to the ground in order to be sure that the tree will be perpendicular to the ground.

24. The length of one side of an equilateral triangle is 12 feet. What is the distance from the centroid of the triangle to a side? (Express the answer in simplest radical form.)

12-10 THE DISTANCE FORMULA

When two points in the coordinate plane are on the same vertical line, they have the same x-coordinate and the distance between them is the absolute value of the difference of their y-coordinates. In the diagram, the coordinates of A are $(4, 8)$ and the coordinates of C are $(4, 2)$.

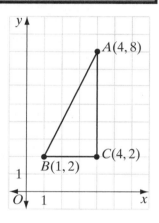

$$CA = |8 - 2|$$
$$= 6$$

When two points in the coordinate plane are on the same horizontal line, they have the same y-coordinate and the distance between them is the absolute value of the difference of their x-coordinates. In the diagram, the coordinates of B are $(1, 2)$ and the coordinates of C are $(4, 2)$.

$$CB = |1 - 4|$$
$$= 3$$

In $\triangle ABC$, $\angle C$ is a right angle and $\overline{AB}$ is the hypotenuse of a right triangle. Using the Pythagorean Theorem, we can find AB:

$$AB^2 = CA^2 + CB^2$$
$$AB^2 = 6^2 + 3^2$$
$$AB^2 = 36 + 9$$
$$AB = \sqrt{45}$$
$$AB = 3\sqrt{5}$$

This example suggests a method that can be used to find a formula for the length of any segment in the coordinate plane.

Let $B(x_1, y_1)$ and $A(x_2, y_2)$ be any two points in the coordinate plane. From A draw a vertical line and from B draw a horizontal line. Let the intersection of these two lines be C. The coordinates of C are (x_2, y_1). Let $AB = c$, $CB = a = |x_2 - x_1|$, and $CA = b = |y_2 - y_1|$. Then,

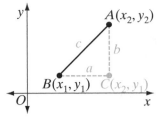

$$c^2 = a^2 + b^2$$

$$c^2 = |x_2 - x_1|^2 + |y_2 - y_1|^2$$

$$c = \sqrt{(x_2 - x_1)^2 + (y_2 - y_1)^2}$$

This result is called the distance formula. If the endpoints of a line segment in the coordinate plane are $B(x_1, y_1)$ and $A(x_2, y_2)$, then:

$$AB = \sqrt{(x_2 - x_1)^2 + (y_2 - y_1)^2}$$

EXAMPLE I

The coordinates of the vertices of quadrilateral $ABCD$ are $A(-1, -3)$, $B(6, -4)$, $C(5, 3)$, and $D(-2, 4)$.

a. Prove that $ABCD$ is a rhombus.

b. Prove that $ABCD$ is not a square.

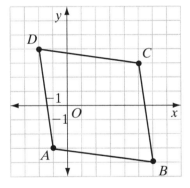

Solution

a.
$$AB = \sqrt{(6 - (-1))^2 + (-4 - (-3))^2}$$
$$= \sqrt{(7)^2 + (-1)^2}$$
$$= \sqrt{49 + 1}$$
$$= \sqrt{50}$$

$$BC = \sqrt{(5 - 6)^2 + (3 - (-4))^2}$$
$$= \sqrt{(-1)^2 + (7)^2}$$
$$= \sqrt{1 + 49}$$
$$= \sqrt{50}$$

$$CD = \sqrt{(-2 - 5)^2 + (4 - 3)^2}$$
$$= \sqrt{(-7)^2 + (1)^2}$$
$$= \sqrt{49 + 1}$$
$$= \sqrt{50}$$

$$DA = \sqrt{(-1 - (-2))^2 + (-3 - 4)^2}$$
$$= \sqrt{(1)^2 + (-7)^2}$$
$$= \sqrt{1 + 49}$$
$$= \sqrt{50}$$

The lengths of the sides of the quadrilateral are equal. Therefore, the quadrilateral is a rhombus.

b. If a rhombus is a square, then it has a right angle.

METHOD I

$$AC = \sqrt{(5 - (-1))^2 + (3 - (-3))^2}$$
$$= \sqrt{(6)^2 + (6)^2}$$
$$= \sqrt{36 + 36}$$
$$= \sqrt{72}$$

If $\angle B$ is a right angle, then

$$AC^2 = AB^2 + BC^2$$
$$\left(\sqrt{72}\right)^2 \overset{?}{=} \left(\sqrt{50}\right)^2 + \left(\sqrt{50}\right)^2$$
$$72 \neq 50 + 50 \;\; ✗$$

Therefore, $\triangle ABC$ is not a right triangle, $\angle B$ is not a right angle and $ABCD$ is not a square.

METHOD 2

$$\text{slope of } \overline{AB} = \frac{-4 - (-3)}{6 - (-1)}$$
$$= \frac{-1}{7}$$
$$= -\frac{1}{7}$$

$$\text{slope of } \overline{BC} = \frac{3 - (-4)}{5 - 6}$$
$$= \frac{7}{-1}$$
$$= -7$$

The slope of $\overline{AB}$ is not equal to the negative reciprocal of the slope of $\overline{BC}$. Therefore, $\overline{AB}$ is not perpendicular to $\overline{BC}$, $\angle B$ is not a right angle and the rhombus is not a square.

EXAMPLE 2

Prove that the midpoint of the hypotenuse of a right triangle is equidistant from the vertices using the distance formula.

Proof   We will use a coordinate proof. The triangle can be placed at any convenient position. Let right triangle ABC have vertices $A(2a, 0)$, $B(0, 2b)$, and $C(0, 0)$. Let M be the midpoint of the hypotenuse $\overline{AB}$. The coordinates of M, the midpoint of $\overline{AB}$, are

$$\left(\tfrac{2a + 0}{2}, \tfrac{0 + 2b}{2}\right) = (a, b)$$

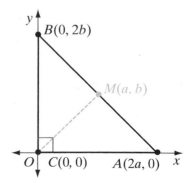

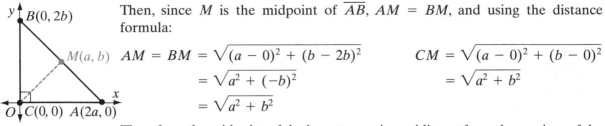

Then, since M is the midpoint of $\overline{AB}$, $AM = BM$, and using the distance formula:

$$AM = BM = \sqrt{(a - 0)^2 + (b - 2b)^2} \qquad CM = \sqrt{(a - 0)^2 + (b - 0)^2}$$
$$= \sqrt{a^2 + (-b)^2} \qquad\qquad\qquad = \sqrt{a^2 + b^2}$$
$$= \sqrt{a^2 + b^2}$$

Therefore, the midpoint of the hypotenuse is equidistant from the vertices of the triangle. ◻

EXAMPLE 3

Prove that the medians to the base angles of an isosceles triangle are congruent.

Given: Isosceles $\triangle ABC$ with vertices $A(-2a, 0)$, $B(0, 2b)$, $C(2a, 0)$. Let M be the midpoint of $\overline{AB}$ and N be the midpoint of $\overline{BC}$.

Prove: $\overline{CM} \cong \overline{AN}$

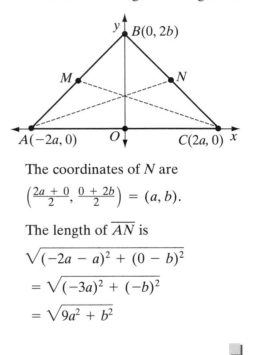

Proof The coordinates of M are

$$\left(\frac{-2a + 0}{2}, \frac{0 + 2b}{2}\right) = (-a, b).$$

The length of $\overline{CM}$ is

$$\sqrt{(-a - 2a)^2 + (b - 0)^2}$$
$$= \sqrt{(-3a)^2 + b^2}$$
$$= \sqrt{9a^2 + b^2}$$

The coordinates of N are

$$\left(\frac{2a + 0}{2}, \frac{0 + 2b}{2}\right) = (a, b).$$

The length of $\overline{AN}$ is

$$\sqrt{(-2a - a)^2 + (0 - b)^2}$$
$$= \sqrt{(-3a)^2 + (-b)^2}$$
$$= \sqrt{9a^2 + b^2}$$

$CM = AN$; therefore, $\overline{CM} \cong \overline{AN}$. ◻

Exercises

Writing About Mathematics

1. Can the distance formula be used to find the length of a line segment when the endpoints of the segment are on the same horizontal line or on the same vertical line? Justify your answer.

2. Explain why $|x_2 - x_1|^2 = (x_2 - x_1)^2$.

Developing Skills

In 3–10, the coordinates of the endpoints of $\overline{AB}$ are given. In each case, find the exact value of AB in simplest form.

3. $A(1, 2), B(4, 6)$ **4.** $A(-1, -6), B(4, 6)$

5. $A(3, -2), B(5, 4)$ **6.** $A(0, 2), B(3, -1)$

7. $A(1, 2), B(3, 4)$ **8.** $A(-5, 2), B(1, -6)$

9. $A(6, 2), B(1, -3)$ **10.** $A(-3, 3), B(3, -3)$

11. The coordinates of A are $(0, 4)$ and the x-coordinate of B is 5. What is the y-coordinate of B if $AB = 13$? (Two answers are possible.)

12. The coordinates of M are $(2, -1)$ and the y-coordinate of N is 5. What is the x-coordinate of N if $MN = 3\sqrt{5}$? (Two answers are possible.)

13. The vertices of a quadrilateral are $A(0, -2), B(5, -2), C(8, 2), D(3, 2)$. Prove that the quadrilateral is a rhombus using the distance formula.

14. The vertices of a triangle are $P(1, -1), Q(7, 1)$, and $R(3, 3)$.

    a. Show that $\triangle PQR$ is an isosceles triangle.

    b. Show that $\triangle PQR$ is a right triangle using the Pythagorean Theorem.

    c. Show that the midpoint of the hypotenuse is equidistant from the vertices.

15. The vertices of a triangle are $L(1, -1), M(7, -3)$, and $N(2, 2)$.

    a. Show that $\triangle LMN$ is a scalene triangle.

    b. Show that $\triangle LMN$ is a right triangle using the Pythagorean Theorem.

    c. Show that the midpoint of $\overline{MN}$ is equidistant from the vertices.

16. The vertices of $\triangle DEF$ are $D(-2, -3), E(5, 0)$, and $F(-2, 3)$. Show that $\overline{DE} \cong \overline{FE}$.

17. The coordinates of the vertices of $\triangle BAT$ are $B(-2, 7), A(2, -1)$, and $T(11, -4)$.

    a. Find the equation of the line that is the altitude from B to $\overline{AT}$.

    b. Find the coordinates of D, the foot of the perpendicular or the **foot of the altitude** from part **a**.

    c. Find the length of the altitude $\overline{BD}$.

18. The coordinates of the vertices of $\triangle EDF$ are $E(-2, 0), D(4, 0)$, and $F\left(1, 3\sqrt{3}\right)$.

    a. Find ED, DF, and FE.

    b. Is $\triangle EDF$ equilateral? Justify your answer.

19. The vertices of $\triangle ABC$ are $A(1, -1), B(4, 1)$, and $C(2, 4)$.

    a. Find the coordinates of the vertices of $\triangle A'B'C'$, the image of $\triangle ABC$ under the transformation D_2.

    b. Show that distance is *not* preserved under the dilation.

c. Show that $\triangle A'B'C' \sim \triangle ABC$ using SSS$\sim$.

d. Use part **c** to show that the angle measures of $\triangle ABC$ are preserved under the dilation.

20. The vertices of quadrilateral $ABCD$ are $A(2,0)$, $B(3,-1)$, $C(4,1)$, and $D(3,2)$.

a. Show that $ABCD$ is a parallelogram using the distance formula.

b. Find the coordinates of the vertices of quadrilateral $A'B'C'D'$, the image of $ABCD$ under the transformation D_3.

c. Show that $A'B'C'D'$ is a parallelogram using the distance formula.

d. Use part **c** to show that the images of the parallel segments of $ABCD$ are also parallel under the dilation.

21. The vertices of quadrilateral $ABCD$ are $A(-2,-2)$, $B(2,0)$, $C(3,3)$, and $D(-1,1)$. Use the distance formula to prove that $ABCD$ is a parallelogram but not a rhombus.

22. The vertices of $\triangle ABC$ are $A(0,-2)$, $B(4,6)$, and $C(-2,4)$. Prove that $\triangle ABC$ is an isosceles right triangle using the Pythagorean Theorem.

Applying Skills

23. Use the distance formula to prove that $(-a,0)$, $(0,b)$ and $(a,0)$ are the vertices of an isosceles triangle.

24. The vertices of square $EFGH$ are $E(0,0)$, $F(a,0)$, $G(a,a)$, and $H(0,a)$. Prove that the diagonals of a square, $\overline{EG}$ and $\overline{FH}$, are congruent and perpendicular.

25. The vertices of quadrilateral $ABCD$ are $A(0,0)$, $B(b,c)$, $C(b+a,c)$, and $D(a,0)$. Prove that if $a^2 = b^2 + c^2$ then $ABCD$ is a rhombus.

26. Prove the midpoint formula using the distance formula. Let P and Q have coordinates (x_1, y_1) and (x_2, y_2), respectively. Let M have coordinates $\left(\frac{x_1 + x_2}{2}, \frac{y_1 + y_2}{2}\right)$. M is the midpoint of $\overline{PQ}$ if and only if $PM = MQ$.

a. Find PM.   **b.** Find MQ.   **c.** Show that $PM = MQ$.

27. The vertices of $\overline{WX}$ are $W(w, y)$ and $X(x, z)$.

a. What are the coordinates of $\overline{W'X'}$, the image of $\overline{WX}$ under the dilation D_k?

b. Show, using the distance formula, that $W'X'$ is k times the length of WX.

CHAPTER SUMMARY

Definitions to Know

- The **ratio of two numbers**, a and b, where b is not zero, is the number $\frac{a}{b}$.
- A **proportion** is an equation that states that two ratios are equal.
- In the proportion $\frac{a}{b} = \frac{c}{d}$, the first and fourth terms, a and d, are the **extremes** of the proportion, and the second and third terms, b and c, are the **means**.
- If the means of a proportion are equal, the **mean proportional** is one of the means.
- Two line segments are **divided proportionally** when the ratio of the lengths of the parts of one segment is equal to the ratio of the lengths of the parts of the other.
- Two polygons are **similar** if there is a one-to-one correspondence between their vertices such that:
  1. All pairs of corresponding angles are congruent.
  2. The ratios of the lengths of all pairs of corresponding sides are equal.
- The **ratio of similitude** of two similar polygons is the ratio of the lengths of corresponding sides.
- A **dilation** of k is a transformation of the plane such that:
  1. The image of point O, the center of dilation, is O.
  2. When k is positive and the image of P is P', then $\overrightarrow{OP}$ and $\overrightarrow{OP'}$ are the same ray and $OP' = kOP$.
  3. When k is negative and the image of P is P', then $\overrightarrow{OP}$ and $\overrightarrow{OP'}$ are opposite rays and $OP' = -kOP$.
- The **projection of a point on a line** is the foot of the perpendicular drawn from that point to the line.
- The **projection of a segment on a line**, when the segment is not perpendicular to the line, is the segment whose endpoints are the projections of the endpoints of the given line segment on the line.
- A **Pythagorean triple** is a set of three integers that can be the lengths of the sides of a right triangle.

Postulates

12.1 Any geometric figure is similar to itself. (Reflexive property)

12.2 A similarity between two geometric figures may be expressed in either order. (Symmetric property)

12.3 Two geometric figures similar to the same geometric figure are similar to each other. (Transitive property)

12.4 For any given triangle there exist a similar triangle with any given ratio of similitude.

Theorems **12.1** In a proportion, the product of the means is equal to the product of the extremes.

12.1a In a proportion, the means may be interchanged.

12.1b In a proportion, the extremes may be interchanged.

12.1c If the products of two pairs of factors are equal, the factors of one pair can be the means and the factors of the other the extremes of a proportion.

12.2 A line segment joining the midpoints of two sides of a triangle is parallel to the third side and its length is one-half the length of the third side.

12.3 Two line segments are divided proportionally if and only if the ratio of the length of a part of one segment to the length of the whole is equal to the ratio of the corresponding lengths of the other segment.

12.4 Two triangles are similar if two angles of one triangle are congruent to two corresponding angles of the other. (AA~)

12.5 Two triangles are similar if the three ratios of corresponding sides are equal. (SSS~)

12.6 Two triangles are similar if the ratios of two pairs of corresponding sides are equal and the corresponding angles included between these sides are congruent. (SAS~)

12.7 A line is parallel to one side of a triangle and intersects the other two sides if and only if the points of intersection divide the sides proportionally.

12.8 Under a dilation, angle measure is preserved.

12.9 Under a dilation, midpoint is preserved.

12.10 Under a dilation, collinearity is preserved.

12.11 If two triangles are similar, the lengths of corresponding altitudes have the same ratio as the lengths of any two corresponding sides.

12.12 If two triangles are similar, the lengths of corresponding medians have the same ratio as the lengths of any two corresponding sides.

12.13 If two triangles are similar, the lengths of corresponding angle bisectors have the same ratio as the lengths of any two corresponding sides.

12.14 Any two medians of a triangle intersect in a point that divides each median in the ratio 2 : 1.

12.15 The medians of a triangle are concurrent.

12.16 The altitude to the hypotenuse of a right triangle divides the triangle into two triangles that are similar to each other and to the original triangle.

12.16a The length of each leg of a right triangle is the mean proportional between the length of the projection of that leg on the hypotenuse and the length of the hypotenuse.

12.16b The length of the altitude to the hypotenuse of a right triangle is the mean proportional between the lengths of the projections of the legs on the hypotenuse.

12.17 A triangle is a right triangle if and only if the square of the length of the longest side is equal to the sum of the squares of the lengths of the other two sides.

Formulas   In the coordinate plane, under a dilation of k with the center at the origin:

$$P(x, y) \rightarrow P'(kx, ky) \quad \text{or} \quad D_k(x, y) = (kx, ky)$$

If x is the length of a leg of an isosceles right triangle and s is the length of the hypotenuse, then:

$$x = \frac{\sqrt{2}}{2}s \quad \text{and} \quad s = x\sqrt{2}$$

If s is the length of a side of an equilateral triangle and h is the length of an altitude then:

$$h = \frac{\sqrt{3}}{2}s$$

If the endpoints of a line segment in the coordinate plane are $B(x_1, y_1)$ and $A(x_2, y_2)$, then:

$$AB = \sqrt{(x_2 - x_1)^2 + (y_2 - y_1)^2}$$

VOCABULARY

12-1 Similar • Ratio of two numbers • Proportion • Extremes • Means • Mean proportional • Geometric mean

12-2 Midsegment theorem • Line segments divided proportionally

12-3 Similar polygons • Ratio of similitude • Constant of proportionality

12-4 Postulate of similarity • AA triangle similarity • SSS similarity theorem • SAS similarity theorem

12-5 Dilation • Enlargement • Contraction • Constant of dilation, k

12-7 Centroid

12-8 Projection of a point on a line • Projection of a segment on a line

12-9 Pythagorean Theorem • Pythagorean triple • 45-45-degree right triangle • 30-60-degree right triangle

12-10 Distance formula • Foot of an altitude

REVIEW EXERCISES

1. Two triangles are similar. The lengths of the sides of the smaller triangle are 5, 6, and 9. The perimeter of the larger triangle is 50. What are the lengths of the sides of the larger triangle?

2. The measure of one angle of right $\triangle ABC$ is 67° and the measure of an angle of right $\triangle LMN$ is 23°. Are the triangles similar? Justify your answer.

3. A line parallel to side $\overline{AB}$ of $\triangle ABC$ intersects $\overline{AC}$ at E and $\overline{BC}$ at F. If $EC = 12$, $AC = 20$, and $AB = 15$, find EF.

4. A line intersects side $\overline{AC}$ of $\triangle ABC$ at E and $\overline{BC}$ at F. If $EC = 4$, $AC = 12$, $FC = 5$, $BC = 15$, prove that $\triangle EFC \sim \triangle ABC$.

5. The altitude to the hypotenuse of a right triangle divides the hypotenuse into two segments. If the length of the altitude is 12 and the length of the longer segment is 18, what is the length of the shorter segment?

6. In $\triangle LMN$, $\angle L$ is a right angle, $\overline{LP}$ is an altitude, $MP = 8$, and $PN = 32$.

   a. Find LP.   **b.** Find MN.   **c.** Find ML.   **d.** Find NL.

7. The length of a side of an equilateral triangle is 18 centimeters. Find, to the nearest tenth of a centimeter, the length of the altitude of the triangle.

8. The length of the altitude to the base of an isosceles triangle is 10.0 centimeters and the length of the base is 14.0 centimeters. Find, to the nearest tenth of a centimeter, the length of each of the legs.

9. The coordinates of the endpoints of $\overline{PQ}$ are $P(2, 7)$ and $Q(8, -1)$.

   a. Find the coordinates of the endpoints of $\overline{P'Q'}$ under the composition $D_2 \circ r_{x\text{-axis}}$.

   b. What is the ratio $\frac{PQ}{P'Q'}$?

   c. What are the coordinates of M, the midpoint of $\overline{PQ}$?

   d. What are the coordinates of M', the image of M under $D_2 \circ r_{x\text{-axis}}$?

   e. What are the coordinates of N, the midpoint of $\overline{P'Q'}$?

   f. Is M' the midpoint of $\overline{P'Q'}$? Justify your answer.

10. If $\overleftrightarrow{AB} \parallel \overleftrightarrow{CD}$ and $\overline{AD}$ and $\overline{BC}$ intersect at E, prove that $\triangle ABE \sim \triangle DCE$.

11. A line intersects $\overline{AC}$ at E and $\overline{BC}$ at F. If $\triangle ABC \sim \triangle EFC$, prove that $\overleftrightarrow{EF} \parallel \overleftrightarrow{AB}$.

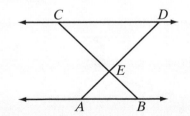

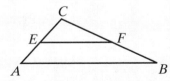

12. Find the length of the altitude to the bases of isosceles trapezoid $KLMN$ if $KL = 20$ cm, $MN = 38$ cm, and $KN = 15$ cm.

13. Find the length of a side of a rhombus if the measures of the diagonals of the rhombus are 30 inches and 40 inches.

14. The length of a side of a rhombus is 26.0 centimeters and the length of one diagonal is 28.0 centimeters. Find to the nearest tenth the length of the other diagonal.

15. The coordinates of the vertices of $\triangle RST$ are $R(4, 1)$, $S(-3, 2)$, and $T(-2, -1)$.

   a. Find the length of each side of the triangle in simplest radical form.

   b. Prove that the triangle is a right triangle.

16. The vertices of $\triangle ABC$ are $A(0, 0)$, $B(4, 3)$, and $C(0, 5)$.

   a. Prove that $\triangle ABC$ is isosceles.

   b. The median to $\overline{BC}$ is $\overline{AD}$. Find the coordinates of D.

   c. Find AD and DB.

   d. Prove that $\overline{AD}$ is the altitude to $\overline{BC}$ using the Pythagorean Theorem.

17. The vertices of $\triangle ABC$ are $A(-2, 1)$, $B(2, -1)$, and $C(0, 3)$.

   a. Find the coordinates of $\triangle A'B'C'$, the image of $\triangle ABC$ under D_3.

   b. Find, in radical form, the lengths of the sides of $\triangle ABC$ and of $\triangle A'B'C'$.

   c. Prove that $\triangle ABC \sim \triangle A'B'C'$.

   d. Find the coordinates of P, the centroid of $\triangle ABC$.

   e. Let M be the midpoint of $\overline{AB}$. Prove that $\frac{CP}{PM} = \frac{2}{1}$.

   f. Find the coordinates of P', the centroid of $\triangle A'B'C'$.

   g. Is P' the image of P under D_3?

   h. Let M' be the midpoint of $\overline{A'B'}$. Prove that $\frac{C'P}{PM'} = \frac{2}{1}$.

18. A right circular cone is cut by a plane through a diameter of the base and the vertex of the cone. If the diameter of the base is 20 centimeters and the height of the cone is 24 centimeters, what is the slant height of the cone?

Exploration

A line parallel to the shorter sides of a rectangle can divide the rectangle into a square and a smaller rectangle. If the smaller rectangle is similar to the given rectangle, then both rectangles are called golden rectangles and the ratio of the lengths of their sides is called the golden ratio. The golden ratio is

$$\left(1 + \sqrt{5}\right) : 2.$$

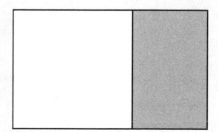

Follow the steps to construct a golden rectangle. You may use compass and straightedge or geometry software.

STEP 1. Draw square $ABCD$.

STEP 2. Construct E, the midpoint of $\overline{AB}$.

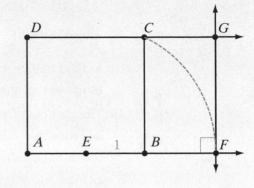

STEP 3. Draw the ray $\overrightarrow{AB}$.

STEP 4. With E as the center and radius EC, draw an arc that intersects $\overrightarrow{AB}$. Call this point F.

STEP 5. Draw the ray $\overrightarrow{DC}$.

STEP 6. Construct the line perpendicular to $\overrightarrow{AB}$ through F. Let the intersection of this line with $\overrightarrow{DC}$ be point G.

a. Let $AB = BC = 2$ and $EB = 1$. Find $EC = EF$, $AF = AE + EF$, and $BF = EF - EB$. Express each length as an irrational number in simplest radical form.

b. Show that $AFGD$ and $FGCB$ are golden rectangles by showing that they are similar, that is, that $\frac{AF}{AD} = \frac{FG}{BF}$.

c. Repeat steps 1 through 6 using a different square. Let $AB = x$. Complete parts **a** and **b**. Do you obtain the same ratio?

d. Research the golden rectangle and share your findings with your classmates.

CUMULATIVE REVIEW   Chapters 1–12

Part I

Answer all questions in this part. Each correct answer will receive 2 credits. No partial credit will be allowed.

1. The length and width of a rectangle are 16 centimeters and 12 centimeters. What is the length of a diagonal of the rectangle?
   (1) 4 cm   (2) 20 cm   (3) 25 cm   (4) $4\sqrt{7}$ cm

2. The measure of an angle is 12 degrees more than twice the measure of its supplement. What is the measure of the angle?
   (1) 26   (2) 56   (3) 64   (4) 124

3. What is the slope of the line through $A(2, 8)$ and $B(-4, -1)$?
   (1) $-\frac{3}{2}$ (2) $\frac{3}{2}$ (3) $-\frac{2}{3}$ (4) $\frac{2}{3}$

4. The measures of the opposite angles of a parallelogram are represented by $2x + 34$ and $3x - 12$. Find the value of x.

(1) 22   (2) 46   (3) 78   (4) 126

5. Which of the following do not determine a plane?

(1) three lines each perpendicular to the other two

(2) two parallel lines   (3) two intersecting lines

(4) a line and a point not on it

6. Which of the following cannot be the measures of the sides of a triangle?

(1) 5, 7, 8   (2) 3, 8, 9   (3) 5, 7, 7   (4) 2, 6, 8

7. Under a rotation of 90° about the origin, the image of the point whose coordinates are $(3, -2)$ is the point whose coordinates are

(1) $(2, -3)$ (2) $(2, 3)$ (3) $(-3, 2)$ (4) $(-2, -3)$

8. If a conditional statement is true, which of the following must be true?

(1) converse   (3) contrapositive

(2) inverse   (4) biconditional

9. In the figure, side $\overline{AB}$ of $\triangle ABC$ is extended through B to D. If $m\angle CBD = 105$ and $m\angle A = 53$, what is the measure of $\angle C$?

(1) 22   (3) 75

(2) 52   (4) 158

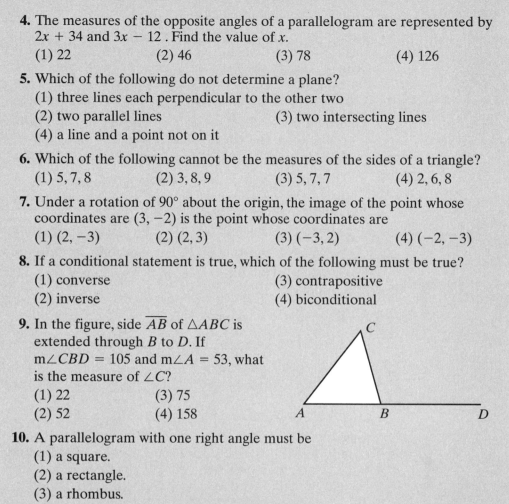

10. A parallelogram with one right angle must be

(1) a square.

(2) a rectangle.

(3) a rhombus.

(4) a trapezoid.

Part II

Answer all questions in this part. Each correct answer will receive 2 credits. Clearly indicate the necessary steps, including appropriate formula substitutions, diagrams, graphs, charts, etc. For all questions in this part, a correct numerical answer with no work shown will receive only 1 credit.

11. *Given:* $\overline{ABCD}$ with $\overline{AB} \cong \overline{CD}$, E not on $\overleftrightarrow{ABCD}$, and $\overline{BE} \cong \overline{CE}$.

Prove: $\overline{AE} \cong \overline{DE}$.

12. *Given:* $\overline{RS}$ perpendicular to plane p at R, points A and B in plane p, and $\overline{RA} \cong \overline{RB}$.

Prove: $\overline{SA} \cong \overline{SB}$

_ _ 1s
_ _ 1e

Part III

Answer all questions in this part. Each correct answer will receive 4 credits. Clearly indicate the necessary steps, including appropriate formula substitutions, diagrams, graphs, charts, etc. For all questions in this part, a correct numerical answer with no work shown will receive only 1 credit.

13. The radius of the base of a right circular cone is one-half the slant height of the cone. The radius of the base is 2.50 feet.

    a. Find the lateral area of the cone to the nearest tenth of a square foot.

    b. Find the volume of the cone to the nearest tenth of a cubic foot.

14. The coordinates of the vertices of $\triangle ABC$ are $A(-1, 0)$, $B(4, 0)$, and $C(2, 6)$.

    a. Write an equation of the line that contains the altitude from C to $\overline{AB}$.

    b. Write an equation of the line that contains the altitude from B to $\overline{AC}$.

    c. What are the coordinates of D, the intersection of the altitudes from C and from B?

    d. Write an equation for $\overleftrightarrow{AD}$.

    e. Is $\overleftrightarrow{AD}$ perpendicular to $\overline{BC}$, that is, does $\overleftrightarrow{AD}$ contain the altitude from A to $\overline{BC}$?

Part IV

Answer all questions in this part. Each correct answer will receive 6 credits. Clearly indicate the necessary steps, including appropriate formula substitutions, diagrams, graphs, charts, etc. For all questions in this part, a correct numerical answer with no work shown will receive only 1 credit.

15. In the diagram, $ABCD$ is a rectangle and $\triangle ADE$ is an isosceles triangle with $\overline{AE} \cong \overline{DE}$. If $\overline{EF}$, a median to $\overline{AD}$ of $\triangle ADE$, is extended to intersect $\overline{BC}$ at G, prove that G is the midpoint of $\overline{BC}$.

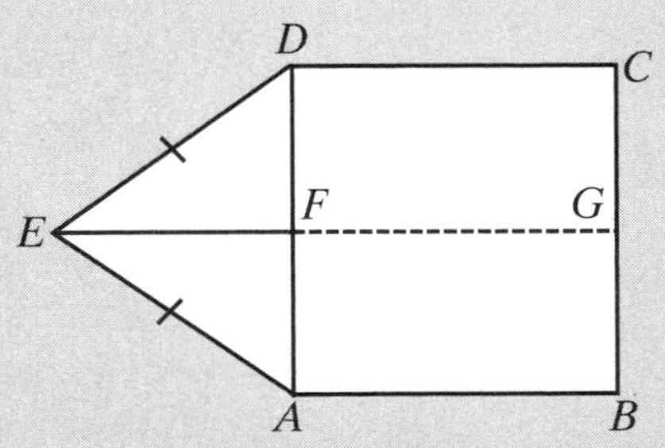

16. The coordinates of the vertices of $\triangle ABC$ are $A(2, 5)$, $B(3, 1)$, and $C(6, 4)$.

    a. Find the coordinates of $\triangle A'B'C'$, the image of $\triangle ABC$ under the composition $r_{y=x} \circ r_{x\text{-axis}}$.

    b. Show that $r_{y=x} \circ r_{x\text{-axis}}(x, y) = R_{90°}(x, y)$.

GEOMETRY OF THE CIRCLE

Early geometers in many parts of the world knew that, for all circles, the ratio of the circumference of a circle to its diameter was a constant. Today, we write $\frac{C}{d} = \pi$, but early geometers did not use the symbol π to represent this constant. Euclid established that the ratio of the area of a circle to the square of its diameter was also a constant, that is, $\frac{A}{d^2} = k$. How do these constants, π and k, relate to one another?

Archimedes (287–212 B.C.) proposed that the area of a circle was equal to the area of a right triangle whose legs have lengths equal to the radius, r, and the circumference, C, of a circle. Thus $A = \frac{1}{2}rC$. He used indirect proof and the areas of inscribed and circumscribed polygons to prove his conjecture and to prove that $3\frac{10}{71} < \pi < 3\frac{1}{7}$. Since this inequality can be written as $3.140845\ldots < \pi < 3.142857\ldots$, Archimedes' approximation was correct to two decimal places.

Use Archimedes' formula for the area of a circle and the facts that $\frac{C}{d} = \pi$ and $d = 2r$ to show that $A = \pi r^2$ and that $4k = \pi$.

13-1 ARCS AND ANGLES

In Chapter 11, we defined a sphere and found that the intersection of a plane and a sphere was a circle. In this chapter, we will prove some important relationships involving the measures of angles and line segments associated with circles. Recall the definition of a circle.

DEFINITION

A **circle** is the set of all points in a plane that are equidistant from a fixed point of the plane called the **center** of the circle.

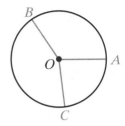

If the center of a circle is point O, the circle is called circle O, written in symbols as $\odot O$.

A **radius** of a circle (plural, *radii*) is a line segment from the center of the circle to any point of the circle. The term *radius* is used to mean both the line segment and the length of the line segment. If A, B, and C are points of circle O, then $\overline{OA}$, $\overline{OB}$, and $\overline{OC}$ are radii of the circle. Since the definition of a circle states that all points of the circle are equidistant from its center, $OA = OB = OC$. Thus, $\overline{OA} \cong \overline{OB} \cong \overline{OC}$ because equal line segments are congruent. We can state what we have just proved as a theorem.

Theorem 13.1

> All radii of the same circle are congruent.

A circle separates a plane into three sets of points. If we let the length of the radius of circle O be r, then:

- Point C is on the circle if $OC = r$.
- Point D is outside the circle if $OD > r$.
- Point E is inside the circle if $OE < r$.

The **interior of a circle** is the set of all points whose distance from the center of the circle is less than the length of the radius of the circle.

The **exterior of a circle** is the set of all points whose distance from the center of the circle is greater than the length of the radius of the circle.

Central Angles

Recall that an angle is the union of two rays having a common endpoint and that the common endpoint is called the vertex of the angle.

DEFINITION

A **central angle of a circle** is an angle whose vertex is the center of the circle.

In the diagram, $\angle LOM$ and $\angle MOR$ are central angles because the vertex of each angle is point O, the center of the circle.

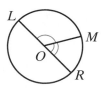

Types of Arcs

An **arc of a circle** is the part of the circle between two points on the circle. In the diagram, A, B, C, and D are points on circle O and $\angle AOB$ intersects the circle at two distinct points, A and B, separating the circle into two arcs.

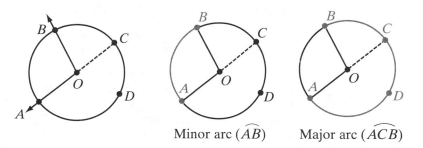

Minor arc ($\widehat{AB}$)   Major arc ($\widehat{ACB}$)

1. If $m\angle AOB < 180$, points A and B and the points of the circle in the interior of $\angle AOB$ make up **minor arc AB**, written as $\widehat{AB}$.

2. Points A and B and the points of the circle not in the interior of $\angle AOB$ make up **major arc AB**. A major arc is usually named by three points: the two endpoints and any other point on the major arc. Thus, the major arc with endpoints A and B is written as $\widehat{ACB}$ or $\widehat{ADB}$.

3. If $m\angle AOC = 180$, points A and C separate circle O into two equal parts, each of which is called a **semicircle**. In the diagram above, $\widehat{ADC}$ and $\widehat{ABC}$ name two different semicircles.

Degree Measure of an Arc

An arc of a circle is called an **intercepted arc**, or an arc intercepted by an angle, if each endpoint of the arc is on a different ray of the angle and the other points of the arc are in the interior of the angle.

DEFINITION

The **degree measure of an arc** is equal to the measure of the central angle that intercepts the arc.

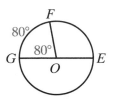

In circle O, $\angle GOE$ is a straight angle, m$\angle GOE = 180$, and m$\angle FOG = 80$. Therefore, the degree measure of $\overset{\frown}{FG}$ is 80°, written as m$\overset{\frown}{FG} = 80$. Also, since m$\overset{\frown}{GFE} = 180$,

$$m\overset{\frown}{FE} = 180 - 80 \qquad\text{and}\qquad m\overset{\frown}{FEG} = 100 + 180$$
$$= 100 \qquad\qquad\qquad\qquad = 280$$

Therefore,

$$m\overset{\frown}{GF} + m\overset{\frown}{FEG} = 80 + 280$$
$$= 360$$

1. The degree measure of a semicircle is 180. Thus:

2. The degree measure of a major arc is equal to 360 minus the degree measure of the minor arc having the same endpoints.

Do not confuse the degree measure of an arc with the length of an arc. The degree measure of every circle is 360 but the circumference of a circle is 2π times the radius of the circle. Example: in circle O, $OA = 1$ cm, and in circle O', $O'A' = 1.5$ cm. In 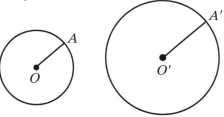 both circles, the degree measure of the circle is 360° but the circumference of circle O is 2π centimeters, and the circumference of circle O' is 3π centimeters.

Congruent Circles, Congruent Arcs, and Arc Addition

DEFINITION
Congruent circles are circles with congruent radii.

If $\overline{OC} \cong \overline{O'C'}$, then circles O' and O are congruent.

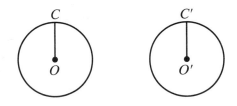

DEFINITION
Congruent arcs are arcs of the same circle or of congruent circles that are equal in measure.

If $\odot O \cong \odot O'$ and $m\widehat{CD} = m\widehat{C'D'} = 60$, then $\widehat{CD} \cong \widehat{C'D'}$. However, if circle O is *not* congruent to circle O'', then $\widehat{CD}$ is *not* congruent to $\widehat{C''D''}$ even if $\widehat{CD}$ and $\widehat{C''D''}$ have the same degree measure.

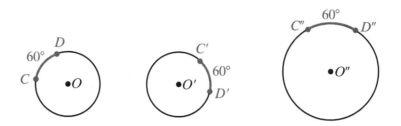

Postulate 13.1

Arc Addition Postulate

If $\widehat{AB}$ and $\widehat{BC}$ are two arcs of the same circle having a common endpoint and no other points in common, then $\widehat{AB} + \widehat{BC} = \widehat{ABC}$ and $m\widehat{AB} + m\widehat{BC} = m\widehat{ABC}$.

The arc that is the sum of two arcs may be a minor arc, a major arc, or semicircle. For example, A, B, C, and D are points of circle O, $m\widehat{AB} = 90$, $m\widehat{BC} = 40$, and $\overrightarrow{OB}$ and $\overrightarrow{OD}$ are opposite rays.

1. *Minor arc:* $\widehat{AB} + \widehat{BC} = \widehat{AC}$

Also, $m\widehat{AC} = m\widehat{AB} + m\widehat{BC}$

$= 90 + 40$

$= 130$

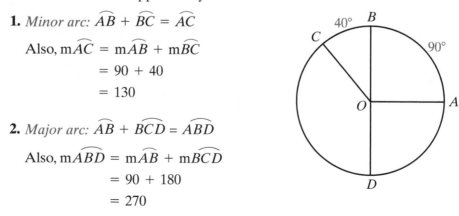

2. *Major arc:* $\widehat{AB} + \widehat{BCD} = \widehat{ABD}$

Also, $m\widehat{ABD} = m\widehat{AB} + m\widehat{BCD}$

$= 90 + 180$

$= 270$

3. *Semicircle:* Since $\overrightarrow{OB}$ and $\overrightarrow{OD}$ are opposite rays, $\angle BOD$ is a straight angle. Thus, $\widehat{BC} + \widehat{CD} = \widehat{BCD}$, a semicircle. Also, $m\widehat{BC} + m\widehat{CD} = m\widehat{BCD} = 180$.

Theorem 13.2a | In a circle or in congruent circles, if central angles are congruent, then their intercepted arcs are congruent

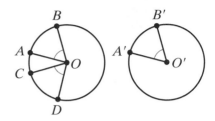

Given   Circle $O \cong$ circle O', $\angle AOB \cong \angle COD$, and $\angle AOB \cong \angle A'O'B'$.

Prove $\overset{\frown}{AB} \cong \overset{\frown}{CD}$ and $\overset{\frown}{AB} \cong \overset{\frown}{A'B'}$.

Proof   It is given that $\angle AOB \cong \angle COD$ and $\angle AOB \cong \angle A'O'B'$. Therefore, $m\angle AOB = m\angle COD = m\angle A'O'B'$ because congruent angles have equal measures. Then since the degree measure of an arc is equal to the degree measure of the central angle that intercepts that arc, $m\overset{\frown}{AB} = m\overset{\frown}{CD} = m\overset{\frown}{A'B'}$. It is also given that circle O and circle O' are congruent circles. Congruent arcs are defined as arcs of the same circle or of congruent circles that are equal in measure. Therefore, since their measures are equal, $\overset{\frown}{AB} \cong \overset{\frown}{CD}$ and $\overset{\frown}{AB} \cong \overset{\frown}{A'B'}$.

The converse of this theorem can be proved by using the same definitions and postulates.

Theorem 13.2b | In a circle or in congruent circles, central angles are congruent if their intercepted arcs are congruent.

Theorems 13.2a and 13.2b can be written as a biconditional.

Theorem 13.2 | In a circle or in congruent circles, central angles are congruent if and only if their intercepted arcs are congruent.

EXAMPLE I

Let $\overrightarrow{OA}$ and $\overrightarrow{OB}$ be opposite rays and $m\angle AOC = 75$. Find:

a. $m\angle BOC$ **b.** $m\overset{\frown}{AC}$ **c.** $m\overset{\frown}{BC}$ **d.** $m\overset{\frown}{AB}$ **e.** $m\overset{\frown}{BAC}$

Solution   **a.** $m\angle BOC = m\angle AOB - m\angle AOC$

$\qquad\qquad = 180 - 75$

$\qquad\qquad = 105$

b. $\text{m}\widehat{AC} = \text{m}\angle AOC = 75$

c. $\text{m}\widehat{BC} = \text{m}\angle BOC = 105$

d. $\text{m}\widehat{AB} = \text{m}\angle AOB = 180$

e. $\text{m}\widehat{BAC} = \text{m}\widehat{BDA} + \text{m}\widehat{AC}$ or $\text{m}\widehat{BAC} = 360 - \text{m}\widehat{BC}$
$$= 180 + 75 \qquad\qquad\qquad = 360 - 105$$
$$= 255 \qquad\qquad\qquad\quad\ = 255$$

Answers **a.** 105 **b.** 75 **c.** 105 **d.** 180 **e.** 255

Exercises

Writing About Mathematics

1. Kay said that if two lines intersect at the center of a circle, then they intercept two pairs of congruent arcs. Do you agree with Kay? Justify your answer.

2. Four points on a circle separate the circle into four congruent arcs: $\widehat{AB}$, $\widehat{BC}$, $\widehat{CD}$, and $\widehat{DA}$. Is it true that $\overleftrightarrow{AC} \perp \overleftrightarrow{BD}$? Justify your answer.

Developing Skills

In 3–7, find the measure of the central angle that intercepts an arc with the given degree measure.

3. 35 **4.** 48 **5.** 90 **6.** 140 **7.** 180

In 8–12, find the measure of the arc intercepted by a central angle with the given measure.

8. 60 **9.** 75 **10.** 100 **11.** 120 **12.** 170

In 13–22, the endpoints of $\overline{AOC}$ are on circle O, $\text{m}\angle AOB = 89$, and $\text{m}\angle COD = 42$. Find each measure.

13. $\text{m}\angle BOC$ **14.** $\text{m}\widehat{AB}$

15. $\text{m}\widehat{BC}$ **16.** $\text{m}\angle DOA$

17. $\text{m}\widehat{DA}$ **18.** $\text{m}\angle BOD$

19. $\text{m}\widehat{BCD}$ **20.** $\text{m}\widehat{DAB}$

21. $\text{m}\angle AOC$ **22.** $\text{m}\widehat{ADC}$

In 23–32, P, Q, S, and R are points on circle O, m$\angle POQ = 100$, m$\angle QOS = 110$, and m$\angle SOR = 35$. Find each measure.

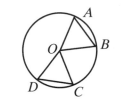

23. m$\overarc{PQ}$

24. m$\overarc{QS}$

25. m$\overarc{SR}$

26. m$\angle ROP$

27. m$\overarc{RP}$

28. m$\overarc{PQS}$

29. m$\angle QOR$

30. m$\overarc{QSR}$

31. m$\overarc{SRP}$

32. m$\overarc{RPQ}$

Applying Skills

33. *Given:* Circle O with $\overarc{AB} \cong \overarc{CD}$.

   Prove: $\triangle AOB \cong \triangle COD$

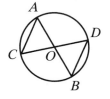

34. *Given:* $\overline{AB}$ and $\overline{CD}$ intersect at O, and the endpoints of $\overline{AB}$ and $\overline{CD}$ are on circle O.

   Prove: $\overline{AC} \cong \overline{BD}$

35. In circle O, $\overline{AOB} \perp \overline{COD}$. Find m$\overarc{AC}$ and m$\overarc{ADC}$.

36. Points A, B, C, and D lie on circle O, and $\overline{AC} \perp \overline{BD}$ at O. Prove that quadrilateral $ABCD$ is a square.

Hands-On Activity

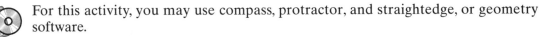

For this activity, you may use compass, protractor, and straightedge, or geometry software.

1. Draw circle O with a radius of 2 inches and circle O' with a radius of 3 inches.

2. Draw points A and B on the circle O so that m$\overarc{AB} = 60$ and points A' and B' on circle O' so that m$\overarc{A'B'} = 60$.

3. Show that $\triangle AOB \sim \triangle A'O'B'$.

13-2 ARCS AND CHORDS

DEFINITION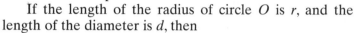

A **chord** of a circle is a line segment whose endpoints are points of the circle.

A **diameter** of a circle is a chord that has the center of the circle as one of its points.

In the diagram, $\overline{AB}$ and $\overline{AOC}$ are chords of circle O. Since O is a point of $\overline{AOC}$, $\overline{AOC}$ is a diameter. Since OA and OC are the lengths of the radius of circle O, $OA = OC$ and O is the midpoint of $\overline{AOC}$.

If the length of the radius of circle O is r, and the length of the diameter is d, then

$$d = AOC$$
$$= OA + OC$$
$$= r + r$$
$$= 2r$$

That is:

$$d = 2r$$

The endpoints of a chord are points on a circle and, therefore, determine two arcs of a circle, a minor arc and a major arc. In the diagram, chord $\overline{AB}$, central $\angle AOB$, minor $\overset{\frown}{AB}$, and major $\overset{\frown}{AB}$ are all determined by points A and B. We proved in the previous section that in a circle, congruent central angles intercept congruent arcs. Now we can prove that in a circle, congruent central angles have congruent chords and that congruent arcs have congruent chords.

Theorem 13.3a

In a circle or in congruent circles, congruent central angles have congruent chords.

Given $\odot O \cong \odot O'$ and
$\angle COD \cong \angle AOB \cong \angle A'O'B'$

Prove $\overline{CD} \cong \overline{AB} \cong \overline{A'B'}$

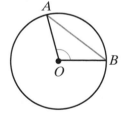

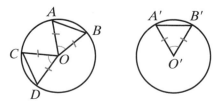

Proof We will show that
$\triangle COD \cong \triangle AOB \cong \triangle A'O'B'$ by SAS.

It is given that $\angle AOB \cong \angle COD$ and $\angle AOB \cong \angle A'O'B'$. Therefore, $\angle AOB \cong \angle COD \cong \angle A'O'B'$ by the transitive property of congruence. Since $\overline{DO}, \overline{CO}, \overline{AO}, \overline{BO}, \overline{A'O'}$, and $\overline{B'O'}$ are the radii of congruent circles, these segments are all congruent:

$$\overline{DO} \cong \overline{CO} \cong \overline{AO} \cong \overline{BO} \cong \overline{A'O'} \cong \overline{B'O'}$$

Therefore, by SAS, $\triangle COD \cong \triangle AOB \cong \triangle A'O'B'$. Since corresponding parts of congruent triangles are congruent, $\overline{CD} \cong \overline{AB} \cong \overline{A'B'}$.

The converse of this theorem is also true.

Theorem 13.3b   In a circle or in congruent circles, congruent chords have congruent central angles.

Given $\odot O \cong \odot O'$ and $\overline{CD} \cong \overline{AB} \cong \overline{A'B'}$

Prove $\angle COD \cong \angle AOB \cong \angle A'O'B'$

Strategy   This theorem can be proved in a manner similar to Theorem 13.3a: prove that $\triangle COD \cong \triangle AOB \cong \triangle A'O'B'$ by SSS.

The proof of Theorem 13.3b is left to the student. (See exercise 23.) Theorems 13.3a and 13.3b can be stated as a biconditional.

Theorem 13.3   In a circle or in congruent circles, two chords are congruent if and only if their central angles are congruent.

Since central angles and their intercepted arcs have equal degree measures, we can also prove the following theorems.

Theorem 13.4a   In a circle or in congruent circles, congruent arcs have congruent chords.

Given $\odot O \cong \odot O'$ and $\overset{\frown}{CD} \cong \overset{\frown}{AB} \cong \overset{\frown}{A'B'}$

Prove $\overline{CD} \cong \overline{AB} \cong \overline{A'B'}$

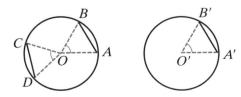

Proof First draw line segments from O to A, B, C, D, A', and B'. Congruent arcs have congruent central angles. Therefore, $\angle COD \cong \angle AOB \cong \angle A'O'B'$. In a circle or in congruent circles, two chords are congruent if and only if their central angles are congruent. Therefore, $\overline{CD} \cong \overline{AB} \cong \overline{A'B'}$. ■

The converse of this theorem is also true.

Theorem 13.4b | In a circle or in congruent circles, congruent chords have congruent arcs.

Given $\odot O \cong \odot O'$ and $\overline{CD} \cong \overline{AB} \cong \overline{A'B'}$

Prove $\overparen{CD} \cong \overparen{AB} \cong \overparen{A'B'}$

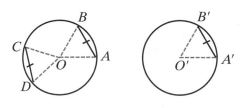

Strategy First draw line segments from O to A, B, C, D, A' and B'.

Prove $\triangle COD \cong \triangle AOB \cong \triangle A'O'B'$ by SSS. Then use congruent central angles to prove congruent arcs.

The proof of Theorem 13.4b is left to the student. (See exercise 24.) Theorems 13.4a and 13.4b can be stated as a biconditional.

Theorem 13.4 | In a circle or in congruent circles, two chords are congruent if and only if their arcs are congruent.

EXAMPLE I

In circle O, $m\overparen{AB} = 35$, $m\angle BOC = 110$, and $\overline{AOD}$ is a diameter.

a. Find $m\overparen{BC}$ and $m\overparen{CD}$.

b. Explain why $AB = CD$.

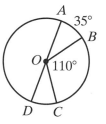

Solution **a.** $\begin{aligned} m\overparen{BC} &= m\angle BOC \\ &= 110 \end{aligned}$ $\begin{aligned} m\overparen{CD} &= 180 - m\overparen{AB} - m\overparen{BC} \\ &= 180 - 35 - 110 \\ &= 35 \end{aligned}$

b. In a circle, arcs with equal measure are congruent. Therefore, since $m\overparen{AB} = 35$ and $m\overparen{CD} = 35$, $\overparen{AB} \cong \overparen{CD}$. In a circle, chords are congruent if their arcs are congruent. Therefore, $\overline{AB} \cong \overline{CD}$ and $AB = CD$. ■

Chords Equidistant from the Center of a Circle

We defined the distance from a point to a line as the length of the perpendicular from the point to the line. The perpendicular is the shortest line segment that can be drawn from a point to a line. These facts can be used to prove the following theorem.

Theorem 13.5

A diameter perpendicular to a chord bisects the chord and its arcs.

Given Diameter $\overline{COD}$ of circle O, chord $\overline{AB}$, and $\overline{AB} \perp \overline{CD}$ at E.

Prove $\overline{AE} \cong \overline{BE}$, $\overset{\frown}{AC} \cong \overset{\frown}{BC}$, and $\overset{\frown}{AD} \cong \overset{\frown}{BD}$.

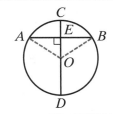

Proof

Statements	Reasons
1. Draw $\overline{OA}$ and $\overline{OB}$.	**1.** Two points determine a line.
2. $\overline{AB} \perp \overline{CD}$	**2.** Given.
3. $\angle AEO$ and $\angle BEO$ are right angles.	**3.** Perpendicular lines intersect to form right angles.
4. $\overline{OA} \cong \overline{OB}$	**4.** Radii of a circle are congruent.
5. $\overline{OE} \cong \overline{OE}$	**5.** Reflexive property of congruence.
6. $\triangle AOE \cong \triangle BOE$	**6.** HL.
7. $\overline{AE} \cong \overline{BE}$	**7.** Corresponding parts of congruent triangles are congruent.
8. $\angle AOE \cong \angle BOE$	**8.** Corresponding parts of congruent triangles are congruent.
9. $\overset{\frown}{AC} \cong \overset{\frown}{BC}$	**9.** In a circle, congruent central angles have congruent arcs.
10. $\angle AOD$ is the supplement of $\angle AOE$. $\angle BOD$ is the supplement of $\angle BOE$.	**10.** If two angles form a linear pair, then they are supplementary.
11. $\angle AOD \cong \angle BOD$	**11.** Supplements of congruent angles are congruent.
12. $\overset{\frown}{AD} \cong \overset{\frown}{BD}$	**12.** In a circle, congruent central angles have congruent arcs.

Since a diameter is a segment of a line, the following corollary is also true:

Corollary 13.5a

A line through the center of a circle that is perpendicular to a chord bisects the chord and its arcs.

An **apothem** of a circle is a perpendicular line segment from the center of a circle to the midpoint of a chord. The term *apothem* also refers to the length of the segment. In the diagram, E is the midpoint of chord $\overline{AB}$ in circle O, $\overline{AB} \perp \overline{CD}$, and $\overline{OE}$, or OE, is the apothem.

Theorem 13.6

The perpendicular bisector of the chord of a circle contains the center of the circle.

Given Circle O, and chord $\overline{AB}$ with midpoint M and perpendicular bisector k.

Prove Point O is a point on k.

Proof In the diagram, M is the midpoint of chord $\overline{AB}$ in circle O. Then, $AM = MB$ and $AO = OB$ (since these are radii). Points O and M are each equidistant from the endpoints of $\overline{AB}$. Two points that are each equidistant from the endpoints of a line segment determine the perpendicular bisector of the line segment. Therefore, $\overleftrightarrow{OM}$ is the perpendicular bisector of $\overline{AB}$. Through a point on a line there is only one perpendicular line. Thus, $\overleftrightarrow{OM}$ and k are the same line, and O is on k, the perpendicular bisector of $\overline{AB}$.

Theorem 13.7a

If two chords of a circle are congruent, then they are equidistant from the center of the circle.

Given Circle O with $\overline{AB} \cong \overline{CD}$, $\overline{OE} \perp \overline{AB}$, and $\overline{OF} \perp \overline{CD}$.

Prove $\overline{OE} \cong \overline{OF}$

Proof A line through the center of a circle that is perpendicular to a chord bisects the chord and its arcs. Therefore, $\overleftrightarrow{OE}$ and $\overleftrightarrow{OF}$ bisect the congruent chords $\overline{AB}$ and $\overline{CD}$. Since halves of congruent segments are congruent, $\overline{EB} \cong \overline{FD}$. Draw $\overline{OB}$ and $\overline{OD}$. Since $\overline{OB}$ and $\overline{OD}$ are radii of the same circle, $\overline{OB} \cong \overline{OD}$. Therefore, $\triangle OBE \cong \triangle ODF$ by HL, and $\overline{OE} \cong \overline{OF}$.

The converse of this theorem is also true.

Theorem 13.7b | If two chords of a circle are equidistant from the center of the circle, then the chords are congruent.

Given Circle O with $\overline{OE} \perp \overline{AB}, \overline{OF} \perp \overline{CD}$, and $\overline{OE} \cong \overline{OF}$.

Prove $\overline{AB} \cong \overline{CD}$

Proof Draw $\overline{OB}$ and $\overline{OD}$. Since $\overline{OB}$ and $\overline{OD}$ are radii of the same circle, $\overline{OB} \cong \overline{OD}$. Therefore, $\triangle OBE \cong \triangle ODF$ by HL, and $\overline{EB} \cong \overline{FD}$. A line through the center of a circle that is perpendicular to a chord bisects the chord. Thus, $\overline{AE} \cong \overline{EB}$ and $\overline{CF} \cong \overline{FD}$. Since doubles of congruent segments are congruent, $\overline{AB} \cong \overline{CD}$.

We can state theorems 13.7a and 13.7b as a biconditional.

Theorem 13.7 | Two chords are equidistant from the center of a circle if and only if the chords are congruent.

What if two chords are not equidistant from the center of a circle? Which is the longer chord? We know that a diameter contains the center of the circle and is the longest chord of the circle. This suggests that the shorter chord is farther from the center of the circle.

(1) Let $\overline{AB}$ and $\overline{CD}$ be two chords of circle O and $AB < CD$.

(2) Draw $\overline{OE} \perp \overline{AB}$ and $\overline{OF} \perp \overline{CD}$.

(3) A line through the center of a circle that is perpendicular to the chord bisects the chord. Therefore, $\frac{1}{2}AB = EB$ and $\frac{1}{2}CD = FD$.

(4) The distance from a point to a line is the length of the perpendicular from the point to the line. Therefore, OE is the distance from the center of the circle to $\overline{AB}$, and OF is the distance from the center of the circle to $\overline{CD}$.

Recall that the squares of equal quantities are equal, that the positive square roots of equal quantities are equal, and that when an inequality is multiplied by a negative number, the inequality is reversed.

(5) Since $AB < CD$:

$$AB < CD$$
$$\tfrac{1}{2}AB < \tfrac{1}{2}CD$$
$$EB < FD$$
$$EB^2 < FD^2$$
$$-EB^2 > -FD^2$$

(6) Since $\triangle OBE$ and $\triangle ODF$ are right triangles and $OB = OD$:

$$OB^2 = OD^2$$
$$OE^2 + EB^2 = OF^2 + FD^2$$

(7) When equal quantities are added to both sides of an inequality, the order of the inequality remains the same. Therefore, adding the equal quantities from step 6 to the inequality in step 5 gives:

$$OE^2 + EB^2 - EB^2 > OF^2 + FD^2 - FD^2$$
$$OE^2 > OF^2$$
$$OE > OF$$

Therefore, the shorter chord is farther from the center of the circle. We have just proved the following theorem:

Theorem 13.8 | In a circle, if the lengths of two chords are unequal, then the shorter chord is farther from the center.

EXAMPLE 2

In circle O, $\text{m}\widehat{AB} = 90$ and $OA = 6$.

a. Prove that $\triangle AOB$ is a right triangle.

b. Find AB.

c. Find OC, the apothem to $\overline{AB}$.

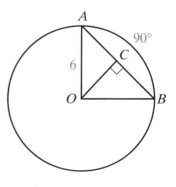

Solution **a.** If $\text{m}\widehat{AB} = 90$, then $\text{m}\angle AOB = 90$ because the measure of an arc is equal to the measure of the central angle that intercepts the arc. Since $\angle AOB$ is a right angle, $\triangle AOB$ is a right triangle.

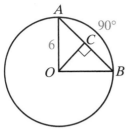

b. Use the Pythagorean Theorem for right $\triangle AOB$. Since $\overline{OA}$ and $\overline{OB}$ are radii, $OB = OA = 6$.

$$AB^2 = OA^2 + OB^2$$
$$AB^2 = 6^2 + 6^2$$
$$AB^2 = 36 + 36$$
$$AB^2 = 72$$
$$AB = \sqrt{72} = \sqrt{36}\sqrt{2} = 6\sqrt{2}$$

c. Since OC is the apothem to $\overline{AB}$, $\overline{OC} \perp \overline{AB}$ and bisects $\overline{AB}$. Therefore, $AC = 3\sqrt{2}$. In right $\triangle OCA$,

$$OC^2 + AC^2 = OA^2$$
$$OC^2 + \left(3\sqrt{2}\right)^2 = 6^2$$
$$OC^2 + 18 = 36$$
$$OC^2 = 18$$
$$OC = \sqrt{18} = \sqrt{9}\sqrt{2} = 3\sqrt{2}$$

Note: In the example, since $\triangle AOB$ is an isosceles right triangle, $m\angle AOB$ is 45. Therefore $\triangle AOC$ is also an isosceles right triangle and $OC = AC$.

Polygons Inscribed in a Circle

If all of the vertices of a polygon are points of a circle, then the polygon is said to be **inscribed** in the circle. We can also say that the circle is **circumscribed** about the polygon.

In the diagram:

1. Polygon $ABCD$ is inscribed in circle O.

2. Circle O is circumscribed about polygon $ABCD$.

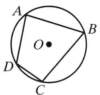

In an earlier chapter we proved that the perpendicular bisectors of the sides of a triangle meet at a point and that that point is equidistant from the vertices of the triangle. In the diagram, $\overleftrightarrow{PL}$, $\overleftrightarrow{PM}$, and $\overleftrightarrow{PN}$ are the perpendicular bisectors of the sides of $\triangle ABC$. Every point on the perpendicular bisectors of a line segment is equidistant from the endpoints of the line segment. Therefore, $PA = PB = PC$ and A, B, and C are points on a circle with center at P, that is, any triangle can be inscribed in a circle.

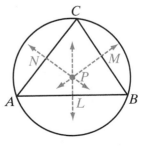

EXAMPLE 3

Prove that any rectangle can be inscribed in a circle.

Proof   Let $ABCD$ be any rectangle. The diagonals of a rectangle are congruent, so $\overline{AC} \cong \overline{BD}$ and $AC = BD$. Since a rectangle is a parallelogram, the diagonals of a rectangle bisect each other. If $\overline{AC}$ and $\overline{BD}$ intersect at E, then $\frac{1}{2}AC = AE = EC$ and $\frac{1}{2}BD = BE = ED$. Halves of equal quantities are equal. Therefore, $AE = EC = BE = ED$ and the vertices of the rectangle are equidistant from E. Let E be the center of a circle with radius AE. The vertices of $ABCD$ are on the circle and $ABCD$ is inscribed in the circle.

Exercises

Writing About Mathematics

1. Daniela said that if a chord is 3 inches from the center of a circle that has a radius of 5 inches, then a 3-4-5 right triangle is formed by the chord, its apothem, and a radius. Additionally, the length of the chord is 4 inches. Do you agree with Daniela? Explain why or why not.

2. Two angles that have the same measure are always congruent. Are two arcs that have the same measure always congruent? Explain why or why not.

Developing Skills

In 3–7, find the length of the radius of a circle whose diameter has the given measure.

3. 6 in.   **4.** 9 cm   **5.** 3 ft   **6.** 24 mm   **7.** $\sqrt{24}$ cm

In 8–12, find the length of the diameter of a circle whose radius has the given measure.

8. 5 in.   **9.** 12 ft   **10.** 7 cm   **11.** 6.2 mm   **12.** $\sqrt{5}$ yd

13. In circle O, $\overline{AOB}$ is a diameter, $AB = 3x + 13$, and $AO = 2x + 5$. Find the length of the radius and of the diameter of the circle.

In 14–21, $\overline{DCOE}$ is a diameter of circle O, $\overline{AB}$ is a chord of the circle, and $\overline{OD} \perp \overline{AB}$ at C.

14. If $AB = 8$ and $OC = 3$, find OB.   **15.** If $AB = 48$ and $OC = 7$, find OB.

16. If $OC = 20$ and $OB = 25$, find AB.   **17.** If $OC = 12$ and $OB = 18$, find AB.

18. If $AB = 18$ and $OB = 15$, find OC.

19. If $AB = 20$ and $OB = 15$, find OC.

20. If $m\angle AOB = 90$, and $AB = 30\sqrt{2}$, find OB and DE.

21. If $m\angle AOB = 60$, and $AB = 30$, find OB and OC.

22. In circle O, chord $\overline{LM}$ is 3 centimeters from the center and chord $\overline{RS}$ is 5 centimeters from the center. Which is the longer chord?

Applying Skills

23. Prove Theorem 13.3b, "In a circle or in congruent circles, congruent chords have congruent central angles."

24. Prove Theorem 13.4b, "In a circle or in congruent circles, congruent chords have congruent arcs."

25. Diameter $\overline{AOB}$ of circle O intersects chord $\overline{CD}$ at E and bisects $\overset{\frown}{CD}$ at B. Prove that $\overline{AOB}$ bisects chord $\overline{CD}$ and is perpendicular to chord $\overline{CD}$.

26. The radius of a spherical ball is 13 centimeters. A piece that has a plane surface is cut off of the ball at a distance of 12 centimeters from the center of the ball. What is the radius of the circular faces of the cut pieces?

27. Triangle ABC is inscribed in circle O. The distance from the center of the circle to $\overline{AB}$ is greater than the distance from the center of the circle to $\overline{BC}$, and the distance from the center of the circle to $\overline{BC}$ is greater than the distance from the center of the circle to $\overline{AC}$. Which is the largest angle of $\triangle ABC$? Justify your answer.

13-3 INSCRIBED ANGLES AND THEIR MEASURES

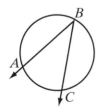

In the diagram, $\angle ABC$ is an angle formed by two chords that have a common endpoint on the circle.

DEFINITION

An **inscribed angle** of a circle is an angle whose vertex is on the circle and whose sides contain chords of the circle.

We can use the fact that the measure of a central angle is equal to the measure of its arc to find the relationship between $\angle ABC$ and the measure of its arc, $\overset{\frown}{AC}$.

CASE 1 *One of the sides of the inscribed angle contains a diameter of the circle.*

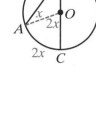

Consider first an inscribed angle, $\angle ABC$, with $\overline{BC}$ a diameter of circle O. Draw $\overline{OA}$. Then $\triangle AOB$ is an isosceles triangle and $m\angle OAB = m\angle OBA = x$. Angle AOC is an exterior angle and $m\angle AOC = x + x = 2x$. Since $\angle AOC$ is a central angle, $m\angle AOC = m\widehat{AC} = 2x$. Therefore, $m\angle ABC = x = \frac{1}{2}m\widehat{AC}$.

We have shown that when one of the sides of an inscribed angle contains a diameter of the circle, the measure of the inscribed angle is equal to one-half the measure of its intercepted arc. Is this true for angles whose sides do not contain the center of the circle?

CASE 2 *The center of the circle is in the interior of the angle.*

Let $\angle ABC$ be an inscribed angle in which the center of the circle is in the interior of the angle. Draw $\overline{BOD}$, a diameter of the circle. Then:

$$m\angle ABD = \tfrac{1}{2}m\widehat{AD} \quad \text{and} \quad m\angle DBC = \tfrac{1}{2}m\widehat{DC}$$

Therefore:

$$m\angle ABC = m\angle ABD + m\angle DBC$$
$$= \tfrac{1}{2}m\widehat{AD} + \tfrac{1}{2}m\widehat{DC}$$
$$= \tfrac{1}{2}(m\widehat{AD} + m\widehat{DC})$$
$$= \tfrac{1}{2}m\widehat{AC}$$

CASE 3 *The center of the circle is not in the interior of the angle.*

Let $\angle ABC$ be an inscribed angle in which the center of the circle is not in the interior of the angle. Draw $\overline{BOD}$, a diameter of the circle. Then:

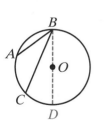

$$m\angle ABD = \tfrac{1}{2}m\widehat{AD} \quad \text{and} \quad m\angle DBC = \tfrac{1}{2}m\widehat{DC}$$

Therefore:

$$m\angle ABC = m\angle ABD - m\angle DBC$$
$$= \tfrac{1}{2}m\widehat{AD} - \tfrac{1}{2}m\widehat{DC}$$
$$= \tfrac{1}{2}(m\widehat{AD} - m\widehat{DC})$$
$$= \tfrac{1}{2}m\widehat{AC}$$

These three possible positions of the sides of the circle with respect to the center the circle prove the following theorem:

Theorem 13.9

> The measure of an inscribed angle of a circle is equal to one-half the measure of its intercepted arc.

There are two statements that can be derived from this theorem.

Corollary 13.9a | An angle inscribed in a semicircle is a right angle.

Proof: In the diagram, $\overline{AOC}$ is a diameter of circle O, and $\angle ABC$ is inscribed in semicircle $\overset{\frown}{ABC}$. Also $\overset{\frown}{ADC}$ is a semicircle whose degree measure is $180°$. Therefore,

$$m\angle ABC = \tfrac{1}{2}m\overset{\frown}{ADC}$$
$$= \tfrac{1}{2}(180)$$
$$= 90$$

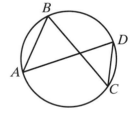

Since any triangle can be inscribed in a circle, the hypotenuse of a triangle can be the diameter of a circle with the midpoint of the hypotenuse the center of the circle.

Corollary 13.9b | If two inscribed angles of a circle intercept the same arc, then they are congruent.

Proof: In the diagram, $\angle ABC$ and $\angle ADC$ are inscribed angles and each angle intercepts $\overset{\frown}{AC}$. Therefore, $m\angle ABC = \tfrac{1}{2}m\overset{\frown}{AC}$ and $m\angle ADC = \tfrac{1}{2}m\overset{\frown}{AC}$. Since $\angle ABC$ and $\angle ADC$ have equal measures, they are congruent.

EXAMPLE

Triangle ABC is inscribed in circle O, $m\angle B = 70$, and $m\overset{\frown}{BC} = 100$. Find:

a. $m\overset{\frown}{AC}$ **b.** $m\angle A$ **c.** $m\angle C$ **d.** $m\overset{\frown}{AB}$

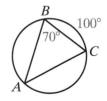

Solution   **a.** If the measure of an inscribed angle is one-half the measure of its intercepted arc, then the measure of the intercepted arc is twice the measure of the inscribed angle.

$$m\angle B = \tfrac{1}{2}m\overset{\frown}{AC}$$
$$2m\angle B = m\overset{\frown}{AC}$$
$$2(70) = m\overset{\frown}{AC}$$
$$140 = m\overset{\frown}{AC}$$

b. $m\angle A = \frac{1}{2}m\widehat{BC}$

$= \frac{1}{2}(100)$

$= 50$

c. $m\angle C = 180 - (m\angle A + m\angle B)$

$= 180 - (50 + 70)$

$= 180 - 120$

$= 60$

d. $m\widehat{AB} = 2m\angle C$

$= 2(60)$

$= 120$

Answers   **a.** $140°$ **b.** $50°$ **c.** $60°$ **d.** $120°$

SUMMARY

Type of Angle	Degree Measure	Example
Central Angle	The measure of a central angle is equal to the measure of its intercepted arc.	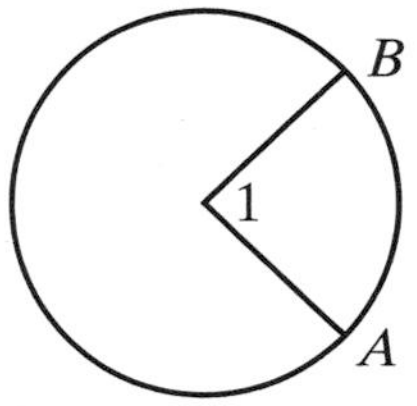 $m\angle 1 = m\widehat{AB}$
Inscribed Angle	The measure of an inscribed angle is equal to one-half the measure of its intercepted arc.	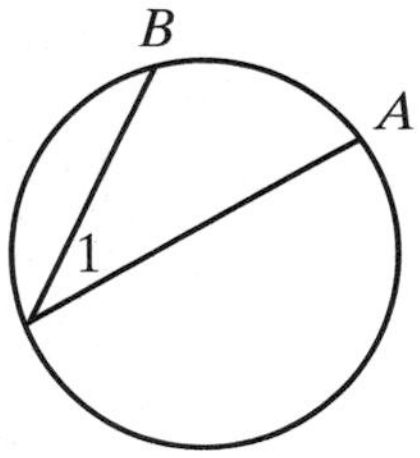 $m\angle 1 = \frac{1}{2}m\widehat{AB}$

Exercises

Writing About Mathematics

1. Explain how you could use Corollary 13.9a to construct a right triangle with two given line segments as the hypotenuse and one leg.

2. In circle O, $\angle ABC$ is an inscribed angle and $m\overarc{AC} = 50$. In circle O', $\angle PQR$ is an inscribed angle and $m\overarc{PR} = 50$. Is $\angle ABC \cong \angle PQR$ if the circles are not congruent circles? Justify your answer.

Developing Skills

In 3–7, B is a point on circle O not on $\overarc{AC}$, an arc of circle O. Find $m\angle ABC$ for each given $m\overarc{AC}$.

3. 88 **4.** 72 **5.** 170 **6.** 200 **7.** 280

In 8–12, B is a point on circle O not on $\overarc{AC}$, an arc of circle O. Find $m\overarc{AC}$ for each given $m\angle ABC$.

8. 12 **9.** 45 **10.** 60 **11.** 95 **12.** 125

13. Triangle ABC is inscribed in a circle, $m\angle A = 80$ and $m\overarc{AC} = 88$. Find:

    a. $m\overarc{BC}$ **b.** $m\angle B$ **c.** $m\angle C$ **d.** $m\overarc{AB}$ **e.** $m\overarc{BAC}$

14. Triangle DEF is inscribed in a circle, $\overline{DE} \cong \overline{EF}$, and $m\overarc{EF} = 100$. Find:

    a. $m\angle D$ **b.** $m\overarc{DE}$ **c.** $m\angle F$ **d.** $m\angle E$ **e.** $m\overarc{DF}$

In 15–17, chords $\overline{AC}$ and $\overline{BD}$ intersect at E in circle O.

15. If $m\angle B = 42$ and $m\angle AEB = 104$, find:

    a. $m\angle A$ **b.** $m\overarc{BC}$ **c.** $m\overarc{AD}$ **d.** $m\angle D$ **e.** $m\angle C$

16. If $\overline{AB} \parallel \overline{DC}$ and $m\angle B = 40$, find:

    a. $m\angle D$ **b.** $m\overarc{AD}$ **c.** $m\overarc{BC}$ **d.** $m\angle A$ **e.** $m\angle DEC$

17. If $m\overarc{AD} = 100$, $m\overarc{AB} = 110$, and $m\overarc{BC} = 96$, find:

    a. $m\overarc{DC}$ **b.** $m\angle A$ **c.** $m\angle B$ **d.** $m\angle AEB$ **e.** $m\angle C$

18. Triangle ABC is inscribed in a circle and $m\overarc{AB} : m\overarc{BC} : m\overarc{CA} = 2 : 3 : 7$. Find:

    a. $m\overarc{AB}$ **b.** $m\overarc{BC}$ **c.** $m\overarc{CA}$ **d.** $m\angle A$ **e.** $m\angle B$ **f.** $m\angle C$

19. Triangle RST is inscribed in a circle and $m\overarc{RS} = m\overarc{ST} = m\overarc{TR}$. Find:

    a. $m\overarc{RS}$ **b.** $m\overarc{ST}$ **c.** $m\overarc{TR}$ **d.** $m\angle R$ **e.** $m\angle S$ **f.** $m\angle T$

Applying Skills

20. In circle O, $\overline{LM}$ and $\overline{RS}$ intersect at P.

    a. Prove that $\triangle LPR \sim \triangle SPM$.

    b. If $LP = 15$ cm, $RP = 12$ cm, and $SP = 10$ cm, find MP.

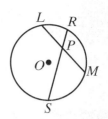

21. Triangle ABC is inscribed in a circle. If $m\overarc{AB} = 100$ and $m\overarc{BC} = 130$, prove that $\triangle ABC$ is isosceles.

22. Parallelogram $ABCD$ is inscribed in a circle.

    a. Explain why $m\overarc{ABC} = m\overarc{ADC}$.

    b. Find $m\overarc{ABC}$ and $m\overarc{ADC}$.

    c. Explain why parallelogram $ABCD$ must be a rectangle.

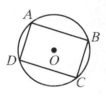

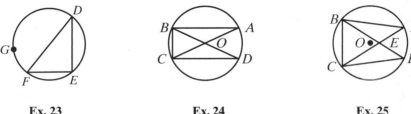

Ex. 23   Ex. 24   Ex. 25

23. Triangle DEF is inscribed in a circle and G is any point not on $\overarc{DEF}$. If $m\overarc{DE} + m\overarc{EF} = m\overarc{FGD}$, show that $\triangle DEF$ is a right triangle.

24. In circle O, $\overline{AOC}$ and $\overline{BOD}$ are diameters. If $\overline{AB} \cong \overline{CD}$, prove that $\triangle ABC \cong \triangle DCB$ by ASA.

25. Chords $\overline{AC}$ and $\overline{BD}$ of circle O intersect at E. If $\overarc{AB} \cong \overarc{CD}$, prove that $\triangle ABC \cong \triangle DCB$.

26. Prove that a trapezoid inscribed in a circle is isosceles.

27. Minor $\overarc{ABC}$ and major $\overarc{ADC}$ are arcs of circle O and $\overline{AB} \parallel \overline{CD}$. Prove that $\overarc{AD} \cong \overarc{BC}$.

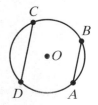

28. Minor $\overarc{ABC}$ and major $\overarc{ADC}$ are arcs of circle O and $\overarc{AD} \cong \overarc{BC}$. Prove that $\overline{AB} \parallel \overline{CD}$.

29. In circle O, $\overline{AOC}$ and $\overline{BOD}$ are diameters. Prove that $\overline{AB} \parallel \overline{CD}$.

30. Points A, B, C, D, E, and F are on circle O, $\overline{AB} \parallel \overline{CD} \parallel \overline{EF}$, and $\overline{CB} \parallel \overline{ED}$. $ABCD$ and $CDEF$ are trapezoids. Prove that $\overarc{CA} \cong \overarc{BD} \cong \overarc{DF} \cong \overarc{EC}$.

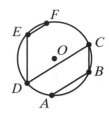

31. Quadrilateral $ABCD$ is inscribed in circle O, and $\overarc{AB}$ is not congruent to $\overarc{CD}$. Prove that $ABCD$ is not a parallelogram.

13-4 TANGENTS AND SECANTS

In the diagram, line p has no points in common with the circle. Line m has one point in common with the circle. Line m is said to be *tangent to the circle*. Line k has two points in common with the circle. Line k is said to be a *secant of the circle*.

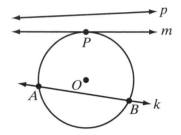

DEFINITION

A **tangent to a circle** is a line in the plane of the circle that intersects the circle in one and only one point.

DEFINITION

A **secant of a circle** is a line that intersects the circle in two points.

Let us begin by assuming that at every point on a circle, there exists exactly one tangent line. We can state this as a postulate.

Postulate 13.2

At a given point on a given circle, one and only one line can be drawn that is tangent to the circle.

Let P be any point on circle O and $\overline{OP}$ be a radius to that point. If line m containing points P and Q is perpendicular to $\overline{OP}$, then $OQ > OP$ because the perpendicular is the shortest distance from a point to a line. Therefore, every point on the line except P is outside of circle O and line m must be tangent to the circle. This establishes the truth of the following theorem.

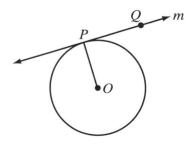

Theorem 13.10a If a line is perpendicular to a radius at a point on the circle, then the line is tangent to the circle.

The converse of this theorem is also true.

Theorem 13.10b If a line is tangent to a circle, then it is perpendicular to a radius at a point on the circle.

Given Line m is tangent to circle O at P.

Prove Line m is perpendicular to $\overline{OP}$.

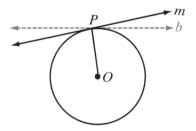

Proof We can use an indirect proof.

Assume that m is not perpendicular to $\overline{OP}$. Then there is some line b that is perpendicular to $\overline{OP}$ at P since, at a given point on a given line, one and only one line can be drawn perpendicular to the given line. Then by Theorem 13.10a, b is a tangent to circle O at P. But this contradicts the postulate that states that at a given point on a circle, one and only one tangent can be drawn. Therefore, our assumption is false and its negation must be true. Line m is perpendicular to $\overline{OP}$.

We can state Theorems 13.10a and 13.10b as a biconditional.

Theorem 13.10 A line is tangent to a circle if and only if it is perpendicular to a radius at its point of intersection with the circle.

Common Tangents

DEFINITION _____
  A **common tangent** is a line that is tangent to each of two circles.

In the diagram, $\overleftrightarrow{AB}$ is tangent to circle O at A and to circle O' at B. Tangent $\overleftrightarrow{AB}$ is said to be a **common internal tangent** because the tangent intersects the line segment joining the centers of the circles.

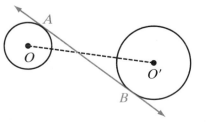

In the diagram, $\overleftrightarrow{CD}$ is tangent to circle P at C and to circle P' at D. Tangent $\overleftrightarrow{CD}$ is said to be a **common external tangent** because the tangent does not intersect the line segment joining the centers of the circles.

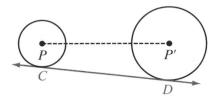

The diagrams below show that two circles can have four, three, two, one, or no common tangents.

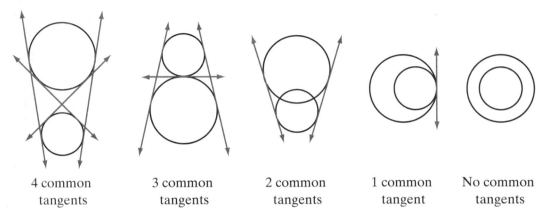

| 4 common tangents | 3 common tangents | 2 common tangents | 1 common tangent | No common tangents |

Two circles are said to be tangent to each other if they are tangent to the same line at the same point. In the diagram, $\overleftrightarrow{ST}$ is tangent to circle O and to circle O' at T. Circles O and O' are **tangent externally** because every point of one of the circles, except the point of tangency, is an external point of the other circle.

In the diagram, $\overleftrightarrow{MN}$ is tangent to circle P and to circle P' at M. Circles P and P' are **tangent internally** because every point of one of the circles, except the point of tangency, is an internal point of the other circle.

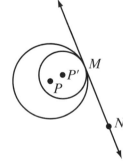

EXAMPLE I

Given: Circles O and O' with a common internal tangent, $\overleftrightarrow{AB}$, tangent to circle O at A and circle O' at B, and C the intersection of $\overline{OO'}$ and $\overleftrightarrow{AB}$.

Prove: $\dfrac{AC}{BC} = \dfrac{OC}{O'C}$

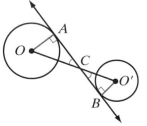

Proof We will use similar triangles to prove the segments proportional.

Line $\overleftrightarrow{AB}$ is tangent to circle O at A and circle O' at B. A line tangent to a circle is perpendicular to a radius drawn to the point of tangency. Since perpendicular lines intersect to form right angles and all right angle are congruent, $\angle OAC \cong \angle O'BC$. Also, $\angle OCA \cong \angle O'CB$ because vertical angles are congruent. Therefore, $\triangle OCA \sim \triangle O'CB$ by AA~. The lengths of corresponding sides of similar triangles are proportional. Therefore, $\frac{AC}{BC} = \frac{OC}{O'C}$.

EXAMPLE 2

Circle O is tangent to $\overleftrightarrow{AB}$ at A, O' is tangent to $\overleftrightarrow{AB}$ at B, and $\overline{OO'}$ intersects $\overleftrightarrow{AB}$ at C.

a. Prove that $\frac{AC}{BC} = \frac{OA}{O'B}$.

b. If $AC = 8$, $AB = 12$, and $OA = 9$, find $O'B$.

Solution **a.** We know that $\angle OAB \cong \angle O'BA$ because they are right angles and that $\angle OCA \cong \angle O'CB$ because they are vertical angles. Therefore, $\triangle OCA \sim \triangle O'CB$ by AA~ and $\frac{AC}{BC} = \frac{OA}{O'B}$.

b.
$$AB = AC + BC$$
$$12 = 8 + BC$$
$$4 = BC$$

$$\frac{AC}{BC} = \frac{OA}{O'B}$$
$$\frac{8}{4} = \frac{9}{O'B}$$
$$8O'B = 36$$
$$O'B = \frac{36}{8} = \frac{9}{2} \; Answer$$

Tangent Segments

DEFINITION

A **tangent segment** is a segment of a tangent line, one of whose endpoints is the point of tangency.

In the diagram, $\overline{PQ}$ and $\overline{PR}$ are tangent segments of the tangents $\overleftrightarrow{PQ}$ and $\overleftrightarrow{PR}$ to circle O from P.

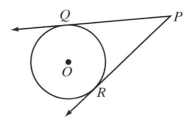

Theorem 13.11 Tangent segments drawn to a circle from an external point are congruent.

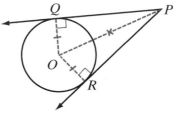

Given $\overleftrightarrow{PQ}$ tangent to circle O at Q and $\overleftrightarrow{PR}$ tangent to circle O at R.

Prove $\overline{PQ} \cong \overline{PR}$

Proof Draw $\overline{OQ}, \overline{OR}$, and $\overline{OP}$. Since $\overline{OQ}$ and $\overline{OR}$ are both radii of the same circle, $\overline{OQ} \cong \overline{OR}$. Since $\overline{QP}$ and $\overline{RP}$ are tangent to the circle at Q and R, $\angle OQP$ and $\angle ORP$ are both right angles, and $\triangle OPQ$ and $\triangle OPR$ are right triangles. Then $\overline{OP}$ is the hypotenuse of both $\triangle OPQ$ and $\triangle OPR$. Therefore, $\triangle OPQ \cong \triangle OPR$ by HL. Corresponding parts of congruent triangles are congruent, so $\overline{PQ} \cong \overline{PR}$.

The following corollaries are also true.

Corollary 13.11a   If two tangents are drawn to a circle from an external point, then the line segment from the center of the circle to the external point bisects the angle formed by the tangents.

Given $\overleftrightarrow{PQ}$ tangent to circle O at Q and $\overleftrightarrow{PR}$ tangent to circle O at R.

Prove $\overrightarrow{PO}$ bisects $\angle RPQ$.

Strategy Use the proof of Theorem 13.11 to show that angles $\angle OPQ$ and $\angle RPO$ are congruent.

Corollary 13.11b   If two tangents are drawn to a circle from an external point, then the line segment from the center of the circle to the external point bisects the angle whose vertex is the center of the circle and whose rays are the two radii drawn to the points of tangency.

Given $\overleftrightarrow{PQ}$ tangent to circle O at Q and $\overleftrightarrow{PR}$ tangent to circle O at R.

Prove $\overrightarrow{OP}$ bisects $\angle QOR$.

Strategy Use the proof of Theorem 13.11 to show that angles $\angle QOP$ and $\angle ROP$ are congruent.

The proofs of Corollaries 13.11a and 13.11b are left to the student. (See exercises 15 and 16.)

A Polygon Circumscribed About a Circle

A polygon is **circumscribed** about a circle if each side of the polygon is tangent to the circle. When a polygon is circumscribed about a circle, we also say that the circle is **inscribed** in the polygon. For example, in the diagram, $\overline{AB}$ is tangent to circle O at E, $\overline{BC}$ is tangent to circle O at F, $\overline{CD}$ is tangent to circle O at G, $\overline{DA}$ is tangent to circle O at H. Therefore, $ABCD$ is circumscribed about circle O and circle O is inscribed in quadrilateral $ABCD$.

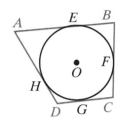

If $\triangle ABC$ is circumscribed about circle O, then we know that $\overline{OA}$, $\overline{OB}$, and $\overline{OC}$ are the bisectors of the angles of $\triangle ABC$, and O is the point at which the angle bisectors of the angles of a triangle intersect.

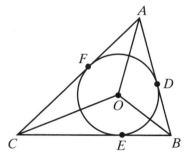

EXAMPLE 3

$\overline{AB}$, $\overline{BC}$, and $\overline{CA}$ are tangent to circle O at D, E, and F, respectively. If $AF = 6$, $BE = 7$, and $CE = 5$, find the perimeter of $\triangle ABC$.

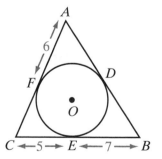

Solution Tangent segments drawn to a circle from an external point are congruent.

$$AD = AF = 6 \qquad BD = BE = 7 \qquad CF = CE = 5$$

Therefore,

$$
\begin{aligned}
AB &= AD + BD & BC &= BE + CE & CA &= CF + AF \\
&= 6 + 7 & &= 7 + 5 & &= 5 + 6 \\
&= 13 & &= 12 & &= 11
\end{aligned}
$$

$$
\begin{aligned}
\text{Perimeter} &= AB + BC + CA \\
&= 13 + 12 + 11 \\
&= 36 \ \textit{Answer}
\end{aligned}
$$

EXAMPLE 4

Point P is a point on a line that is tangent to circle O at R, P is 12.0 centimeters from the center of the circle, and the length of the tangent segment from P is 8.0 centimeters.

a. Find the exact length of the radius of the circle.

b. Find the length of the radius to the nearest tenth.

Solution **a.** P is 12 centimeters from the center of circle O; $OP = 12$.

The length of the tangent segment is 8 centimeters; $RP = 8$.

A line tangent to a circle is perpendicular to the radius drawn to the point of tangency; $\triangle OPR$ is a right triangle.

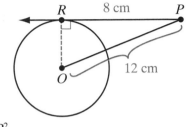

$$RP^2 + OR^2 = OP^2$$
$$8^2 + OP^2 = 12^2$$
$$64 + OP^2 = 144$$
$$OP^2 = 80$$
$$OP = \sqrt{80} = \sqrt{16}\sqrt{5} = 4\sqrt{5}$$

b. Use a calculator to evaluate $4\sqrt{5}$.

ENTER: 4 **2nd** $\sqrt{}$ 5 **ENTER**

DISPLAY: 8.94427191

To the nearest tenth, $OP = 8.9$.

Answers **a.** $4\sqrt{5}$ cm **b.** 8.9 cm

Exercises

Writing About Mathematics

1. Line l is tangent to circle O at A and line m is tangent to circle O at B. If $\overline{AOB}$ is a diameter, does l intersect m? Justify your answer.

2. Explain the difference between a polygon inscribed in a circle and a circle inscribed in a polygon.

Developing Skills

In 3 and 4, $\triangle ABC$ is circumscribed about circle O and D, E, and F are points of tangency.

3. If $AD = 5$, $EB = 5$, and $CF = 10$, find the lengths of the sides of the triangle and show that the triangle is isosceles.

4. If $AF = 10$, $CE = 20$, and $BD = 30$, find the lengths of the sides of the triangle and show that the triangle is a right triangle.

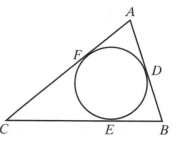

In 5–11, $\overline{PQ}$ is tangent to circle O at P, $\overline{SQ}$ is tangent to circle O at S, and $\overline{OQ}$ intersects circle O at T and R.

5. If $OP = 15$ and $PQ = 20$, find: **a.** OQ **b.** SQ **c.** TQ

6. If $OQ = 25$ and $PQ = 24$, find: **a.** OP **b.** RT **c.** RQ

7. If $OP = 10$ and $OQ = 26$, find: **a.** PQ **b.** RQ **c.** TQ

8. If $OP = 6$ and $TQ = 13$, find: **a.** OQ **b.** PQ **c.** SQ

9. If $OS = 9$ and $RQ = 32$, find: **a.** OQ **b.** SQ **c.** PQ

10. If $PQ = 3x$, $SQ = 5x - 8$, and $OS = x + 1$, find: **a.** PQ **b.** SQ **c.** OS **d.** OQ

11. If $SQ = 2x$, $OS = 2x + 2$, and $OQ = 3x + 1$, find: **a.** x **b.** SQ **c.** OS **d.** OQ

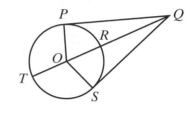

12. The sides of $\triangle ABC$ are tangent to a circle at D, E, and F. If $DB = 4$, $BC = 7$, and the perimeter of the triangle is 30, find:

    a. BE **b.** EC **c.** CF **d.** AF **e.** AC **f.** AB

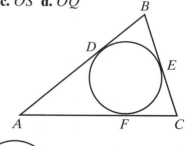

13. Line $\overleftrightarrow{RP}$ is tangent to circle O at P and $\overline{OR}$ intersects the circle at M, the midpoint of $\overline{OR}$. If $RP = 3.00$ cm, find the length of the radius of the circle:

    a. in radical form **b.** to the nearest hundredth

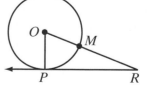

14. Points E, F, G, and H are the points of tangency to circle O of $\overline{AB}$, $\overline{BC}$, $\overline{CD}$, and $\overline{DA}$, respectively. The measure of $\overset{\frown}{EF}$ is $80°$, of $\overset{\frown}{FG}$ is $70°$, and of $\overset{\frown}{GH}$ is $50°$. Find:

    a. m$\angle EOF$ **b.** m$\angle FOG$ **c.** m$\angle GOH$ **d.** m$\angle HOE$ **e.** m$\angle AOE$

    f. m$\angle EAO$ **g.** m$\angle EAH$ **h.** m$\angle FBE$ **i.** m$\angle GCF$ **j.** m$\angle HDG$

    k. the sum of the measures of the angles of quadrilateral $ABCD$

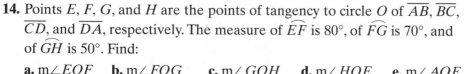

Applying Skills

15. Prove Corollary 13.11a, "If two tangents are drawn to a circle from an external point, then the line segment from the center of the circle to the external point bisects the angle formed by the tangents."

16. Prove Corollary 13.11b, "If two tangents are drawn to a circle from an external point, then the line segment from the center of the circle to the external point bisects the angle whose vertex is the center of the circle and whose rays are the two radii drawn to the points of tangency."

17. Lines $\overleftrightarrow{PQ}$ and $\overleftrightarrow{PR}$ are tangent to circle O at Q and R.

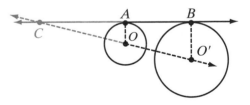

   a. Prove that $\angle PQR \cong \angle PRQ$.

   b. Draw $\overline{OP}$ intersecting $\overline{RQ}$ at S and prove that $QS = RS$ and $\overline{OP} \perp \overline{QR}$.

   c. If $OP = 10$, $SQ = 4$ and $OS < SP$, find OS and SP.

18. Tangents $\overline{AC}$ and $\overline{BC}$ to circle O are perpendicular to each other at C. Prove:

   a. $\overline{AC} \cong \overline{AO}$

   b. $OC = \sqrt{2}OA$

   c. $AOBC$ is a square.

19. Isosceles $\triangle ABC$ is circumscribed about circle O. The points of tangency of the legs, $\overline{AB}$ and $\overline{AC}$, are D and F, and the point of tangency of the base, $\overline{BC}$, is E. Prove that E is the midpoint of $\overline{BC}$.

20. Line $\overleftrightarrow{AB}$ is a common external tangent to circle O and circle O'. $\overleftrightarrow{AB}$ is tangent to circle O at A and to circle O' at B, and $OA < O'B$.

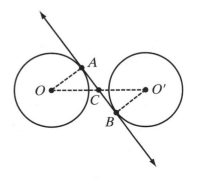

   a. Prove that $\overleftrightarrow{OO'}$ is not parallel to $\overleftrightarrow{AB}$.

   b. Let C be the intersection of $\overleftrightarrow{OO'}$ and $\overleftrightarrow{AB}$. Prove that $\triangle OAC \sim \triangle O'BC$.

   c. If $\frac{OA}{O'B} = \frac{2}{3}$, and $BC = 12$, find AC, AB, OC, $O'C$, and OO'.

21. Line $\overleftrightarrow{AB}$ is a common internal tangent to circles O and O'. $\overleftrightarrow{AB}$ is tangent to circle O at A and to circle O' at B, and $OA = O'B$. The intersection of $\overline{OO'}$ and $\overleftrightarrow{AB}$ is C.

   a. Prove that $OC = O'C$.

   b. Prove that $AC = BC$.

Hands-On Activity

Consider any regular polygon. Construct the angle bisectors of each interior angle. Since the interior angles are all congruent, the angles formed are all congruent. Since the sides of the regular polygon are all congruent, congruent isosceles triangles are formed by ASA. Any two adjacent triangles share a common leg. Therefore, they all share the same vertex. Since the legs of the triangles formed are all congruent, the vertex is equidistant from the vertices of the regular polygon. This common vertex is the **center** of the regular polygon.

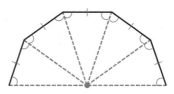

 In this Hands-On Activity, we will use the center of a regular polygon to inscribe a circle in the polygon.

a. Using geometry software or compass, protractor, and straightedge, construct a square, a regular pentagon, and a regular hexagon. For each figure:

(1) Construct the center P of the regular polygon. (The center is the intersection of the angle bisectors of a regular polygon.)

(2) Construct an **apothem** or perpendicular from P to one of the sides of regular polygon.

(3) Construct a circle with center P and radius equal to the length of the apothem.

b. Prove that the circles constructed in part **a** are inscribed inside of the polygon. Prove:

(1) The apothems of each polygon are all congruent.

(2) The foot of each apothem is on the circle.

(3) The sides of the regular polygon are tangent to the circle.

c. Let r be the distance from the center to a vertex of the regular polygon. Since the center is equidistant from each vertex, it is possible to circumscribe a circle about the polygon with radius r. Let a be the length of an apothem and s be the length of a side of the regular polygon. How is the radius, r, of the circumscribed circle related to the radius, a, of the inscribed circle?

13-5 ANGLES FORMED BY TANGENTS, CHORDS, AND SECANTS

Angles Formed by a Tangent and a Chord

In the diagram, $\overleftrightarrow{AB}$ is tangent to circle O at A, $\overline{AD}$ is a chord, and $\overline{AC}$ is a diameter. When $\overline{CD}$ is drawn, $\angle ADC$ is a right angle because it is an angle inscribed in a semicircle, and $\angle ACD$ is the complement of $\angle CAD$. Also, $\overline{CA} \perp \overline{AB}$, $\angle BAC$ is a right angle, and $\angle DAB$ is the complement of $\angle CAD$. Therefore, since complements of the same angle are congruent, $\angle ACD \cong \angle DAB$. We can conclude that since $m\angle ACD = \frac{1}{2}m\overset{\frown}{AD}$, then $m\angle DAB = \frac{1}{2}m\overset{\frown}{AD}$.

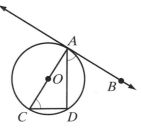

We can state what we have just proved on page 567 as a theorem.

Theorem 13.12

The measure of an angle formed by a tangent and a chord that intersect at the point of tangency is equal to one-half the measure of the intercepted arc.

Angles Formed by Two Intersecting Chords

We can find how the measures of other angles and their intercepted arcs are related. For example, in the diagram, two chords $\overline{AB}$ and $\overline{CD}$ intersect in the interior of circle O and $\overline{DB}$ is drawn. Angle AED is an exterior angle of $\triangle DEB$. Therefore,

$$\text{m}\angle AED = \text{m}\angle BDE + \text{m}\angle DBE$$
$$= \tfrac{1}{2}\text{m}\widehat{BC} + \tfrac{1}{2}\text{m}\widehat{DA}$$
$$= \tfrac{1}{2}(\text{m}\widehat{BC} + \text{m}\widehat{DA})$$

Notice that $\widehat{BC}$ is the arc intercepted by $\angle BEC$ and $\widehat{DA}$ is the arc intercepted by $\angle AED$, the angle vertical to $\angle BEC$. We can state this relationship as a theorem.

Theorem 13.13

The measure of an angle formed by two chords intersecting within a circle is equal to one-half the sum of the measures of the arcs intercepted by the angle and its vertical angle.

Angles Formed by Tangents and Secants

We have shown how the measures of angles whose vertices are on the circle or within the circle are related to the measures of their intercepted arcs. Now we want to show how angles formed by two tangents, a tangent and a secant, or two secants, all of which have vertices outside the circle, are related to the measures of the intercepted arcs.

A Tangent Intersecting a Secant

In the diagram, $\overleftrightarrow{PRS}$ is a tangent to circle O at R and $\overleftrightarrow{PTQ}$ is a secant that intersects the circle at T and at Q. Chord $\overline{RQ}$ is drawn. Then $\angle SRQ$ is an exterior angle of $\triangle PRQ$.

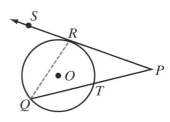

$$m\angle RQP + m\angle P = m\angle SRQ$$
$$m\angle P = m\angle SRQ - m\angle RQP$$
$$m\angle P = \tfrac{1}{2}m\overset{\frown}{RQ} - \tfrac{1}{2}m\overset{\frown}{RT}$$
$$m\angle P = \tfrac{1}{2}(m\overset{\frown}{RQ} - m\overset{\frown}{RT})$$

Two Intersecting Secants

In the diagram, $\overleftrightarrow{PTR}$ is a secant to circle O that intersects the circle at R and T, and $\overleftrightarrow{PQS}$ is a secant to circle O that intersects the circle at Q and S. Chord $\overline{RQ}$ is drawn. Then $\angle RQS$ is an exterior angle of $\triangle RQP$.

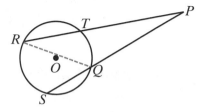

$$m\angle PRQ + m\angle P = m\angle RQS$$
$$m\angle P = m\angle RQS - m\angle PRQ$$
$$m\angle P = \tfrac{1}{2}m\overset{\frown}{RS} - \tfrac{1}{2}m\overset{\frown}{QT}$$
$$m\angle P = \tfrac{1}{2}(m\overset{\frown}{RS} - m\overset{\frown}{QT})$$

Two Intersecting Tangents

In the diagram, $\overleftrightarrow{PRS}$ is tangent to circle O at R, $\overleftrightarrow{PQ}$ is tangent to the circle at Q, and T is a point on major $\overset{\frown}{RQ}$. Chord $\overline{RQ}$ is drawn. Then $\angle SRQ$ is an exterior angle of $\triangle RQP$.

$$m\angle PQR + m\angle P = m\angle SRQ$$
$$m\angle P = m\angle SRQ - m\angle PQR$$
$$m\angle P = \tfrac{1}{2}m\overset{\frown}{RTQ} - \tfrac{1}{2}m\overset{\frown}{RQ}$$
$$m\angle P = \tfrac{1}{2}(m\overset{\frown}{RTQ} - m\overset{\frown}{RQ})$$

For each pair of lines, a tangent and a secant, two secants, and two tangents, the steps necessary to prove the following theorem have been given:

Theorem 13.14

> The measure of an angle formed by a tangent and a secant, two secants, or two tangents intersecting outside the circle is equal to one-half the difference of the measures of the intercepted arcs.

EXAMPLE 1

A tangent and a secant are drawn to circle O from point P. The tangent intersects the circle at Q and the secant at R and S. If $m\overset{\frown}{QR}:m\overset{\frown}{RS}:m\overset{\frown}{SQ} = 3:5:7$, find:

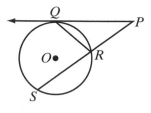

a. $m\overset{\frown}{QR}$ **b.** $m\overset{\frown}{RS}$ **c.** $m\overset{\frown}{SQ}$

d. $m\angle QRS$ **e.** $m\angle RQP$ **f.** $m\angle P$

Solution   Let $m\overset{\frown}{QR} = 3x$, $m\overset{\frown}{RS} = 5x$, and $m\overset{\frown}{SQ} = 7x$.

$$3x + 5x + 7x = 360$$
$$15x = 360$$
$$x = 24$$

a. $m\overset{\frown}{QR} = 3x$ **b.** $m\overset{\frown}{RS} = 5x$ **c.** $m\overset{\frown}{SQ} = 7x$

$\phantom{\text{a. }m\overset{\frown}{QR}} = 3(24)$ $\phantom{\text{b. }m\overset{\frown}{RS}} = 5(24)$ $\phantom{\text{c. }m\overset{\frown}{SQ}} = 7(24)$

$\phantom{\text{a. }m\overset{\frown}{QR}} = 72$ $\phantom{\text{b. }m\overset{\frown}{RS}} = 120$ $\phantom{\text{c. }m\overset{\frown}{SQ}} = 168$

d. $m\angle QRS = \frac{1}{2}m\overset{\frown}{SQ}$ **e.** $m\angle RQP = \frac{1}{2}m\overset{\frown}{QR}$ **f.** $m\angle P = \frac{1}{2}(m\overset{\frown}{SQ} - m\overset{\frown}{QR})$

$\phantom{\text{d. }m\angle QRS} = \frac{1}{2}(168)$ $\phantom{\text{e. }m\angle RQP} = \frac{1}{2}(72)$ $\phantom{\text{f. }m\angle P} = \frac{1}{2}(168 - 72)$

$\phantom{\text{d. }m\angle QRS} = 84$ $\phantom{\text{e. }m\angle RQP} = 36$ $\phantom{\text{f. }m\angle P} = 48$

Answers   **a.** $72°$ **b.** $120°$ **c.** $168°$ **d.** $84°$ **e.** $36°$ **f.** $48°$

Note: $m\angle QRS = m\angle RQP + m\angle P$
$$= 36 + 48$$
$$= 84$$

EXAMPLE 2

Two tangent segments, $\overline{RP}$ and $\overline{RQ}$, are drawn to circle O from an external point R. If $m\angle R$ is 70, find the measure of the minor arc $\overset{\frown}{PQ}$ and of the major arc $\overset{\frown}{PSQ}$ into which the circle is divided.

Solution The sum of the minor arc and the major arc with the same endpoints is 360.

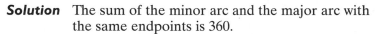

Let $x = m\overset{\frown}{PQ}$.

Then $360 - x = m\overset{\frown}{PSQ}$.

$$m\angle R = \tfrac{1}{2}(m\overset{\frown}{PSQ} - m\overset{\frown}{PQ})$$
$$70 = \tfrac{1}{2}(360 - x - x)$$
$$70 = \tfrac{1}{2}(360 - 2x)$$
$$70 = 180 - x$$
$$x = 110$$
$$360 - x = 360 - 110$$
$$= 250$$

Answer $m\overset{\frown}{PQ} = 110$ and $m\overset{\frown}{PSQ} = 250$

SUMMARY

Type of Angle	Degree Measure	Example
Formed by a Tangent and a Chord	The measure of an angle formed by a tangent and a chord that intersect at the point of tangency is equal to one-half the measure of the intercepted arc.	$m\angle 1 = \tfrac{1}{2}m\overset{\frown}{AB}$
Formed by Two Intersecting Chords	The measure of an angle formed by two intersecting chords is equal to one-half the sum of the measures of the arcs intercepted by the angle and its vertical angle.	$m\angle 1 = \tfrac{1}{2}(m\overset{\frown}{AB} + m\overset{\frown}{CD})$ $m\angle 2 = \tfrac{1}{2}(m\overset{\frown}{AB} + m\overset{\frown}{CD})$

(Continued)

SUMMARY (*Continued*)

Type of Angle	Degree Measure	Example
Formed by Tangents and Secants	The measure of an angle formed by a tangent and a secant, two secants, or two tangents intersecting outside the circle is equal to one-half the difference of the measures of the intercepted arcs.	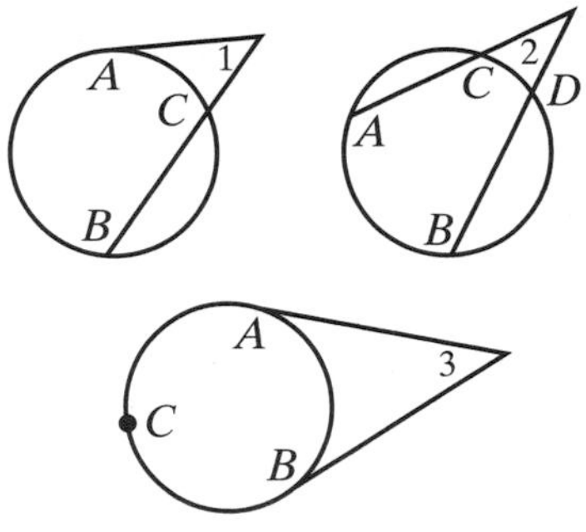 $$m\angle 1 = \tfrac{1}{2}(m\widehat{AB} - m\widehat{AC})$$ $$m\angle 2 = \tfrac{1}{2}(m\widehat{AB} - m\widehat{CD})$$ $$m\angle 3 = \tfrac{1}{2}(m\widehat{ACB} - m\widehat{AB})$$

Exercises

Writing About Mathematics

1. Nina said that a radius drawn to the point at which a secant intersects a circle cannot be perpendicular to the secant. Do you agree with Nina? Explain why or why not.

2. Two chords intersect at the center of a circle forming four central angles. Aaron said that the measure of one of these angles is one-half the sum of the measures of the arcs intercepted by the angle and its vertical angle. Do you agree with Aaron? Explain why or why not.

Developing Skills

In 3–8, secants $\overleftrightarrow{PQS}$ and $\overleftrightarrow{PRT}$ intersect at P.

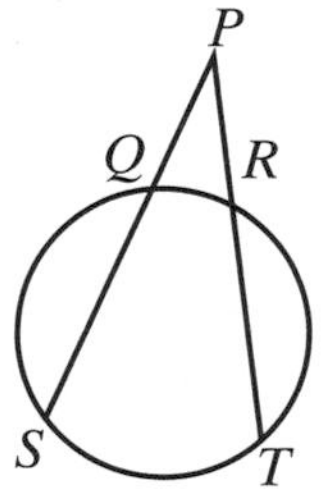

3. If $m\widehat{ST} = 160$ and $m\widehat{QR} = 90$, find $m\angle P$.

4. If $m\widehat{ST} = 100$ and $m\widehat{QR} = 40$, find $m\angle P$.

5. If $m\widehat{ST} = 170$ and $m\widehat{QR} = 110$, find $m\angle P$.

6. If $m\angle P = 40$ and $m\widehat{QR} = 86$, find $m\widehat{ST}$.

7. If $m\angle P = 60$ and $m\widehat{QR} = 50$, find $m\widehat{ST}$.

8. If $m\angle P = 25$ and $m\widehat{ST} = 110$, find $m\widehat{QR}$.

In 9–14, tangent $\overleftrightarrow{QP}$ and secant $\overleftrightarrow{PRT}$ intersect at P.

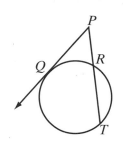

9. If m$\widehat{QT}$ = 170 and m$\widehat{QR}$ = 70, find m∠P.

10. If m$\widehat{QT}$ = 120 and m$\widehat{QR}$ = 30, find m∠P.

11. If m$\widehat{QR}$ = 70 and m$\widehat{RT}$ = 120, find m∠P.

12. If m$\widehat{QR}$ = 50 and m∠P = 40, find m$\widehat{QT}$.

13. If m$\widehat{QR}$ = 60 and m∠P = 35, find m$\widehat{QT}$.

14. If m∠P = 30 and m$\widehat{QR}$ = 120, find m$\widehat{QT}$.

In 15–20, tangents $\overleftrightarrow{RP}$ and $\overleftrightarrow{QP}$ intersect at P and S is on major arc $\widehat{QR}$.

15. If m$\widehat{RQ}$ = 160, find m∠P.

16. If m$\widehat{RQ}$ = 80, find m∠P.

17. If m$\widehat{RSQ}$ = 260, find m∠P.

18. If m$\widehat{RSQ}$ = 210, find m∠P.

19. If m$\widehat{RSQ}$ = 2m$\widehat{RQ}$, find m∠P.

20. If m∠P = 45, find m$\widehat{RQ}$ and m$\widehat{RSQ}$.

In 21–26, chords $\overline{AB}$ and $\overline{CD}$ intersect at E in the interior of a circle.

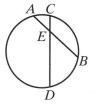

21. If m$\widehat{AC}$ = 30 and m$\widehat{BD}$ = 80, find m∠AEC.

22. If m$\widehat{DA}$ = 180 and m$\widehat{BC}$ = 100 find m∠AED.

23. If m$\widehat{AC}$ = 25 and m$\widehat{BD}$ = 45, find m∠DEB.

24. If m$\widehat{AC}$ = 20 and m$\widehat{BD}$ = 60, find m∠AED.

25. If m$\widehat{AC}$ = 30 and m∠AEC = 50, find m$\widehat{BD}$.

26. If m$\widehat{BC}$ = 80 and m∠AEC = 30, find m$\widehat{DA}$.

27. In the diagram, $\overleftrightarrow{PA}$ and $\overleftrightarrow{PB}$ are tangent to circle O at A and B. Diameter $\overline{BD}$ and chord $\overline{AC}$ intersect at E, m$\widehat{CB}$ = 125 and m∠P = 55. Find:

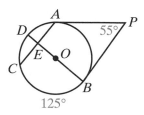

   a. m$\widehat{AB}$ **b.** m$\widehat{AD}$ **c.** m$\widehat{CD}$

   d. m∠DEC **e.** m∠PBD **f.** m∠PAC

   g. Show that $\overline{BD}$ is perpendicular to $\overline{AC}$ and bisects $\overline{AC}$.

28. Tangent segment $\overline{PA}$ and secant segment $\overline{PBC}$ are drawn to circle O and $\overline{AB}$ and $\overline{AC}$ are chords. If $m\angle P = 45$ and $m\widehat{AC}:m\widehat{AB} = 5:2$, find:

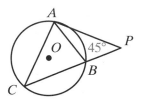

    a. $m\widehat{AC}$ **b.** $m\widehat{BC}$ **c.** $m\angle ACB$

    d. $m\angle PAB$ **e.** $m\angle CAB$ **f.** $m\angle PAC$

Applying Skills

29. Tangent $\overleftrightarrow{PC}$ intersects circle O at C, chord $\overline{AB} \parallel \overleftrightarrow{CP}$, diameter $\overline{COD}$ intersects $\overline{AB}$ at E, and diameter $\overline{AOF}$ is extended to P.

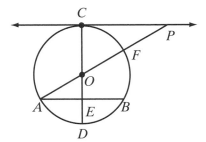

    a. Prove that $\triangle OPC \sim \triangle OAE$.

    b. If $m\angle OAE = 30$, find $m\widehat{AD}$, $m\widehat{CF}$, $m\widehat{FB}$, $m\widehat{BD}$, $m\widehat{AC}$, and $m\angle P$.

30. Tangent $\overleftrightarrow{ABC}$ intersects circle O at B, secant $\overleftrightarrow{AFOD}$ intersects the circle at F and D, and secant $\overleftrightarrow{CGOE}$ intersects the circle at G and E. If $m\widehat{EFB} = m\widehat{DGB}$, prove that $\triangle AOC$ is isosceles.

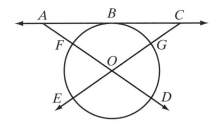

31. Segments $\overline{AP}$ and $\overline{BP}$ are tangent to circle O at A and B, respectively, and $m\angle AOB = 120$. Prove that $\triangle ABP$ is equilateral.

32. Secant $\overleftrightarrow{ABC}$ intersects a circle at A and B. Chord $\overline{BD}$ is drawn. Prove that $m\angle CBD \neq \frac{1}{2}m\widehat{BD}$.

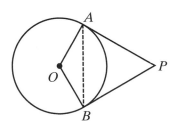

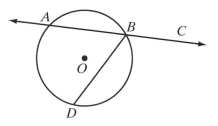

13-6 MEASURES OF TANGENT SEGMENTS, CHORDS, AND SECANT SEGMENTS

Segments Formed by Two Intersecting Chords

We have been proving theorems to establish the relationship between the measures of angles of a circle and the measures of the intercepted arcs. Now we will study the measures of tangent segments, secant segments, and chords. To do this, we will use what we know about similar triangles.

Theorem 13.15 | If two chords intersect within a circle, the product of the measures of the segments of one chord is equal to the product of the measures of the segments of the other.

Given Chords $\overline{AB}$ and $\overline{CD}$ intersect at E in the interior of circle O.

Prove $(AE)(EB) = (CE)(ED)$

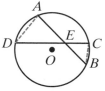

Proof

Statements	Reasons
1. Draw $\overline{AD}$ and $\overline{CB}$.	**1.** Two points determine a line.
2. $\angle A \cong \angle C$ and $\angle D \cong \angle B$	**2.** Inscribed angles of a circle that intercept the same arc are congruent.
3. $\triangle ADE \sim \triangle CBE$	**3.** AA~.
4. $\frac{AE}{CE} = \frac{ED}{EB}$	**4.** The lengths of the corresponding sides of similar triangles are in proportion.
5. $(AE)(EB) = (CE)(ED)$	**5.** In a proportion, the product of the means is equal to the product of the extremes.

Segments Formed by a Tangent Intersecting a Secant

Do similar relationships exist for tangent segments and secant segments? In the diagram, tangent segment $\overline{PA}$ is drawn to circle O, and secant segment $\overline{PBC}$ intersects the circle at B and C.

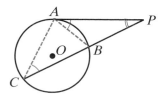

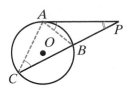

We will call $\overline{PB}$, the part of the secant segment that is outside the circle, the **external segment** of the secant. When chords $\overline{AB}$ and $\overline{AC}$ are drawn, $\angle C \cong \angle PAB$ because the measure of each is one-half the measure of the intercepted arc, $\overarc{AB}$. Also $\angle P \cong \angle P$ by the reflexive property. Therefore, $\triangle BPA \sim \triangle APC$ by AA~. The length of the corresponding sides of similar triangles are in proportion. Therefore:

$$\frac{PB}{PA} = \frac{PA}{PC} \quad \text{and} \quad (PA)^2 = (PC)(PB)$$

We can write what we have just proved as a theorem:

Theorem 13.16

> If a tangent and a secant are drawn to a circle from an external point, then the square of the length of the tangent segment is equal to the product of the lengths of the secant segment and its external segment.

Note that both means of the proportion are PA, the length of the tangent segment. Therefore, we can say that the length of the tangent segment is the mean proportional between the lengths of the secant and its external segment. Theorem 13.16 can be stated in another way.

Theorem 13.16

> If a tangent and a secant are drawn to a circle from an external point, then the length of the tangent segment is the mean proportional between the lengths of the secant segment and its external segment.

Segments Formed by Intersecting Secants

What is the relationship of the lengths of two secants drawn to a circle from an external point? Let $\overline{ABC}$ and $\overline{ADE}$ be two secant segments drawn to a circle as shown in the diagram. Draw $\overline{AF}$ a tangent segment to the circle from A. Since

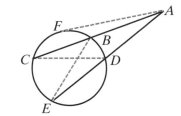

$$AF^2 = (AC)(AB) \quad \text{and} \quad AF^2 = (AE)(AD),$$

then

$$(AC)(AB) = (AE)(AD)$$

Note: This relationship could also have been proved by showing that $\triangle ABE \sim \triangle ADC$.

We can state this as a theorem:

Theorem 13.17 — If two secant segments are drawn to a circle from an external point, then the product of the lengths of one secant segment and its external segment is equal to the product of the lengths of the other secant segment and its external segment.

EXAMPLE I

Two secant segments, $\overline{PAB}$ and $\overline{PCD}$, and a tangent segment, $\overline{PE}$, are drawn to a circle from an external point P. If $PB = 9$ cm, $PD = 12$ cm, and the external segment of $\overline{PAB}$ is 1 centimeter longer than the external segment of $\overline{PCD}$, find: **a.** PA **b.** PC **c.** PE

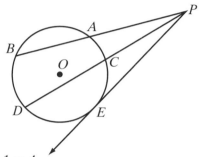

Solution Let $x = PC$ and $x + 1 = PA$.

$$(PB)(PA) = (PD)(PC)$$

$$9PA = 12PC$$

$$9(x + 1) = 12x$$

$$9x + 9 = 12x$$

$$9 = 3x$$

$$3 = x$$

a. $PA = x + 1 = 4$

b. $PC = x = 3$

c. $(PE)^2 = (PB)(PA)$

$$(PE)^2 = (9)(4)$$

$$(PE)^2 = 36$$

$$PE = 6 \text{ (Use the positive square root.)}$$

Answers **a.** $PA = 4$ cm **b.** $PC = 3$ cm **c.** $PE = 6$ cm

EXAMPLE 2

In a circle, chords $\overline{PQ}$ and $\overline{RS}$ intersect at T. If $PT = 2$, $TQ = 10$, and $SR = 9$, find ST and TR.

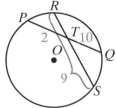

Solution   Since $ST + TR = SR = 9$, let $ST = x$ and $TR = 9 - x$.

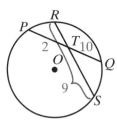

$$(PT)(TQ) = (RT)(TS)$$
$$(2)(10) = x(9 - x)$$
$$20 = 9x - x^2$$
$$x^2 - 9x + 20 = 0$$
$$(x - 5)(x - 4) = 0$$

$x - 5 = 0$	$x - 4 = 0$
$x = 5$	$x = 4$
$9 - x = 4$	$9 - x = 5$

Answer $ST = 5$ and $TR = 4$, or $ST = 4$ and $TR = 5$

EXAMPLE 3

Find the length of a chord that is 20 centimeters from the center of a circle if the length of the radius of the circle is 25 centimeters.

Solution   Draw diameter $\overline{COD}$ perpendicular to chord $\overline{AB}$ at E. Then OE is the distance from the center of the circle to the chord.

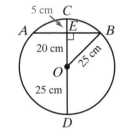

$OE = 20$ $DE = OD + OE$ $CE = OC - OE$
$$= 25 + 20 \qquad\qquad = 25 - 20$$
$$= 45 \qquad\qquad\quad\ = 5$$

A diameter perpendicular to a chord bisects the chord. Therefore, $AE = EB$.

Let $AE = EB = x$.

$$(AE)(EB) = (DE)(CE)$$
$$(x)(x) = (45)(5)$$
$$x^2 = 225$$
$$x = 15 \text{ (Use the positive square root.)}$$

Therefore,

$$AB = AE + EB$$
$$= x + x$$
$$= 30 \text{ cm } \textit{Answer}$$

SUMMARY

Type of Segment	Length	Example
Formed by Two Intersecting Chords	If two chords intersect, the product of the measures of the segments of one chord is equal to the product of the measures of of the segments of the other.	 $(AE)(EB) = (CE)(ED)$
Formed by a Tangent Intersecting a Secant	If a tangent and a secant are drawn to a circle from an external point, then the square of the length of the tangent segment is equal to the product of the lengths of the secant segment and its external segment.	 $(PA)^2 = (PC)(PB)$
Formed by Two Intersecting Secants	If two secant segments are drawn to a circle from an external point, then the product of the lengths of one secant segment and its external segment is equal to the product of the lengths of the other secant segment and its external segment.	 $(PB)(PA) = (PD)(PC)$

Exercises

Writing About Mathematics

1. The length of chord $\overline{AB}$ in circle O is 24. Vanessa said that any chord of circle O that intersects $\overline{AB}$ at its midpoint, M, is separated by M into two segments such that the product of the lengths of the segments is 144. Do you agree with Vanessa? Justify your answer.

2. Secants $\overline{ABP}$ and $\overline{CDP}$ are drawn to circle O. If $AP > CP$, is $BP > DP$? Justify your answer.

Developing Skills

In 3–14, chords $\overline{AB}$ and $\overline{CD}$ intersect at E.

3. If $CE = 12$, $ED = 2$, and $AE = 3$, find EB.

4. If $CE = 16$, $ED = 3$, and $AE = 8$, find EB.

5. If $AE = 20$, $EB = 5$, and $CE = 10$, find ED.

6. If $AE = 14$, $EB = 3$, and $ED = 6$, find CE.

7. If $CE = 10$, $ED = 4$, and $AE = 5$, find EB.

8. If $CE = 56$, $ED = 14$, and $AE = EB$, find EB.

9. If $CE = 12$, $ED = 2$, and AE is 2 more than EB, find EB.

10. If $CE = 16$, $ED = 12$, and AE is 3 times EB, find EB.

11. If $CE = 8$, $ED = 5$, and AE is 6 more than EB, find EB.

12. If $CE = 9$, $ED = 9$, and AE is 24 less than EB, find EB.

13. If $CE = 24$, $ED = 5$, and $AB = 26$, find AE and EB.

14. If $AE = 7$, $EB = 4$, and $CD = 16$, find CE and ED.

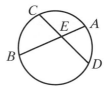

In 15–22, $\overleftrightarrow{AF}$ is tangent to circle O at F and secant $\overline{ABC}$ intersects circle O at B and C.

15. If $AF = 8$ and $AB = 4$, find AC.

16. If $AB = 3$ and $AC = 12$, find AF.

17. If $AF = 6$ and $AC = 9$, find AB.

18. If $AB = 4$ and $BC = 12$, find AF.

19. If $AF = 12$ and BC is 3 times AB, find AC, AB, and BC.

20. If $AF = 10$ and AC is 4 times AB, find AC, AB, and BC.

21. If $AF = 8$ and $CB = 12$, find AC, AB, and BC.

22. If $AF = 15$ and $CB = 16$, find AC, AB, and BC.

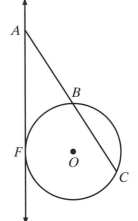

In 23–30, secants $\overrightarrow{ABC}$ and $\overrightarrow{ADE}$ intersect at A outside the circle.

23. If $AB = 8$, $AC = 25$, and $AD = 10$, find AE.

24. If $AB = 6$, $AC = 18$, and $AD = 9$, find AE.

25. If $AD = 12$, $AE = 20$, and $AB = 8$, find AC.

26. If $AD = 9$, $AE = 21$, and AC is 5 times AB, find AB and AC.

27. If $AB = 3$, $AD = 2$, and $DE = 10$, find AC.

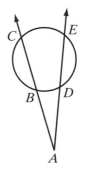

28. If $AB = 4$, $BC = 12$, and $AD = DE$, find AE.

29. If $AB = 2$, $BC = 7$, and $DE = 3$, find AD and AE.

30. If $AB = 6$, $BC = 8$, and $DE = 5$, find AD and AE.

31. In a circle, diameter $\overline{AB}$ is extended through B to P and tangent segment $\overline{PC}$ is drawn. If $BP = 6$ and $PC = 9$, what is the measure of the diameter of the circle?

13-7 CIRCLES IN THE COORDINATE PLANE

In the diagram, a circle with center at the origin and a radius with a length of 5 units is drawn in the coordinate plane. The points (5, 0), (0, 5), (−5, 0) and (0, −5) are points on the circle. What other points are on the circle and what is the equation of the circle?

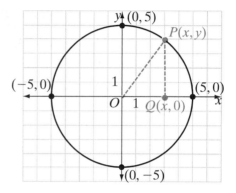

Let $P(x, y)$ be any other point on the circle. From P, draw a vertical line segment to the x-axis. Let this be point Q. Then $\triangle OPQ$ is a right triangle with $OQ = x$, $PQ = y$, and $OP = 5$. We can use the Pythagorean Theorem to write an equation for the circle:

$$OQ^2 + PQ^2 = OP^2$$
$$x^2 + y^2 = 5^2$$

The points (3, 4), (4, 3), (−3, 4), (−4, 3), (−3, −4), (−4, −3), (3, −4), and (4, −3) appear to be points on the circle and all make the equation $x^2 + y^2 = 5^2$ true. The points (5, 0), (0, 5), (−5, 0), and (0, −5) also make the equation true, as do points such as $\left(1, \sqrt{24}\right)$ and $\left(−2, \sqrt{21}\right)$.

If we replace 5 by the length of any radius, r, the equation of a circle whose center is at the origin is:

$$x^2 + y^2 = r^2$$

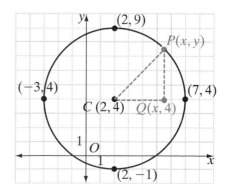

How does the equation change if the center is not at the origin? For example, what is the equation of a circle whose center, C, is at (2, 4) and whose radius has a length of 5 units? The points (7, 4), (−3, 4), (2, 9), and (2, −1) are each 5 units from (2, 4) and are therefore points on the circle. Let $P(x, y)$ be any other point on the circle. From P, draw a vertical line and from C, a horizontal line.

Let the intersection of these two lines be Q. Then $\triangle CPQ$ is a right triangle with:

$$CQ = |x - 2| \quad PQ = |y - 4| \quad CP = 5$$

We can use the Pythagorean Theorem to write an equation for the circle.

$$CQ^2 \quad + \quad PQ^2 \quad = CP^2$$
$$(x - 2)^2 + (y - 4)^2 = 5^2$$

The points $(5, 8)$, $(6, 7)$, $(-1, 8)$, $(-2, 7)$, $(-1, 0)$, $(-2, 1)$ $(5, 0)$, and $(6, 1)$ appear to be points on the circle and all make $(x - 2)^2 + (y - 4)^2 = 5^2$ true. The points $(7, 4)$, $(-3, 4)$, $(2, 9)$, and $(2, -1)$ also make the equation true, as do points whose coordinates are not integers.

We can write a general equation for a circle with center at $C(h, k)$ and radius r. Let $P(x, y)$ be any point on the circle. From P draw a vertical line and from C draw a horizontal line. Let the intersection of these two lines be Q. Then $\triangle CPQ$ is a right triangle with:

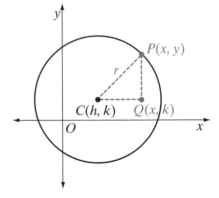

$$CQ = |x - h| \quad PQ = |y - k| \quad CP = r$$

We can use the Pythagorean Theorem to write an equation for the circle.

$$CQ^2 \quad + \quad PQ^2 \quad = CP^2$$
$$(x - h)^2 + (y - k)^2 = r^2$$

In general, the **center-radius equation of a circle** with radius r and center (h, k) is

$$(x - h)^2 + (y - k)^2 = r^2$$

A circle whose diameter $\overline{AB}$ has endpoints at $A(-3, -1)$ and $B(5, -1)$ is shown at the right. The center of the circle, C, is the midpoint of the diameter. Recall that the coordinates of the midpoint of the segment whose endpoints are (a, b) and (c, d) are $\left(\frac{a + c}{2}, \frac{b + d}{2}\right)$. The coordinates of C are

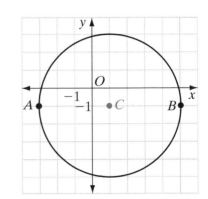

$$\left(\frac{5 + (-3)}{2}, \frac{-1 + (-1)}{2}\right) = (1, -1).$$

The length of the radius is the distance from C to any point on the circle. The distance between two points on the same vertical line, that is, with the same x-coordinates, is the absolute value of the difference of the y-coordinates. The length of the radius is the distance from $C(1, -1)$ to $A(-3, -1)$. The length of the radius is

$$|1 - (-3)| = 4.$$

In the equation of a circle with center at (h, k) and radius r, we use $(x - h)^2 + (y - k)^2 = r^2$. For this circle with center at $(1, -1)$ and radius 4, $h = 1, k = -1$, and $r = 4$. The equation of the circle is:

$$(x - 1)^2 + (y - (-1))^2 = 4^2 \quad \text{or} \quad (x - 1)^2 + (y + 1)^2 = 16$$

The equation of a circle is a rule for a set of ordered pairs, that is, for a relation. For the circle $(x - 1)^2 + (y + 1)^2 = 16$, $(1, -5)$ and $(1, 3)$ are two ordered pairs of the relation. Since these two ordered pairs have the same first element, this relation is not a function.

EXAMPLE I

a. Write an equation of a circle with center at $(3, -2)$ and radius of length 7.

b. What are the coordinates of the endpoints of the horizontal diameter?

Solution   **a.** The center of the circle is $(h, k) = (3, -2)$. The radius is $r = 7$.

The general form of the equation of a circle is $(x - h)^2 + (y - k)^2 = r^2$.

The equation of the given circle is:

$$(x - 3)^2 + (y - (-2))^2 = 7^2 \quad \text{or} \quad (x - 3)^2 + (y + 2)^2 = 49$$

b. METHOD I

If this circle were centered at the origin, then the endpoints of the horizontal diameter would be $(-7, 0)$ and $(7, 0)$. However, the circle is centered at $(3, -2)$. Shift these endpoints using the translation $T_{3, -2}$:

$$(-7, 0) \rightarrow (-7 + 3, 0 - 2) = (-4, -2)$$
$$(7, 0) \rightarrow (7 + 3, 0 - 2) \quad = (10, -2)$$

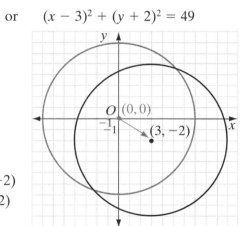

METHOD 2

Since the center of the circle is $(3, -2)$, the y-coordinates of the endpoints are both -2. Substitute $y = -2$ into the equation and solve for x:

$$(x - 3)^2 + (y + 2)^2 = 49$$
$$(x - 3)^2 + (-2 + 2)^2 = 49$$
$$x^2 - 6x + 9 + 0 = 49$$
$$x^2 - 6x - 40 = 0$$
$$(x - 10)(x + 4) = 0$$
$$x = 10 \mid x = -4$$

The coordinates of the endpoints are $(10, -2)$ and $(-4, -2)$.

Answers   **a.** $(x - 3)^2 + (y + 2)^2 = 49$ **b.** $(10, -2)$ and $(-4, -2)$

EXAMPLE 2

The equation of a circle is $(x - 1)^2 + (y - 5)^2 = 36$.

a. What are the coordinates of the center of the circle?

b. What is the length of the radius of the circle?

c. What are the coordinates of two points on the circle?

Solution   Compare the equation $(x - 1)^2 + (y - 5)^2 = 36$ to the general form of the equation of a circle:

$$(x - h)^2 + (y - k)^2 = r^2$$

Therefore, $h = 1, k = 5, r^2 = 36$, and $r = 6$.

a. The coordinates of the center are $(1, 5)$.

b. The length of the radius is 6.

c. Points 7 units from $(1, 5)$ on the same horizontal line are $(8, 5)$ and $(-6, 5)$.

Points 7 units from $(1, 5)$ on the same vertical line are $(1, 12)$ and $(1, -2)$.

Answers   **a.** $(1, 5)$ **b.** 6   **c.** $(8, 5)$ and $(-6, 5)$ or $(1, 12)$ and $(1, -2)$

EXAMPLE 3

The equation of a circle is $x^2 + y^2 = 50$. What is the length of the radius of the circle?

Solution   Compare the given equation to $x^2 + y^2 = r^2$.

$$r^2 = 50$$
$$r = \pm\sqrt{50}$$
$$r = \pm\sqrt{25}\sqrt{2}$$
$$r = \pm 5\sqrt{2}$$

Since a length is always positive, $r = 5\sqrt{2}$. *Answer*

Exercises

Writing About Mathematics

1. Cabel said that for every circle in the coordinate plane, there is always a diameter that is a vertical line segment and one that is a horizontal line segment. Do you agree with Cabel? Justify your answer.

2. Is $3x^2 + 3y^2 = 12$ the equation of a circle? Explain why or why not.

Developing Skills

In 3–8, write an equation of each circle that has the given point as center and the given value of r as the length of the radius.

3. $(0, 0), r = 3$ **4.** $(1, 3), r = 5$ **5.** $(-2, 0), r = 6$

6. $(4, -2), r = 10$ **7.** $(6, 0), r = 9$ **8.** $(-3, -3), r = 2$

In 9–16, write an equation of each circle that has a diameter with the given endpoints.

9. $(-2, 0)$ and $(2, 0)$ **10.** $(0, -4)$ and $(0, 4)$

11. $(2, 5)$ and $(2, 13)$ **12.** $(-5, 3)$ and $(3, 3)$

13. $(5, 12)$ and $(-5, 12)$ **14.** $(-5, 9)$ and $(-7, -7)$

15. $(-7, 3)$ and $(9, 10)$ **16.** $(2, 2)$ and $(18, -4)$

In 17–22, write an equation of each circle.

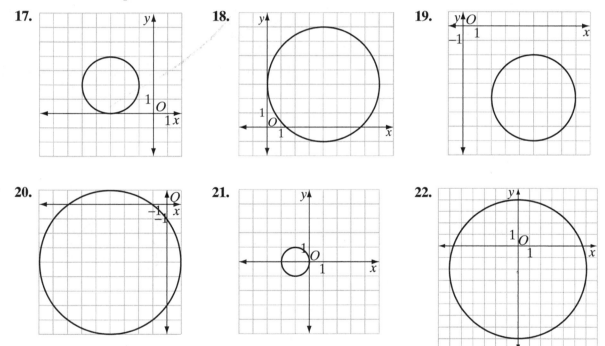

17.   **18.**   **19.**

20.   **21.**   **22.**

In 23–28, find the center of each circle and graph each circle.

23. $(x - 2)^2 + (y + 5)^2 = 4$

24. $(x + 4)^2 + (y - 4) = 36$

25. $\left(x + \frac{3}{2}\right)^2 + (y - 1)^2 = 25$

26. $\left(x - \frac{5}{2}\right)^2 + \left(y + \frac{3}{4}\right)^2 = \frac{81}{25}$

27. $2x^2 + 2y^2 = 18$

28. $5(x - 1)^2 + 5(y - 1)^2 = 245$

29. Point $C(2, 3)$ is the center of a circle and $A(-3, -9)$ is a point on the circle. Write an equation of the circle.

30. Does the point $(4, 4)$ lie on the circle whose center is at the origin and whose radius is $\sqrt{32}$? Justify your answer.

31. Is $x^2 + 4x + 4 + y^2 - 2y + 1 = 25$ the equation of a circle? Explain why or why not.

Applying Skills

32. In the figure on the right, the points $A(2, 6)$, $B(-4, 0)$, and $C(4, 0)$ appear to lie on a circle.

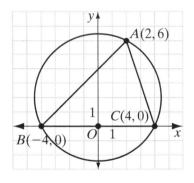

  a. Find the equation of the perpendicular bisector of $\overline{AB}$.

  b. Find the equation of the perpendicular bisector of $\overline{BC}$.

  c. Find the equation of the perpendicular bisector of $\overline{AC}$.

  d. Find the circumcenter of $\triangle ABC$, the point of intersection of the perpendicular bisectors.

  e. From what you know about perpendicular bisectors, why is the circumcenter equidistant from the vertices of $\triangle ABC$?

  f. Do the points A, B, C lie on a circle? Explain.

33. In the figure on the right, the circle with center at $C(3, -1)$ appears to be inscribed in $\triangle PQR$ with vertices $P(-1, 2)$, $Q(3, -12)$, and $R(7, 2)$.

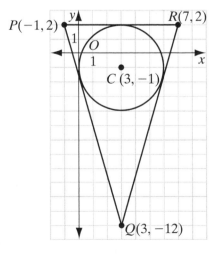

  a. If the equations of the angle bisectors of $\triangle PQR$ are $6x + 8y = 10, x = 3$, and $3x - 4y = 13$, is C the incenter of $\triangle PQR$?

  b. From what you know about angle bisectors, why is the incenter equidistant from the sides of $\triangle PQR$?

  c. If $S(3, 2)$ is a point on the circle, is the circle inscribed in $\triangle PQR$? Justify your answer.

  d. Write the equation of the circle.

34. In the figure on the right, the circle is circumscribed about $\triangle ABC$ with vertices $A(-1, 3)$, $B(-5, 1)$, and $C(-5, -3)$. Find the equation of the circle. Justify your answer algebraically.

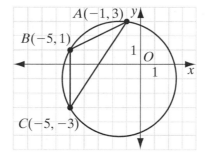

35. Bill Bekebrede wants to build a circular pond in his garden. The garden is in the shape of an equilateral triangle. The length of the altitude to one side of the triangle is 18 feet. To plan the pond, Bill made a scale drawing on graph paper, letting one vertex of the equilateral triangle OAB be $O(0, 0)$ and another vertex be $A(2s, 0)$. Therefore, the length of a side of the triangle is $2s$. Bill knows that an inscribed circle has its center at the intersection of the angle bisectors of the triangle. Bill also knows that the altitude, median, and angle bisector from any vertex of an equilateral triangle are the same line.

    a. What is the exact length, in feet, of a side of the garden?

    b. In terms of s, what are the coordinates of B, the third vertex of the triangle?

    c. What are the coordinates of C, the intersection of the altitudes and of the angle bisectors of the triangle?

    d. What is the exact distance, in feet, from C to the sides of the garden?

    e. What should be the radius of the largest possible pond?

36. The director of the town park is planning walking paths within the park. One is to be a circular path with a radius of 1,300 feet. Two straight paths are to be perpendicular to each other. One of these straight paths is to be a diameter of the circle. The other is a chord of the circle. The two straight paths intersect 800 feet from the circle. Draw a model of the paths on graph paper letting 1 unit = 100 feet. Place the center of the circle at $(13, 13)$ and draw the diameter as a horizontal line and the chord as a vertical line.

    a. What is the equation of the circle?

    b. What are all the possible coordinates of the points at which the straight paths intersect the circular path?

    c. What are all the possible coordinates of the point at which the straight paths intersect?

    d. What are the lengths of the segments into which the point of intersection separates the straight paths?

13-8 TANGENTS AND SECANTS IN THE COORDINATE PLANE

Tangents in the Coordinate Plane

The circle with center at the origin and radius 5 is shown on the graph. Let l be a line tangent to the circle at $A(3, 4)$. Therefore, $l \perp \overline{OA}$ since a tangent is perpendicular to the radius drawn to the point of tangency. The slope of l is the negative reciprocal of the slope of $\overline{OA}$.

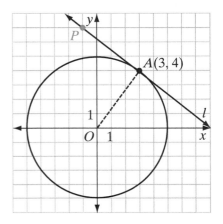

$$\text{slope of } \overline{OA} = \frac{4 - 0}{3 - 0}$$
$$= \frac{4}{3}$$

Therefore, the slope of $l = -\frac{3}{4}$. We can use the slope of l and the point $A(3, 4)$ to write the equation of l.

$$\frac{y - 4}{x - 3} = -\frac{3}{4}$$
$$4(y - 4) = -3(x - 3)$$
$$4y - 16 = -3x + 9$$
$$3x + 4y = 25$$

The point $P(-1, 7)$ makes the equation true and is therefore a point on the tangent line $3x + 4y = 25$.

Secants in the Coordinate Plane

A secant intersects a circle in two points. We can use an algebraic solution of a pair of equations to show that a given line is a secant. The equation of a circle with radius 10 and center at the origin is $x^2 + y^2 = 100$. The equation of a line in the plane is $x + y = 2$. Is the line a secant of the circle?

How to Proceed

(1) Solve the pair of equations algebraically:

$$x^2 + y^2 = 100$$
$$x + y = 2$$

(2) Solve the linear equation for y in terms of x:

$$y = 2 - x$$

(3) Substitute the resulting expression for y in the equation of the circle:

$$x^2 + (2 - x)^2 = 100$$

(4) Square the binomial:

$$x^2 + 4 - 4x + x^2 = 100$$

(5) Write the equation in standard form:

$$2x^2 - 4x - 96 = 0$$

(6) Divide by the common factor, 2:

$$x^2 - 2x - 48 = 0$$

(7) Factor the quadratic equation:

$$(x - 8)(x + 6) = 0$$

(8) Set each factor equal to zero:

$$x - 8 = 0 \quad | \quad x + 6 = 0$$

(9) Solve each equation for x:

$$x = 8 \quad | \quad x = -6$$

(10) For each value of x find the corresponding value of y:

$$y = 2 - x \quad | \quad y = 2 - x$$
$$y = 2 - 8 \quad | \quad y = 2 - (-6)$$
$$y = -6 \quad | \quad y = 8$$

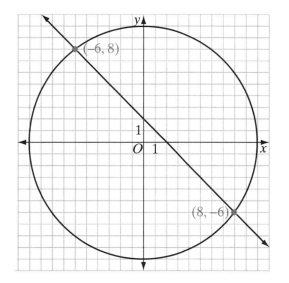

The common solutions are $(8, -6)$ and $(-6, 8)$. The line intersects the circle in two points and is therefore a secant. In the diagram, the circle is drawn with its center at the origin and radius 10. The line $y = 2 - x$ is drawn with a y-intercept of 2 and a slope of -1. The line intersects the circle at $(8, -6)$ and $(-6, 8)$.

EXAMPLE 1

Find the coordinates of the points at which the line $y = 2x - 1$ intersects a circle with center at $(0, -1)$ and radius of length $\sqrt{20}$.

Solution In the equation $(x - h)^2 + (y - k)^2 = r^2$, let $h = 0, k = -1$, and $r = \sqrt{20}$. The equation of the circle is:

$$(x - 0)^2 + (y - (-1))^2 = \left(\sqrt{20}\right)^2 \quad \text{or} \quad x^2 + (y + 1)^2 = 20.$$

Find the common solution of $x^2 + (y + 1)^2 = 20$ and $y = 2x - 1$.

(1) The linear equation is solved for y in terms of x. Substitute, in the equation of the circle, the expression for y and simplify the result.

$$x^2 + (y + 1)^2 = 20$$
$$x^2 + (2x - 1 + 1)^2 = 20$$
$$x^2 + (2x)^2 = 20$$

(2) Square the monomial:

$$x^2 + 4x^2 = 20$$

(3) Write the equation in standard form:

$$5x^2 - 20 = 0$$

(4) Divide by the common factor, 5:

$$x^2 - 4 = 0$$

(5) Factor the left side of the equation:

$$(x - 2)(x + 2) = 0$$

(6) Set each factor equal to zero:

$$x - 2 = 0 \qquad x + 2 = 0$$

(7) Solve each equation for x:

$$x = 2 \qquad x = -2$$

(8) For each value of x find the corresponding value of y:

$$\begin{array}{c|c} y = 2x - 1 & y = 2x - 1 \\ y = 2(2) - 1 & y = 2(-2) - 1 \\ y = 3 & y = -5 \end{array}$$

Answer The coordinates of the points of intersection are $(2, 3)$ and $(-2, -5)$.

EXAMPLE 2

The line $x + y = 2$ intersects the circle $x^2 + y^2 = 100$ at $A(8, -6)$ and $B(-6, 8)$. The line $y = 10$ is tangent to the circle at $C(0, 10)$.

a. Find the coordinates of P, the point of intersection of the secant $x + y = 2$ and the tangent $y = 10$.

b. Show that $PC^2 = (PA)(PB)$.

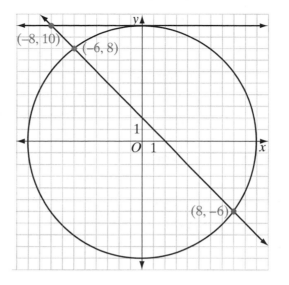

Solution   **a.** Use substitution to find the intersection: If $x + y = 2$ and $y = 10$, then $x + 10 = 2$ and $x = -8$. The coordinates of P are $(-8, 10)$.

b. Use the distance formula, $d = \sqrt{(x_2 - x_1)^2 + (y_2 - y_1)^2}$, to find the lengths of PC, PA, and PB.

$$PA = \sqrt{(-8 - 8)^2 + (10 - (-6))^2} \qquad PB = \sqrt{(-8 - (-6))^2 + (10 - 8)^2}$$
$$= \sqrt{256 + 256} \qquad\qquad\qquad = \sqrt{4 + 4}$$
$$= \sqrt{256}\sqrt{2} \qquad\qquad\qquad\quad = \sqrt{4}\sqrt{2}$$
$$= 16\sqrt{2} \qquad\qquad\qquad\qquad = 2\sqrt{2}$$

$$PC = \sqrt{(-8 - 0)^2 + (10 - 10)^2}$$
$$= \sqrt{64 + 0}$$
$$= 8$$

Then:

$$PC^2 = 8^2 \qquad \text{and} \qquad (PA)(PB) = (16\sqrt{2})(2\sqrt{2})$$
$$= 64 \qquad\qquad\qquad\qquad\quad = (32)(2)$$
$$= 64$$

Therefore, $PC^2 = (PA)(PB)$.

Exercises

Writing About Mathematics

1. Ron said that if the x-coordinate of the center of a circle is equal to the length of the radius of the circle, then the y-axis is tangent to the circle. Do you agree with Ron? Explain why or why not.

2. At A, $\overleftrightarrow{AB}$ intersects a circle with center at C. The slope of $\overleftrightarrow{AB}$ is m and the slope of $\overline{CA}$ is $-m$. Is $\overleftrightarrow{AB}$ tangent to the circle? Explain your answer.

Developing Skills

In 3–14: **a.** Find the coordinates of the points of intersection of the circle and the line. **b.** Is the line a secant or a tangent to the circle?

3. $x^2 + y^2 = 36$
$y = 6$

4. $x^2 + y^2 = 100$
$x + y = 14$

5. $x^2 + y^2 = 25$
$x + y = 7$

6. $x^2 + y^2 = 10$
$y = 3x$

7. $x^2 + y^2 = 9$
$y = x - 3$

8. $x^2 + y^2 = 8$
$x = y$

9. $x^2 + y^2 = 25$
$y = x - 1$

10. $x^2 + y^2 = 20$
$x + y = 6$

11. $x^2 + y^2 = 18$
$y = x + 6$

12. $x^2 + y^2 = 50$
$x + y = 10$

13. $x^2 + y^2 = 8$
$x + y = 4$

14. $x^2 + (y + 2)^2 = 4$
$y = x - 4$

In 15–18, write an equation of the line tangent to the given circle at the given point.

15. $x^2 + y^2 = 9$ at $(0, 3)$

16. $x^2 + y^2 = 16$ at $(-4, 0)$

17. $x^2 + y^2 = 8$ at $(2, -2)$

18. $x^2 + y^2 = 20$ at $(4, -2)$

Applying Skills

19. a. Write an equation of the secant that intersects $x^2 + y^2 = 25$ at $A(3, 4)$ and $B(0, -5)$.

   b. Write an equation of the secant that intersects $x^2 + y^2 = 25$ at $D(0, 5)$ and $E(0, -5)$.

   c. Find the coordinates of P, the intersection of $\overleftrightarrow{AB}$ and $\overleftrightarrow{DE}$.

   d. Show that $(PA)(PB) = (PD)(PE)$.

20. a. Write an equation of the secant that intersects $x^2 + y^2 = 100$ at $A(6, 8)$ and $B(-8, -6)$.

   b. Write an equation of the tangent to $x^2 + y^2 = 100$ at $D(0, 10)$.

   c. Find the coordinates of P, the intersection of $\overleftrightarrow{AB}$ and the tangent line at D.

   d. Show that $(PA)(PB) = (PD)^2$.

21. a. Write an equation of the tangent to $x^2 + y^2 = 18$ at $A(3, 3)$.

    b. Write an equation of the tangent to $x^2 + y^2 = 18$ at $B(3, -3)$.

    c. Find the point P at which the tangent to $x^2 + y^2 = 18$ at A intersects the tangent to $x^2 + y^2 = 18$ at B.

    d. Show that $PA = PB$.

22. Show that the line whose equation is $x + 2y = 10$ is tangent to the circle whose equation is $x^2 + y^2 = 20$.

23. a. Show that the points $A(-1, 7)$ and $B(5, 7)$ lie on a circle whose radius is 5 and whose center is at $(2, 3)$.

    b. What is the distance from the center of the circle to the chord $\overline{AB}$?

24. Triangle ABC has vertices $A(-7, 10)$, $B(2, -2)$, and $C(2, 10)$.

    a. Find the coordinates of the points where the circle with equation $(x + 1)^2 + (y - 7)^2 = 9$ intersects the sides of the triangle.

    b. Show that the sides of the triangle are tangent to the circle.

    c. Is the circle inscribed in the triangle? Explain.

CHAPTER SUMMARY

Definitions to Know

- A **circle** is the set of all points in a plane that are equidistant from a fixed point of the plane called the **center** of the circle.
- A **radius** of a circle (plural, *radii*) is a line segment from the center of the circle to any point of the circle.
- A **central angle of a circle** is an angle whose vertex is the center of the circle.
- An **arc of a circle** is the part of the circle between two points on the circle.
- An arc of a circle is called an **intercepted arc**, or an arc intercepted by an angle, if each endpoint of the arc is on a different ray of the angle and the other points of the arc are in the interior of the angle.
- The **degree measure of an arc** is equal to the measure of the central angle that intercepts the arc.
- **Congruent circles** are circles with congruent radii.
- **Congruent arcs** are arcs of the same circle or of congruent circles that are equal in measure.
- A **chord** of a circle is a line segment whose endpoints are points of the circle.
- A **diameter** of a circle is a chord that has the center of the circle as one of its points.
- An **inscribed angle** of a circle is an angle whose vertex is on the circle and whose sides contain chords of the circle.

- A **tangent to a circle** is a line in the plane of the circle that intersects the circle in one and only one point.
- A **secant of a circle** is a line that intersects the circle in two points.
- A **common tangent** is a line that is tangent to each of two circles.
- A **tangent segment** is a segment of a tangent line, one of whose endpoints is the point of tangency.

Postulates   **13.1**   If $\overset{\frown}{AB}$ and $\overset{\frown}{BC}$ are two arcs of the same circle having a common endpoint and no other points in common, then $\overset{\frown}{AB} + \overset{\frown}{BC} = \overset{\frown}{ABC}$ and $m\overset{\frown}{AB} + m\overset{\frown}{BC} = m\overset{\frown}{ABC}$. (**Arc Addition Postulate**)

13.2   At a given point on a given circle, one and only one line can be drawn that is tangent to the circle.

Theorems and   **13.1**   All radii of the same circle are congruent.
Corollaries   **13.2**   In a circle or in congruent circles, central angles are congruent if and only if their intercepted arcs are congruent.

13.3   In a circle or in congruent circles, two chords are congruent if and only if their central angles are congruent.

13.4   In a circle or in congruent circles, two chords are congruent if and only if their arcs are congruent.

13.5   A diameter perpendicular to a chord bisects the chord and its arcs.

13.5a A line through the center of a circle that is perpendicular to a chord bisects the chord and its arcs.

13.6   The perpendicular bisector of the chord of a circle contains the center of the circle.

13.7   Two chords are equidistant from the center of a circle if and only if the chords are congruent.

13.8   In a circle, if the lengths of two chords are unequal, the shorter chord is farther from the center.

13.9   The measure of an inscribed angle of a circle is equal to one-half the measure of its intercepted arc.

13.9a An angle inscribed in a semicircle is a right angle.

13.9b If two inscribed angles of a circle intercept the same arc, then they are congruent.

13.10 A line is tangent to a circle if and only if it is perpendicular to a radius at its point of intersection with the circle.

13.11 Tangent segments drawn to a circle from an external point are congruent.

13.11a If two tangents are drawn to a circle from an external point, then the line segment from the center of the circle to the external point bisects the angle formed by the tangents.

13.11b If two tangents are drawn to a circle from an external point, then the line segment from the center of the circle to the external point bisects the angle whose vertex is the center of the circle and whose rays are the two radii drawn to the points of tangency.

13.12 The measure of an angle formed by a tangent and a chord that intersect at the point of tangency is equal to one-half the measure of the intercepted arc.

13.13 The measure of an angle formed by two chords intersecting within a circle is equal to one-half the sum of the measures of the arcs intercepted by the angle and its vertical angle.

13.14 The measure of an angle formed by a tangent and a secant, two secants, or two tangents intersecting outside the circle is equal to one-half the difference of the measures of the intercepted arcs.

13.15 If two chords intersect within a circle, the product of the measures of the segments of one chord is equal to the product of the measures of the segments of the other.

13.16 If a tangent and a secant are drawn to a circle from an external point, then the square of the length of the tangent segment is equal to the product of the lengths of the secant segment and its external segment.

13.16 If a tangent and a secant are drawn to a circle from an external point, then the length of the tangent segment is the mean proportional between the lengths of the secant segment and its external segment.

13.17 If two secant segments are drawn to a circle from an external point, then the product of the lengths of one secant segment and its external segment is equal to the product of the lengths of the other secant segment and its external segment.

Formulas

Type of Angle	Degree Measure	Example
Central Angle	The measure of a central angle is equal to the measure of its intercepted arc.	$m\angle 1 = m\overset{\frown}{AB}$
Inscribed Angle	The measure of an inscribed angle is equal to one-half the measure of its intercepted arc.	$m\angle 1 = \tfrac{1}{2}m\overset{\frown}{AB}$

(Continued)

Formulas (Continued)

Type of Angle	Degree Measure	Example
Formed by a Tangent and a Chord	The measure of an angle formed by a tangent and a chord that intersect at the point of tangency is equal to one-half the measure of the intercepted arc.	 $m\angle 1 = \frac{1}{2}m\widehat{AB}$
Formed by Two Intersecting Chords	The measure of an angle formed by two intersecting chords is equal to one-half the sum of the measures of the arcs intercepted by the angle and its vertical angle.	 $m\angle 1 = \frac{1}{2}(m\widehat{AB} + m\widehat{CD})$ $m\angle 2 = \frac{1}{2}(m\widehat{AB} + m\widehat{CD})$
Formed by Tangents and Secants	The measure of an angle formed by a tangent and a secant, two secants, or two tangents intersecting outside the circle is equal to one-half the difference of the measures of the intercepted arcs.	 $m\angle 1 = \frac{1}{2}(m\widehat{AB} - m\widehat{AC})$ $m\angle 2 = \frac{1}{2}(m\widehat{AB} - m\widehat{CD})$ $m\angle 3 = \frac{1}{2}(m\widehat{ACB} - m\widehat{AB})$

Formulas (*Continued*)

Type of Segment	Length	Example
Formed by Two Intersecting Chords	If two chords intersect, the product of the measures of the segments of one chord is equal to the product of the measures of the segments of the other.	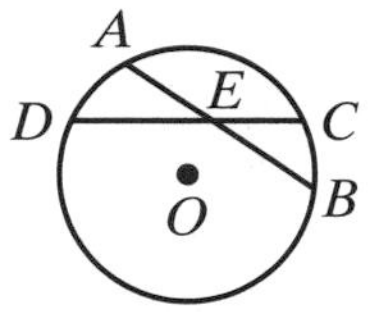 $(AE)(EB) = (CE)(ED)$
Formed by a Tangent Intersecting a Secant	If a tangent and a secant are drawn to a circle from an external point, then the square of the length of the tangent segment is equal to the product of the lengths of the secant segment and its external segment.	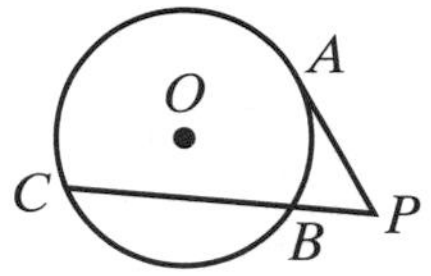 $(PA)^2 = (PC)(PB)$
Formed by Two Intersecting Secants	If two secant segments are drawn to a circle from an external point, then the product of the lengths of one secant segment and its external segment is equal to the product of the lengths of the other secant segment and its external segment.	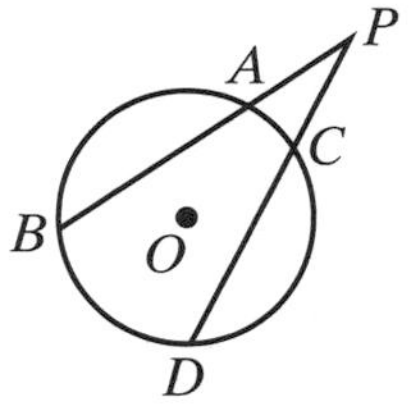 $(PB)(PA) = (PD)(PC)$

The equation of a circle with radius *r* and center (*h*, *k*) is

$$(x - h)^2 + (y - k)^2 = r^2$$

VOCABULARY

13-1 Circle • Center • Radius • Interior of a circle • Exterior of a circle • Central angle of a circle • Arc of a circle • Minor arc • Major arc • Semicircle • Intercepted arc • Degree measure of an arc • Congruent circles • Congruent arcs

13-2 Chord • Diameter • Apothem • Inscribed polygon • Circumscribed circle

13-3 Inscribed angle

13-4 Tangent to a circle • Secant of a circle • Common tangent • Common internal tangent • Common external tangent • Tangent externally • Tangent internally • Tangent segment • Circumscribed polygon • Inscribed circle • Center of a regular polygon

13-6 External segment

13-7 Center-radius equation of a circle

REVIEW EXERCISES

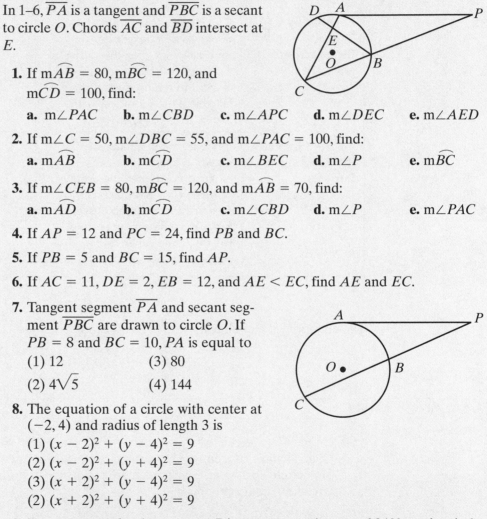

In 1–6, $\overline{PA}$ is a tangent and $\overline{PBC}$ is a secant to circle O. Chords $\overline{AC}$ and $\overline{BD}$ intersect at E.

1. If m$\overset{\frown}{AB}$ = 80, m$\overset{\frown}{BC}$ = 120, and m$\overset{\frown}{CD}$ = 100, find:

    a. m∠PAC **b.** m∠CBD **c.** m∠APC **d.** m∠DEC **e.** m∠AED

2. If m∠C = 50, m∠DBC = 55, and m∠PAC = 100, find:

    a. m$\overset{\frown}{AB}$ **b.** m$\overset{\frown}{CD}$ **c.** m∠BEC **d.** m∠P **e.** m$\overset{\frown}{BC}$

3. If m∠CEB = 80, m$\overset{\frown}{BC}$ = 120, and m$\overset{\frown}{AB}$ = 70, find:

    a. m$\overset{\frown}{AD}$ **b.** m$\overset{\frown}{CD}$ **c.** m∠CBD **d.** m∠P **e.** m∠PAC

4. If AP = 12 and PC = 24, find PB and BC.

5. If PB = 5 and BC = 15, find AP.

6. If AC = 11, DE = 2, EB = 12, and $AE < EC$, find AE and EC.

7. Tangent segment $\overline{PA}$ and secant segment $\overline{PBC}$ are drawn to circle O. If PB = 8 and BC = 10, PA is equal to

    (1) 12   (3) 80

    (2) $4\sqrt{5}$ (4) 144

8. The equation of a circle with center at $(-2, 4)$ and radius of length 3 is

    (1) $(x - 2)^2 + (y - 4)^2 = 9$

    (2) $(x - 2)^2 + (y + 4)^2 = 9$

    (3) $(x + 2)^2 + (y - 4)^2 = 9$

    (2) $(x + 2)^2 + (y + 4)^2 = 9$

9. Two tangents that intersect at P intercept a major arc of 240° on the circle. What is the measure of ∠P?

10. A chord that is 24 centimeters long is 9 centimeters from the center of a circle. What is the measure of the radius of the circle?

11. Two circles, O and O', are tangent externally at P, $OP = 5$, and $O'P = 3$. Segment $\overline{ABC}$ is tangent to circle O' at B and to circle O at C, $\overline{AO'O}$ intersects circle O' at D and P, and circle O at P. If $AD = 2$, find AB and AC.

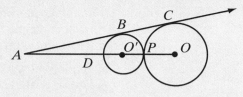

12. Isosceles $\triangle ABC$ is inscribed in a circle. If the measure of the vertex angle, $\angle A$, is 20 degrees less than twice the measure of each of the base angles, find the measures of $\overset{\frown}{AB}$, $\overset{\frown}{BC}$, and $\overset{\frown}{CA}$.

13. Prove that a trapezoid inscribed in a circle is isosceles.

14. In circle O, chords $\overline{AB}$ and $\overline{CD}$ are parallel and $\overline{AD}$ intersects $\overline{BC}$ at E.

    a. Prove that $\triangle ABE$ and $\triangle CDE$ are isosceles triangles.

    b. Prove that $\overset{\frown}{AC} \cong \overset{\frown}{BD}$.

    c. Prove that $\triangle ABE \sim \triangle CDE$.

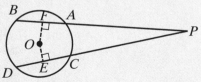

15. Prove that if A, B, C, and D separate a circle into four congruent arcs, then quadrilateral $ABCD$ is a square.

16. Prove, using a circumscribed circle, that the midpoint of the hypotenuse of a right triangle is equidistant from the vertices of the triangle.

17. Secant segments $\overline{PAB}$ and $\overline{PCD}$ are drawn to circle O. If $\overline{PAB} \cong \overline{PCD}$, prove that $\overline{AB}$ and $\overline{CD}$ are equidistant from the center of the circle.

18. An equilateral triangle is inscribed in a circle whose radius measures 12 centimeters. How far from the center of the circle is the centroid of the triangle?

Exploration

A regular polygon can be constructed by constructing the congruent isosceles triangles into which it can be divided. The measure of each base angle of the isosceles triangles is one-half the measure of an interior angle of the polygon. However, the interior angles of many regular polygons are not angles that can be constructed using compass and straightedge. For example, a regular polygon with nine sides has angles that measure 140 degrees. Each of the nine isosceles triangles of this polygon has base angles of 70 degrees which cannot be constructed with straightedge and compass.

In this exploration, we will construct a regular triangle (equilateral triangle), a regular hexagon, a regular quadrilateral (a square), a regular octagon, and a regular dodecagon (a polygon with 12 sides) inscribed in a circle.

a. Explain how a compass and a straightedge can be used to construct an equilateral triangle. Prove that your construction is valid.

b. Explain how the construction in part **a** can be used to construct a regular hexagon. Prove that your construction is valid.

c. Explain how a square, that is, a regular quadrilateral, can be inscribed in a circle using only a compass and a straightedge. (*Hint:* What is true about the diagonals of a square?) Prove that your construction is valid.

d. Bisect the arcs determined by the chords that are sides of the square from the construction in part **c**. Join the endpoints of the chords that are formed to draw a regular octagon. Prove that this construction is valid.

e. A regular octagon can also be constructed by constructing eight isosceles triangles. The interior angles of a regular octagon measure 135 degrees. Bisect a right angle to construct an angle of 45 degrees. The complement of that angle is an angle of 135 degrees. Bisect this angle to construct the base angle of the isosceles triangles needed to construct a regular octagon.

f. Explain how a regular hexagon can be inscribed in a circle using only a compass and a straightedge. (*Hint:* Recall how a regular polygon can be divided into congruent isosceles triangles.)

g. Bisect the arcs determined by the chords that are sides of the hexagon from part **f** to draw a regular dodecagon.

h. A regular dodecagon can also be constructed by constructing twelve isosceles triangles. The interior angles of a regular dodecagon measure 150 degrees. Bisect a 60-degree angle to construct an angle of 30 degrees. The complement of that angle is an angle of 150 degrees. Bisect this angle to construct the base angle of the isosceles triangles needed to construct a regular dodecagon.

CUMULATIVE REVIEW   Chapters 1–13

Part I

Answer all questions in this part. Each correct answer will receive 2 credits. No partial credit will be allowed.

1. The measure of $\angle A$ is 12 degrees more than twice the measure of its complement. The measure of $\angle A$ is

(1) 26   (2) 39   (3) 64   (4) 124

2. The coordinates of the midpoint of a line segment with endpoints at $(4, 9)$ and $(-2, 15)$ are

(1) $(1, 12)$ (2) $(3, -3)$ (3) $(2, 24)$ (4) $(6, -6)$

3. What is the slope of a line that is perpendicular to the line whose equation is $2x + y = 8$?

(1) -2 (2) 2 (3) $-\frac{1}{2}$ (4) $\frac{1}{2}$

4. The altitude to the hypotenuse of a right triangle separates the hypotenuse into segments of length 6 and 12. The measure of the altitude is

(1) 18 (2) $3\sqrt{2}$ (3) $6\sqrt{2}$ (4) $6\sqrt{3}$

5. The diagonals of a quadrilateral bisect each other. The quadrilateral cannot be a

(1) trapezoid   (2) rectangle   (3) rhombus   (4) square

6. Two triangles, $\triangle ABC$ and $\triangle DEF$, are similar. If $AB = 12$, $DE = 18$, and the perimeter of $\triangle ABC$ is 36, then the perimeter of $\triangle DEF$ is

(1) 24 (2) 42 (3) 54 (4) 162

7. Which of the following do not always lie in the same plane?

(1) two points   (3) two lines

(2) three points   (4) a line and a point not on the line

8. At A, the measure of an exterior angle of $\triangle ABC$ is 110 degrees. If the measure of $\angle B$ is 45 degrees, what is the measure of $\angle C$?

(1) 55 (2) 65 (3) 70 (4) 135

9. Under the composition $r_{x\text{-axis}} \circ T_{2,3}$, what are the coordinates of the image of $A(3, -5)$?

(1) $(5, 2)$ (2) $(-5, -2)$ (3) $(5, 8)$ (4) $(-1, -2)$

10. In the diagram, $\overleftrightarrow{AB} \parallel \overleftrightarrow{CD}$ and $\overleftrightarrow{EF}$ intersects $\overleftrightarrow{AB}$ at E and $\overleftrightarrow{CD}$ at F. If $m\angle AEF$ is represented by $3x$ and $m\angle CFE$ is represented by $2x + 20$, what is the value of x?

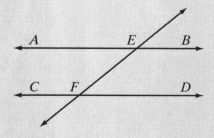

(1) 4 (3) 32

(2) 12 (4) 96

Part II

Answer all questions in this part. Each correct answer will receive 2 credits. Clearly indicate the necessary steps, including appropriate formula substitutions, diagrams, graphs, charts, etc. For all questions in this part, a correct numerical answer with no work shown will receive only 1 credit.

11. The coordinates of the vertices of $\triangle ABC$ are $A(8, 0)$, $B(-4, 4)$, and $C(0, -4)$.

    a. Find the coordinates of D, the midpoint of $\overline{AB}$.

    b. Find the coordinates of E, the midpoint of $\overline{BC}$.

    c. Is $\overline{DE} \parallel \overline{AC}$? Justify your answer.

    d. Are $\triangle ABC$ and $\triangle DBE$ similar triangles? Justify your answer.

12. $ABCD$ is a quadrilateral, $\overline{AC} \cong \overline{BD}$ and $\overline{AC}$ and $\overline{BD}$ bisect each other at E. Prove that $ABCD$ is a rectangle.

Part III

Answer all questions in this part. Each correct answer will receive 4 credits. Clearly indicate the necessary steps, including appropriate formula substitutions, diagrams, graphs, charts, etc. For all questions in this part, a correct numerical answer with no work shown will receive only 1 credit.

13. In the diagram, $\overleftrightarrow{AB}$ and $\overleftrightarrow{CD}$ intersect at E in plane p, $\overleftrightarrow{EF} \perp \overleftrightarrow{AB}$, and $\overleftrightarrow{EF} \perp \overleftrightarrow{CD}$. If $\overline{EA} \cong \overline{EC}$, prove that $\overline{FA} \cong \overline{FC}$.

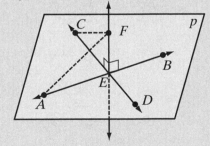

14. The length of the hypotenuse of a right triangle is 2 more than the length of the longer leg. The length of the shorter leg is 7 less than the length of the longer leg. Find the lengths of the sides of the right triangle.

Part IV

Answer all questions in this part. Each correct answer will receive 6 credits. Clearly indicate the necessary steps, including appropriate formula substitutions, diagrams, graphs, charts, etc. For all questions in this part, a correct numerical answer with no work shown will receive only 1 credit.

15. a. Find the coordinates of A', the image of $A(5, 2)$ under the composition $R_{90} \circ r_{y=x}$.

   b. What single transformation is equivalent to $R_{90} \circ r_{y=x}$?

   c. Is $R_{90} \circ r_{y=x}$ a direct isometry? Justify your answer.

16. In the diagram, D is a point on $\overline{ADC}$ such that $AD : DC = 1 : 3$, and E is a point on $\overline{BEC}$ such that $BE : EC = 1 : 3$.

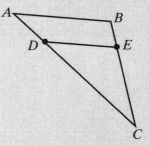

   a. Show that $AC : DC = BC : EC$.

   b. Prove that $\triangle ABC \sim \triangle DEC$.

CHAPTER 14

LOCUS AND CONSTRUCTION

CHAPTER TABLE OF CONTENTS

Classical Greek construction problems limit the solution of the problem to the use of two instruments: the straightedge and the compass. There are three construction problems that have challenged mathematicians through the centuries and have been proved impossible:

- the duplication of the cube
- the trisection of an angle
- the squaring of the circle

The duplication of the cube requires that a cube be constructed that is equal in volume to twice that of a given cube. The origin of this problem has many versions. For example, it is said to stem from an attempt at Delos to appease the god Apollo by doubling the size of the altar dedicated to Apollo.

The trisection of an angle, separating the angle into three congruent parts using only a straightedge and compass, has intrigued mathematicians through the ages.

The squaring of the circle means constructing a square equal in area to the area of a circle. This is equivalent to constructing a line segment whose length is equal to $\sqrt{\pi}$ times the radius of the circle.

Although solutions to these problems have been presented using other instruments, solutions using only straightedge and compass have been proven to be impossible.

14-1 CONSTRUCTING PARALLEL LINES

In Chapter 5, we developed procedures to construct the following lines and rays:

1. a line segment congruent to a given line segment

2. an angle congruent to a given angle

3. the bisector of a given line segment

4. the bisector of a given angle

5. a line perpendicular to a given line through a given point on the line

6. a line perpendicular to a given line through a given point not on the line

Then, in Chapter 9, we constructed parallel lines using the theorem that if two coplanar lines are each perpendicular to the same line, then they are parallel.

Now we want to use the construction of congruent angles to construct parallel lines.

Two lines cut by a transversal are parallel if and only if the corresponding angles are congruent. For example, in the diagram, the transversal $\overleftrightarrow{EF}$ intersects $\overleftrightarrow{AB}$ at G and $\overleftrightarrow{CD}$ at H. If $\angle EGB \cong \angle GHD$, then $\overleftrightarrow{AB} \parallel \overleftrightarrow{CD}$. Therefore, we can construct parallel lines by constructing congruent corresponding angles.

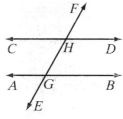

Construction 7	**Construct a Line Parallel to a Given Line at a Given Point.**

Given $\overleftrightarrow{AB}$ and point P not on $\overleftrightarrow{AB}$

Construct   A line through P that is parallel to $\overleftrightarrow{AB}$

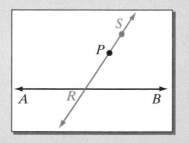

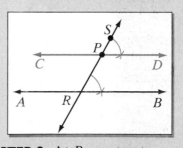

STEP 1. Through P, draw any line intersecting $\overleftrightarrow{AB}$ at R. Let S be any point on the ray opposite $\overrightarrow{PR}$.

STEP 2. At P, construct $\angle SPD \cong \angle PRB$. Draw $\overrightarrow{PC}$, the opposite ray of $\overrightarrow{PD}$, forming $\overleftrightarrow{CD}$.

Continued

Construction 7	Construct a Line Parallel to a Given Line at a Given Point. (*continued*)

Conclusion $\overleftrightarrow{CPD} \parallel \overleftrightarrow{AB}$

Proof Corresponding angles, $\angle SPD$ and $\angle PRB$, are congruent. Therefore, $\overleftrightarrow{CPD} \parallel \overleftrightarrow{AB}$.

This construction can be used to construct the points and lines that satisfy other conditions.

EXAMPLE I

Given: $\overleftrightarrow{AB}$ and $\overline{PQ}$

Construct: $\overleftrightarrow{CD} \parallel \overleftrightarrow{AB}$ at a distance $2PQ$ from $\overleftrightarrow{AB}$.

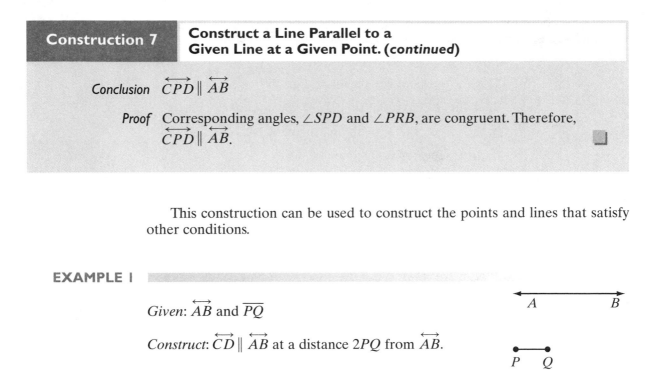

Construction The distance from a point to a line is the length of the perpendicular from the point to the line. Therefore, we must locate points at a distance of $2PQ$ from a point on $\overleftrightarrow{AB}$ at which to draw a parallel line.

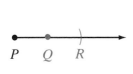 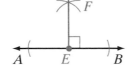 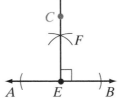 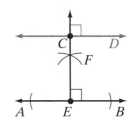

1. Extend $\overline{PQ}$. Locate point R on $\overrightarrow{PQ}$ such that $QR = PQ$, making $PR = 2PQ$.

2. Choose any point E on $\overleftrightarrow{AB}$. At E, construct the line $\overrightarrow{EF}$ perpendicular to $\overleftrightarrow{AB}$.

3. On $\overleftrightarrow{EF}$, locate point C at a distance PR from E.

4. At C, construct $\overleftrightarrow{CD}$ perpendicular to $\overleftrightarrow{EF}$ and therefore parallel to $\overleftrightarrow{AB}$.

Conclusion $\overleftrightarrow{CD}$ is parallel to $\overleftrightarrow{AB}$ at a distance $2PQ$ from $\overleftrightarrow{AB}$.

Exercises

Writing About Mathematics

1. In the example, every point on $\overleftrightarrow{CD}$ is at a fixed distance, $2PQ$, from $\overleftrightarrow{AB}$. Explain how you know that this is true.

2. Two lines, $\overleftrightarrow{AB}$ and $\overleftrightarrow{CD}$, are parallel. A third line, $\overleftrightarrow{EF}$, is perpendicular to $\overleftrightarrow{AB}$ at G and to $\overleftrightarrow{CD}$ at H. The perpendicular bisector of $\overline{GH}$ is the set of all points equidistant from $\overleftrightarrow{AB}$ and $\overleftrightarrow{CD}$. Explain how you know that this is true.

Developing Skills

 In 3–9, complete each required construction using a compass and straightedge, or geometry software. Draw each given figure and do the required constructions. Draw a separate figure for each construction. Enlarge the given figure for convenience.

3. *Given*: Parallel lines l and m and point P.

Construct:

a. a line through P that is parallel to l.

b. a line that is parallel to l and to m and is equidistant from l and m.

c. a line n that is parallel to l and to m such that l is equidistant from m and n.

d. Is the line that you constructed in **a** parallel to m? Justify your answer.

4. *Given*: Line segments of length a and b and $\angle A$

Construct:

a. a rectangle whose length is a and whose width is b.

b. a square such that the length of a side is a.

c. a parallelogram that has sides with measures a and b and an angle congruent to $\angle A$.

d. a rhombus that has sides with measure a and an angle congruent to $\angle A$.

5. *Given*: $\overline{AB}$

a. Divide $\overline{AB}$ into four congruent parts.

b. Construct a circle whose radius is AB.

c. Construct a circle whose diameter is AB.

6. *Given*: △*ABC*

Construct:

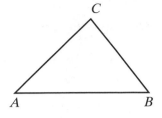

a. a line parallel to $\overleftrightarrow{AB}$ at *C*.

b. the median to $\overline{AC}$ by first constructing the midpoint of $\overline{AC}$.

c. the median to $\overline{BC}$ by first constructing the midpoint of $\overline{BC}$.

d. If the median to $\overline{AC}$ and the median to $\overline{BC}$ intersect at *P*, can the median to $\overline{AB}$ be drawn without first locating the midpoint of $\overline{AB}$? Justify your answer.

7. *Given*: Obtuse triangle *ABC*

Construct:

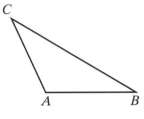

a. the perpendicular bisector of $\overline{AC}$.

b. the perpendicular bisector of $\overline{BC}$.

c. *M*, the midpoint of $\overline{AB}$.

d. If the perpendicular bisector of $\overline{AC}$ and the perpendicular bisector of $\overline{BC}$ intersect at *P*, what two points determine the perpendicular bisector of $\overline{AB}$? Justify your answer.

8. *Given*: △*ABC*

Construct:

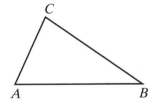

a. the altitude to $\overline{AC}$.

b. the altitude to $\overline{BC}$.

c. If the altitude to $\overline{AC}$ and the altitude to $\overline{BC}$ intersect at *P*, what two points determine the altitude to $\overline{AB}$? Justify your answer.

9. *Given*: △*ABC*

Construct:

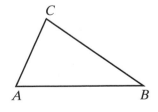

a. the angle bisector from *B* to $\overline{AC}$.

b. the angle bisector from *A* to $\overline{BC}$.

c. If the angle bisector of ∠*B* and the angle bisector of ∠*A* intersect at *P*, what two points determine the angle bisector from *C* to $\overline{AB}$? Justify your answer.

10. a. Draw any line segment, $\overline{AB}$. Draw any ray, $\overrightarrow{AC}$, forming ∠*BAC*.

b. Choose any point, *D*, on $\overrightarrow{AC}$. Construct $\overline{DE}$ on $\overrightarrow{AC}$ such that *DE* = 2*AD*.

c. Draw $\overline{EB}$ and construct a line through *D* parallel to $\overline{EB}$ and intersecting $\overrightarrow{AB}$ at *F*.

d. Prove that *AD* : *DE* = *AF* : *FB* = 1 : 2.

11. a. Draw an angle, $\angle LPR$.

   b. Choose point N on $\overrightarrow{PL}$. Construct Q on $\overrightarrow{PL}$ such that $PQ = 8PN$. Locate S on $\overline{PQ}$ such that $PS : SQ = 3 : 5$.

   c. Draw $\overline{QR}$ and construct a line through S parallel to $\overline{QR}$ and intersecting $\overrightarrow{PR}$ at T.

   d. Prove that $\triangle PST \sim \triangle PQR$ with a constant of proportionality of $\frac{3}{8}$.

14-2 THE MEANING OF LOCUS

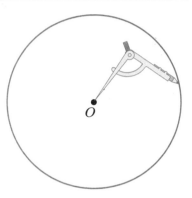

In a construction, the opening between the pencil and the point of the compass is a fixed distance, the length of the radius of a circle. The point of the compass determines a fixed point, point O in the diagram. If the length of the radius remains unchanged, all of the points in the plane that can be drawn by the compass form a circle, and any points that cannot be drawn by the compass do not lie on the circle. Thus, the circle is the set of all points at a fixed distance from a fixed point. This set is called a *locus*.

DEFINITION _____

A **locus** is the set of all points that satisfy a given condition or set of conditions.

The example of the circle given above helps us to understand what the definition means. Every definition can be written as a biconditional:

       p: A point is on the locus.
       q: A point satisfies the given conditions.

1. $(p \rightarrow q)$: If a point is on the locus, then the point satisfies the given conditions. All points on the circle are at a given fixed distance from the center.

2. $(\sim p \rightarrow \sim q)$: If a point is not on the locus, then the point does not satisfy the given conditions. Any point that is *not* on the circle is *not* at the given distance from the center.

Recall that the statement $(\sim p \rightarrow \sim q)$ is the inverse of the statement $(p \rightarrow q)$. A locus is correct when both statements are true: the conditional and its inverse.

We can restate the definition of locus in biconditional form:

▶ **A point P is a point of the locus if and only if P satisfies the given conditions of the locus.**

Discovering a Locus

Procedure

To discover a probable locus:

1. Make a diagram that contains the fixed lines or points that are given.

2. Decide what condition must be satisfied and locate one point that meets the given condition.

3. Locate several other points that satisfy the given condition. These points should be sufficiently close together to develop the shape or the nature of the locus.

4. Use the points to draw a line or smooth curve that appears to be the locus.

5. Describe in words the geometric figure that appears to be the locus.

Note: In this chapter, we will assume that all given points, segments, rays, lines, and circles lie in the same plane and the desired locus lies in that plane also.

EXAMPLE I

What is the locus of points equidistant from the endpoints of a given line segment?

Solution Apply the steps of the procedure for discovering a probable locus.

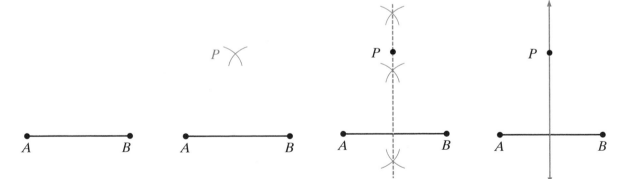

1. Make a diagram: $\overline{AB}$ is the given line segment.

2. Decide the condition to be satisfied: P is to be equidistant from A and B. Use a compass opened to any convenient radius to locate one such point, P.

3. Locate several other points equidistant from A and B, using a different opening of the compass for each point.

4. Through these points, draw the straight line that appears to be the locus.

5. Describe the locus in words: The locus is a straight line that is the perpendicular bisector of the given line segment. *Answer*

Note that in earlier chapters, we proved two theorems that justify these results:

- If a point is equidistant from the endpoints of a line segment, then it lies on the perpendicular bisector of the segment.

- If a point lies on the perpendicular bisector of a line segment, then it is equidistant from the endpoints of the segment.

EXAMPLE 2

Construct the locus of points in the interior of an angle equidistant from the rays that form the sides of the given angle.

Construction Corollaries 9.13b and 9.15a together state: A point is equidistant from the sides of an angle if and only if it lies on the bisector of the angle. Therefore, the required locus is the bisector of the angle.

1. Make a diagram: ∠*ABC* is the given angle.

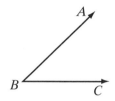

2. Decide the condition to be satisfied: *P* is to be equidistant from $\overrightarrow{BA}$ and $\overrightarrow{BC}$, the rays that are the sides of ∠*ABC*. Construct the angle bisector.

Use a compass to draw an arc with center *B* that intersects $\overrightarrow{BA}$ at *R* and $\overrightarrow{BC}$ at *S*. Then, with the compass open to a convenient radius, draw arcs from *R* and *S* that intersect in the interior of ∠*ABC*. Call the intersection *P*. *PR = PS*.

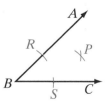

3. Through points *P* and *B*, draw the ray that is the locus.

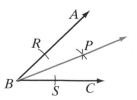

Exercises

Writing About Mathematics

1. Are all of the points that are equidistant from the endpoints of a line segment that is 8 centimeters long 4 centimeters from the endpoints? Explain your answer.

2. What line segment do we measure to find the distance from a point to a line or to a ray?

Developing Skills

3. What is the locus of points that are 10 centimeters from a given point?

4. What is the locus of points equidistant from two points, A and B, that are 8 meters apart?

5. What is the locus of points equidistant from two parallel lines 8 meters apart?

6. What is the locus of points 4 inches away from a given line, $\overleftrightarrow{AB}$?

7. What is the locus of points 3 inches from each of two parallel lines that are 6 inches apart?

8. What is the locus of points that are equidistant from two opposite sides of a square?

9. What is the locus of points that are equidistant from the vertices of two opposite angles of a square?

10. What is the locus of points that are equidistant from the four vertices of a square? (A locus can consist of a single point or no points.)

11. What is the locus of points in the interior of a circle whose radius measures 3 inches if the points are 2 inches from the circle?

12. What is the locus of points in the exterior of a circle whose radius measures 3 inches if the points are 2 inches from the circle?

13. What is the locus of points 2 inches from a circle whose radius measures 3 inches?

14. Circle O has a radius of length r, and it is given that $r > m$.

    a. What is the locus of points in the exterior of circle O and at a distance m from the circle?

    b. What is the locus of points in the interior of circle O and at a distance m from the circle?

    c. What is the locus of points at a distance m from circle O?

15. Concentric circles have the same center. What is the locus of points equidistant from two concentric circles whose radii measure 10 centimeters and 18 centimeters?

16. A series of isosceles triangles are drawn, each of which has a fixed segment, $\overline{AB}$, as its base. What is the locus of the vertices of the vertex angles of all such isosceles triangles?

17. Triangle ABC is drawn with a fixed base, $\overline{AB}$, and an altitude to $\overline{AB}$ whose measure is 3 feet. What is the locus of points that can indicate vertex C in all such triangles?

Applying Skills

18. What is the locus of the tip of the hour hand of a clock during a 12-hour period?

19. What is the locus of the center of a train wheel that is moving along a straight, level track?

20. What is the locus of the path of a car that is being driven down a street equidistant from the two opposite parallel curbs?

21. A dog is tied to a stake by a rope 6 meters long. What is the boundary of the surface over which he can move?

22. A boy walks through an open field that is bounded on two sides by straight, intersecting roads. He walks so that he is always equidistant from the two intersecting roads. Determine his path.

23. There are two stationary floats on a lake. A girl swims so that she is always equidistant from both floats. Determine her path.

24. A dime is rolled along a horizontal line so that the dime always touches the line. What is the locus of the center of the dime?

25. The outer edge of circular track is 40 feet from a central point. The track is 10 feet wide. What is path of a runner who runs on the track, 2 feet from the inner edge of the track?

14-3 FIVE FUNDAMENTAL LOCI

There are five fundamental loci, each based on a different set of conditions. In each of the following, a condition is stated, and the locus that fits the condition is described in words and drawn below. Each of these loci has been shown as a construction in preceding sections.

1. *Equidistant from two points:* Find points equidistant from points A and B.

Locus: The locus of points equidistant from two fixed points is the perpendicular bisector of the segment determined by the two points.

2. *Equidistant from two intersecting lines:* Find points equidistant from the intersecting lines $\overleftrightarrow{AB}$ and $\overleftrightarrow{CD}$.

Locus: The locus of points equidistant from two intersecting lines is a pair of lines that bisect the angles formed by the intersecting lines.

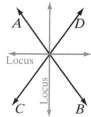

3. *Equidistant from two parallel lines:* Find points equidistant from the parallel lines $\overleftrightarrow{AB}$ and $\overleftrightarrow{CD}$.

Locus: The locus of points equidistant from two parallel lines is a third line, parallel to the given lines and midway between them.

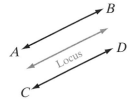

4. *At a fixed distance from a line:* Find points that are at distance d from the line $\overleftrightarrow{AB}$.

Locus: The locus of points at a fixed distance from a line is a pair of lines, each parallel to the given line and at the fixed distance from the given line.

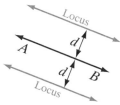

5. *At a fixed distance from a point:* Find points that are at a distance d from the fixed point A.

Locus: The locus of points at a fixed distance from a fixed point is a circle whose center is the fixed point and whose radius is the fixed distance.

These loci are often combined to find a point or set of points that satisfy two or more conditions. The resulting set is called a **compound locus**.

Procedure

To find the locus of points that satisfy two conditions:

1. Determine the locus of points that satisfy the first condition. Sketch a diagram showing these points.

2. Determine the locus of points that satisfy the second condition. Sketch these points on the diagram drawn in step 1.

3. Locate the points, if any exist, that are common to both loci.

Steps 2 and 3 can be repeated if the locus must satisfy three or more conditions.

EXAMPLE 1

Quadrilateral $ABCD$ is a parallelogram. What is the locus of points equidistant from $\overleftrightarrow{AB}$ and $\overleftrightarrow{CD}$ and also equidistant from $\overleftrightarrow{AB}$ and $\overleftrightarrow{BC}$?

Solution   Follow the procedure for finding a compound locus.

(1) Since *ABCD* is a parallelogram, $\overleftrightarrow{AB} \parallel \overleftrightarrow{CD}$. The locus of points equidistant from two parallel lines is a third line parallel to the given lines and midway between them. In the figure, $\overleftrightarrow{EF}$ is equidistant from $\overleftrightarrow{AB}$ and $\overleftrightarrow{CD}$.

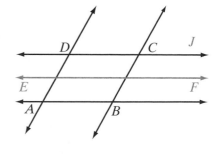

(2) The lines $\overleftrightarrow{AB}$ and $\overleftrightarrow{BC}$ are intersecting lines. The locus of points equidistant from intersecting lines is a pair of lines that bisect the angles formed by the given lines. In the figure, $\overleftrightarrow{GH}$ and $\overleftrightarrow{JK}$ are equidistant from $\overleftrightarrow{AB}$ and $\overleftrightarrow{BC}$.

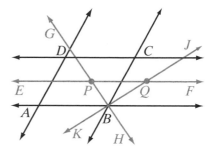

(3) The point *P* at which $\overleftrightarrow{EF}$ intersects $\overleftrightarrow{GH}$ and the point *Q* at which $\overleftrightarrow{EF}$ intersects $\overleftrightarrow{JK}$ are equidistant from $\overleftrightarrow{AB}$ and $\overleftrightarrow{CD}$ and also equidistant from $\overleftrightarrow{AB}$ and $\overleftrightarrow{BC}$.

Answer   *P* and *Q*

Note that only point *P* is equidistant from the three segments that are sides of the parallelogram, but both *P* and *Q* are equidistant from the lines of which these three sides are segments.

Exercises

Writing About Mathematics

1. If *PQRS* is a square, are the points that are equidistant from $\overleftrightarrow{PQ}$ and $\overleftrightarrow{RS}$ also equidistant from *P* and *S*? Explain your answer.

2. Show that the two lines that are equidistant from two intersecting lines are perpendicular to each other.

Developing Skills

In 3–10, sketch and describe each required locus.

3. The locus of points equidistant from two points that are 4 centimeters apart.

4. The locus of points that are 6 inches from the midpoint of a segment that is 1 foot long.

5. The locus of points equidistant from the endpoints of the base of an isosceles triangle.

6. The locus of points equidistant from the legs of an isosceles triangle.

7. The locus of points equidistant from the diagonals of a square.

8. The locus of points equidistant from the lines that contain the bases of a trapezoid.

9. The locus of points 4 centimeters from the midpoint of the base of an isosceles triangle if the base is 8 centimeters long.

10. The locus of points that are 6 centimeters from the altitude to the base of an isosceles triangle if the measure of the base is 12 centimeters

11.   **a.** Sketch the locus of points equidistant from two parallel lines that are 4 centimeters apart.

   b. On the diagram drawn in **a**, place point P on one of the given parallel lines. Sketch the locus of points that are 3 centimeters from P.

   c. How many points are equidistant from the two parallel lines and 3 centimeters from P?

12. **a.** Construct the locus of points equidistant from the endpoints of a line segment $\overline{AB}$.

   b. Construct the locus of points at a distance $\frac{1}{2}(AB)$ from M, the midpoint of $\overline{AB}$.

   c. How many points are equidistant from A and B and at a distance $\frac{1}{2}(AB)$ from the midpoint of $\overline{AB}$?

   d. Draw line segments joining A and B to the points described in **c** to form a polygon. What kind of a polygon was formed? Explain your answer.

14-4 POINTS AT A FIXED DISTANCE IN COORDINATE GEOMETRY

We know that the locus of points at a fixed distance from a given point is a circle whose radius is the fixed distance. In the coordinate plane, the locus of points r units from (h, k) is the circle whose equation is $(x - h)^2 + (y - k)^2 = r^2$.

For example, the equation of the locus of points $\sqrt{10}$ units from $(2, -3)$ is:

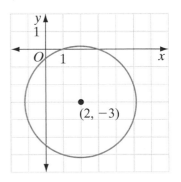

$$(x - 2)^2 + (y - (-3))^2 = (\sqrt{10})^2$$

or

$$(x - 2)^2 + (y + 3)^2 = 10$$

We also know that the locus of points at a fixed distance from a given line is a pair of lines parallel to the given line.

For example, to write the equations of the locus of points 3 units from the horizontal line $y = 2$, we need to write the equations of two horizontal lines, one 3 units above the line $y = 2$ and the other 3 units below the line $y = 2$. The equations of the locus are $y = 5$ and $y = -1$.

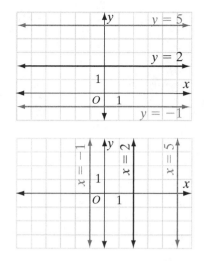

To write the equations of the locus of points 3 units from the vertical line $x = 2$, we need to write the equations of two vertical lines, one 3 units to the right of the line $x = 2$ and the other 3 units to the left of the line $x = 2$. The equations of the locus are $x = 5$ and $x = -1$.

From these examples, we can infer the following:

▶ **The locus of points d units from the horizontal line $y = a$ is the pair of lines $y = a + d$ and $y = a - d$.**

▶ **The locus of points d units from the vertical line $x = a$ is the pair of lines $x = a + d$ and $x = a - d$.**

EXAMPLE 1

What is the equation of the locus of points at a distance of $\sqrt{20}$ units from $(0, 1)$?

Solution The locus of points at a fixed distance from a point is the circle with the given point as center and the given distance as radius. The equation of the locus is

$$(x - 0)^2 + (y - 1)^2 = \left(\sqrt{20}\right)^2 \quad \text{or} \quad x^2 + (y - 1)^2 = 20 \; \text{Answer}$$

EXAMPLE 2

What are the coordinates of the points on the line $y = -\frac{1}{2}x + 1$ at a distance of $\sqrt{20}$ from $(0, 1)$?

Solution From Example 1, we know that the set of all points at a distance of $\sqrt{20}$ from $(0, 1)$ lie on the circle whose equation is $x^2 + (y - 1)^2 = 20$. Therefore, the points on the line $y = -\frac{1}{2}x + 1$ at a distance of $\sqrt{20}$ from $(0, 1)$ are the intersections of the circle and the line.

METHOD 1 Solve the system of equations graphically. Sketch the graphs and read the coordinates from the graph.

METHOD 2 Solve the system of equations algebraically. Substitute the value of y from the linear equation in the equation of the circle.

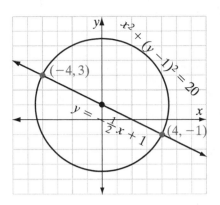

$$x^2 + (y - 1)^2 = 20$$
$$x^2 + \left(-\tfrac{1}{2}x + 1 - 1\right)^2 = 20$$
$$x^2 + \tfrac{1}{4}x^2 = 20$$
$$\tfrac{5}{4}x^2 = 20$$
$$x^2 = 16$$
$$x = \pm 4$$

If $x = 4$:

$$y = -\tfrac{1}{2}(4) + 1$$
$$y = -1$$

If $x = -4$:

$$y = -\tfrac{1}{2}(-4) + 1$$
$$y = 3$$

Answer The points are $(4, -1)$ and $(-4, 3)$.

Exercises

Writing About Mathematics

1. In Example 2, is the line $y = -\tfrac{1}{2}x + 1$ a secant of the circle $x^2 + (y - 1)^2 = 20$? Justify your answer.

2. $\overleftrightarrow{PA}$ and $\overleftrightarrow{PB}$ are tangent to circle O. Martin said that point O is in the locus of points equidistant from $\overleftrightarrow{PA}$ and $\overleftrightarrow{PB}$. Do you agree with Martin? Explain why or why not.

Developing Skills

In 3–8, write an equation of the locus of points at the given distance d from the given point P.

3. $d = 4, P(0, 0)$ 4. $d = 1, P(-1, 0)$ 5. $d = 3, P(0, -2)$

6. $d = 7, P(1, 1)$ 7. $d = \sqrt{10}, P(3, -1)$ 8. $d = \sqrt{18}, P(-3, 5)$

In 9–12, find the equations of the locus of points at the given distance d from the given line.

9. $d = 5, x = 7$ 10. $d = 1, x = 1$ 11. $d = 4, y = 2$ 12. $d = 6, y = 7$

In 13–16, find the coordinates of the points at the given distance from the given point and on the given line.

13. 5 units from $(0, 0)$ on $y = x + 1$

14. 13 units from $(0, 0)$ on $y = x - 7$

15. 10 units from $(0, 1)$ on $y = x + 3$

16. $\sqrt{10}$ units from $(1, -1)$ on $y = x + 2$

In 17–22, write the equation(s) or coordinates and sketch each locus.

17. a. The locus of points that are 3 units from $y = 4$.

   b. The locus of points that are 1 unit from $x = 2$.

   c. The locus of points that are 3 units from $y = 4$ and 1 unit from $x = 2$.

18. a. The locus of points that are 3 units from $(2, 2)$.

   b. The locus of points that are 3 units from $y = -1$.

   c. The locus of points that are 3 units from $(2, 2)$ and 3 units from $y = -1$.

19. a. The locus of points that are 5 units from the origin.

   b. The locus of points that are 3 units from the x-axis.

   c. The locus of points that are 5 units from the origin and 3 units from the x-axis.

20. a. The locus of points that are 10 units from the origin.

   b. The locus of points that are 8 units from the y-axis.

   c. The locus of points that are 10 units from the origin and 8 units from the y-axis.

21. a. The locus of points that are 2 units from $(x + 4)^2 + y^2 = 16$.

   b. The locus of points that are 2 units from the y-axis.

   c. The locus of points that are 2 units from $(x + 4)^2 + y^2 = 16$ and 2 units from the y-axis.

22. a. The locus of points that are 3 units from $(x + 1)^2 + (y - 5)^2 = 4$.

   b. The locus of points that are $\frac{5}{2}$ units from $x = \frac{3}{2}$.

   c. The locus of points that are 3 units from $(x + 1)^2 + (y - 5)^2 = 4$ and $\frac{5}{2}$ units from $x = \frac{3}{2}$.

14-5 EQUIDISTANT LINES IN COORDINATE GEOMETRY

Equidistant from Two Points

The locus of points equidistant from two fixed points is the perpendicular bisector of the line segment joining the two points. For example, the locus of points equidistant from $A(2, -1)$ and $B(8, 5)$ is a line perpendicular to $\overline{AB}$ at its midpoint.

$$\text{midpoint of } \overline{AB} = \left(\tfrac{2+8}{2}, \tfrac{5+(-1)}{2}\right)$$
$$= (5, 2)$$

$$\text{slope of } \overline{AB} = \tfrac{5-(-1)}{8-2}$$
$$= \tfrac{6}{6}$$
$$= 1$$

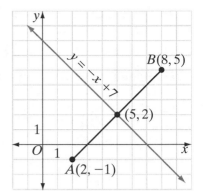

Therefore, the slope of a line perpendicular to $\overline{AB}$ is -1. The perpendicular bisector of $\overline{AB}$ is the line through $(5, 2)$ with slope -1. The equation of this line is:

$$\tfrac{y-2}{x-5} = -1$$
$$y - 2 = -x + 5$$
$$y = -x + 7$$

EXAMPLE I

Describe and write an equation for the locus of points equidistant from $A(-2, 5)$ and $B(6, 1)$.

Solution (1) Find the midpoint, M, of $\overline{AB}$:

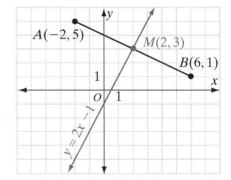

$$M = \left(\tfrac{-2+6}{2}, \tfrac{5+1}{2}\right) = (2, 3)$$

(2) Find the slope of $\overline{AB}$:

$$\text{slope of } \overline{AB} = \tfrac{5-1}{-2-6} = \tfrac{4}{-8} = -\tfrac{1}{2}$$

(3) The slope of a line perpendicular to $\overline{AB}$ is 2.

(4) Write an equation of the line through $(2, 3)$ with slope 2.

$$\tfrac{y-3}{x-2} = 2$$
$$y - 3 = 2x - 4$$
$$y = 2x - 1$$

Answer The locus of points equidistant from $A(-2, 5)$ and $B(6, 1)$ is the perpendicular bisector of $\overline{AB}$. The equation of the locus is $y = 2x - 1$.

Equidistant from Two Parallel Lines

The locus of points equidistant from two parallel lines is a line parallel to the two lines and midway between them.

For example, the locus of points equidistant from the vertical lines $x = -2$ and $x = 6$ is a vertical line midway between them. Since the given lines intersect the x-axis at $(-2, 0)$ and $(6, 0)$, the line midway between them intersects the x-axis at $(2, 0)$ and has the equation $x = 2$.

EXAMPLE 2

Write an equation of the locus of points equidistant from the parallel lines $y = 3x + 2$ and $y = 3x - 6$.

Solution The locus is a line parallel to the given lines and midway between them.

The slope of the locus is 3, the slope of the given lines.

The y-intercept of the locus, b, is the average of the y-intercepts of the given lines.

$$b = \frac{2 + (-6)}{2} = \frac{-4}{2} = -2$$

The equation of the locus is $y = 3x - 2$. *Answer*

Note that in Example 2, we have used the midpoint of the y-intercepts of the given lines as the y-intercept of the locus. In Exercise 21, you will prove that the midpoint of the segment at which the two given parallel lines intercept the y-axis is the point at which the line equidistant from the given lines intersects the y-axis.

Equidistant from Two Intersecting Lines

The locus of points equidistant from two intersecting lines is a pair of lines that are perpendicular to each other and bisect the angles at which the given lines intersect. We will consider two special cases.

1. *The locus of points equidistant from the axes*

The x-axis and the y-axis intersect at the origin to form right angles. Therefore, the lines that bisect the angles between the axes will also go through the origin and will form angles measuring 45° with the axes. One bisector will have a positive slope and one will have a negative slope.

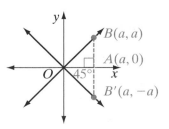

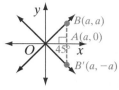

Let $A(a, 0)$ be a point on the x-axis and B be a point on the bisector with a positive slope such that $\overline{AB}$ is perpendicular to the x-axis. The triangle formed by A, B, and the origin O is a 45-45 right triangle. Since 45-45 right triangles are isosceles, $OA = AB = |a|$, and the coordinates of B are (a, a). The line through (a, a) and the origin is $y = x$.

Similarly, if B' is a point on the bisector with a negative slope, then the coordinates of B' are $(a, -a)$. The line through $(a, -a)$ and the origin is $y = -x$.

We have shown that the lines $y = x$ and $y = -x$ are the locus of points equidistant from the axes. These lines are perpendicular to each other since their slopes, 1 and -1, are negative reciprocals.

▶ **The locus of points in the coordinate plane equidistant from the axes is the pair of lines $y = x$ and $y = -x$.**

2. *The locus of points equidistant from two lines with slopes m and $-m$ that intersect at the origin*

Let $O(0, 0)$ and $A(a, ma)$ be any two points on $y = mx$ and $B(0, ma)$ be a point on the y-axis. Under a reflection in the y-axis, the image of $O(0, 0)$ is $O(0, 0)$ and the image of $A(a, ma)$ is $A'(-a, ma)$. The points O and A' are on the line $y = -mx$. Therefore, the image of the line $y = mx$ is the line $y = -mx$ since collinearity is preserved under a line reflection. Also, $m\angle AOB = m\angle A'OB$ since angle measure is preserved under a line reflection. Therefore, the y-axis bisects the angle between the lines $y = mx$ and $y = -mx$.

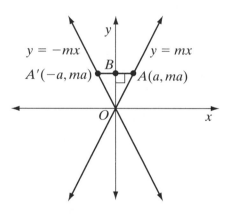

In a similar way, it can be shown that the x-axis bisects the other pair of angles between $y = mx$ and $y = -mx$. Therefore, the y-axis, together with the x-axis, is the locus of points equidistant from the lines $y = mx$ and $y = -mx$.

▶ **The locus of points in the coordinate plane equidistant from two lines with slopes m and $-m$ that intersect at the origin is the pair of lines $y = 0$ and $x = 0$, that is, the x-axis and the y-axis.**

Exercises

Writing About Mathematics

1. In the coordinate plane, are the x-axis and the y-axis the locus of points equidistant from the intersecting lines $y = x$ and $y = -x$? Justify your answer.

2. Ryan said that if the locus of points equidistant from $y = x + 2$ and $y = x + 10$ is $y = x + 6$, then the distance from $y = x + 6$ to $y = x + 2$ and to $y = x + 10$ is 4. Do you agree with Ryan? Justify your answer.

Developing Skills

In 3–8, find the equation of the locus of points equidistant from each given pair of points.

3. $(1, 1)$ and $(9, 1)$ **4.** $(3, 1)$ and $(3, -3)$ **5.** $(0, 2)$ and $(2, 0)$

6. $(0, 6)$ and $(4, -2)$ **7.** $(-2, -2)$ and $(0, 2)$ **8.** $(-4, -5)$ and $(-2, 1)$

In 9–14, find the equation of the locus of points equidistant from each given pair of parallel lines

9. $x = -1$ and $x = 7$ **10.** $y = 0$ and $y = -6$

11. $y = x + 3$ and $y = x + 9$ **12.** $y = -x - 2$ and $y = -x + 6$

13. $y = 2x + 1$ and $y = 2x + 5$ **14.** $2x + y = 7$ and $y = -2x + 9$

15. Find the coordinates of the locus of points equidistant from $(-2, 3)$ and $(4, 3)$, and 3 units from $(1, 3)$.

16. Find the coordinates of the locus of points equidistant from $(2, -5)$ and $(2, 3)$, and 4 units from $(0, -1)$.

Applying Skills

17. a. Find the equation of the locus of points equidistant from $(3, 1)$ and $(5, 5)$.

    b. Prove that the point $(-2, 6)$ is equidistant from the points $(3, 1)$ and $(5, 5)$ by showing that it lies on the line whose equation you wrote in **a**.

    c. Prove that the point $(-2, 6)$ is equidistant from $(3, 1)$ and $(5, 5)$ by using the distance formula.

18. a. Find the equation of the locus of points equidistant from the parallel lines $y = x + 3$ and $y = x - 5$.

    b. Show that point $P(3, 2)$ is equidistant from the given parallel lines by showing that it lies on the line whose equation you wrote in **a**.

    c. Find the equation of the line that is perpendicular to the given parallel lines through point $P(3, 2)$.

    d. Find the coordinates of point A at which the line whose equation you wrote in **c** intersects $y = x + 3$.

    e. Find the coordinates of point B at which the line whose equation you wrote in **c** intersects $y = x - 5$.

    f. Use the distance formula to show that $PA = PB$, that is, that P is equidistant from $y = x + 3$ and $y = x - 5$.

19. Show that the locus of points equidistant from the line $y = x + 1$ and the line $y = -x + 1$ is the y-axis and the line $y = 1$.

20. Show that the locus of points equidistant from the line $y = 3x - 2$ and the line $y = -3x - 2$ is the y-axis and the line $y = -2$.

21. Prove that the midpoint of the segment at which two given parallel lines intercept the y-axis is the point at which the line equidistant from the given lines intersects the y-axis. That is, if the y-intercept of the first line is b, and the y-intercept of the second line is c, then the y-intercept of the line equidistant from them is $\frac{b + c}{2}$.

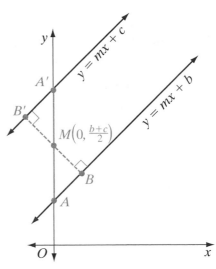

a. Find the coordinates of A, the point where the first line intersects the y-axis.

b. Find the coordinates of A', the point where the second line intersects the y-axis.

c. Show that $M\left(0, \frac{b + c}{2}\right)$ is the midpoint of $\overline{AA'}$.

d. Let B and B' be points on the lines such that M is a point on $\overline{BB'}$ and $\overline{BB'}$ is perpendicular to both lines. Show that $\triangle ABM \cong \triangle A'B'M$.

e. Using part **d**, explain why M is the point at which the line equidistant from the given lines intersects the y-axis.

14-6 POINTS EQUIDISTANT FROM A POINT AND A LINE

We have seen that a straight line is the locus of points equidistant from two points or from two parallel lines. What is the locus of points equidistant from a given point and a given line?

Consider a fixed horizontal line $\overleftrightarrow{AB}$ and a fixed point F above the line. The point P_m, that is, the midpoint of the vertical line from F to $\overleftrightarrow{AB}$, is on the locus of points equidistant from the point and the line. Let P_n be any other point on the locus. As we move to the right or to the left from P_m along $\overleftrightarrow{AB}$,

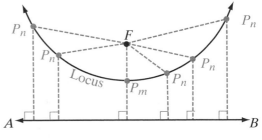

the distance from $\overleftrightarrow{AB}$ to P_n continues to be the length of a vertical line, but the distance from F to P_n is along a slant line. The locus of points is a curve.

Consider a point and a horizontal line that are at a distance d from the origin.

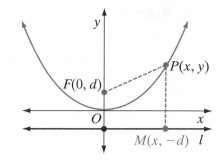

- Let $F(0, d)$ be the point and l be the line. The equation of l is $y = -d$.

- Let $P(x, y)$ be any point equidistant from F and l.

- Let $M(x, -d)$ be the point at which a vertical line from P intersects l.

The distance from P to M is equal to the distance from P to F.

$$PM = PF$$
$$|y - (-d)| = \sqrt{(x - 0)^2 + (y - d)^2}$$
$$(y + d)^2 = x^2 + (y - d)^2$$
$$y^2 + 2dy + d^2 = x^2 + y^2 - 2dy + d^2$$
$$4dy = x^2$$

For instance, if $d = \frac{1}{4}$, that is, if the given point is $\left(0, \frac{1}{4}\right)$ and the given line is $y = -\frac{1}{4}$, then the equation of the locus is $y = x^2$. Recall that $y = x^2$ is the equation of a parabola whose turning point is the origin and whose axis of symmetry is the y-axis.

Recall that under the translation $T_{h, k}$, the image of $4dy = x^2$ is

$$4d(y - k) = (x - h)^2.$$

For example, if the fixed point is $F(2, 1)$ and the fixed line is $y = -1$, then $d = 1$ and the turning point of the parabola is $(2, 0)$, so $(h, k) = (2, 0)$. Therefore, the equation of the parabola is

$$4(1)(y - 0) = (x - 2)^2 \quad \text{or} \quad 4y = x^2 - 4x + 4.$$

This equation can also be written as $y = \frac{1}{4}x^2 - x + 1$.

For any horizontal line and any point not on the line, the equation of the locus of points equidistant from the point and the line is a parabola whose equation can be written as $y = ax^2 + bx + c$. If the given point is above the line, the coefficient a is positive and the parabola opens upward. If the given point is below the line, the coefficient a is negative and the parabola opens downward.

The axis of symmetry is a vertical line whose equation is $x = \frac{-b}{2a}$. Since the turning point is on the axis of symmetry, its x-coordinate is $\frac{-b}{2a}$.

EXAMPLE I

a. Draw the graph of $y = x^2 - 4x - 1$ from $x = -1$ to $x = 5$.

b. Write the coordinates of the turning point.

c. Write an equation of the axis of symmetry.

d. What are the coordinates of the fixed point and the fixed line for this parabola?

Solution **a.** (1) Make a table of values using integral values of x from $x = -1$ to $x = 5$.

(2) Plot the points whose coordinates are given in the table and draw a smooth curve through them.

x	$x^2 - 4x - 1$	y
-1	$1 + 4 - 1$	4
0	$0 - 0 - 1$	-1
1	$1 - 4 - 1$	-4
2	$4 - 8 - 1$	-5
3	$9 - 12 - 1$	-4
4	$16 - 16 - 1$	-1
5	$25 - 20 - 1$	4

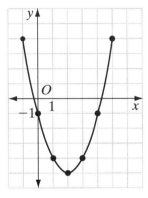

b. From the graph or from the table, the coordinates of the turning point appear to be $(2, -5)$. We can verify this algebraically:

$$x = \frac{-b}{2a} \qquad\qquad y = x^2 - 4x - 1$$
$$= \frac{-(-4)}{2(1)} \qquad\qquad = (2)^2 - 4(2) - 1$$
$$= 2 \qquad\qquad = -5$$

c. The axis of symmetry is the vertical line through the turning point, $x = 2$.

d. Note that the turning point of the parabola is $(2, -5)$. When the turning point of the parabola $y = x^2$ has been moved 2 units to the right and 5 units down, the equation becomes the equation of the graph that we drew:

$$y - (-5) = (x - 2)^2$$
$$y + 5 = x^2 - 4x + 4$$
$$y = x^2 - 4x - 1$$

The turning point of the parabola is the midpoint of the perpendicular segment from the fixed point to the fixed line. Since the coefficient of y in the equation of the parabola is 1, $4d = 1$ or $d = \frac{1}{4}$. The parabola opens upward so the fixed point is $\frac{1}{4}$ unit above the turning point and the fixed line is $\frac{1}{4}$ unit below the turning point. The coordinates of the fixed point are $\left(2, -4\frac{3}{4}\right)$ and the equation of the fixed line is $y = -5\frac{1}{4}$. ◾

EXAMPLE 2

a. Draw the graph of $y = -x^2 - 2x + 8$ from $x = -4$ to $x = 2$.

b. Write the coordinates of the turning point.

c. Write an equation of the axis of symmetry.

Solution a. (1) Make a table of values using integral values of x from $x = -4$ to $x = 2$.

(2) Plot the points whose coordinates are given in the table and draw a smooth curve through them.

x	$-x^2 - 2x + 8$	y
−4	−16 + 8 + 8	0
−3	−9 + 6 + 8	5
−2	−4 + 4 + 8	8
−1	−1 + 2 + 8	9
0	0 − 0 + 8	8
1	−1 − 2 + 8	5
2	−4 − 4 + 8	0

b. From the graph or from the table, the coordinates of the turning point appear to be $(-1, 9)$. We can verify this algebraically:

$$x = \frac{-b}{2a} \qquad\qquad y = -x^2 - 2x + 8$$
$$= \frac{-(-2)}{2(-1)} \qquad\qquad = -(-1)^2 - 2(-1) + 8$$
$$= -1 \qquad\qquad\qquad = 9$$

c. The axis of symmetry is the vertical line through the turning point, $x = -1$.

Here the parabola $y = x^2$ has been reflected in the x-axis so that the equation becomes $-y = x^2$. Then that parabola has been moved 1 unit to the left and 9 units up so that the equation becomes $-(y - 9) = (x - (-1))^2$ or $-y + 9 = x^2 + 2x + 1$, which can be written as $-y = x^2 + 2x - 8$ or $y = -x^2 - 2x + 8$. ◾

EXAMPLE 3

A parabola is equidistant from a given point and a line. How does the turning point of the parabola relate to the given point and line?

Solution The *x*-coordinate of the turning point is the same as the *x*-coordinate of the given point and is halfway between the given point and the line. ◻

EXAMPLE 4

Solve the following system of equations graphically and check:

$$y = x^2 - 2x + 1$$
$$y = -x + 3$$

Solution (1) Make a table of values using at least three integral values of *x* that are less than that of the turning point and three that are greater. The *x*-coordinate of the turning point is:

$$\frac{-b}{2a} = \frac{-(-2)}{2(1)} = \frac{2}{2} = 1$$

(2) Plot the points whose coordinates are given in the table and draw a smooth curve through them.

x	*x*² − 2*x* + 1	*y*
−2	4 + 4 + 1	9
−1	1 + 2 + 1	4
0	0 + 0 + 1	1
1	1 − 2 + 1	0
2	4 − 4 + 1	1
3	9 − 6 + 1	4
4	16 − 8 + 1	9

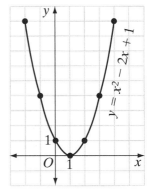

(3) On the same set of axes, sketch the graph of $y = -x + 3$.

The *y*-intercept is 3. Start at the point $(0, 3)$.

The slope is −1 or $\frac{-1}{1}$. Move 1 unit down and 1 unit to the right to find a second point of the line. From this point, again move 1 unit down and 1 unit to the right to find a third point. Draw a line through these three points.

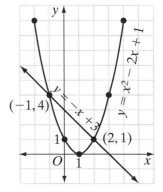

(4) Read the coordinates of the points of intersection from the graph. The common solutions are $(-1, 4)$ and $(2, 1)$.

Answer $(-1, 4)$ and $(2, 1)$ or $x = -1, y = 4$ and $x = 2, y = 1$

Exercises

Writing About Mathematics

1. Luis said that the solutions to the equation $-x^2 - 2x + 8 = 0$ are the x-coordinates of the points at which the graph of $y = -x^2 - 2x + 8$ intersects the y-axis. Do you agree with Luis? Explain why or why not.

2. Amanda said that if the turning point of a parabola is $(1, 0)$, then the x-axis is tangent to the parabola. Do you agree with Amanda? Explain why or why not.

Developing Skills

In 3–8, find the coordinates of the turning point and the equation of the axis of symmetry of each parabola.

3. $y = x^2 - 6x + 1$ **4.** $y = x^2 - 2x + 3$ **5.** $y = x^2 + 4x - 1$

6. $y = -x^2 + 2x + 5$ **7.** $y = -x^2 - 8x + 4$ **8.** $y = x^2 - 5x + 2$

In 9–16, find the common solution of each system of equations graphically and check your solution.

9. $y = x^2 - 2x + 2$ **10.** $y = x^2 - 1$

    $y = x + 2$ $x + y = 1$

11. $y = x^2 - 4x + 3$ **12.** $y = x^2 + 2x - 3$

    $y = x - 1$ $y = 1 - x$

13. $y = x^2 - 4x + 2$ **14.** $y = -x^2 + 2x + 2$

    $y = 2x - 3$ $y = 2x - 2$

15. $y = -x^2 + 6x - 5$ **16.** $y = 2x - x^2$

    $y = 7 - 2x$ $y = 2x - 4$

In 17–20 write the equation of the parabola that is the locus of points equidistant from each given point and line.

17. $F\left(0, \frac{1}{4}\right)$ and $y = -\frac{1}{4}$ **18.** $F\left(0, -\frac{1}{4}\right)$ and $y = \frac{1}{4}$

19. $F(0, 2)$ and $y = -2$ **20.** $F(3, -3)$ and $y = 3$

Hands-On Activity

If the graph of an equation is moved h units in the horizontal direction and k units in the vertical direction, then x is replaced by $x - h$ and y is replaced by $y - k$ in the given equation.

1. The turning point of the parabola $y = ax^2$ is $(0, 0)$. If the parabola $y = ax^2$ is moved so that the coordinates of the turning point are $(3, 5)$, what is the equation of the parabola?

2. If the parabola $y = ax^2$ is moved so that the coordinates of the turning point are (h, k), what is the equation of the parabola?

CHAPTER SUMMARY

Loci
- A **locus of points** is the set of all points, and only those points, that satisfy a given condition.
- The **locus of points equidistant from two fixed points** that are the endpoints of a segment is the perpendicular bisector of the segment.
- The **locus of points equidistant from two intersecting lines** is a pair of lines that bisect the angles formed by the intersecting lines.
- The **locus of points equidistant from two parallel lines** is a third line, parallel to the given lines and midway between them.
- The **locus of points at a fixed distance from a line** is a pair of lines, each parallel to the given line and at the fixed distance from the given line.
- The **locus of points at a fixed distance from a fixed point** is a circle whose center is the fixed point and whose radius is the fixed distance.
- The **locus of points equidistant from a fixed point and a line** is a parabola.

Loci in the Coordinate Plane
- The locus of points in the coordinate plane r units from (h, k) is the circle whose equation is $(x - h)^2 + (y - k)^2 = r^2$.
- The locus of points d units from the horizontal line $y = a$ is the pair of lines $y = a + d$ and $y = a - d$.
- The locus of points d units from the vertical line $x = a$ is the pair of lines $x = a + d$ and $x = a - d$.
- The locus of points equidistant from $A(a, c)$ and $B(b, d)$ is a line perpendicular to $\overline{AB}$ at its midpoint, $\left(\frac{a + b}{2}, \frac{c + d}{2}\right)$.
- The locus of points equidistant from the axes is the pair of lines $y = x$ and $y = -x$.
- The locus of points equidistant from the lines $y = mx$ and $y = -mx$ is the pair of lines $y = 0$ and $x = 0$, that is, the x-axis and the y-axis.
- The locus of points equidistant from $(h, k + d)$ and $y = k - d$ is the parabola whose equation is $4d(y - k) = (x - h)^2$.

14-2 Locus • Concentric circles

14-3 Compound locus

1. *Construct*:

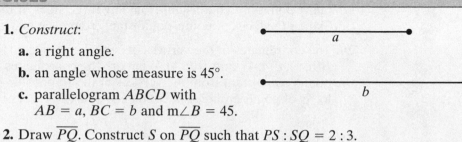

   a. a right angle.

   b. an angle whose measure is 45°.

   c. parallelogram *ABCD* with
       $AB = a$, $BC = b$ and m∠$B = 45$.

2. Draw $\overline{PQ}$. Construct *S* on $\overline{PQ}$ such that $PS : SQ = 2 : 3$.

In 3–6, sketch and describe each locus.

3. Equidistant from two points that are 6 centimeters apart.

4. Four centimeters from *A* and equidistant from *A* and *B*, the endpoints of a line segment that is 6 centimeters long.

5. Equidistant from the endpoints of a line segment that is 6 centimeters long and 2 centimeters from the midpoint of the segment.

6. Equidistant from parallel lines that are 5 inches apart and 4 inches from a point on one of the given lines.

In 7–12, sketch the locus of points on graph paper and write the equation or equations of the locus.

7. 3 units from $(-1, 2)$.

8. 2 units from $(2, 4)$ and 2 units from $x = 2$.

9. Equidistant from $x = -1$ and $x = 5$.

10. Equidistant from $y = 2$ and $y = 8$.

11. Equidistant from $y = 2x$ and $y = -2x$.

12. Equidistant from $(1, -3)$ and $(3, 1)$.

13. Find the coordinates of the points on the line $x + y = 7$ that are 5 units from the origin.

14. a. Draw the graph of $y = x^2 - 4x - 1$.

   b. On the same set of axes, draw the graph of $y = -2x + 2$.

   c. What are the coordinates of the points of intersection of the graphs drawn in **a** and **b**?

In 15–18, solve each system of equations graphically.

15. $x^2 + y^2 = 25$

$y = x - 1$

16. $(x - 2)^2 + (y + 1)^2 = 4$

$x = 2$

17. $y = x^2 + 2x - 1$

$y = x + 1$

18. $y = -x^2 + 4x + 2$

$x + y = 6$

19. A field is rectangular in shape and measures 80 feet by 120 feet. How many points are equidistant from any three sides of the field? (*Hint:* Sketch the locus of points equidistant from each pair of sides of the field.)

20. The coordinates of the vertices of isosceles trapezoid $ABCD$ are $A(0, 0)$, $B(6, 0)$, $C(4, 4)$, and $D(2, 4)$. What are the coordinates of the locus of points equidistant from the vertices of the trapezoid? (*Hint:* Sketch the locus of points equidistant from each pair of vertices.)

Exploration

An **ellipse** is the locus of points such that the sum of the distances from two fixed points F_1 and F_2 is a constant, k. Use the following procedure to create an ellipse. You will need a piece of string, a piece of thick cardboard, and two thumbtacks.

STEP 1. Place two thumbtacks in the cardboard separated by a distance that is less than the length of the string. Call the thumbtacks F_1 and F_2. Attach one end of the string to F_1 and the other to F_2. The length of the string represents the sum of the distances from a point on the locus to the fixed points.

STEP 2. Place your pencil in the loop of string and pull the string taut to locate some point P. Keeping the string taut, slowly trace your pencil around the fixed points until you have created a closed figure.

a. Prove that the closed figure you created is an ellipse. (*Hint:* Let k be the length of the string.)

b. If F_1 and F_2 move closer and closer together, how is the shape of the ellipse affected?

A **hyperbola** is the locus of points such that the difference of the distances from fixed points F_1 and F_2 is a constant, k.

c. Explain why a hyperbola is not a closed figure.

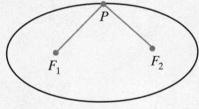

Part I

Answer all questions in this part. Each correct answer will receive 2 credits. No partial credit will be allowed.

1. Which of the following has a midpoint?

(1) $\overleftrightarrow{AB}$ (2) $\overline{AB}$ (3) $\overrightarrow{AB}$ (4) $\angle ABC$

2. What is the inverse of the statement "If two lines segments are congruent, then they have equal measures?"

(1) If two line segments are not congruent, then they do not have equal measures.

(2) If two line segments do not have equal measures, then they are not congruent.

(3) If two line segments have equal measures, then they are congruent.

(4) Two line segments are not congruent if they have unequal measures.

3. In triangle ABC, $\overline{AD}$ is the altitude to $\overline{BC}$. If D is the midpoint of $\overline{BC}$, then which of the following may be false?

(1) $AB = AC$ (3) $AB = AD$

(2) $BD = CD$ (4) $m\angle B = m\angle C$

4. If two angles of a triangle are congruent and complementary, then the triangle is

(1) isosceles and right   (3) isosceles and obtuse

(2) scalene and right   (4) scalene and acute

5. The equation of a line that is perpendicular to the line $x + 3y = 6$ is

(1) $y = 3x + 1$ (3) $y = -\frac{1}{3}x + 1$

(2) $y = -3x + 1$ (4) $y = \frac{1}{3}x + 1$

6. Under a line reflection, which of the following is *not* preserved?

(1) angle measure   (3) orientation

(2) collinearity   (4) midpoint

7. Which of the following is *not* sufficient to prove that quadrilateral $ABCD$ is a parallelogram?

(1) $\overline{AB} \cong \overline{CD}$ and $\overline{BC} \cong \overline{DA}$ (3) $\overline{AC}$ and $\overline{BD}$ bisect each other

(2) $\overline{AB} \cong \overline{CD}$ and $\overline{AB} \parallel \overline{CD}$ (4) $\overline{AC} \perp \overline{BD}$

8. A prism with bases that are equilateral triangles has a height of 12.0 centimeters. If the length of each side of a base is 8.00 centimeters, what is the number of square centimeters in the lateral area of the prism?

(1) 48.0 (2) 96.0 (3) 288 (4) 384

9. The altitude to the hypotenuse of a right triangle divides the hypotenuse into segments of lengths 4 and 45. The length of the shorter leg is

(1) $6\sqrt{5}$ (2) 14   (3) $\sqrt{2{,}009}$ (4) $\sqrt{2{,}041}$

10. Triangle ABC is inscribed in circle O. If $m\widehat{AB} = 110$ and $m\widehat{BC} = 90$, what is the measure of $\angle ABC$?

(1) 45   (2) 55   (3) 80   (4) 110

Part II

Answer all questions in this part. Each correct answer will receive 2 credits. Clearly indicate the necessary steps, including appropriate formula substitutions, diagrams, graphs, charts, etc. For all questions in this part, a correct numerical answer with no work shown will receive only 1 credit

11. D is a point on side $\overline{AB}$ and E is a point on side $\overline{AC}$ of $\triangle ABC$ such that $\overline{DE} \parallel \overline{BC}$. If $AD = 6$, $DB = 9$, and $AC = 20$, find AE and EC.

12. A pile of gravel is in the form of a cone. The circumference of the pile of gravel is 75 feet and its height is 12 feet. How many cubic feet of gravel does the pile contain? Give your answer to the nearest hundred cubic feet.

Part III

Answer all questions in this part. Each correct answer will receive 4 credits. Clearly indicate the necessary steps, including appropriate formula substitutions, diagrams, graphs, charts, etc. For all questions in this part, a correct numerical answer with no work shown will receive only 1 credit.

13. a. On graph paper, sketch the graph of $y = x^2 - x + 2$.

    b. On the same set of axes, sketch the graph of $y = x + 5$.

    c. What are the common solutions of the equations $y = x^2 - x + 2$ and $y = x + 5$?

14. a. Show that the line whose equation is $x + y = 4$ intersects the circle whose equation is $x^2 + y^2 = 8$ in exactly one point and is therefore tangent to the circle.

    b. Show that the radius to the point of tangency is perpendicular to the tangent.

Part IV

Answer all questions in this part. Each correct answer will receive 6 credits. Clearly indicate the necessary steps, including appropriate formula substitutions, diagrams, graphs, charts, etc. For all questions in this part, a correct numerical answer with no work shown will receive only 1 credit.

15. a. The figure below shows the construction of the perpendicular bisector, $\overleftrightarrow{DE}$, of segment $\overline{AB}$. Identify all congruent segments and angles in the construction and state the theorems that prove that $\overleftrightarrow{DE}$ is the perpendicular bisector of $\overline{AB}$.

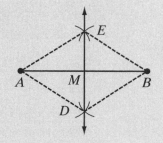

b. The figure below shows the construction of $\angle FDE$ congruent to $\angle CAB$. Identify all congruent lines and angles in the construction and state the theorems that prove that $\angle FDE \cong \angle CAB$.

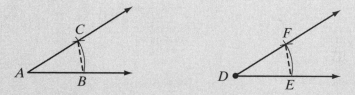

16. a. Quadrilateral $ABCD$ is inscribed in circle O, $\overline{AB} \parallel \overline{CD}$, and $\overline{AC}$ is a diameter. Prove that $ABCD$ is a rectangle.

b. In quadrilateral $ABCD$, diagonals $\overline{AC}$ and $\overline{BD}$ intersect at E. If $\triangle ABE$ and $\triangle CDE$ are similar but not congruent, prove that the quadrilateral is a trapezoid.

c. Triangle ABC is equilateral. From D, the midpoint of $\overline{AB}$, a line segment is drawn to E, the midpoint of $\overline{BC}$, and to F, the midpoint of $\overline{AC}$. Prove that $DECF$ is a rhombus.

INDEX